CAMBRIDGE LIBRARY COLLECTION
Books of enduring scholarly value

Mathematical Sciences

From its pre-historic roots in simple counting to the algorithms powering modern desktop computers, from the genius of Archimedes to the genius of Einstein, advances in mathematical understanding and numerical techniques have been directly responsible for creating the modern world as we know it. This series will provide a library of the most influential publications and writers on mathematics in its broadest sense. As such, it will show not only the deep roots from which modern science and technology have grown, but also the astonishing breadth of application of mathematical techniques in the humanities and social sciences, and in everyday life.

Scientific Papers

As Cavendish Professor of Experimental Physics at Cambridge University, Lord Rayleigh focussed his considerable energies on the study of electricity – building on the work of his illustrious predecessor, James Clerk Maxwell. This second volume of his papers, covering 1881-7, includes a series of four major contributions from 1881 and 1883 concerning the absolute determination of the ohm. Related reports include the measurement of current, and the electrical properties of various materials. A note from 1884 pessimistically predicts an absolute practical limit of less than 50 miles for a working telephone cable. In 1884, Rayleigh stepped down from his post as Cavendish Professor but continued his research work from his private laboratory. He proposed the existence of surface waves in a paper of 1885. These 'Rayleigh waves' roll along the surface of the earth and are responsible for producing most of the shaking experienced in an earthquake.

Cambridge University Press has long been a pioneer in the reissuing of out-of-print titles from its own backlist, producing digital reprints of books that are still sought after by scholars and students but could not be reprinted economically using traditional technology. The Cambridge Library Collection extends this activity to a wider range of books which are still of importance to researchers and professionals, either for the source material they contain, or as landmarks in the history of their academic discipline.

Drawing from the world-renowned collections in the Cambridge University Library, and guided by the advice of experts in each subject area, Cambridge University Press is using state-of-the-art scanning machines in its own Printing House to capture the content of each book selected for inclusion. The files are processed to give a consistently clear, crisp image, and the books finished to the high quality standard for which the Press is recognised around the world. The latest print-on-demand technology ensures that the books will remain available indefinitely, and that orders for single or multiple copies can quickly be supplied.

The Cambridge Library Collection will bring back to life books of enduring scholarly value across a wide range of disciplines in the humanities and social sciences and in science and technology.

Scientific Papers

VOLUME 2: 1881–1887

BARON JOHN WILLIAM STRUTT RAYLEIGH

CAMBRIDGE
UNIVERSITY PRESS

CAMBRIDGE UNIVERSITY PRESS

Cambridge New York Melbourne Madrid Cape Town Singapore São Paolo Delhi

Published in the United States of America by Cambridge University Press, New York

www.cambridge.org
Information on this title: www.cambridge.org/9781108005432

This edition first published 1900
This digitally printed version 2009

ISBN 978-1-108-00543-2

SCIENTIFIC PAPERS

London: C. J. CLAY AND SONS,
CAMBRIDGE UNIVERSITY PRESS WAREHOUSE,
AVE MARIA LANE.

Glasgow: 50, WELLINGTON STREET.

Leipzig: F. A. BROCKHAUS.
New York: THE MACMILLAN COMPANY.
Bombay: E. SEYMOUR HALE.

SCIENTIFIC PAPERS

BY

JOHN WILLIAM STRUTT,

BARON RAYLEIGH,

D.Sc., F.R.S.,

HONORARY FELLOW OF TRINITY COLLEGE, CAMBRIDGE,
PROFESSOR OF NATURAL PHILOSOPHY IN THE ROYAL INSTITUTION.

VOL. II.

1881—1887.

CAMBRIDGE:

AT THE UNIVERSITY PRESS.

1900

Cambridge:

PRINTED BY J. AND C. F. CLAY,
AT THE UNIVERSITY PRESS.

PREFACE.

AMONG the Papers here reprinted several, relating to the Electrical Units, were written conjointly with Prof. Schuster and Mrs Sidgwick. It may perhaps be well to remind the reader that at the time of these researches the ohm was uncertain to the extent of 4 per cent., and that the silver equivalent then generally accepted differed 2 per cent. from the value arrived at by us.

TERLING PLACE, WITHAM,
   October 1900.

The works of the Lord are great,
Sought out of all them that have pleasure therein.

CONTENTS.

CONTENTS.

79.

ON THE DETERMINATION OF THE OHM [B. A. UNIT] IN ABSOLUTE MEASURE. By Lord RAYLEIGH, F.R.S., AND ARTHUR SCHUSTER, Ph.D., F.R.S.

[*Proceedings of the Royal Society*, XXXII. pp. 104—141, 1881.]

Part I.—By Lord RAYLEIGH.

It is generally felt that considerable uncertainty still attaches to the real value of the ohm, or British Association unit of resistance. The ohm was constructed to represent 10^9 C.G.S. absolute units, but according to Kohlrausch* it is nearly 2 per cent. too great, and according to Rowland† nearly 1 per cent. too small. On the other hand, H. Weber‡ has obtained by more than one method results very nearly in harmony with those of the British Association Committee. Influenced partly by the fact that the original apparatus (though a good deal out of repair) and the standard coils themselves were in the Cavendish Laboratory, I determined last June to repeat the measurement by the method of the Committee, which has been employed by no subsequent experimenter, and sought permission from the Council of the British Association to make the necessary alterations in the apparatus. In this way I hoped not merely to obtain an independent result, but also to form an opinion upon the importance of certain criticisms which have been passed upon the work of the Committee.

The method, it will be remembered, consists in causing a coil of insulated wire, forming a closed circuit, to revolve about a vertical axis, and in observing the deflection from the magnetic meridian of a magnet suspended at its centre, the deflection being due to the currents developed in the coil under the influence of the earth's magnetism. The amount of the deflection

* *Phil. Mag.* vol. XLVII. p. 294, 1874.
† *American Journal of Science and Arts*, 1878.
‡ *Phil. Mag.* vol. V. p. 30, 1878.

is independent of the intensity of the earth's magnetic force, and it varies inversely as the resistance of the circuit. The theory of the experiment is explained very fully in the reports of the Committee* and in Maxwell's *Electricity and Magnetism*, section 763. For the sake of distinctness, and as affording an opportunity for one or two minor criticisms, a short statement in the original notation will be convenient :—

H = horizontal component of earth's magnetism.

γ = strength of current in coil at time t.

G = total area inclosed by all the windings of the wire.

ω = angular velocity of rotation.

$\theta = \omega t$ = angle between plane of coil and magnetic meridian.

M = magnetic moment of suspended magnet.

ϕ = angle between the axis of the magnet and the magnetic meridian.

K = magnetic force at the centre of the coil due to unit current in the wire.

L = coefficient of self-induction of coil.

R = resistance of coil in absolute measure.

$MH\tau$ = force of torsion of fibre per unit of angular rotation.

The equation determining the current is—

$$L \frac{d\gamma}{dt} + R\gamma = HG\omega \cos \omega t + MK\omega \cos (\omega t - \phi), \quad \ldots\ldots\ldots\ldots(1)$$

whence

$$\gamma = \frac{\omega}{R^2 + L^2\omega^2} \{GH (R \cos \theta + L\omega \sin \theta) + KM (R \cos (\theta - \phi) + L\omega \sin (\theta - \phi))\}. \quad \ldots\ldots(2)$$

If L were zero, or if the rotation were extremely slow, the current would (apart from KM) be greatest when the coil is passing through the meridian. In consequence of self-induction, the phase of the current is retarded, and its maximum value is diminished. At the higher speeds used by the Committee, the retardation of phase amounted to 20°.

To find the effect of (2) upon the suspended needle, we have to introduce MK and the resolving factor $\cos (\theta - \phi)$, and then to take the average. This, on the supposition that the needle remains on the whole balanced at ϕ, must be equal to the force of restitution due to the direct action of the earth's magnetism and to torsion, i.e., $MH \sin \phi + MH\tau \phi$. Thus—

$$\frac{\frac{1}{2}MK\omega}{R^2 + L^2\omega^2} \{GH (R \cos \phi + L\omega \sin \phi) + KMR\} - MH (\sin \phi + \tau\phi) = 0.$$

* Collected in one volume. Spon, London, 1873.

In the actual experiment τ is a very small quantity, say $\frac{1}{1000}$; and the distinction between $\tau\phi$ and $\tau \sin \phi$ may be neglected.

$$R^2 - R\frac{\frac{1}{2}GK\omega \cot \phi}{1+\tau}\left(1+\frac{MK}{GH}\sec \phi\right) + L^2\omega^2 - \frac{\frac{1}{2}GKL\omega^2}{1+\tau} = 0. \ \ldots\ldots(3)$$

If we omit the small terms depending upon τ and upon MK/GH, we get on solution and expansion of the radical—

$$R = \tfrac{1}{2}GK\omega \cot \phi \left\{1 - \frac{2L}{GK}\left(\frac{2L}{GK} - 1\right)\tan^2 \phi - \left(\frac{2L}{GK}\right)^2\left(\frac{2L}{GK} - 1\right)^2\tan^4\phi\ldots\right\}.$$

$$\ldots\ldots(4)$$

The term in $\tan^4 \phi$ is not given in the report of the Committee; but, as I learn from Mr Hockin through Dr Schuster, it was included in the actual reductions. But the next term in $\tan^6 \phi$, and one arising from a combination of the correction for self-induction with that depending on M, are not altogether insensible, so that probably the direct use of the quadratic is more convenient than the expansion. At the high speeds used by the Committee the correction for self-induction amounted to some 8 per cent., and therefore cannot be treated as very small.

If the axis of rotation be not truly vertical, a correction for level is necessary. In the case of coincidence with the line of dip, no currents, due to the earth's magnetism, would be developed. If the upper end of the axis deviate from the vertical by a small angle β towards the north, the electromotive forces are increased in the ratio $\cos(I+\beta) : \cos I$, i.e., in the ratio $1 + \tan I . \beta$, I being the angle of dip. A deviation in the east and west plane will have an effect of the second order only. The magnetic forces due to the currents will not act upon the needle precisely as if the plane of the coil were always vertical, but the difference is of the second order, so that the whole effect of a small error of level may be represented by writing $G(1 + \tan I . \beta)$ for G in (3) or (4).

The next step is to express GK in terms of the measurements of the coil. In order that there may be a passage for the suspending fibre and its enveloping tube, it is necessary that the coil be double, or if we prefer so to express it, that there be a gap in the middle. If [see figure]

a = mean radius of each coil,

n = whole number of windings,

b = axial dimension of section of each coil,

c = radial dimension of section of each coil,

b' = distance of mean plane of each coil from the axis of motion,

α = angle subtended at centre by radius of each coil, so that $\cot \alpha = b'/a$,

1—2

then—

$$G = \pi n a^2 \left(1 + \tfrac{1}{12}\frac{c^2}{a^2}\right), \quad \dots\dots\dots\dots\dots (5)$$

$$K = \frac{2\pi n}{a}\sin^3\alpha\left\{1 + \tfrac{1}{24}\frac{c^2}{a^2}(2 - 15\sin^2\alpha\cos^2\alpha) + \tfrac{1}{24}\frac{b^2}{a^2}(15\sin^2\alpha\cos^2\alpha - 3\sin^2\alpha)\right\},$$
$$\dots\dots (6)$$

so that

$$GK = 2\pi^2 n^2 a\sin^3\alpha\left\{1 + \tfrac{1}{6}\frac{c^2}{a^2} + \tfrac{5}{8}\frac{b^2 - c^2}{a^2}\sin^2\alpha\cos^2\alpha - \tfrac{1}{8}\frac{b^2}{a^2}\sin^2\alpha\right\}. \quad \dots\dots (7)$$

The correction due to the finiteness of b and c is in practice extremely small, but the factor $\sin^3\alpha$ must be determined with full accuracy.

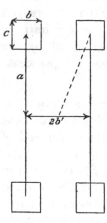

In order to arrive at the value of MK/GH, which occurs in (3), we observe that the approximate value of K/G is $2\sin^3\alpha/a^3$; so that MK/GH is equal to $\tan\mu$, where μ is the angle through which the needle of a magnetometer is deflected when the suspended magnet (M) is placed at a distance from it $a/\sin\alpha$ to the east or west, with the magnetic axis pointing east or west. In practice the difference of readings when M is reversed is taken in order to double the effect, and any convenient distance is used in lieu of $a/\sin\alpha$, allowance being easily made by the law of cubes.

The correction for torsion is determined by giving the suspended magnet one (or more) complete turns, and observing the displacement. If this be δ_1, reckoned in divisions of the scale, i.e., in millimetres, and D be the distance from the mirror to the scale reckoned in millimetres,

$$\tau = \frac{\delta_1}{4\pi D}. \quad \dots\dots\dots\dots\dots\dots (8)$$

The correction for scale reading, necessary in order to pass from $\tfrac{1}{2}\tan 2\phi$ to $\tan\phi$, will be explained under the head of reductions.

Corrections depending upon irregularity in the magnetic field, and in the adjustment of the magnet to the centre of the coil, are given in the report. They are exceedingly small. The same may be said of errors due to imperfect adjustment of the coil with respect to the axis of rotation.

In remounting the apparatus the first point for consideration was the driving gear. The Committee used a Huyghens' gearing, driven by hand, in conjunction with a governor. This, it appeared to me, might advantageously be replaced by a water-motor; and Bailey's "Thirlmere" engine, which acts by the impulse of a jet of water upon revolving 'cups, was chosen as suitable for the purpose. As the pressure in the public water pipes is not sufficiently uniform, it was at first intended to introduce a reducing valve; but on reflection it seemed simpler to obtain a constant head of water by connecting the engine with a small cistern at the top of the building. This cistern is just big enough to hold the ball-tap by which it is supplied, and gives at the engine a head of about 50 feet.

The success of this arrangement depends upon attention to principles, as to which it may be well to say a few words. The work done by many prime movers is within practical limits proportional to the speed. If the work necessary to be done in order to overcome resistances, as in overcoming solid friction, or in pulling up weights, be also proportional to the speed, there is nothing to determine the rate of the engine, and in the absence of an effective governor the motion will be extremely unsteady. In general the resistance function will be of the form—

$$Bv + Cv^2 + Dv^3 + ...,$$

in which the above-mentioned resistances are included under B. The term in C will represent resistances of the nature of viscosity, and that in D a resistance such as is incurred in setting fluids in motion by a fan or otherwise. By these resistances, if present, the speed of working will be determined.

In the water impulse engine, however, the work is not proportional to the speed. At zero speed no work is done; neither is any work done at a speed such that the cups retreat with the full velocity of the jet. The speed of maximum efficiency is the half of the last, and the curve representing work as a function of speed is a parabola with vertex directed upwards. If we draw upon the same diagram the curve of work and the curve of resistance, the actual speed will correspond to the point of intersection, and will be well or ill defined according as the angle of intersection is great or small. At the higher speeds of the coil (four to six revolutions per second) so much air is set in motion that the resistance curve is highly convex downwards, and no difficulty is experienced in obtaining a nearly uniform motion. But when the speed of rotation is as slow as once a second, the principal resistance is due to solid friction, and the requisite curvature in the diagrams must be obtained in the curve of work. It was necessary in order to obtain a satisfactory performance at low speeds to introduce an additional reducing pulley, so that the engine might run fast, although the coil was running slow.

The revolving coil with its frame, and the apparatus for suspending the magnet, were at first arranged as described by the Committee. This description, with drawings, is to be found in the report, and it is reproduced in Gordon's *Electricity and Magnetism*, vol. I. The water engine was ready about the middle of June, and towards the end of the month the apparatus was mounted by Mr Horace Darwin. During July and August preliminary trials were made by Mr Darwin, Mrs Sidgwick, and myself, and various troubles were encountered.

The only point in which the arrangement adopted by the Committee was intentionally departed from was in the connexion of the magnet and mirror. The magnet is necessarily placed at the centre of the revolving coil, but in their arrangement the mirror is on the top of the frame and is connected to the magnet by a brass wire. In order to save weight, I preferred to have the magnet and mirror close together, not anticipating any difficulty from the periodic and very brief interruption caused by the passage of the coil across the line of sight. A box was, therefore, prepared with a glass front, through which the mirror could be observed, and was attached to the end of a brass tube coming through the hollow axle of the coil. This tube itself was supported on screws resting on the top of the frame. The upper end of the suspension fibre was carried by a tall tripod resting independently on the floor.

The first matter for examination was the behaviour of the magnet and mirror when the coil was spinning with circuit open. At low speeds the result was fairly satisfactory, but at six or more revolutions per second a violent disturbance set in. This could not be attributed to the direct action of wind, as the case surrounding the suspended parts was nearly air-tight, except at the top. It was noticed by Mr Darwin that even at low speeds a disturbance was caused at every stroke of the bell. This observation pointed to mechanical tremor, communicated through the frame, as the cause of the difficulty, and the next step was to support the case surrounding the suspended parts independently. A rough trial indicated some improvement, but at this point the experiments had to be laid aside for a time.

From the fact that the disturbance in question was produced by the slightest touch (as by a tap of the finger nail) upon the box, while the upper parts of the tube could be shaken with impunity, it appeared that it must depend upon a reaction between the air included in the box and the mirror. It is known that a flat body tends to set itself across the direction of any steady current of the fluid in which it is immersed, and we may fairly suppose than an effect of the same character will follow from an alternating current. At the moment of the tap upon the box the air inside is made to move past the mirror, and probably executes several vibrations. While these vibrations last, the mirror is subject to a twisting force tending to set it at right angles to the direction of vibration. The whole action being over in a time very small compared with that of the free vibrations of the magnet and mirror, the observed effect is as if an impulse had been given to the suspended parts.

In order to illustrate this effect I contrived the following experiment *. A small disk of paper, about the size of a sixpence, was hung by a fine silk fibre across the mouth of a resonator of pitch 128. When a sound of this pitch is excited in the neighbourhood, there is a powerful rush of air in and out of the resonator, and the disk sets itself promptly across the passage. A fork of pitch 128 may be held near the resonator, but it is better to use a second resonator at a little distance in order to avoid any possible disturbance due to the neighbourhood of the vibrating prongs. The experiment, though rather less striking, was also successful with forks and resonators of pitch 256.

It will be convenient here to describe the method adopted for regulating and determining the speed of rotation, which has proved thoroughly satisfactory. In the experiments of the Committee a governor was employed, and the speed was determined by means of the bell already referred to. This bell received a stroke every 100 revolutions, and the times were taken with a chronometer. In this method rather long spinnings (ten or twenty minutes) are necessary in order to get the speed with sufficient accuracy, much longer than are required to take the readings at the telescope. Desirous, if possible, of making the observations more quickly, I determined to try the stroboscopic method. On the axis of the instrument a stout card of 14 inches diameter was mounted, divided into concentric circles of black and white teeth. The black and white spaces were equal, and the black only were counted as teeth. There were five circles, containing 60, 32, 24, 20, 16 teeth respectively, the outside circle having the largest number of teeth.

This disk was observed from a distance through a telescope, and an arrangement for affording an intermittent view. An electric tuning-fork of frequency about $63\frac{1}{2}$ was maintained in regular vibration in the usual way by means of a Grove cell. To the ends of the prongs are attached thin plates of metal, perforated with somewhat narrow slits parallel to the prongs. In the position of equilibrium these slits overlap so as to allow an unobstructed view, but in other positions of the fork the disk cannot be seen. When the fork vibrates, the disk is seen intermittently 127 times a second; and if the speed be such that on any one of the circles 127 teeth a second pass a fixed pointer, that circle is seen as if it were at rest.

By means of the various circles it is possible to observe correspondingly varied speeds without any change in the frequency of the fork's vibration. A further step in this direction may be taken by modifying the arrangement for intermittent view. If the eye be placed at the top or bottom of one of the vibrating plates, a view is obtained once only, instead of twice, during

---

* *Proc. Camb. Phil. Soc.* Nov. 8, 1880. [1899. For a lecture experiment the paper disc may be replaced by a magnet and mirror, such as are used for galvanometers. See also *Phil. Mag.* vol. XIV. p. 186, 1882.]

each vibration of the fork. This plan was adopted for the slowest rotation, and allowed 60 teeth to take the place of 120, which would otherwise have been necessary.

The performance of the fork was very satisfactory. It would go for hours without the smallest attention, except an occasional renewal of the alcohol in the mercury cup. Pure (not methylated) alcohol was used for this purpose, and a *platinum* point made and broke the contacts. Although, as it turned out, this fork vibrated with great regularity, dependence was not placed upon it, but repeated comparisons by means of beats were made between it and a standard fork of Kœnig's construction, of pitch (about) 128. These beats, at pitch 128, were about 48 per minute, and scarcely varied perceptibly during the course of the experiments. They could have been counted for an even longer time, but this was not necessary. It was intended at first to make the comparisons of the forks simultaneous with the other observations, but this was given up as a needless refinement.

Some care was necessary in the optical arrangements to obviate undue fatigue of the eyes in a long series of observations. In daylight the illumination of the card was sufficient without special provision, but at night, when the actual observations were made, the image of an Argand gas flame was thrown upon the pointer and the part of the card near it. On account of the necessity of removing the electric fork and its appliances to a distance, the card, if looked at directly, would appear too much fore-shortened, and a looking-glass was therefore introduced. The eyepiece of the telescope, close in front of the slits, was adjusted to the exact height, and the eye was placed immediately behind the slits. By cutting off stray light as completely as possible, the observation may be made without fatigue and with slits narrow enough to give good definition when the speed is correct.

As governor I had originally intended to employ an electro-magnetic contrivance, invented a few years ago by La Cour and myself*, in which a revolving wheel is made to take its time from a vibrating fork, and it was partly for this reason that the water engine was placed at a considerable distance from the revolving coil. I was, however, not without hopes that a governor would be found unnecessary, and a few trials with the stroboscopic apparatus were very encouraging. It appeared that by having the water power a little in excess, the observer looking through the vibrating slits could easily control the speed by applying a slight friction to the cord connecting the engine and coil. For this purpose the cord was allowed to run lightly through the fingers, and after a little practice there was no difficulty in so regulating the speed that a tooth was never allowed finally to pass the pointer, however long the observation was continued. If from a

* *Nature*, May 23, 1878. [Art. 56; vol. I. p. 355.]

momentary inadvertence or from some slight disturbance a tooth passed, it could readily be brought back again. The power of control thus obtained will be appreciated when it is remembered that the passage of a tooth *per second* would correspond to less than one per cent. on the speed. In many of the observations the pointer covered the same tooth all the while, so that the introduction of a governor could only have done harm.

Another, and perhaps still more important, improvement on the original method related to the manner of making correction for the changes of declination which usually occur during the progress of the experiments. The Committee relied for this purpose upon comparisons with the photographic records made at Kew, and they recognise that considerable disturbances arose from the passage of steamers, &c. All difficulty of this kind is removed by the plan which we adopted of taking simultaneous readings of a second magnetometer, called the auxiliary magnetometer, placed at a sufficient distance from the revolving coil to be sensibly unaffected by it, but near enough to be similarly influenced by changes in the earth's magnetism, and by other disturbances having their origin at a moderate distance. The auxiliary magnetometer was of very simple construction, and was read with a telescope and a millimetre scale, the distance between mirror and scale (about $2\frac{1}{2}$ metres) being adjusted to approximate equality with that used for the principal magnet, so that disturbances were eliminated by simple comparisons of the scale readings. During a magnetic storm it was very interesting to watch the simultaneous movements of the magnets.

In the month of September the apparatus was remounted under the direction of Professor Stuart, to whose advice we have often been indebted. In order to examine whether any errors were caused by the circulation of currents in the frame, as has been suggested by more than one critic, insulating pieces were inserted, mercury cups at the same time being provided, so that the contacts could be restored at pleasure. But the principal changes related to the manner of suspending the fibre and supporting the box and tube. In order to eliminate tremor, as far as possible, these parts were supported by a massive wooden stand, resting on the floor and overhanging, but without contact, the top of the metal frame of the coil. The upper end of the fibre was fastened to a rod sliding in a metal cap, which formed the upper extremity of a 2-inch glass tube. Near the other end this tube was attached to a triangular piece of brass, resting on three screws, by which the whole could be raised or lowered bodily and levelled. Rigidly attached to this tube, and forming a continuation of it, a second glass tube, narrow enough to pass freely through the hollow axle of the coil, protected the fibre as far as the box in which the mirror and magnet were hung. This box was cylindrical and about 3 inches in diameter. The top fitted stiffly to the lower end of the narrow glass tube, and the body of the box could be unscrewed, so as to give access to the interior. The window necessary for observation of the mirror was made of a piece of worked glass, and was fitted air-tight.

On my return to Cambridge in October the apparatus was tested, but without the full success that had been hoped for. At high speeds there was still unsteadiness enough to preclude the use of these speeds for measurement. Since it is impossible to suppose that the tremor is propagated with sufficient intensity through the floor and massive brickwork on which the coil is supported, the cause must be looked for in the fanning action of the revolving coil, aggravated no doubt by the somewhat pendulous character of the box, and perhaps by the nearness of the approach between the coil and its frame at three points of the revolution.

At this time the experiment was in danger of languishing, as other occupations prevented Mr Darwin from taking any further part; but on Dr Schuster's return to Cambridge he offered his valuable assistance. Encouraged by Sir W. Thomson, we determined to proceed with the measurements, inasmuch as no disturbance, due to the rotation of the coil with circuit open, could be detected until higher speeds were approached than it was at all necessary to use.

One of the first points submitted to examination was the influence of currents induced in the frame. Without altering the speed or making any other change, readings were taken alternately with the contact-pieces in and out. Observations made on several days agreed in showing a small effect, due to the currents in the frame, in the direction of a *diminished* deflection. The whole deflection being 516 divisions of the scale, the mean diminution on making the top contacts was ·86 division. When the coil was at rest no difference in the zero could be detected on moving the contact-pieces.

In these preliminary experiments very consistent results were obtained at constant speeds, whether the rotation was in one direction or the other; but when deflections at various speeds were compared, we were startled to find the larger deflections falling very considerably short of proportionality to the speeds. There are only two corrections which tend to disturb this proportionality—(1) the correction for scale-reading, (2) the correction for self-induction. The effect of the first is to make the readings too high, and of the second to make the readings too low at the greater speeds. According to the figures given by the Committee (Report, p. 106), the aggregate effect is to increase the readings, on account of the preponderance of (1) over (2), whereas our results were consistently of the opposite character. Everything that could be thought of as a possible explanation was examined theoretically and experimentally, but without success. The coil was dismounted and the wire unwound, in order to see whether there was any false contact which might be supposed to vary with the speed and so account for the discrepancy. After much vexation and delay, it was discovered that the error was in the statement in the Report, the effect of self-induction being given at nearly ten times less than its real value. The correction for scale-reading, instead of preponderating over the correction for self-induction, is in reality quite a small part of the whole.

At this stage, as time was running short, we determined to proceed at once to a complete series of readings at sufficiently varied speeds, postponing the measurement of the coil to the end. The wire had been rewound without extreme care to secure the utmost attainable evenness, and the condition of the groove was such that a thoroughly satisfactory coil could not have been obtained, even with extreme care. It appeared, however, on examination that irregularities of this sort were not likely to affect the final result more than one or two parts in a thousand, if so much; and many points of interest could be decided altogether independently of this measurement.

The details of the experiments and reductions are given below by Dr Schuster, who took all the readings of the principal magnetometer. Mrs Sidgwick observed the auxiliary magnetometer; while the regulation of the speed by stroboscopic observation fell to myself. Dr Schuster also undertook the labour of the reductions and the final comparisons of our arbitrary German silver coil with the standard ohms.

The observations were very satisfactory, and at constant speeds agreed better than we had expected. The only irregularity that we met with was a slight disturbance of the zero, due to convection currents in the air surrounding the mirror, the effect of which, however, almost entirely disappears in the means. This disturbance could be magnified by bringing a paraffin lamp into the neighbourhood of the mirror. After about half a minute, apparently the time occupied in conduction through the box and in starting the current, the readings began to move off. Complete recovery would occupy twenty or thirty minutes. In future experiments this kind of disturbance will be very much reduced by increasing the moment of the magnet five or six times, and by diminishing the size of the mirror, both of which may be done without objection.

The comparison of the results at various speeds requires a knowledge of the coefficient of self-induction L. Nothing is said in the Report as to the value of L for the second year's experiments, but the missing information is supplied in Maxwell's paper on the "Electro-magnetic Field[*]," together with an indication of the process followed in calculating it. The first approximation to the value of L, in which the dimensions of the section are neglected in comparison with the radius of the coil, is 437,440

[*] *Phil. Trans.* 1865.

metres, but this is reduced by corrections to 430,165. The value which best satisfies the observations is considerably greater, viz., 456,748. A rough experiment with the electric balance gave 410,000; but Professor Maxwell remarks that the value calculated from the dimensions of the coil is probably much the more accurate, and was used in the actual reductions. I had supposed at one time that the discrepancy between the results at various speeds and the calculated value of L was due to the omission of the term in $\tan^4 \phi$, given above, which would have the same general effect as an under-estimate of L; but, as has been already mentioned, this term was in fact included in the reductions made by Mr Hockin, in conjunction, moreover, with the value $L = 437,440$.

A rough preliminary reduction of our observations showed at once that they could not be satisfied by any such value of L as 437,000, but pointed rather to 454,000, and we began to suspect that the influence of self-induction had been seriously under-estimated by the Committee. Preliminary trials by Maxwell's method with the electric balance giving promise of results trustworthy within one per cent., we proceeded to apply it with care to the determination of L, but the galvanometer at our command—a single needle Thomson of 2,000 ohms resistance—was not specially suitable for ballistic work. As this method is not explained in any of the usual text-books, it may be convenient here to give a statement of it.

The arrangement is identical with that adopted to measure the resistance of the coil in the usual way by the bridge. If P be the resistance of the copper coil, Q, R, S, nearly inductionless resistances from resistance-boxes, balance is obtained at the galvanometer when $PS = QR$. This is a resistance balance, and to observe it the influence of induction must be eliminated by making the battery contact a second or two before making the galvanometer contact. Let us now suppose that P is altered to $P + \delta P$. The effect of this change would be annulled by the operation of an electromotive force in branch P of magnitude $\delta P \cdot x$, where x denotes the magnitude of the current in this branch before the change. Since electromotive forces act independently, the effect upon the galvanometer of the change from P to $P + \delta P$ is the same as would be caused by $\delta P \cdot x$ acting in branch P, if there be no E.M.F. in the battery branch at all*.

Returning now to resistance P, let us make the galvanometer contact before making the battery contact. There is no permanent current through the galvanometer (G), but, at the moment of make, self-induction opposes an obstacle to the development of the current in P, which causes a transient current through G, showing itself by a throw of the needle. The integral

* [1899. A slight error should here be corrected. The electromotive force should be reckoned as $\delta P \cdot x'$, where x' is the actual current flowing through δP. The ratio of x' to x is very near unity in practice. See *Phil. Trans.* vol. CLXXIII. p. 677, 1882; Art. 80 below.]

magnitude of this opposing E.M.F. is simply Lx, and it produces the same effect upon G as if it acted by itself. We have now to compare the effects of a transient and of a permanent E.M.F. upon G. This is merely a question of galvanometry. If T be the time of half a complete vibration of the needle, θ the permanent deflection due to the steady E.M.F., α the throw due to the transient E.M.F., then the ratio of the electromotive forces, or of the currents, is

$$\frac{T}{\pi} \frac{2 \sin \frac{1}{2}\alpha}{\tan \theta} .$$

If, instead of the permanent deflection θ, we observe the first throw (β) of the galvanometer needle, this becomes

$$\frac{T}{\pi} \frac{2 \sin \frac{1}{2}\alpha}{\tan \frac{1}{2}\beta} .$$

In the present case, the ratio in question is, by what has been shown above, $\delta P . x : Lx$, or $\delta P : L$; so that

$$\frac{L}{P} = \frac{\delta P}{P} \frac{T}{\pi} \frac{2 \sin \frac{1}{2}\alpha}{\tan \frac{1}{2}\beta} , \quad \dots\dots\dots\dots\dots\dots\dots(9)$$

a formula which exhibits the time-constant of the coil P in terms of the period of the galvanometer needle. Further to deduce the value of L in absolute measure from the formula requires a knowledge of resistances in absolute measure.

In carrying out the experiment the principal difficulty arose from want of permanence of the resistance balance, due to changes of temperature in the copper coil. The error from this source was, however, diminished by protecting the coil with flannel, and was in great measure eliminated in the reductions. The result was $L = 455,000$ metres. This is on the supposition that the ohm is correct. If, as we consider more probable, the ohm is one per cent. too small, the result would be $L' = 450,000$.

Without attributing too great importance to this determination, there were now three independent arguments pointing to the higher value of L: first, from the experiments of the Committee; secondly, and more distinctly, from our experiments; and thirdly, from the special determination; and I entertained little doubt that a direct calculation from the dimensions of the coil would lead to a similar conclusion.

This direct calculation proved no very easy matter. Mr W. D. Niven (whom I was fortunately able to interest in the question) and myself had no difficulty in verifying independently the formulæ given in Maxwell's *Electricity and Magnetism*, §§ 692, 705, from which the self-induction of a simple coil of rectangular section can be found, on the supposition that the dimensions of the section are very small in comparison with the radius. In the notation of the paper on the electro-magnetic field, if r be the diagonal of the section, and θ the angle between it and the plane of the coil,

$$L = 4\pi n^2 a \left[\log_e \frac{8a}{r} + \tfrac{1}{12} - \tfrac{4}{3}(\theta - \tfrac{1}{4}\pi)\cot 2\theta - \tfrac{1}{3}\pi \operatorname{cosec} 2\theta \right.$$
$$\left. - \tfrac{1}{6}\cot^2\theta \log_e \cos\theta - \tfrac{1}{6}\tan^2\theta \log_e \sin\theta \right]. \quad \dots\dots(10)$$

In the paper itself, probably by a misprint, $\cos 2\theta$ appears, instead of $\operatorname{cosec} 2\theta$, in (10). The expression is, as it evidently ought to be, unchanged when $\tfrac{1}{2}\pi - \theta$ is written for θ. By an ingenious process, explained in the paper, the formula is applied to calculate the self-induction of a double coil*.

The whole self-induction of the double coil is found by adding together twice the self-induction of each part and twice the mutual induction of the two parts. The self-induction of each part is found (to this approximation) by a simple application of (10). For twice this quantity Mr Niven found 301,802, and I found 301,920 metres. For twice the mutual induction of the two parts I found, by Maxwell's method, 145,820 metres. Adding 301,920 and 145,820, we get 447,740 metres as the value of the whole self-induction, on the supposition that the curvature may be neglected. This corresponds to the value 437,440 given in the paper.

As to the origin of the discrepancy I am not able to offer any satisfactory explanation. It should be noticed, however, that owing to his peculiar use of the words "depth" and "breadth" as applied to coils, Maxwell has inter-changed what, to avoid any possible ambiguity, I have called the *axial* and *radial* dimensions of the section. Thus the depth, *i.e.*, in his use of the word, the axial dimension, is given as ·01608, but this is really the radial dimension, as appears clearly enough from the Report of the Committee, as well as from our recent measurements. The real value of the axial dimension is ·01841 metre. But I do not think that this interchange will explain the difference in the results of the calculation.

When we proceed to apply corrections for the finite size of the section, further discrepancies develope themselves. The second term in the expression for L given in the paper (p. 508) does not appear to be correct, and the final numerical correction for curvature ($-7,345$ metres) differs in sign from that which we obtain. Mr Niven has attacked the problem of determining the correction for curvature in the general case of a single coil of rectangular section, and (subject to a certain difficulty of interpretation) has obtained a solution†. The application of the result to the actual case of a double coil

---

* The following misprints may be noticed:—Page 509, line 11, for B read C; line 13, for $L(AC)$ read $M(AC)$; line 13, for $L(B)$ read $L(C)$. Attention must be directed to the peculiar meaning attached to *depth*.

† [1899. On this subject see Stefan (*Wied. Ann.* XXII. p. 107, 1884).]

would, however, be a very troublesome matter. For the two particular cases in which only one of the two dimensions of the section of a simple coil is considered to be finite, Mr Niven and myself have independently obtained tolerably simple results. Thus, if the axial dimension be zero $(b = 0)$,

$$L = 4\pi n^2 a \left[\log \frac{8a}{c} - \tfrac{1}{2} + \frac{c^2}{96a^2} \left(\log \frac{8a}{c} + \tfrac{43}{12} \right) \right]; \quad \dots\dots\dots\dots(11)$$

and if the radial dimension be zero $[c = 0]$,

$$L = 4\pi n^2 a \left[\log \frac{8a}{b} - \tfrac{1}{2} + \frac{b^2}{32a^2} \left(\log \frac{8a}{b} + \tfrac{1}{4} \right) \right]. \quad \dots\dots\dots\dots(12)$$

Again, for a circular section of radius ρ,

$$L = 4\pi n^2 a \left[\log \frac{8a}{\rho} - \tfrac{7}{4} + \frac{\rho^2}{8a^2} \left(\log \frac{8a}{\rho} + \tfrac{1}{3} \right) \right] \dots\dots\dots\dots(13)$$

In all these cases we see that the correction increases the value of L, and there can be no doubt that the same is true for the double coil.

I have applied (13) to estimate the correction for curvature in the self-induction of each part of the double coil. For reasons which it would take too long to explain, I arrived at the conclusion that the value of the small term must be very nearly the same for a circular section as for a square section of the same area, and the actual section is nearly enough square to allow of the use of this principle. The necessary addition to the originally calculated self-induction of each part, in order to take account of curvature, comes out 119·5 metres; so that the final value of L for the double coil will on this account be increased 239 metres. This is a small quantity, but a much larger correction for curvature must be expected in the mutual induction of the two parts. By a sufficiently approximate method I find as the correction to twice the mutual induction 3,469 metres, giving altogether for twice the mutual induction 149,289 metres. This added to 302,159 $(= 301,920 + 239)$ metres gives as the final calculated value of L for the double coil, $L = 451,448$ metres. This result is confirmed by calculation of the mutual induction by means of a table founded on elliptic functions. In this way, and with a suitable formula for quadrature, we find, $2M = 149,394$ metres, agreeing nearly enough with the value found by Maxwell's method, viz., 149,289 metres*. When all the evidence is taken into consideration, there can remain, I suppose, little doubt that the value 451,000 is substantially correct, and that the reductions of the Committee are affected by a serious under-estimate.

* The arithmetical calculations were made from the data given by the Committee (Reprint, p. 115), $a = \cdot153194$, $2b' = \cdot03851$, $b = \cdot01841$ (not ·1841), $c = \cdot01608$, all in metres. $n = 313$. The whole number of turns (313) was supposed to be equally divided between the two parts.

Professor Rowland, in ignorance apparently of Maxwell's previous calculation, has shown that if in the original experiments we assume an unknown cause of error proportional to the square of the speed, and eliminate it, we shall arrive at a value of the ohm differing very appreciably from that adopted by the Committee. In this way he finds that

$$1 \text{ ohm [B.A. unit]} = \cdot 9926 \frac{\text{earth quadrant}}{\text{second}} .$$

Rowland is himself disposed to attribute the error to currents induced in the frame. Our experiments prove these currents had not much effect, though they may explain the difference between the value of L which best satisfies our experiments (where the currents could not exist), i.e., 451,000, and the higher value 457,000 calculated by Maxwell as most in harmony with the original experiments. The process adopted by Rowland is evidently equivalent to determining the coefficient of self-induction from the deflections themselves, and his result, rather than that given by the Committee, must be regarded as the one supported by the evidence of the original experiments.

Rowland's own determination, by a wholly distinct method, gives—

$$1 \text{ ohm [B.A. unit]} = \cdot 9911 \frac{\text{earth quadrant}}{\text{second}} ;$$

and according to our experiments the ohm is even smaller—

$$1 \text{ ohm [B.A. unit]} = \cdot 9893 \frac{\text{earth quadrant}}{\text{second}} .$$

The question, therefore, arises whether any further explanation can be given of the different result obtained by the Committee. The value of GK employed in calculating the experiments according to (4) was $GK = 299,775$ metres. For the principal term in GK, as given by (7), we require the values of n, a, and α. From p. 115 of the Reprint we find $a = \cdot 158194$ metre, $n = 313$. The angle α must be recalculated, as the value of $\log \sin^3 \alpha$ ($1 \cdot 9624955$) is evidently incorrect. From $2b' = \cdot 03851$ metre, by means of $\sin \alpha = a/\sqrt{(a^2 + b'^2)}$, we find $\log \sin^3 \alpha = \overline{1} \cdot 99043$. From these data the final value is $GK = 299,290$ metres, differing appreciably from that used by the Committee. The further discussion of the question is a matter of difficulty at this distance of time. There may have been some reason for the value adopted, which it is now impossible to trace, so that I desire to be understood as merely throwing out a suggestion with all reserve. But I think it right to point out a possible explanation, depending upon the interchange of the axial and radial dimensions in the paper on the Electro-magnetic Field. The data there given are the mean radius, the two dimensions of the sections, and the distance between the coils ($\cdot 02010$). This distance is correct, being equal to $2b' - b$, that is, to ·03851 − ·01841. The distance between the mean planes of the coils is not given, but could, of course, be calculated by addition of ·02010 and ·01841. If, however, the radial dimension ·01608 were substituted for the axial dimension ·01841, an erroneous value would be obtained for $2b'$, that is, ·03618 instead of ·03851. Using ·03618 to calculate α, I find $GK = 299,860$ metres, agreeing much more nearly with the value used in the reductions.

If it be thought probable that the value of GK was really 299,290, a still further reduction of nearly two parts in a thousand must be made in the number which expresses the ohm in absolute measure, and we should get—

$$1 \text{ ohm [B.A. unit]} = ·9910 \; \frac{\text{earth quadrant}}{\text{second}},$$

coinciding practically with the value obtained by Rowland from his own experiments.

In the course of our experiments various doubts suggested themselves, and were subjected to examination. It may be well to say a few words about some of these, though the results are for the most part negative.

The energy of the currents circulating in the coil is expended in heating the copper, and a rise of temperature affects the resistance. Calculation shows that the disturbance from this cause is utterly insensible. If at the highest speeds of rotation all the heat were retained, the rise of temperature would be only at the rate of $3·2 \times 10^{-8°}$ C. per second.

Much more heating may be looked for during the operation of taking the resistance. Under the actual circumstances a rise of resistance of about one part in 30,000 might be expected as the effect of the battery current in one minute. The aggregate duration of the battery contact in each of the resistance measurements was probably less than a minute.

Another question related to the possible effect of a want of rigidity in the magnetism of the needle. It is known that galvanometers will some-times, when it is certain that there is no average current passing through the coils, show a powerful effect as a consequence of fluctuating magnetism corresponding to the fluctuating magnetic field. In the present experiment the magnetic field is fluctuating, and the magnet is expected to integrate the effect as if its own magnetism were constant. It is unlikely that any appreciable error arises in this way, as I find by calculation that a theoretically soft iron needle would point in the same direction as a theoretically hard needle when placed at the centre of the revolving coil.

From the details given the reader will be in a position to judge for himself as to the accuracy of our experiments. If, as we believe, the principal error to be feared is in the measurement of the coil, there is little to be gained by further experimenting with the present apparatus. Accordingly a new apparatus has been ordered, from which superior results may be expected. In designing this several questions presented themselves for solution.

All corrections being omitted, the effect—

$$\tan \phi \propto n^2 a \omega / R \, ;$$

and, if σ denote the section of the wire, and $S (= n\sigma)$, the aggregate section of the coil—

$$R \propto na/\sigma \propto n^2a/S \, ;$$

so that if S be given, $\tan \phi$ is independent both of the number of turns n and of the mean radius a. If ϕ be given, the correction for self-induction depends upon L/GK, while both L and GK vary approximately as n^2a. So far, therefore, there is nothing to help us in determining n and a. The following considerations, however, tell in favour of a rather large radius :—

(1)   Easier measurement of coil.

(2)   Smaller correction for moment of suspended magnet.

(3)   Smaller errors from maladjustment to centre, and from size of magnet.

The question of insulation is important. During the rotation the electromotive force acts independently in every turn, and there is no strain upon the insulation; but in taking the resistance, when a battery is employed, the circumstances are materially different. Any leakage from one turn to another would, therefore, be a direct source of error. It is proposed to use triply covered wire.

In order to obtain room for the tube encasing the fibre, it is necessary to use a double coil. In the new apparatus there will be opportunity for a much larger diameter, by which it is hoped to obtain an advantage in respect of stiffness; but the further question presents itself, whether the interval between the coils should be increased so as to obtain a very uniform field, as in Helmholtz's arrangement of galvanometer. The advantages of this plan would be considerable in several respects, but on the whole I decided against it, mainly on the ground that it would magnify the errors due to imperfect measurement. If we call the effect (so far as it depends upon the quantities now under consideration) u, we have, in previous notation,

$$u = a \sin^3 \alpha = a^4 (a^2 + b'^2)^{-\frac{3}{2}},$$

so that

$$\frac{du}{u} = \frac{4\,da}{a} - 3\,\frac{a\,da + b'\,db'}{a^2 + b'^2} \, .$$

If $b' = 0$, $\qquad\qquad\qquad du/u = da/a$;

but if, as in Helmholtz's arrangement, $b' = \tfrac{1}{2}a$,

$$\frac{du}{u} = \frac{8}{5}\frac{da}{a} - \frac{6}{5}\frac{db'}{a}.$$

The increase of b' from 0 to $\tfrac{1}{2}a$ not only introduces a new source of error in the measurement of b', but also magnifies the effect of an error in the measurement of a. If $b' = \tfrac{1}{10}a$, we have nearly

$$\frac{du}{u} = \frac{da}{a} - \frac{3}{10}\frac{db'}{a},$$

showing that an absolute error in b' has about $\tfrac{1}{3}$ of the importance of an equal absolute error in a.

As will be evident from what has been said already, the treatment of the correction for self-induction is a very important matter. It is probable that L may be best determined from the deflections themselves with the use of sufficiently varied speeds. If L be arrived at by calculation, or by independent experiments, it is important to keep down the amount of the correction. We have seen, however, that L/GK is almost independent of n, a, and S, so that if we regard $\tan\phi$ as given, the magnitude of the correction cannot be controlled so long as a single pair of coils is used. An improvement in this respect would result from the employment of two pairs of coils in perpendicular planes, giving two distinct and independent circuits. In virtue of the conjugate character, the currents in each double coil would be the same as if the other did not exist, and the effects of both would conspire in deflecting the suspended magnet. This doubled deflection would be obtained without increase of the correction for self-induction, such as would arise if the same deflection were arrived at by increasing the speed of rotation with a single pair of coils. A second advantage of this arrangement is to be found in the production of a field of force uniform with respect to time.

However the correction for self-induction be treated, it is important to obtain trustworthy observations at low speeds. In order to get a zero sufficiently independent of air currents, it will be advantageous largely to increase the moment of the suspended magnet. Preliminary experiments have, however, shown that there is some difficulty in getting the necessary moment in a very small space, in consequence of the interference with each other of neighbouring magnets, and thus the question presents itself as to the most advantageous arrangement for a compound magnet.

A sphere of steel, as used by the Committee, has the advantage that if uniformly magnetised it exercises the same action as an infinitely small magnet at its centre. But the weight of such a sphere is considerable in

2—2

proportion to its moment, and it is probable that a combination of detached magnets is preferable. It is possible so to choose the proportions as to imitate pretty closely the action of an infinitely small magnet. Thus, if the magnet consist of a piece of sheet steel bent into a cylinder and uniformly magnetised parallel to the axis, the length of the cylinder should be to the diameter as $\sqrt{3}$ to $\sqrt{2}$. In this case the action is the same as of an infinitely small magnet as far as the fourth term inclusive of the harmonic expansion. Without loss of this property the cylinder may be replaced by four equal line magnets, coinciding with four symmetrically situated generating lines. Thus, if we make a compound magnet by placing four equal thin magnets along the parallel edges of a cube, the length of the magnets should be $\sqrt{3}$ times the side of the cube. This is on the supposition that the thin magnets are uniformly magnetised, as is never the case in practice. To allow for the distance between the poles and the ends of the bars, we may take the length of the bars 2·3 times the side of the cube.

With the new apparatus, and with the precautions pointed out by experience, we hope to arrive at very accurate results, competing on at least equal terms with those obtained by other methods. Most of the determinations hitherto made depend upon the use of a ballistic galvanometer, and the element of time is introduced as the time of swing of the galvanometer needle. There is no reason to doubt that very good results may be thus obtained; but it is, to say the least, satisfactory to have them confirmed by a method in which the element of time enters in a wholly different manner.

Part II.—By Arthur Schuster.

Adjustment of the Instruments and Determination of Constants.

The only adjustments to be made consist in—

1. The levelling of the coil.

2. The suspension of the magnet in the centre of the coil.

3. The proper disposition of the scale and telescope by means of which the angles of deflection are read off.

Level.—The first of these presents no difficulty, and any small error can be easily taken account of in the calculations. It was found that the upper end of the axis of rotation was inclined towards the north by an angle of ·0003 circular measure. Hence, as has already been explained [p. 3], we must in the reductions write throughout $G(1 + ·0003 \tan I)$ or $1·0008\,G$ for G. This correction is small, but a little uncertain, as the coil was not very steadily fixed in its bearings, and small variations in the inclination of the axis could be produced by slightly pressing on one side or the other of the coil. When left to itself the coil seemed, however, very nearly to return to the same position.

The Magnet.—The magnet, which was suspended in the centre of the coil, consisted of four separate magnetised needles, each about 0·5 centim. long. These were mounted on four parallel edges of a small cube of cork. A needle attached to the back of the mirror went through a small hole in the cork, and was kept in its place by means of shellac, to prevent any slipping between the magnets and the mirror. The proper suspension of the magnet is a point of some delicacy and importance. As regards the vertical adjustment, the distance of the cube of magnets from the highest and lowest points of the circular frame was measured, and the magnet raised or lowered until the distances became equal. A pointer was next fixed to the frame, reaching very nearly to the centre of the coil. As the coil was rotated, the pointer described a small circle round the axis of revolution, and the position of the magnet could be easily altered until it occupied the centre of the small circle. It is supposed that this adjustment was made to within less than 1 millim., and could, therefore, for all practical purposes, be considered as perfect. The magnetic moment of the magnet was measured in the usual way. Two closely agreeing sets of measurements showed that at a distance of 1 foot it deflected a suspended needle through an angle, the tangent of which was ·000298. Hence at the mean distance of the coil (15·85 centims.) the deflection would have been ·0021. This number is equal to MK/GH, and will be referred to as tan μ in the discussion of the calculations. The magnetic moment was determined a few days after the last spinnings had been taken; but on each day on which experiments were made, the time of vibration of the magnet was determined, and we thus assured ourselves that no appreciable change in the magnetic moment had taken place while the experiments were going on. The time of one complete vibration was 14·6 seconds.

Adjustment of Scale and Telescope.—The telescope which served to read the angle of deflection rested on a small table to which it could be clamped. In front of the table and below the telescope, the scale could be raised or lowered and fixed when the proper position had been found. It was levelled by deflecting the magnet successively towards both sides, and observing the point of the scale at which the cross wires of the telescope seemed to cut the scale. If in both positions of the mirror the scale was intersected at the same height, it was considered to be sufficiently levelled. It remained to place the scale at right angles to the line joining its centre to the mirror. This was done by measuring the distance of both ends to the mirror by means of a deal rod, with metallic adjustable pointers (presently to be described), and altering the position until these distances were equal. It is supposed that considerable accuracy was thus obtained. A small remaining error would be eliminated by observing deflections on both sides of the zero. To adjust the telescope we had now only to point it to the centre of the mirror, and at the same time to place it in such a position that its optic axis passed vertically over the centre of the scale. By suspending a plumb-line from the telescope so as to divide its objective into two equal parts, and focussing alternately on the mirror and on the image of the scale, both points could be simultaneously attended to.

To measure the distance of the scale from the mirror the deal rod used for the adjustment of the scale was cut down so as to have nearly the required length. The two brass pointers attached to the two ends made an angle of about 45° with the rod. One of the pointers was fixed, but the other could be moved round a fixed point in the rod by means of a screw. As it moved, the distance of the two points changed, and by properly supporting the rod and leaning one point against the centre of the scale at a known height from the ground, while the moveable point was made to touch the centre of the mirror, the distance could be accurately found. It was compared with the scale itself, in order that the calculation of the angles of deflection should be independent of the absolute length of a scale division. The length required is the shortest line between the centre of the mirror and the plane of the scale, and this can be calculated if the difference in height of the centre of the mirror and the point to which the distance was measured, is known. These heights were determined by means of a cathetometer. The height of the centre of the objective was measured at the same time; so that all data required to find the inclination of the normal of the mirror to the horizontal are known. The following numbers were obtained; each division of the scale is for simplicity supposed to be equal to 1 millim., which is very nearly correct, but as has been said, its absolute value is of no importance.

Distance of mirror from scale in centims. .... 252·28

As the position of the magnet was always read off through a glass plate, a small correction equal to the thickness of the glass (3·2 millims.) multiplied into $(1 - \mu^{-1})$, where μ is the refractive index, has to be applied. This correction is subtractive and equal to   0·11

Hence, $D =$ .... 252·17

It was also found that the mirror pointed downwards, and made an angle of ·004 with the horizontal. A small correction due to this cause will be discussed in another place.

Torsion.—The torsion was as much as possible taken out of the silk fibre, which was about 4 feet long, before the magnet was attached to the mirror. The coefficient of torsion was determined by turning the magnet through five whole revolutions and observing the displacement of the magnet. It was calculated from the numbers obtained that one revolution shifted the position of rest through 5·6 scale divisions, corresponding to an angle of ·001107.

Another experiment in which the magnet was turned in the opposite direction gave ·001117.

Hence $\tau = ·00111/2\pi = ·00018.$

The correction due to torsion is best applied to the value of G at the same time as the correction for level by writing everywhere

$$G\frac{1 + \beta \tan I}{1 + \tau} \text{ for } G.$$

Constants of the Coil.—The accurate determination of the constants of the coil forms the most difficult part of the measurements. Unwinding the coil, the outer circumference of the successive layers was measured by means of a steel tape. Each measurement was repeated several times, and the agreement was satisfactory. The thickness of the wire was found to be 1·37 millims., which, on the circumference of the successive layers, should produce a constant difference of 2·74 π or 8·62 millims. Owing, however, to defective winding, each layer enters a little into the grooves of the subjacent layer, and the differences in circumference of successive layers were therefore always smaller than they ought to have been. The differences varied between 7·7 millims. and 8·6 millims., but generally were about 8·1 millims. The wire was a little too thick, and as it had been taken off the coil and rewound, small irregularities were formed which prevented a satisfactory winding. Each coil had 156·5 windings. Of these 156 were in one coil regularly distributed over twelve layers of thirteen windings each; while half a turn was outside. In the second coil the twelve layers only contained 155 windings, and one turn and a half was lying outside. In the calculation for mean radius it was assumed that each complete layer contained the same number of turns. Let S be the sum of all measurements for one coil, also C the circumference of the layer containing the loose extra turns; then we find the mean circumference of the first coil,

$$\frac{13S + 0·5C}{156·5} \quad\ldots\ldots\ldots\ldots\ldots\ldots = \quad 99·680$$

and for the second,

$$\frac{(13 - 1/12)S + 1·5C}{156·5} \quad\ldots\ldots\ldots\ldots = \quad 99·651$$

Or as the circumference of the outside of the mean turn ... =   99·666

From this is to be subtracted a correction equal $\pi \times$ thickness

   of tape .... =   ·031

                                                                              99·635

To obtain the circumference of the axis of the mean winding we have to subtract π × thickness of wire .... =   ·431

Hence the final value of the mean circumference ....β =   99·204

Or for the mean radius ....a =   15·789

The grooves of the coils and their distance was also measured, and it was found that—

       b = axial dimension of coil .... =   1·833

       b' = distance of mean plane from axis of motion .... =   1·918

All measurements are given in centimetres.

    We calculate—

α = angle subtended at axis by mean radius = $\cot^{-1}(b'/a)$..... =   83° 04'

And $\log \sin^3 \alpha$ .... = $\bar{1}$·990458

The principal term in the expansion of GK is $\pi n^2 \beta \sin^3 \alpha$... = 29,869,300

To this is to be added a small correction given on [p. 4]... =   100

The final value of GK being 29,869,200

    Applying the corrections for level and torsion to GK as explained, and writing 𝕲𝕶 for the value so corrected, we find

$$\mathfrak{GK} = 29,887,600.$$

The Observations.

    The observations consisted of two parts : the comparison of the resistance of the rotating copper coil with that of a standard German silver coil, and the observation of the deflections during the spinnings. The comparison of resistance was made by a resistance balance devised by Mr Fleming*, to whom we are indebted for advice and assistance in all questions concerning the comparison of resistances. In this balance, which only differs by a more convenient arrangement from an ordinary Wheatstone's bridge, Professor Carey Foster's method of comparing resistances is used. The method consists in interchanging the resistances in the two arms of the balance which contain the graduated wire, and thus finding the difference between these two resistances in terms of that of a certain length of the bridge wire. The balance was placed on a table near the rotating coil, and could be electrically connected with it by means of two stout copper rods. The German silver coil which served as the standard of comparison was prepared so as to have a resistance nearly equal to that of the copper coil, that is about 4·6 ohms. Any error due to thermo-electric currents, which have sometimes been found to be generated at the moveable contact of the galvanometer circuit with the bridge wire, is eliminated in Foster's

* *Phil. Mag.* vol. IX. p. 109, 1880.

method, but to ensure greater accuracy and safety all measurements were repeated with reversed battery current. The whole comparison seldom occupied more than five minutes; and the readings obtained with the battery current in different directions closely agreed with each' other.

The spinnings were always taken in sets of four at the same speed, and the comparison of resistance was made at the beginning and end of each set. During the time of spinning the resistance was found to have altered owing to a rise of temperature which always took place during the time of experimentation. The corrections for the change of resistance and the possible errors introduced owing to the uncertainty of this correction will be described further on.

After the resistance of the coil had been measured, it was disconnected from the balance and set into rotation with open circuit, so that no current could pass. While the water supply was adjusted so as to give approximately the required speed, the magnet in the centre of the coil, which had been strongly disturbed during the measurement of resistance, was brought to rest either by means of an outside magnet or more often by means of a small coil and LeClanché cell, which was always placed in the neighbourhood of the rotating coil. A key within reach of the observer served to make and break contact at the proper time. When the speed had been approximately regulated and the magnet was vibrating through a small arc only, its position of rest was determined, while at the same time the auxiliary magnetometer placed in the adjoining room was observed. The two ends of the rotating coil were now connected together, by means of a stout piece of copper, the well amalgamated ends of which were pressed into cups containing a little mercury, into which they tightly fitted.

As the coil was set into rotation the magnet slowly moved towards one side, and a proper use of the damping coil brought it quickly to approximate rest near its new position of equilibrium. When the swings extended through no more than ten or twenty divisions of the scale, the coil was kept, as nearly as possible, at the proper speed, by the observer at the tuning-fork (Lord Rayleigh, see p. 8). Readings of the successive elongations were taken for about two minutes, and a signal given at the beginning and end of each set of readings enabled the observer at the auxiliary magnetometer (Mrs Sidgwick) to note its position as well as any changes in the direction of the earth's magnetic force during the time of observation. The direction of rotation was now reversed, and the deflection observed in the same manner; the whole process being twice repeated, so that four sets of readings were obtained. When all the observations for the given speed had been completed, the position of rest of the magnet, when no current passed through the coil, was again determined and compared with the auxiliary magnetometer. A recomparison of resistance with the standard completed the set.

The magnet in the centre of the coil should, when no current is passing through the coil, always go through the same changes as the magnet of the auxiliary magnetometer. If this could be ensured, the two might be compared once for all, or the comparison might even be omitted altogether, for the difference between the deflections of positive and negative rotations, when corrected for changes in the earth's magnetism, would give the double deflection independently of the actual zero position. Unfortunately, however, and this was our greatest trouble, the comparison between the magnet and the auxiliary magnetometer showed that we had to deal with a disturbing cause, which rendered a frequent comparison between the two instruments necessary. This disturbing cause, which we traced to air currents circulating in the box containing the magnet, will be discussed presently.

The observations were taken on three different evenings and one afternoon. The evenings (8 h. P.M. to 11 h. P.M.) were chosen on account of the absence of disturbances, which, during the usual working hours, are almost unavoidable in a laboratory. We may give, as an example for the regularity with which the magnet vibrated round its position of rest, a set of readings which were taken while the coil revolved about four times in one second, the circuit being closed.

$$T = 9^h \ 36^m. \qquad t = 13°·0 \ C.$$

Rotation.		Negative.
374·4	....	362·1
373·3	....	362·8
372·2	....	362·0
373·9	....	361·4
372·8	....	362·0
372·8	....	362·0
372·4	....	363·8
371·8	....	364·0
371·1	....	364·0
370·5	....	

Mean.... 372·52   ....   362·68

Position of rest, 367·60.

$$T = 9^h \ 38^m·5. \qquad t = 13°·0 \ C.$$

The number of readings taken was not always the same, but varied generally between sixteen and twenty.

We used, in the course of our experiments, four different speeds. The method of obtaining and regulating these has been explained by Lord Rayleigh. For simplicity we generally denoted the speed by means of the number of teeth on the circle which seemed stationary when looked at through the tuning-fork; thus we spoke of a speed 24 teeth, 32 teeth, and 60 teeth. To obtain the lowest speed the circle containing 60 teeth was looked at over the top of the tuning-fork, so that only one view for each complete vibration was obtained; this was equivalent to a circle of 120 teeth in the ordinary arrangement, which allowed a view for each half vibration, and, consequently, the lowest speed was called 120 teeth. The velocity of rotation depends, of course, on the frequency of the fork, which varied only within narrow limits, and was always very near 63·69. If f denote this frequency and N the number of teeth on the stationary card, the velocity of rotation ω is given by the equation $\omega = 4\pi f/N$. In the "British Association Report" the speed is always indicated by means of the time occupied by 100 revolutions. If T is this time, we find $T = 50N/f$. The following table gives the comparison of ω, T, and N, on the supposition that the frequency of the fork was always the same and equal to 63·69.

N.		ω.		T.		Number of turns per second.
120	......	6·670	......	94·206	......	1·06
60	......	13·339	......	47·103	......	2·12
32	......	25·011	......	25·122	......	3·98
24	......	33·348	......	18·841	......	5·31

The last column gives the number of turns per second.

Three speeds were taken on each of the three nights, and one set of observations with the lowest speed was secured in the course of one afternoon. We obtained in this way three sets for each of the two intermediate speeds and two sets for the lowest and highest speeds. A comparison with the Report of the British Association Committee shows that we do not go up quite to their highest speeds, but that on the other hand our lowest speed was considerably below the one used by them. In the Report for the year 1863, it is mentioned that the forced vibration of the magnet, depending on the rotation of the coil, could not be noticed, and it is calculated that the amplitude of this vibration was less than $\frac{1}{100}$ of a millimetre on the scale. No mention is made of this forced vibration in the Report for 1864, although much lower speeds were used during that year. In our lowest speed a slight shake of the needle, due to the varying action of the currents in the coil, was distinctly seen; but as calculation showed that the amplitude was only the eighth part of a millimetre on the scale, no appreciable error is supposed to have been introduced by it. The moment of inertia of the suspended parts was higher in the experiments made by the British Association, and this, no doubt, is partly the reason why this forced vibration escaped their notice.

Air Currents.

It has already been noticed that air currents in the box containing the magnet effected its position to some extent, and we had to investigate in how far our final results might be affected by this disturbance. During the first night (December 2) our attention had not been drawn so much as it was afterwards to the effect of these air currents. We had previously ascertained, by a series of careful measurements, that the rotation of the coil with open circuit did not sensibly affect the zero position of the magnet, and we considered it sufficient to note the zero as short a time as possible before each set of four spinnings. The comparison of these zeros with the auxiliary magnetometer showed that during the two hours of experimenting, the needle had kept its zero within two divisions of the scale, so that the changes during two successive spinnings (generally about five minutes) must have been very small. On the second night (December 6), however, the zeros were taken at the beginning and end of each set of four spinnings, and the disturbance due to air currents was found to be of more importance. The following table reveals the fact that during a set of spinnings the magnet seems to have moved in one direction, but that during the time the coil was at rest and the comparison of resistance was made, it went in the opposite direction. The numbers given are corrected for changes in the direction of the earth's magnetic force as shown by the auxiliary magnetometer.

December 6.

Number of teeth on stationary circle.	Time. h. m.	Position of rest.	Approximate deflection.
60	8 53	763·60	218
	9 12	766·35	
32	9 31	764·88	397
	9 56	765·78	
24	10 9	762·67	514
	10 38	766·48	

Here, then, we have a gradual rise in the zero from one to over three divisions during a set of four spinnings. The approximate deflection is given in order to give an idea what amount of error the uncertainty of the zero might introduce.

Special experiments were now made, and it was found that by placing a lamp about a foot and a half from the magnet box, changes amounting to eighteen divisions of the scale would be observed; greater precautions were taken, in consequence of the experience thus gained, to secure the box from the radiation of the lamp and gas-jets, which could not be dispensed with in the course of the experiments. The magnet box was covered with gold-leaf so as to reflect the heat as much as possible. On the last night of work frequent determinations of the position of rest were made, but in spite of all precautions an unknown cause produced a sudden displacement of five scale divisions. The exact time at which this change took place could not be determined, and two spinnings were therefore rejected. During the remainder of the evening the magnet gradually came back to its original position. With the exception of the two spinnings just mentioned we have not rejected any observations.

When we come to inquire into the amount of uncertainty to which our results are liable, owing to the effects of these air currents, we find that it cannot be greater than the more dangerous, because less evident, errors to which the determination of our constants (mean radius and distance of mirror from scale) are subject. As long as the changes of the position of rest take place irregularly, the error would tend to disappear in the mean, and the probable error deduced from our experiments would give a fair idea of the uncertainty due to this cause. This probable error, as we shall see, is very small. A regular displacement of zero in one direction would, however, produce a constant error which would not disappear in the final mean. We have some evidence that such a regular displacement has to some extent taken place. The comparison of zeros on December 6, as quoted above, for instance, shows the position of rest in the course of the spinnings shifted towards increasing numbers. Such a shift, if not taken into account, would tend to make the deflections towards increasing numbers (positive rotation) appear larger than those towards decreasing numbers. This, indeed, was observed. Supposing the shift takes place regularly during the time of spinnings we might have taken it into account. But the correction which we should have had to apply is so small and uncertain that it is doubtful whether we should have improved our final result, and it would certainly not have altered it within the limits within which we consider it accurate; for we find that reducing the deflections on the supposition, 1st, that the zero has kept constant; and 2nd, that it has changed uniformly during each set of spinnings; the two results agree to within about one and a-half tenths of a division, which, even at the lowest speeds, would only make a difference of about 1 in 750, and on the highest speeds four times less. The fact that a regular shift in the zero position of the magnet affects the mean of four spinnings is due to the arrangement of experiments, adopted during the first two nights, in which four rotations succeeded each other in alternate directions. If, after a rotation in the positive direction, two negative rotations, followed again by a positive one, had been taken, a regular displacement such as that we are discussing would not have affected the mean. This latter plan was adopted on the last night. In the measurements undertaken by the British Association Committee, the deflections in one direction were generally greater than in the other, and the difference was ascribed to a considerable torsion in the fibre of suspension, which, in order to explain the discrepancy, must, as pointed out by Rowland, have displaced the magnetic axis considerably out of the meridian. The differences in the readings taken when the coil was spinning in opposite directions were, on the average, about 3 per cent., and amounted in one case to 8 per cent. They show no regularity dependent on the speed of rotation. We also observed some slight differences of the same nature; but they are very small, and always remain within such limits that they may easily have been produced by a displacement due to air currents. On the last night, when more frequent zero readings were taken, the differences were, it is true, not much reduced in amount, but their sign was reversed. It is, perhaps, worth remarking that, owing to the absence of any controlling instrument equivalent to our auxiliary magnetometer, the Committee of the British Association had no opportunity of discovering the presence of these air currents, as any changes in the zero position would naturally have been ascribed by them to a casual change in the direction of the earth's magnetic force. Owing to the different shape and material of the box containing the mirror, it seems possible that the effect of air currents may have been considerably larger than it has been in our experiments.

Reduction of Observations.

Scale Corrections.—The first step in reducing the observations consists in calculating the value of $2 \tan \phi$ from the observed deflection. The correction to be applied to the reading in order to obtain numbers proportional to the tangents of deflection, if the position of rest of the magnet is at the centre of the scale, would be $-d^3/4D^3$; d being the observed reading, and D the distance of the mirror from the scale. If the zero, however, is at a point δ of the scale, the correction becomes $-(d-\delta)(d+\delta)^2/4D^2$, where δ is to be reckoned positive when in the same direction as d. For the higher speeds a second correction, to $+d^5/8D^4$, was applied, which comes just within the limits of accuracy aimed at in the actual readings. The corrected deflections so obtained are equal to $2D \tan \phi$. Small errors, due to a faulty division of the scale, ought also to be applied. It is difficult to obtain a proper scale in one piece of sufficient length to be used in these experiments; and the one in actual use consisted of three parts, cemented with caoutchouc cement to a thick piece of deal. No appreciable error was introduced by a very slight warping of the wood, and the scales were found to be very accurately divided, but the small errors existing were corrected; small corrections had also to be introduced, which are due to the imperfect joining of the different pieces. The combined correction never amounted to more than ·15 of a division. Each division, as has already been stated, being very nearly equal to 1 millim.

It has already been noticed that the normal to the mirror pointed slightly downwards. The correction due to this want of adjustment seems to have been generally neglected, yet it is not altogether unimportant. If p is the vertical distance between the centre of the objective and that point of the scale where it is cut by the normal to the mirror; also if α is the inclination between the normal to the mirror and the horizontal, the correction to be applied to a deflection d is $dp\alpha/D$, where D is the distance of the mirror from the scale. In our experiments the correction amounted to $d \times 0\cdot00014$, although the angle between the normal and the horizontal was only about 14 minutes of arc. The correction is positive only if the normal lies between the horizontal through the mirror and the line joining the mirror to the cross wires of the telescope.

Correction for Temperature.—We have now to discuss a series of corrections which have to be applied in order to make a comparison of the results obtained in different spinnings possible. It has already been noticed that four spinnings at the same speed were always taken into one set. The comparison of resistance at the beginning and end of each set showed that during the time of spinning the temperature had altered; before combining the mean within each set we had, therefore, to correct for the change of temperature. We endeavoured to keep the room as much as possible at a constant temperature during the experiments; the lamps used were always lighted nearly two hours before beginning, but, in spite of all precautions, the temperature always rose after we had entered the room and begun to work. The thermometer rose at first pretty rapidly through about 1 degree, and then rose slowly until at the end of the evening it stood generally nearly 2 degrees higher than at first. When the first set of spinnings commenced, the rapid rise, as shown by the thermometer in the room, had already subsided; but, as was to be expected, the temperature of the coil was lagging somewhat behind that of the room, and its resistance still rose appreciably. Thus, during the first night, the resistance of the copper coil rose almost ·4 per cent. during the course of the first set of four spinnings. If the curve of temperature of the coil is known, there is of course no difficulty in applying the proper correction. This curve can be obtained approximately by plotting down the measured resistances as ordinates with the time as abscissæ. This was done for all observations made on December 2; but during the succeeding nights it was found that the curve could not be sufficiently well determined by the observations, and that the assumption of a uniform rise in resistance during the time elapsing between two successive measurements would give as good results as the experiments themselves would allow us to obtain. The proper determination of this correction is a subject to which we shall have to give some attention in the more accurate measurements which we have in view. At present it will suffice to point out that, as far as we can judge, the error due to uncertainty of temperature is not more than ·05 per cent. during the first set of spinnings on each night. It is much smaller in the succeeding sets. It may increase the clearness of this account if at this point we give a specimen, worked out in detail, of one set of deflections. Let the resistance of our standard German silver coil, which we always have assumed to be at the temperature of the air, be called S, and the resistance of the rotating coil C. A comparison by means of the balance shows that

$$C = S + a,$$

where a is the resistance of a certain length of the bridge wire, differing slightly at the beginning and end of the experiment.

December 6.

Number of teeth on stationary circle, 32.

Comparison of resistance, $C = S + \cdot 0225$. Time $= 9^h\ 17^m$.

Position of rest.....................766·48.   Time $= 9^h\ 32^m$.

Auxiliary magnetometer......... 26·9.

Time =	$9^h\ 37^m$	...	$9^h\ 42^m$	...	$9^h\ 47^m$	...	$9^h\ 53^m$
$t =$	$13°\cdot0$	...	$13°\cdot0$	...	$13°\cdot0$	...	$13°\cdot0$
Rotation	negative...		positive...		negative...		positive
Deflected reading	367·60	...	1166·40...		366·23	...	1166·09
Auxiliary magnetometer	27·55	...	28·24...		28·50	...	28·30

Auxiliary magnetometer .... 27·2. Time $= 9^h\ 57^m$.

Position of rest .... 767·08.

Comparison of resistance, $C = S + \cdot 0272$. Time $= 10^h\ 0^m$.

From the comparison of zeros with the auxiliary magnetometer at the beginning and end of the experiments, we find for the corresponding readings during the experiments, 766·78 and 27·05. Considering that increased readings, if the magnet in the coil correspond to decreased readings in the auxiliary magnetometer, we find the following numbers for the positions of rest during the experiments:—

Position of rest	766·28	765·59	765·33	765·53
Deflected reading...........	367·60	1166·40	366·23	1166·09
Deflections	− 398·61	+ 400·81	− 399·10	+ 400·56
Scale correction	+ 2·08	− 2·93	+ 2·08	− 2·94
Reduction of temperature to Time $= 9^h\ 37^m$.........	+ 0·05	+ 0·05	− 0·21	+ 0·35
Corrected deflection ...	− 396·55	+ 397·93	− 397·23	+ 397·97

Mean deflection...... 397·42.

$$C = S + 0\cdot0248.$$

When all the spinnings had been reduced in this way, the final calculations for the actual resistance were made. The determination of all
quantities involved has been explained, with the exception of the measurement of the absolute pitch of the tuning-fork.

Rate of Vibration of Tuning-fork.—As has already been explained, the
tuning-fork which was used to regulate the speed was on every night
compared with a standard fork, and our determinations, therefore, all depend
on the absolute pitch of this standard fork. The method used to determine
that pitch has been described by Lord Rayleigh*.

A fork, vibrating about 32 times a second, maintained by means of an
electric current, and driving a second fork of fourfold frequency, was
compared directly with the clock. The vibrations of the driven fork were
simultaneously compared with the standard by counting the number of
beats in a given time. A few experiments have to be made in order to see
whether the fork gains on the clock, or *vice versâ*, and also whether the
standard vibrates quicker or slower than the driven fork. This can be done
by gradually shifting weights on the driver. The difference in the time of
vibration of the clock and driving fork was generally such as to give one
cycle in between 20 or 30 seconds. The driven fork gave at the same time
from 5 to 11 beats per minute.

The experiments agreed well with each other, and both the rate of
vibration and the temperature variation are in close agreement with the
determinations made by Professor McLeod and Mr G. S. Clarke† of other
tuning-forks which, like ours, were made by König.

The following series of determinations was made at a temperature of
about 13° C. :—

128·179	128·181	128·174	128·189
128·180	128·179	128·180	128·185

The small discrepancies would very likely be still further reduced if
greater care was taken to ascertain the exact temperature of the fork. As a
mean of different sets we find

Number of vibrations in 1 second $= 128\cdot180$ for $t = 13^\circ\cdot0$ C.

$$128\cdot161 \qquad t = 14^\circ\cdot6 \text{ C.}$$

From these data and the number of beats counted during each course of
experiments we can, with the necessary accuracy, determine the number of
vibrations of the fork, which served to regulate the velocity of the revolving
coil.

* *Nature*, vol. xvii. p. 12, 1877. [Art. 49, vol. i. p. 331.]
† *Phil. Trans.*, vol. clxxi. p. 1, 1880.

Calculation of Results.—For accurate calculation, the expansion given in the Report of the British Association is not sufficient. Instead of taking into account a greater number of terms, we may with equal facility have recourse to the original quadratic equation for the resistance. We find

$$R = \frac{2f\,GK\cot\phi}{N}\,[\tfrac{1}{2}(1+\tan\mu\sec\phi)+\sqrt{\tfrac{1}{4}(1+\tan\mu\sec\phi)^2-U\tan^2\phi}].$$

In this equation, f, as before, is written for the frequency of the electrically maintained fork, and N for the number of the teeth on the apparently stationary circle.

U is written for

$$\frac{2L}{GK}\left(\frac{2L}{GK}-1\right).$$

The equation is correct if the torsion and deviation from level are taken into account in the value of GK as has been explained. The only approximation used in the equation is that of writing $\tan\mu$ for KM/GH.

Results.

The results of the calculation are collected in the following table. The first column contains the date on which the experiments were made; the second, the speed in terms of the number of teeth on the stationary card; the third column gives the deflection corrected for all scale errors and variations of temperature during each set; the fourth column shows the value of resistance in absolute measure as obtained by calculation on the assumption that the coefficient of self-induction of the coil is $4\cdot51\times10^7$ centims. This absolute resistance refers to the German silver coil, and a small length of the bridge wire at a given temperature. As both the temperature and this length of bridge wire varied in different experiments, the different results cannot be directly compared, but we can easily apply a correction which shall reduce the numbers to the absolute resistance of the German silver coil at a fixed temperature. The temperature chosen was $11°\cdot5$ C., which was approximately the lowest temperature observed in the course of the experiments. The fifth column contains the corrected values, which now can be compared together, and give the absolute resistance of the standard coil as observed on different occasions, and with different speeds. In the last column the mean value for the different speeds is given. In taking these, as well as the final mean, it must be observed that the set of observations made on December 10 with speed 60 teeth contained only two spins, or half the usual number.

Date	Speed. No. of teeth on stationary card	Deflection	$R \times 10^{-9}$	$R \times 10^{-9}$ corrected	Mean
Dec. 7 ...	120	110·42	4·5486	4·5419	4·5364
10 ...		110·22	4·5568	4·5309	
Dec. 2 ...	60	218·61	4·5580	4·5487	4·5467
6 ...		218·30	4·5620	4·5471	
10 ...		218·72	4·5531	4·5422	
Dec. 2 ...	32	397·75	4·5639	4·5417	4·5427
6 ...		397·39	4·5672	4·5415	
10 ...		397·26	4·5687	4·5448	
Dec. 2 ...	24	513·73	4·5719	4·5446	4·5442
6 ...		513·58	4·5734	4·5438	

The mean of all the observations is—

$$R = 4·5427 \ \frac{\text{earth quadrant}}{\text{second}}.$$

The value of the self-induction which was adopted in these calculations is slightly smaller than the values calculated by Lord Rayleigh and Mr Niven. A comparison of the results obtained with different speeds shows that the value must be very nearly correct, for there is no decided difference between the results. Nevertheless, it seemed of interest to calculate the value of the self-induction which best agreed with the experiments, and to see whether that value gave an appreciably different result for R.

We may, in fact, treat both R and L as unknown quantities, and employ the methods of least squares to find out the most probable values. We use for this purpose the approximate values already found, and find the most probable corrections to them. Neglecting the small corrections for torsion, magnetic moment, and level, and writing $P = 2R/GK\omega$, we find for the quadratic which determines R—

$$P^2 - P \cot \phi + U = 0,$$

where U as before is written for

$$\frac{2L}{GK} \left(\frac{2L}{GK} - 1 \right).$$

We find approximately by differentiation, remembering that

$$dP/P = dR/R,$$

$$d(\tan \phi) = -\frac{dR}{R} \left(\frac{1}{P} - \frac{3U}{P^3} \right) - \frac{dU}{P^3}.$$

We may consider dR/R and dU to be the unknown quantities to be determined. The coefficients with which they are multiplied are known with sufficient accuracy. $d \tan \phi$ is found for each observation by putting $dU = 0$ and dR equal to the difference between the value of R calculated by means of that observation, and the value of R provisionally adopted. The usual methods to form the normal equations were employed. We find in this way—

$$R = 10^9 \times (4\cdot5433 \pm 0\cdot0019),$$

$$L = 10^7 \times (4\cdot5130 \pm \cdot0110).$$

It is satisfactory to note that the final value of R derived with the aid of the theory of probability is practically identical with the mean value directly calculated from our experiments with $4\cdot51 \times 10^7$ as coefficient of self-induction. A remarkable agreement is shown between the value of this coefficient of self-induction best fitting our experiments ....…...... $4\cdot5130 \times 10^7$

and the value calculated from the dimensions of
the coil... $4\cdot5145 \times 10^7$.

The large probable error, however, shows that the agreement is partly accidental.

To give an idea of the accuracy with which R has been determined by means of our experiments independently of constant errors, it may be mentioned that the probability of our value being wrong by one in a thousand is only one in ten, while the experiments made by the British Association give an even chance for the same deviation.

It only remains to determine the resistance of the German silver standard in ohms [B.A. units] at a temperature of $11°\cdot5$ C.

We had at our disposal the original standards prepared by the Committee of the British Association. These are very nearly equal at the temperature at which they are supposed to be correct, and the ohm as determined by the Committee is, of course, uncertain within the limits within which the standards differ, but for our present purpose these may be considered identical. The coils were carefully compared by Mr Fleming, who also determined their temperature coefficients. One coil in a flat case (hence called the "flat coil"), which had the smallest temperature coefficient, and supposed to be right at $14°\cdot8$ C., was taken at that temperature as the true ohm. Six of the standards were so arranged and joined together by means of mercury cups, that four were in a row, and the remaining two in double circuit, the whole system of coils being thus equivalent to about $4\cdot5$ ohms. Our standard German silver was nearly $4\cdot6$ ohms. As the difference was too great to allow a direct comparison by means of Mr Fleming's bridge, a piece of German silver wire was prepared so as to have a resistance of $\cdot1$ ohm ;

this could easily be done within the required limits of accuracy by means of a set of resistance coils belonging to the Laboratory. Having thus a set of resistances very nearly equal to the one to be measured, a series of experiments was made on two successive days. Knowing all the temperature coefficients, we could easily reduce the measurements to ohms. Four different experiments gave for the German silver standard at 8°·5 C.—

4·5902,   4·5896,   4·5869,   4·5879,   4·5890.   Mean = 4·5887.

Assuming the German silver to vary 4·4 per cent. for 100° C., we find for our standard at 11°·5 C. 4·595 ohms [B.A. units]. We have hitherto neglected to take account of the resistance of the two stout copper rods which connected the rotating coil with the resistance bridge. This resistance was determined by Mr Fleming to be ·003 ohm. To make matters equal, we ought to have added that resistance to the British Association standards in comparing them with the standard used by us, and we should then have found that the absolute resistance found by us to be equal to

$$4·543 \frac{\text{earth quadrant}}{\text{second}},$$

was equal to 4·592 ohms [B.A. units].

From this we calculate that the ohm [B.A. unit] as fixed by the Committee of the British Association is

$$·9893 \frac{\text{earth quadrant}}{\text{second}},$$

this being the final result of our experiments.

80.

EXPERIMENTS TO DETERMINE THE VALUE OF THE BRITISH ASSOCIATION UNIT OF RESISTANCE IN ABSOLUTE MEASURE.

[*Phil. Trans.* CLXXIII. pp. 661—697, 1882.]

THE present paper relates to the same subject as that entitled " On the Determination of the Ohm in Absolute Measure," communicated to the Society by Dr Schuster and myself, and published in the *Proceedings* for April, 1881 [Art. 79]—referred to in the sequel as the former paper. The title has been altered to bring it into agreement with the resolutions of the Paris Electrical Congress, who decided that the ohm was to mean in future the absolute unit (10^9 C. G. S.), and not, as has usually been the intention, the unit issued by the Committee of the British Association, called for brevity the B.A. unit. Much that was said in the former paper applies equally to the present experiments, and will not in general be repeated, except for correction or additional emphasis.

The new apparatus [fig. 0] was constructed by Messrs Elliott on the same general plan as that employed by the original Committee, the principal difference being an enlargement of the linear dimensions in the ratio of about 3 : 2. The frame by which the revolving parts are supported is provided with insulating pieces to prevent the formation of induced electric currents, and more space is allowed than before between the frame and those parts of the ring which most nearly approach it during the revolution. It is supported on three levelling screws, and is clamped by bolts and nuts to the stone table upon which it stands. The ring is firmly fastened by nuts to two gun-metal pieces which penetrate it at the ends of the vertical diameter, and which form the shaft on which it rotates. The lower end of the bottom piece is rounded, and bears upon a plate of agate, on which the weight of the revolving parts is taken. A little above this comes the driving pulley (9 inches in diameter), and above this again the screw and nut by which the divided card is held. The top piece is hollow, forming a tube with an aperture of $1\frac{1}{4}$ inches, and is held by a well-fitting brass collar attached to the upper part of the frame. On this bearing the force is very small, so that the considerable relative velocity of the sliding surfaces has no ill effect. Notwithstanding its great weight, the ring ran very lightly, and the principal resistance to be overcome was that due to setting air in motion.

Fig. 0.

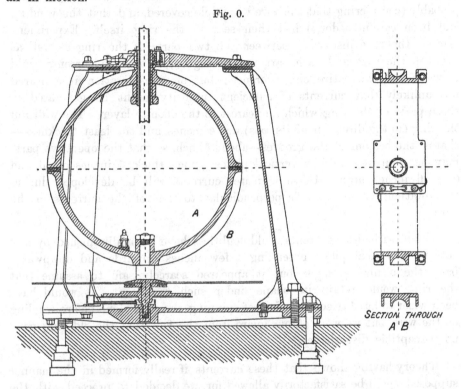

SECTION THROUGH
A'B

In the original apparatus the ring is very light, in fact scarcely strong enough to stand the forces to which it is subjected in winding on the wire. In order to avoid this defect, and also on account of its larger size, the new ring was made very massive. Cast solid, with lugs at the ends of what was to be in use the horizontal diameter, it was cut into two equal parts along a horizontal plane. The two parts were then insulated from one another by a layer of ebonite, and firmly joined together again at the lugs by bolts and nuts, after which the grooves, &c., were carefully turned. As it was intended to use two coils of wire in perpendicular planes, two rings were prepared. The smaller ring fitted into the larger, the end pieces passing through holes along the vertical diameters of both. But for a reason that will presently be given, only the larger ring was used in the present experiments.

In the spring of 1881 the larger ring was wound in Messrs Elliott's shop under the superintendence of Dr Schuster and myself, and the necessary measurements were taken. On mounting the apparatus a few days later in the magnetical room of the Cavendish Laboratory, and making preliminary trials, we were annoyed by finding a very perceptible effect upon the suspended magnet even when the wire circuit was open. The currents thus indicated might have been due to a short circuit in the wire, or more probably (considering that the wire was triple covered, and that the winding had been carefully done) had their seat in the ring itself. Experiment showed that the insulation between the two parts of the ring as well as between the wire and each part, was very good, so that no currents could travel round the entire circumference; but on consideration it appeared not unlikely that currents of sufficient intensity might be generated in those parts of the ring which lie nearest to the ebonite layer. The width of the ring (in the direction of its axis) was 4 inches, and the least thickness— that at the bottom of the grooves—about $\frac{3}{8}$ inch, so that the operative parts may be compared to four vertical plates $\frac{3}{8}$ inch thick, 4 inches broad, and (say) 6 inches high. In these plates currents will be developed during the rotation, whose plane is perpendicular to that of the currents in the wire.

The unwished-for currents could doubtless have been diminished by saw cuts in a vertical plane extending a few inches upwards and downwards from the insulating layer, but it appeared scarcely safe to assume that the ring would retain its shape under such treatment. It would have been wiser to have tried the effect of spinning the ring alone before winding on the wire, but we were off our guard from the fact that the old ring gave no perceptible disturbance.

Theory having shown that these currents, if really formed in the manner supposed, could be satisfactorily allowed for, we decided to proceed with the experiment. At the worst, the differential effect between wire circuit closed and wire circuit open could only be in error by a quantity depending upon the square of the speed, and therefore capable of elimination upon the evidence of the spinnings themselves; while if the view were correct that the disturbing currents were principally in a plane perpendicular to that of the wire, even the correction for induction would not be much affected. A special experiment, in which the ring (with wire circuit open) was oscillated backwards and forwards through a small angle in time with the natural vibrations of the magnet, allowed us to verify the plane of the currents. A marked effect was produced when the plane of the ring was east and west, but nothing could be detected with certainty when it was north and south—the opposite of what would happen with the wire circuit closed. After this, no doubt could remain but that most of the disturbance was due to currents in the ring, and subsequent spinnings after the removal of the wire have proved that no sensible part of it was caused by leakage through the silk insulation. The existence of this disturbance, however, so far modified our original plan as to induce us to omit the second ring as giving rise to too great a complication.

The suspended magnet was made of four pieces of steel attached to the edges of a cube of pith and of such length (about $\frac{1}{2}$ inch) as to be equivalent in their action to an infinitely small magnet at the centre of the cube. Before the pieces were put together the approximate equality of their magnetic moments was ascertained. The resultant moment was between six and seven times as great as that used in our former experiments. In virtue of the greater radius of the coil, this important advantage was obtained without undue increase of the correction for magnetic moment, which amounted to about ·004, only twice as great as before. The effect of mechanical disturbances, such as air currents, was still further reduced by diminishing the size of the mirror, particularly in its horizontal dimension. On both accounts the influence of air currents was probably lessened about 15 times, and, in fact, no marked disturbance was now caused by the proximity of a lamp to the magnet box*. In consequence of these changes, however, it was found necessary to introduce an inertia ring in order to bring the time of vibration up to the amount (about $5\frac{1}{2}$ seconds from rest to rest) necessary for convenient observation. The diameter of the ring was about $\frac{3}{4}$ inch, and the whole weight of the suspended parts was not too great to be borne easily by a single fibre of silk. A brass wire passing between the spokes of the ring prevented the needle from making a complete revolution.

The enlarged scale of the apparatus allowed us to introduce a great improvement into the arrangement of the case necessary for screening the suspended parts from the mechanical disturbance of the air caused by the revolution of the coil. A brass tube of an inch in diameter was not too large to pass freely through the hollow axis. At its lower extremity (fig. 1) it was provided with an outside screw, to which the magnet box was attached air-tight. By unscrewing the box, whose aperture was large enough to allow the inertia ring to pass, the suspended parts could be exposed to view, and by drawing up the brass tube they could be removed altogether, so as to allow the coil to be dismounted, without breaking the fibre. The upper end of the fibre was attached to a brass rod sliding in a socket at the upper end of the tube, by which the height of the magnet could readily be adjusted. The whole was supported on three screws passing through the corners of a brass triangle attached to the tube not far above the place

---

* See pp. 115, 132 [vol. ɪɪ. pp. 11, 28] of the former paper.

where it emerged from the hollow axis. The points of the screws rested upon the same overhanging stand as in the former experiments [p. 9]*.

The larger diameter of the tube made the system so rigid that no mechanical disturbance of the kind formerly met with was to be detected at the highest speed to which we could drive the coil. Even a tap with the finger-nail upon the magnet box produced but a small disturbance.

Fig. 1.

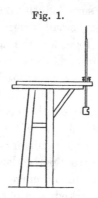

No change was required in the arrangements for regulating and determining the speed of the coil, which worked, if possible, more perfectly than before, in consequence of the greater inertia of the revolving parts. The divided card was, however, on an enlarged scale, and the numbers of the teeth in the various circles were so arranged that each circle was available for a distinct pair of speeds according as it was observed through the slits in the plates carried by the electric fork or over the top of the upper plate. The speeds actually used corresponded to 80, 60, 45, 35, and 30 teeth, seen through the slits, i.e., about 127 times per second.

The greater resistance of the copper coil (23 instead of 4·6) rendered necessary a modification in the method of making the comparisons with the standard. The whole value of the divided platinum-iridium wire on Fleming's bridge being only $\frac{1}{20}$ ohm, a change of temperature in the copper of not much more than a degree would exhaust the range of the instrument. To meet this difficulty it was only necessary to add resistances to the copper circuit so as to compensate approximately the temperature variations, for it is evident that it can make no difference whether the change of resistance of the entire revolving circuit is due to a rise of temperature, or to the insertion of an additional piece. The platinum-silver standard was therefore prepared so as to have a resistance (about 24 ohms) greater than any which we were likely to meet with in the copper, and the additional pieces were relied upon to bring the total within distance. As at first arranged, the additional resistance was inserted at the mercury cups, instead of a contact piece of

* June, 1882. The general disposition of the apparatus is shown in fig. 2.

no appreciable resistance.   During the comparison with the standard it was transferred to another part of the circuit.

In the course of May, 1881, a complete series of spinnings were taken, the arrangements and adjustments being (except as above-mentioned) in all respects the same as with the old apparatus.   Five different speeds were used, and each of them on three different evenings.   The work of observing was also distributed as before, Dr Schuster taking the readings of the principal magnetometer, and Mrs Sidgwick the simultaneous readings of the auxiliary magnetometer, while I observed the divided card and regulated the speed.   At each speed on each evening four readings were taken with

Fig. 2.

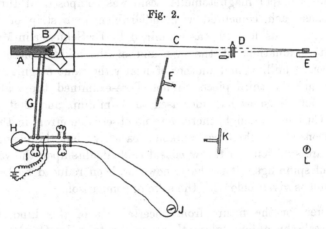

A.	Stand for suspended parts.	H.	Fleming's bridge.
B.	Frame of revolving coil.	I.	Platinum-silver standard.
C.	Driving cord.	J.	Bridge galvanometer.
D.	Electro-magnetic fork and telescope.	K.	Telescope and scale of auxiliary magneto-
E.	Water engine.		meter.
F.	Principal telescope and scale.	L.	Auxiliary magnetometer needle and
G.	Copper connecting bars.		mirror.

wire circuit closed, two with positive and two with negative rotation, and in like manner four readings were taken with the wire circuit open.   Observations on the zero with the coil at rest were for the most part dispensed with, as it was thought that the time could be better employed otherwise ; in fact, the mean of the two not very different positions of equilibrium obtained with positive and negative rotation when the wire circuit was open, gives all that is wanted in this respect.   In the actual reductions we only require the *difference* of readings with positive and negative rotations.

It was hoped that these observations would have been sufficient, but on the introduction by Dr Schuster of the various corrections for temperature, for the beats between the two forks, and for the outstanding bridge-wire divisions, the necessity for which disguises the significance of the numbers first obtained, it was found that the agreement of the results corresponding to a given speed was by no means so good as we had expected in view of the precautions taken and the accuracy of the readings. What was worse, there was evidence of a decided progression, as if the absolute resistance of the standard had gradually diminished during the time occupied by the spinnings.

It is not impossible that there really was some change in the standard which had then been newly prepared; but the discrepancies were not, as according to this view they ought to have been, proportional to the speeds of rotation. I am inclined rather to attribute them to shiftings of the paper scales. The principal magnetometer scale was composed of three lengths of 50 centims. each, cemented with indiarubber to a strip of deal. The compound scale thus formed was examined by Dr Schuster in March, 1881. Between the graduations of the first and of the middle piece there was a gap of about $\frac{1}{4}$ millim., and another of nearly the same magnitude between the middle and the third piece. When I re-examined the scale in July, the gap at 500 divisions had increased to $\frac{9}{10}$ millim., and that at 1000 to $\frac{1}{2}$ millim. Curiously enough, there were no observable errors in the equality of the divisions of the three parts taken separately; but the changes above-mentioned are sufficient to throw considerable doubts upon the value of the first series of spinnings. They have, however, been reduced by Dr Schuster, and the result is given below for the sake of comparison.

To be free for the future from uncertainties of this kind, I replaced the paper scale by a long glass thermometer tube by Casella, graduated into millimetres. The divisions were fine and accurately placed, but the imperfect straightness of the tube has rendered necessary certain small corrections in the final results. Probably a straight strip of flat opal would have been an improvement.

The second series of spinnings was made in August, 1881, and this, it was fondly hoped, would be final. To guard against possible change in the platinum-silver coil a careful comparison with the standard units was previously instituted by Mrs Sidgwick, of which the details are given later. As we had unfortunately lost the advantage of Dr Schuster's assistance, the observations at the principal magnetometer devolved upon Mrs Sidgwick. The much easier post at the auxiliary magnetometer was usually occupied by Lady Rayleigh; occasional assistance has been rendered by Mr A. Mallock and by Mr J. J. Thomson.

In the conduct of the second series one or two minor changes were introduced. In order to know the temperature of the standard tuning-fork more accurately, a thermometer was placed between its prongs and read at the same time as the number of beats was taken. The insertion of the small resistances necessary to bring the copper coil within range of the standard was also arranged in a different manner. Some trouble had been experienced in getting a sufficiently good fit between the contact pieces used in the first series and the mercury cups. It is necessary that the stout copper terminals should press down closely upon the bottoms of the cups, and also that the mercury should not be liable to escape at high speeds from the effect of centrifugal force. Bits of indiarubber tubing were placed round the copper legs, by which a fair fit with the sides of the cups was effected; but I thought that it would be an improvement to revert to a single contact piece for the mercury cups of no sensible resistance, whose fit could be carefully adjusted, and to insert the extra resistances at the connexion of the other (outer) ends of the component coils. For this purpose binding screws were employed, pressing firmly together the flat copper terminals of the copper wire and of the German-silver resistance pieces. It is almost unnecessary to say that these short lengths of German-silver wire were doubled upon themselves before being coiled, and that the pieces were not touched between a spinning and the associated resistance comparisons. Used in this way the screwed up contacts seemed unobjectionable, even though the surfaces were not amalgamated.

On each night and for each speed a set of twelve spinnings was made, six with wire circuit open, and six with wire circuit closed. It was usual to take, first, two of the former (one with positive and one with negative rotation); secondly, to compare the resistances of the revolving circuit and the standard; thirdly, after inserting the contact piece and adjusting the indiarubber strap by which it was held down, to make the six closed contact spinnings; fourthly, to compare the resistances again; and lastly, to complete the open contact readings. Each spinning, it will be understood, involved the reading of several elongations (about six for the open contact and ten for the closed), from which the position of equilibrium was deduced.

Table II. [p. 70] gives all the results of the second series, except one for 35 teeth on August 27th, which was rejected on the ground that it exhibited such large *internal* discrepancies, as to force us to the conclusion that the contact piece had been inserted improperly. It will be seen that the agreement is good except on August 29th, in which case the deflections are as much as four or five tenths of a millimetre too small. These discrepancies, though not very important in themselves, gave me a good deal of anxiety, as they were much too large to be attributed to mere errors of reading, and seemed to indicate a source of disturbance against which we were not on our guard.

The least unlikely explanations seemed to be (1) a change in the distance of the mirror from the scale, which unfortunately had not been remeasured at the close of the spinnings, though this would require to reach 3 millims.; (2) imperfect action of the contact piece from displacement of mercury or otherwise; (3) a change of level in the axis of rotation. The anomalous result of August 27th seemed to favour (2), while on behalf of (1) it must be said that the stand of the telescope and scale as well as the support for the suspended parts of the principal magnetometer were of wood. It was just conceivable that under the influence of heat or moisture some bending might have occurred.

On my return to Cambridge in October we proceeded to investigate these questions with the closest attention. As repeated direct measurements of the distance of the mirror and scale were inconvenient, measuring rods (like beam compasses) were provided to check the relative positions of the telescope stand and of the upper end of the suspending fibre with regard to fixed points on the walls of the room. But no changes comparable with 3 millims. were detected, even under much greater provocation than could have existed during the August spinnings. The next step was to examine the action of the contact piece. For this purpose the coil was balanced against the standard as usual, except that the contact piece was inserted and connexion with the bridge made at the other ends of the double coil. It was presently found that the resistance *did* depend upon the manner in which the contact piece was pressed, and that to an extent sufficient to account for the August discrepancies. Eventually it was discovered that one of the legs of the contact piece, which by a mistake had been merely rivetted and not soldered in, was shaky.

After this there could be no reasonable doubt that the faulty contact piece was the cause of our troubles. In all probability the leg became loose on the 27th, in which case the earlier results would be correct. Moreover, the final means are not very different, whether the spinnings of August 29th are retained or not. This being the case, we might perhaps have been content to let the matter rest here; but in view of the importance of the determination, and the desirability as far as possible of convincing others as well as ourselves, we thought that it would be more satisfactory to make a third and completely independent series of spinnings.

In this series the faulty composite contact piece was replaced by a horse shoe of continuous copper, and a check was instituted upon the distance between mirror and scale. The opportunity was also taken to make a minor improvement in connexion with the auxiliary magnetometer. The somewhat unsteady table on which the telescope and scale had stood was replaced by one of stone, and the arrangements for illumination were improved by throwing an image of a gas flame on the part of the scale under observation. The same number of readings were made as in the second series, but we found it more expeditious to take the six open contact spinnings together. At the beginning of the evening it was desirable to commence with these open contact spinnings in order to give more time for the coil to acquire the temperature of the room, which always rose somewhat, although the lamps and gas were lit a couple of hours beforehand. Later in the evening we sometimes took the closed contact readings for two speeds consecutively, in order that the intermediate resistance comparison might serve for both. In other respects the arrangements were unaltered.

Full details of the observations and reductions are given below. It will be sufficient here to mention that the maximum discrepancy between any two deflections at the same speed amounts only to $\frac{7}{100}$ of a millimetre, so that the agreement on different nights is more perfect than could have reasonably been expected. At the lowest speed the above-mentioned discrepancy is less than one part in 3000, and at the highest speed less than one part in 6000. No spinnings in the third series were rejected, except on one or two occasions when it appeared *at the time of observation*, from the behaviour of the auxiliary magnetometer, that there was too much earth disturbance. The spinnings were then suspended, and the observations already obtained were not reduced.

At the close of the spinnings, Mrs Sidgwick made a further comparison of our platinum-silver coil with the standard units.

The value arrived at for the B.A. unit ($\cdot 9865$ ohm) differs nearly three parts in a thousand from that which we obtained with the original apparatus. This difference is not very great, and may possibly be accounted for by errors in the measurement of the coil (see [p. 11] of former paper). If a coil be imperfectly wound, the mean radius, as determined by a tape, is liable to be too great. At any rate, this discrepancy sinks into insignificance in comparison with that which exists between either of these determinations and that of Professor Kohlrausch*, according to whom the B.A. unit would be as much as $1\cdot 0196$ ohms. With respect to the method employed by Kohlrausch, I agree with Rowland† in thinking it difficult, and unlikely to give the highest accuracy; but how in the hands of a skilful experimenter it could lead to a result 3 per cent. in error, is difficult to understand. The only suggestion I have to make is that possibly sufficient care was not taken in levelling the earth-inductor. Although estimates are given of the probable errors due to uncertainties in the various data, nothing is said upon this subject. In consequence, however, of the occurrence of the horizontal intensity as a square in the final formula, in conjunction with the largeness of the angle of dip, the method is especially sensitive to a maladjustment of

---

* *Pogg. Ann.*, Ergänzungband vi. *Phil. Mag.*, April, 1874.
† *American Journal*, April, 1878.

this kind. I calculate that a deviation of the axis of rotation from the vertical through 21′ in the plane of the meridian, would alter the final result by 3 per cent.*

According to Rowland's determination, the value of the B.A. unit is ·9912 ohm. The method consists essentially in comparing the integral current in a secondary circuit, due to the reversal of the battery in a primary circuit, with the magnitude of the primary current itself. The determination of the secondary current involves the use of a ballistic galvanometer, whose damping is small, and whose time of vibration can be ascertained with full accuracy; and it is here, I think, that the weakest point in the method is to be found. The logarithmic decrement is obtained by observation of a long series of vibrations, and it is assumed that the value so arrived at is applicable to the correction of the observed throw. I am not aware whether the origin of damping in galvanometers has ever been fully investigated, but the effect is usually supposed to be represented by a term in the differential equation of motion proportional to the momentary velocity. This mode of representation is no doubt applicable to that part of the damping which depends upon the induction of currents in the galvanometer coil under the influence of the swinging magnet. If this were all, a correction for damping would be accurately effected on the basis of a determination of the logarithmic decrement, made with the galvanometer circuit closed in the same manner as when the throw is taken. In all galvanometers, however, a very sensible damping remains in operation even when the circuit is open, of which the greatest part is doubtless due to aerial viscosity; and it is certain that the retarding force arising from viscosity is not simply proportional to the velocity at the moment, without regard to the state of things immediately preceding.

In particular, the force acting upon the suspended parts as they start suddenly from rest in the observation of the throw, must be immensely greater than in subsequent passages through the position of equilibrium, when the vibrations have assumed their ultimate character. I calculate that in the first quarter vibration (i.e., from the position of equilibrium to the first elongation) of a disc vibrating in its own plane and started impulsively from rest, the loss of energy from aerial viscosity would be 1·373 times that undergone in subsequent motion between the same phases. From this it might at first appear that in this ideal case the logarithmic decrement observed in the usual manner would need to be increased by more than a third part in order to make it applicable to the correction of a throw from rest; but in order to carry out this view consistently we should have to employ in the formula the time in which the needle would vibrate if the aerial forces were non-existent, instead of the actually observed time of vibration. Now since the action of

* See p. [63].

viscosity is to increase the time of vibration, the second effect is antagonistic to the first, so that probably the error arising from the complete neglect of these considerations is very small.

There is another point in which it appears to me that the theory of the ballistic galvanometer is incomplete. It is assumed that the magnetism of the needle in the direction of its axis is the same at the moment of the impulse as during regular vibrations. Can we be sure of this? The impulse is due to a momentary but very intense magnetic force in the perpendicular direction, and it seems not impossible that there may be in consequence a temporary loss of magnetism along the axis. If this were so, the actual impulse and subsequent elongation would be less than is supposed in the calculation, and too high a value would be obtained of the resistance of the secondary circuit in absolute measure. In making these remarks I desire merely to elicit discussion, and not to imply that Rowland's value is certainly four parts in a thousand too high.

Determinations of the absolute unit have been made also by H. Weber*, whose results indicate that the B.A. unit is substantially correct. In the absence of sufficient detail it is difficult to compare this determination with others, so as to assign their relative weights.

The value of the B.A. unit in absolute measure is involved in the two series of experiments executed by Joule on the mechanical equivalent of heat†. The result from the agitation of water is 24868, while that derived from the passage of a known absolute current through a resistance compared with the B.A. unit was 25187. The latter result is on the supposition that the B.A. unit is really 10^9 C.G.S. If we inquire what value of the B.A. unit will reconcile the two results, we find—

$$1 \text{ B.A. unit} = \cdot 9873 \text{ ohm,}$$

in very close agreement with the measurement described in the present paper. It should be remarked that in the comparison of the two thermal results some of the principal causes of error are eliminated; and it is not improbable that an experiment in which heat should be simultaneously developed in one calorimeter by friction, and in a second similar calorimeter by electric currents, would lead to a very accurate determination of resistance, more especially if care were taken so to adjust matters that the rise of temperature in the two vessels was nearly the same, and a watch were kept upon the resistance of the wire while the development of heat was in progress.

[*June*, 1882.—Since this paper was sent to the Society, Mr Glazebrook has worked out the results of a determination of the B.A. unit in absolute

---

* *Phil. Mag.*, Jan., Feb., March, 1878.

† *Phil. Trans.*, Part II., 1878. Brit. Ass. Rep., 1867; Reprint, p. 175.

measure by a method not essentially different from that adopted by Rowland. The final number is practically identical with that of the present paper; and the agreement tends to show that the difference between ourselves and Rowland is not to be attributed to the use of a ballistic galvanometer.

Reference should have been made to the results of Lorenz*. He finds as the value of the mercury unit *defined* by Siemens

$$1 \text{ mercury unit} = \cdot 9337 \ \frac{\text{earth quadrant}}{\text{second}}.$$

The corresponding number calculated from the results of the present paper with use of the value of the specific resistance of mercury lately found (*Proc. Roy. Soc.*, May 4, 1882) is ·9413. If we invert the calculation, we find that according to Lorenz the value of the B.A. unit would be ·9786 absolute measure. The method of Lorenz is ingenious, and apparently capable with good apparatus of giving a result to much within 1 per cent. Mrs Sidgwick and myself are at present making a trial of it.]

It will be desirable here to consider briefly some of the criticisms of Kohlrausch and Rowland upon the method of the original British Association Committee, which has been adopted in the present investigation without fundamental alteration. The difficulty, remarked upon by Kohlrausch, of obtaining a rapid and uniform rotation, has not been found serious, and I believe that no appreciable error can be due either to irregularity of rotation or to faulty determination of its rapidity. It has also been brought as an objection to the method that a correction is necessary on account of the magnetic influence of the suspended magnet upon the revolving circuit. The theory of this action is, however, perfectly simple, and the application of the correction requires only a knowledge of the ratio of the magnetic moment to the earth's horizontal force. If the magnetic moment is very small, the correction is unimportant; if larger, it can on that very account be determined with the greater ease and accuracy. It is probable that in the original experiments too feeble a magnetic moment was used, and that in consequence the suspended parts were too easily disturbed by non-magnetic causes; but this might have been remedied without increasing objectionably the correction in question. At any rate the larger coil of the new apparatus allows the use of any reasonable magnetic moment.

Perhaps the least advantageous feature in the method is the necessity for creating a violent aerial disturbance in the immediate neighbourhood of a delicately suspended magnet and mirror. If, however, any deflection occurs in this way, very little error can remain when the open contact effect is subtracted from the closed contact effect. The difficulty of avoiding a

---

* *Pogg. Ann.*, 1873.

sensible deflection, due to currents in the ring, when the wire circuit is open, is connected with a special advantage—*i.e.*, the possibility of assuring ourselves that there is no leakage from turn to turn of the coil. In the method followed by Rowland, for instance, such a leakage would lead to error, and could not be submitted to any direct test.

The correction for self-induction cannot be made very small without a disadvantageous reduction of the whole angular deflection; but so far as the wire is concerned it can be calculated *à priori*, or determined by independent experiment, with the necessary accuracy. There is reason, however, to think that the best method of treatment is to determine this correction from the spinnings themselves, combining the results of widely different speeds so as to obtain what would have been observed at a small speed. At small speeds it is certain that all effects of self-induction and of mutual induction between the wire circuit and other circuits in the ring will disappear.

Measurements of coil.

The mean radius of the coil, being the fundamental linear measurement of the investigation, must be found with full accuracy. There has been some difference of opinion as to the best method of effecting this. The greatest accuracy is probably attained by the use of the cathetometer. The measurement of the circumference of every layer by a steel tape has the advantage that the subject of measurement is three times as large, and is much less troublesome. The disadvantage is that if a layer be not quite even, there is danger of measuring the maximum rather than the mean outside circumference. In the present investigation the coil was so large that the tape could be employed without fear*.

Each of the component coils marked A and B had $18 \times 16 = 288$ windings, but in consequence of variations in the thickness of the triple silk covering, there was a difficulty in getting exactly 18 turns into each layer. In the eleventh layer of A it was necessary to be content with 17 turns, and to place an extra turn on the outside, so as to form the commencement of a seventeenth layer—a circumstance which of course was taken into account in calculating the mean. The number thus arrived at, after correction for the thickness of tape, is the mean *outside* circumference. What we require is the mean circumference of the axis of the wire; it may be derived from the first by subtraction of half the difference between the tape readings for the first layer, and for the bottom of the gun-metal groove.

* The original Committee also employed the tape method. Their measurement of the length of the wire when unwound was not in order to find the mean radius, as Siemens and Kohlrausch suppose, but to verify the number of turns.

4—2

The results obtained by Dr Schuster and myself when the coils were wound are:

	Coil A.	Coil B.
Mean of readings in millims.	1489·3	1487·5
Correction for tape	·6	·6
Mean outside circumference	1488·7	1486·9
Correction for thickness of wire	3·4	3·4
Mean circumference	1485·3	1483·5
Mean radius	236·39	236·11
Mean circumference of A and B	1484·4	
Mean radius of A and B (a)	236·25	
Axial dimension of section in millims.	19·9	19·9
Radial	15·9	15·4
Distance of mean planes (2b′)	65·95	

Two or three readings were taken of the circumference of every layer, and to prevent mistakes in the number of turns, the plan described by Maxwell*, of simultaneously winding string on wooden rods, was followed. Without some such device, there is great risk of confusion.

In estimating the degree of accuracy obtainable, we must remember that the circumference of each layer is measured before the outer layers are wound on; any change produced by the pressure of these outer layers is a source of error. We had already observed a tendency in the measurements to be less during the unwinding of a coil than during the winding, and we fully intended to remeasure the coil after the spinnings were completed. This was done on December 6, 1881, by Mrs Sidgwick and myself. As we expected, somewhat smaller readings (by about ¾ millim.) were obtained for the circumference of the middle layers. The results were:

	Coil A.	Coil B.
Mean radius	236·31	236·02
Mean of both	236·16	

or nearly one part in 2000 less than before. Of the two values, it would appear that the latter is more likely to represent the actual condition of the coil during the spinnings, and is therefore entitled to greater weight. If we give weights in the proportion of two to one, we get

$$\text{Mean radius} = 23\cdot619 \text{ centims.}†$$

* *Electricity and Magnetism*, II. § 708.

† August, 1882. At the time of use the tape was compared with a measuring rod, which again has been compared with a standard metre verified by the Standards Department of the Board of Trade. For the purposes of this investigation the differences observed are altogether negligible. I may add that the clock with which the standard tuning-fork was compared (see p. [33] of former paper) was rated from astronomical observations.

Calculation of GK.

We have

$$GK = 2\pi^2 n^2 a \sin^3\alpha \left\{1 + \tfrac{1}{6}\frac{c^2}{a^2} + \tfrac{5}{8}\frac{b^2 - c^2}{a^2}\sin^2\alpha\,\cos^2\alpha - \tfrac{1}{8}\frac{h^2}{a^2}\sin^2\alpha\right\},$$

in which

a = mean radius   = 23·625 (1st measurement)

b = axial dimension of section = 1·990

c = radial dimension of section = 1·565

n = total number of turns   = 576

$2b'$ = distance of mean planes = 6·595

$$\sin\alpha = a \div \sqrt{(a^2 + b'^2)}.$$

From these data we find

$$\log 2\pi^2 n^2 = 6·81617$$
$$\log a \quad = 1·37337$$
$$\log \sin^3\alpha = \bar{1}·98744$$
$$\log \{...\} = \bar{1}·99995$$
$$\overline{\log GK = 8·17693}$$

But if we substitute the adopted value of a, i.e., 23·619 centims., we have by subtraction of ·00011

$$\log GK = 8·17682.$$

Calculation of L.

We may write

$$L = 16^2 \times 18^2 (L_1 + L_2 + 2M),$$

where L_1, L_2 are the coefficients of self-induction of the two parts, and M the coefficient of mutual induction without regard to the number of turns. L_1 and L_2 may be calculated from the formula

$$L = 4\pi a [\log_e (8a/r) + \tfrac{1}{12} - \tfrac{4}{3}(\theta - \tfrac{1}{4}\pi)\cot 2\theta - \tfrac{1}{3}\pi \operatorname{cosec} 2\theta$$
$$- \tfrac{1}{6}\cot^2\theta \log_e \cos\theta - \tfrac{1}{4}\tan^2\theta \log_e \sin\theta],$$

in which r is the diagonal of the section, and θ the angle between it and the plane of the coil. With this formula and with the dimensions as measured when the coil was wound, we get

L_1 (for A) = 1029·3 centims. L_2 (for B) = 1031·9 centims.

It would not be difficult to calculate an approximate correction for the curvature of the coil, but this is scarcely necessary. (See p. [15] of former paper.) Adding the above, we have

$$L_1 + L_2 = 2061·2 \text{ centims.}$$

The value of M was found from the tables given as Appendix I. to § 706 of the new edition of Maxwell's *Electricity*. If we suppose each coil condensed into the centre of its section, we find $M = 4\pi \times 33 \cdot 061$. A more exact calculation by the formula of interpolation explained in Appendix II. gives $M = 4\pi \times 33 \cdot 140$, so that

$$2M = 832 \cdot 88 \text{ centims.}$$

The final result is accordingly

$$L = 16^2 \times 18^2 \times 2894 \cdot 1 = 2 \cdot 4004 \times 10^8 \text{ centims.}$$

These calculations of the coefficients of induction have been made independently by Mr Niven and myself, and are so far reliable; but we must not forget that the accuracy of the result depends upon the accuracy of the data, and that in the present case the diagonal of the section (r) on which the most important part of L depends is an element subject to considerable relative uncertainty. It is probable that the effective axial dimension of the section is somewhat less than the width of the groove, and therefore that the real value of L may be a little greater than would appear from the preceding calculation.

Theory of the ring currents.

If the circuits are conjugate, the currents in the wire and in the ring are formed in complete independence of one another, a circumstance which simplifies the theory very materially. In the same notation as was used in the former paper (p. 105) [p. 2], and with dashed letters for the ring circuit, we have as the equation determining the angle of deflection (ϕ) when the wire circuit is closed,

$$\tan \phi + \tau \frac{\phi}{\cos \phi} = \frac{\frac{1}{2} G K \omega}{R^2 + L^2 \omega^2} \{R + L\omega \tan \phi + R \tan \mu \sec \phi\}$$

$$+ \frac{\frac{1}{2} G' K' \omega}{R'^2 + L'^2 \omega^2} \{R' + L'\omega \tan \phi + R' \tan \mu \sec \phi\}.$$

When the wire circuit is open, the equation determining the angle of deflection (ϕ_0) is

$$\tan \phi_0 + \tau \frac{\phi_0}{\cos \phi_0} = \frac{\frac{1}{2} G' K' \omega}{R'^2 + L'^2 \omega^2} \{R' + L'\omega \tan \phi_0 + R' \tan \mu \sec \phi_0\}.$$

Since τ is an extremely small quantity it is unnecessary to keep up the distinction between $\tau \phi / \cos \phi$ and $\tau \tan \phi$. By subtraction

$$(1 + \tau)(\tan \phi - \tan \phi_0)$$

$$= \frac{\frac{1}{2} G K \omega}{R^2 + L^2 \omega^2} \{R + L\omega \tan \phi + R \tan \mu \sec \phi\}$$

$$+ \frac{\frac{1}{2} G' K' \omega}{R'^2 + L'^2 \omega^2} \{L'\omega (\tan \phi - \tan \phi_0) + R' \tan \mu (\sec \phi - \sec \phi_0)\}.$$

The last term is small, and we may neglect $(\sec \phi - \sec \phi_0)$ in combination with $R' \tan \mu$.

Moreover

$$\frac{\frac{1}{2}G'K'\omega}{R'^2 + L'^2\omega^2} = \frac{(1+\tau)\tan \phi_0}{R' + L'\omega \tan \phi_0},$$

so that

$$(1+\tau)(\tan \phi - \tan \phi_0) = \frac{\frac{1}{2}GK\omega}{R^2 + L^2\omega^2}\{R + L\omega \tan \phi + R \tan \mu \sec \phi\}$$

$$+ (1+\tau)(\tan \phi - \tan \phi_0)\frac{L'\omega \tan \phi_0}{R' + L'\omega \tan \phi_0}.$$

If now we write (GK) for $GK/(1+\tau)$, we get

$$\tan \phi - \tan \phi_0 = \frac{\frac{1}{2}(GK)\omega}{R^2 + L^2\omega^2}\{R + L\omega \tan \phi + R \tan \mu \sec \phi\}\left\{1 + \frac{L'\omega}{R'}\tan \phi_0\right\}.$$

The effect of L' would therefore be to *increase* disproportionately the deflections at high speeds, *i.e.*, contrary to the effect of L. It appears, however, that in these experiments it could not have been sensible. At the highest speed $\tan \phi_0$ was about $\frac{1}{670}$, and ω about 26 per second, so that $\omega \tan \phi_0$ would be about $\frac{1}{26}$. The value of L'/R' is difficult to estimate with any accuracy. But the value of L/R for the wire circuit is about ·01 second, and that for the ring circuit must be much less, so that the terms involving L' may safely be omitted.

The quadratic in R then becomes

$$R^2 - R\frac{\frac{1}{2}(GK)\omega(1 + \tan \mu \sec \phi)}{\tan \phi - \tan \phi_0} + L^2\omega^2 - \frac{1}{2}(GK)L\omega^2\frac{\tan \phi}{\tan \phi - \tan \phi_0} = 0,$$

whence

$$R = \frac{\frac{1}{2}(GK)\omega}{\tan \phi - \tan \phi_0}[\frac{1}{2}(1 + \tan \mu \sec \phi)$$

$$+ \sqrt{\{\frac{1}{4}(1 + \tan \mu \sec \phi)^2 - U(\tan \phi - \tan \phi_0)^2\}}]$$

where

$$U = \frac{2L}{(GK)}\left\{\frac{2L}{(GK)} - \frac{\tan \phi}{\tan \phi - \tan \phi_0}\right\}.$$

L by direct experiment*.

Although the calculated value of L was the result of two independent computations, I considered that it would be satisfactory still further to

---

* In consequence of the necessity which ultimately appeared of introducing an arbitrary correction proportional to the square of the speed of rotation, the result of the present section does not influence the final number expressing the B.A. unit in absolute measure. The method, however, is of some interest, and (it is believed) has not been carried out before with the precautions necessary to secure a satisfactory result.

verify it by an experiment with Wheatstone's balance. The statement of this method and the final formula, as given on pp. [12, 13] of the former paper, being approximate only, it will be convenient here to repeat them with the necessary corrections.

The four resistances in the balance are two equal resistances (10 units each), that of the copper coil P, and a fourth resistance Q (nearly equal to P) taken from resistance boxes, of which P is the only one associated with sensible self-induction. When P and Q are equal, there is no permanent current through the galvanometer; but if the galvanometer circuit be first closed and then the battery current be made, broken, or reversed, the needle receives an impulse, whose magnitude depends upon L.

If x denote the change of current in the branch P, the action of self-induction is the same as that of an electromotive impulse in that branch of magnitude Lx, and the effect upon the galvanometer is that due to this electromotive impulse acting independently of the electromotive force in the battery branch.

In order now to get a second quantity with which to compare the induction throw, the resistance balance is upset in a known manner. If while Q remains unaltered, P be increased to $P + \delta P$, there is a steady current through the galvanometer, which we may regard as due to an electromotive force $\delta P \cdot x'$ in the branch $P + \delta P$, x' being the current through the branch. If θ be the deflection of the needle under the action of the steady current, α the angular throw, and T the time of swing from rest to rest, we have by the theory of the ballistic galvanometer as the ratio of the instantaneous to the steady electromotive force

$$\frac{T}{\pi} \frac{2 \sin \frac{1}{2}\alpha}{\tan \theta},$$

subject to a correction for damping; so that this expression represents the ratio of $Lx : \delta P \cdot x'$. If the induction throw be due to the make or break of the battery circuit, x represents simply the current in the branch P. In the case where the battery current is reversed, we may write $2Lx$ for Lx, understanding by x the same as before. As this method was the one actually adopted, we will write the result in the appropriate form

$$\frac{L \cdot x}{\delta P \cdot x'} = \frac{T}{\pi} \frac{\sin \frac{1}{2}\alpha}{\tan \theta}.$$

In the formula as originally given by Maxwell, and as stated in the former paper, the distinction between x and x' (the currents before and after the resistance balance is upset) was neglected. This step is legitimate if δP be taken small enough, to which course however there are experimental objections. In order that $\tan \theta$ might be of suitable magnitude, it was found necessary to make the ratio of $\delta P : P$ equal to about $\frac{1}{300}$, a fraction too large to be neglected.

In carrying out the experiment it was found more convenient to insert the additional resistance in the branch Q, leaving P unaltered. By the symmetry of the arrangement it is evident that this alteration is immaterial, and that we may take the formula in the form

$$\frac{L}{Q} = \frac{L}{P} = \frac{\delta Q \cdot x'}{Q \cdot x} \frac{T}{\pi} \frac{\sin \frac{1}{2}\alpha}{\tan \theta},$$

x being the current in the branch Q when the resistance balance is perfect, x' the diminished current when the additional resistance δQ is inserted.

The principal difficulties in carrying out the experiment arose from variation in the battery and in the resistance balance. From these causes the results of two days' experiments were rejected, as unlikely to repay the trouble of reduction. On the last day (December 3, 1881) the first difficulty was overcome by using three large Daniell cells (charged with zinc sulphate) in multiple arc. As precautions against rapid change of temperature the copper coils were wrapped thickly round with strips of blanket and deposited in a closed box. The delicacy of our arrangements was such that about $\frac{1}{1000}$ of a degree centigrade would manifest itself, so that it was hopeless to try to maintain the resistance balance absolutely undisturbed. The mode of applying a suitable correction will presently be explained. On December 3, partly by good luck, the necessary correction remained small throughout. In order to avoid a direct action of the current upon the galvanometer needle, the coil was placed at a considerable distance, at the same level, and with its plane horizontal. Any outstanding effect of the kind would, however, be eliminated from the final result by the reversals practised.

The induction throws were always taken by reversal of the battery current. A reversal has two advantages over a simple make or break. In the first place the effect is doubled and is therefore more easily measured; and in the second the battery is more likely to work in a uniform manner, the circuit being always closed except for a fraction of a second at the moment of reversal. The key was of the usual rocker and mercury cup pattern.

The galvanometer was one belonging to the laboratory of about 80 ohms resistance. It was set up by Mr Glazebrook for his experiments by an allied method, and with its appurtenances was ready for use at the time that this determination of self-induction was undertaken. The scale was divided into millimetres, and was placed at a distance of 218 centims. from the galvanometer mirror. The instrument was adapted for ballistic work, as the vibrations were subject to a logarithmic decrement of only about ·0142.

The electric balance was provided for by a resistance box from Messrs Elliotts. The battery current after leaving the reversing key divides itself on entering the box, each part traversing 10 ohms. At the ends of these resistances come the galvanometer electrodes. The first part of the current now traverses the copper coil, and the second part other resistances, after which the two parts reunite and pass back to the battery. In the use of the "other resistances," a special arrangement was adopted which I must now explain. The resistance of the copper coil being somewhat under 24 ohms, the most obvious way to obtain a balance was to add to it a piece of adjustable wire until the whole would balance 24 ohms from the box. The objection to this plan is that the smallest known disturbance which we can then introduce, i.e., by the addition or subtraction of a single unit, is much too great for the purpose.

The difficulty thus arising is completely met by the use of high resistances, taken from a second box, in multiple arc with the 24 ohms.

In order to balance the copper coil and its leading wires at the actual temperature (about 14°), 753 ohms were required in multiple arc with the 24. To calculate the resultant resistance we have

$$\tfrac{1}{24} + \tfrac{1}{753} = \cdot041666667 + \cdot001328021 = \cdot042994688 = \tfrac{1}{23\cdot25869},$$

so that the resistance of the copper coil in terms of the units of the box is 23·25869. A suitable deflection θ was obtained by the substitution of 853 for 753 in the auxiliary box. In this case

$$\tfrac{1}{24} + \tfrac{1}{853} = \cdot041666667 + \cdot001172333 = \cdot042839000 = \tfrac{1}{23\cdot34322};$$

so that the additional resistance was

$$\delta Q = \cdot08453 \text{ unit}.$$

It may be remarked that if the copper coil had been about 1° warmer, its resistance would have been greater by $\tfrac{1}{300}$th part, and the balance would have required 853 instead of 753 in multiple arc with the 24.

On account of the progressive changes already mentioned, it was advisable to alternate the observations of α and θ as rapidly as possible, and to occupy no more time than was really necessary in taking the readings. A good deal of time may be saved by working the key suitably, and by opening and closing the galvanometer branch (at a mercury cup provided for the purpose) so as to avoid producing unnecessary swings, and to stop those due to induction when done with; but it is unnecessary to go into detail in this part of the subject. After a little practice two induction throws, starting with opposite directions of the current, and two observations of steady deflection, one in each position of the reversing key, could be made in about seven minutes. The vibrations of the galvanometer needle were damped by the operation of a current in a neighbouring coil, the current being excited by a Leclanché cell and controlled by a key within reach of the observer at the telescope. The readings were taken by Mrs Sidgwick, while I reversed the battery current, shifted the resistances, and recorded the results.

In the simple theory of the method the induction throw is supposed to be taken when the needle is at rest and when the resistance balance is perfect. Instead of waiting to reduce the free swing to insignificance, it was much better to observe its actual amount and to allow for it. The first step is, therefore, to read two successive elongations, and this should be taken as soon as the needle is fairly quiet. The battery current is then reversed, to a signal, as the needle passes the position of equilibrium, and a note made whether the free swing is in the same or in the opposite direction to the induction throw. We have also to bear in mind that the zero about which the vibrations take place is different after reversal from what it was before reversal, in consequence of imperfection in the resistance balance. At the moment after reversal we are therefore to regard the needle as displaced from its position of equilibrium, and as affected with a velocity due jointly to the induction impulse and to the free swing previously existing. If the arc of vibration (*i.e.*, the difference of successive elongations) be a_0 before reversal, the arc due to induction be a, and if b be the difference of zeros, the subsequent vibration is expressed by

$$\tfrac{1}{2}(a \pm a_0)\sin nt + b \cos nt,$$

in which t is measured from the moment of reversal, and the damping is for the present neglected. The actually observed arc of vibration is therefore

$$2\sqrt{\{\tfrac{1}{4}(a \pm a_0)^2 + b^2\}},$$

or with sufficient approximation

$$a \pm a_0 + 2b^2/a,$$

so that

$$a = \text{observed arc} \mp a_0 - 2b^2/a.$$

In most cases the correction depending upon b was very small, if not insensible. The "observed arc" was the difference of the readings at the two elongations immediately following the reversal. As a check against mistakes the two next elongations also were observed, but were not used further in the reductions. The needle was then brought nearly to rest, and two elongations observed in the now reversed position of the key, giving with the former ones the data for determining the imperfection of the resistance balance. As the needle next passed the position of equilibrium, it was acted upon by the induction impulse (in the opposite direction to that observed before), and the four following elongations were read.

These observations of the throw were followed as quickly as possible by observations of the effect of substituting 853 for 753 units in the auxiliary arc. As soon as the vibrations could be reduced to a moderate amplitude, readings of three or four consecutive elongations were taken. The galvanometer contact was then broken, and the battery key reversed. When the needle had swung over to the other side, the galvanometer contact was renewed, and four elongations were observed. The difference between the two positions of equilibrium represented the disturbance of the resistance balance.

The whole of this disturbance, however, was not due to the additional 100 introduced, but required correction for the corresponding effect observed even with 753 units in the auxiliary arc. For this purpose it was only necessary to add or subtract the difference between the equilibrium positions of the needle with the key in the two positions, as deduced from the observations immediately preceding the induction throws; and in order to eliminate the influence of the progressive change, the mean of these differences as found before and *after* the insertion of the extra 100 units was employed. This result was compared with the mean of the four induction throws contiguous to it, two preceding and two following, and in this way a ratio obtained which was independent of the gradual but unavoidable changes in the battery current and in the copper resistance. After about half the readings had been taken the galvanometer connexions were reversed.

A specimen set of observations will now be given.

$3^h 36^m$ [753]	L	....	264·4
	Induction	....	246·6
$3^h 38^m$	R	....	262·5
$3^h 38^m$ Induction		....	245·9
$3^h 40^m$ [853]	R	....	182·3
$3^h 41^m$	L	....	344·7
$3^h 44^m$ [753]	L	....	264·4
$3^h 44^m$ Induction		....	245·7
$3^h 45^m$	R	....	263·1
	Induction	....	245·6

At 3^h 36^m with 753 units in the auxiliary arc and with battery key to the left, the position of equilibrium, as deduced from two elongations, was 264·4 on the galvanometer scale. The arc of vibration due to induction consequent on shifting the key from left to right, corrected for the free swing, but uncorrected for damping, was 246·6. In like manner with key to the right, the equilibrium position at 3^h 38^m was 262·5 and the arc due to induction was 245·9. The difference 1·9 between 264·4 and 262·5 represented the defect of balance. In the second set of induction throws the corresponding difference is 1·3, showing that the changes of temperature in progress were (at this stage) improving the balance of resistances. The difference between the readings R and L with 853 units is 162·4, the reading L being the higher. Since the reading L is also higher with 753 units, we have to *subtract* from 162·4 the mean of 1·9 and 1·3, *i.e.*, 1·6. The corrected value is thus 160·8. With this we have to compare the mean of 246·6, 245·9, 245·7, 245·6, *i.e.*, 245·9, and we thus obtain as the ratio of the two effects

$$245·9/160·8, \text{ or } 1·529.$$

The numbers obtained in this way were 1·535, 1·532, 1·529, 1·528, mean 1·5310; and with galvanometer reversed 1·534, 1·529, 1·530, 1·530, 1·532, mean 1·5310. The reversal of the galvanometer appears to have made no difference, and we have as the mean of all 1·5310. The comparison of the partial results shows that during the hour and a half over which the readings extended the battery current fell slowly about one part in 120, and that the resistance of the copper gradually increased, until the balance was perfect, and afterwards became too great, the whole change being about one part in 6000, which would correspond to about one-twentieth of a degree centigrade.

A small correction is required in identifying the above determined ratio with $2 \sin \frac{1}{2}\alpha/\tan \theta$. If A be the induction arc and B be difference of equilibrium positions with 853 units when the commutator is reversed,

$$\tan 2\alpha = \tfrac{1}{2}A/D, \qquad \tan 2\theta = \tfrac{1}{2}B/D,$$

where $D =$ distance of mirror from scale $= 218$ centims.

From these we get

$$\frac{2 \sin \frac{1}{2}\alpha}{\tan \theta} = \frac{A}{B} \frac{1 - \tfrac{11}{32}A^2/4D^2}{1 - \tfrac{1}{4}B^2/4D^2},$$

or in the present case with $A = 24·5$, $B = 16·0$,

$$\frac{2 \sin \frac{1}{2}\alpha}{\tan \theta} = \frac{A}{B} \times ·99925,$$

and

$$A/B = 1·5310.$$

So far we have omitted to consider the effect of damping, which must necessarily cause the observed value of A to be too small. If λ be the logarithmic decrement, the correcting factor is $(1+\lambda)$. The throw from zero to the first elongation is diminished by the fraction $\frac{1}{2}\lambda$, and the distance from zero to the second elongation is too small by the fraction $\frac{3}{2}\lambda$. Observations made in the usual manner after the other readings were concluded gave with considerable accuracy

$$\lambda = \cdot0142.$$

The time of vibration was taken simultaneously. It appeared that

$$T = 11\cdot693 \text{ seconds.}$$

A sufficient approximation to the ratio of currents $x':x$ can be obtained by neglecting in both cases the current through the galvanometer, whose resistance (80 units) was considerable in comparison with the other resistances. On account of the small resistance of the battery, the difference of potentials at the battery electrodes may be regarded as given. On these suppositions we get at once

$$\frac{x'}{x} = \frac{10+23\cdot25869}{10+23\cdot34322}, \quad \text{whence} \quad \log(x'/x) = \bar{1}\cdot99891.$$

A more elaborate calculation, in which the finite conductivity of the galvanometer was taken into account, gave a practically identical result,

$$\log(x'/x) = \bar{1}\cdot99886.$$

We may now enter the numbers in the formula

$$L = \delta Q \, \frac{x'}{x} \cdot \frac{T'}{2\pi} \cdot \frac{A}{B} \, (\cdot99925)(1+\lambda),$$

in which we must remember that δQ is to be expressed in absolute measure. Now the value given before, viz. $\delta Q = \cdot08453$, is expressed in B.A. units. What this would be in absolute units involves the entire question to whose solution this paper is directed. We will suppose that

$$\text{1 B.A. unit} = \cdot987 \text{ ohm,}$$

δQ	log $\cdot08453 \times 10^9$	$= 7\cdot92701$
Correction to absolute units	log $\cdot987$	$= \bar{1}\cdot99432$
$A:B$	log $1\cdot5310$	$= \cdot18498$
Correction for finite arcs	log $\cdot99925$	$= \bar{1}\cdot99967$
Correction for damping	log $1\cdot0142$	$= \cdot00612$
Time of vibration	log $11\cdot693$	$= 1\cdot06793$
Ratio of currents	log (x'/x)	$= \bar{1}\cdot99886$
		$9\cdot17889$
	log $2\pi =$	$\cdot79818$
	log $L \ =$	$8\cdot38071$

whence
$$L = 2{\cdot}4028 \times 10^8 \text{ centims.}$$

The value by *à priori* calculation is
$$L = 2{\cdot}400 \times 10^8 \text{ centims.}$$

about one part in a thousand lower*.

Correction for level.

If the axis of rotation deviate from the vertical in the plane of the meridian a corresponding correction is required. If I be the angle of dip, and β the deviation of the axis from the vertical towards the north, the electromotive forces are increased in the ratio $(1 + \beta \tan I) : 1$, in which proportion we must suppose GK increased. (See pp. 106, 124 [pp. 3, 20] of former paper.) The angle of dip at Greenwich for 1881 is about $67° \ 30'$, so that
$$\tan I = 2{\cdot}414.$$

The correction for an error in level is thus of the first order, and is magnified by the largeness of the angle of dip in these latitudes. If the experiments were made at the magnetic equator, we should not only reduce the correction for level to the second order, but also obtain the advantage of a nearly doubled horizontal force.

Observations on the level were made by Dr Schuster on June 1, by myself on August 30, and by Mrs Sidgwick on October 13, and on November 11 and 23. The August observations gave $\beta = {\cdot}26'$; the October observations gave $\beta = {\cdot}30'$; and the November observations gave $\beta = {\cdot}25'$. The position of the axis is necessarily to a slight extent indefinite, and the differences are probably accidental. The same level was used throughout, and the value of its graduations was tested. We may take
$$\beta = + {\cdot}27' = + {\cdot}000079 \text{ circular measure}$$
and
$$1 + \beta \tan I = 1{\cdot}00019.$$

---

* A further small correction is called for by the fact that at actual temperature of the room (about 14°) the resistances given by the boxes were not exactly multiples of the B.A. unit. The difference in the case of the principal box, which is marked as correct at 14°·2, may be neglected, but the resistances taken from the auxiliary box (marked 18°·3) must have been smaller than their nominal value, to the extent of a little over one part in a thousand. By the same fraction δQ, and consequently L, must be *greater* than is supposed in the above calculation. The corrected value of L will be
$$L = 2{\cdot}4052 \times 10^8.$$
It is about *two* parts in a thousand greater than that found from the measured dimensions, and is, in my opinion, quite as likely to be correct.

Correction for torsion.

To determine τ, about five complete turns in either direction were given to the upper end of the fibre. The difference of reading for one turn was found to be in June 2·58, and in August 2·45. If we take as the mean 2·51, we get

$$\tau = \frac{2\cdot 51}{4\pi \times 2670} = \cdot 000075.$$

Value of GK corrected for level and torsion.

Calling the corrected value 𝕲𝕶, we have

$$\mathfrak{G}\mathfrak{K} = \frac{GK\,(1 + \beta \tan I)}{1 + \tau} = \frac{1\cdot 00019}{1\cdot 000075}\,GK,$$

so that

$$\log \mathfrak{G}\mathfrak{K} + 8\cdot 17686.$$

The corrections are in fact almost insensible.

Calculation of U.

In this we take for 𝕲𝕶 the value just found. For L we take the mean of the values found by *à priori* calculation and by direct experiment, *i.e.*,

$$L = 2\cdot 4026 \times 10^8.$$

Thus

$$\log \frac{2L}{\mathfrak{G}\mathfrak{K}} = \cdot 50485, \quad \frac{2L}{\mathfrak{G}\mathfrak{K}} = 3\cdot 1978.$$

For the values of $\tan \phi$ and $\tan \phi_0$ we must anticipate a little. The ratio is itself in some degree a function of the speed, but it will suffice to take the values applicable to the highest speed, for which

$$\tan \phi_0 : \tan \phi = 7\cdot 81 : 439\cdot 41.$$

Thus

$$\frac{\tan \phi}{\tan \phi - \tan \phi_0} = 1\cdot 0181,$$

and

$$\log U = \cdot 84325.$$

Measurement of $\tan \mu$.

This is the tangent of the angle through which a suspended magnetic needle would be turned when the principal magnet is presented to it at a distance $\sqrt{(a^2 + b'^2)}$ to the east or west, the axis of the principal magnet lying east and west. Actual measurements with the aid of the auxiliary magnetometer were made in April, June, and November; and as a check upon the constancy of the magnetic moment frequent observations were taken of the time of vibration.

To explain the procedure it will be sufficient to take the data of the November measurement. Two positions were chosen for the principal magnet, nearly equidistant from the suspended magnet, to the east and west. The length of the line joining the two positions was 695 millims., and it passed horizontally about 36 millims. below the suspended magnet. In each position the magnet was reversed backwards and forwards several times and readings taken. When the principal magnet was to the east, the mean difference of readings due to reversal was 13·55 divisions on the millimetre scale. When the principal magnet was in the westerly position, the corresponding difference of readings was 14·61. We are to take the mean of these, i.e., 14·08, as the difference of readings due to reversal at a distance of 347·5 millims. The half of this, or 7·04, corresponds to the simple presentation or removal of the magnet. The distance from mirror to scale was 2670 millims., so that the tangent of the angle of deflection was $\frac{7·04}{2 \times 2670}$. This result has to be adjusted to correspond with the distance $\sqrt{(a^2 + b'^2)}$, in place of 347·5. Hence

$$\tan \mu = \frac{(347·5)^3 \times 7·04}{(238·5)^3 \times 2 \times 2670} = ·00408.$$

In this calculation the error due to the principal magnet having been necessarily placed at a different level from that of the suspended magnet is ignored. As a matter of fact a relatively considerable correction is required. If θ be the altitude of one magnet as seen from the other, the observed effect is too small in the ratio $(1 - 3\theta^2) : 1$. The above written value of $\tan \mu$ requires to be increased about 3 per cent.; so that we take

$$\tan \mu = ·00420.$$

Measurement of D.

For the first and second series of spinnings the distance from mirror to scale was measured exactly as described by Dr Schuster (p. [22] of former paper). The value adopted for the second series, after correction for the thickness of the glass window in the magnet box, was

$$D = 2669·0 \text{ millims.}$$

The same method of measurement was applied at the beginning of the third set, and a watch was kept by means of the measuring rods already spoken of [p. 46]. Slight movements were in fact observed, principally of the nature of a recovery of the telescope stand from the rather violent treatment to which it had been subjected as a test. Minute corrections are accordingly introduced into the tabular statement [p. 72], so as to make the results of different days comparable. At the close of the spinnings the direct measurement was repeated, when there appeared a slight discrepancy between the results obtained by Mrs Sidgwick and myself. It is in fact rather a difficult matter to say exactly when the pointer has advanced to the equilibrium position of the centre of a suspended mirror, which cannot be prevented from swinging. Although the amount in question was not important, I thought it might be more satisfactory to check the result by another method, and therefore arranged a travelling microscope focussed alternately upon the centre of the mirror and upon a scratch on the window of the magnet box, by which the distance between these two points was determined. The remaining distance between the scratch and the scale was easily measured with the rod. The result tended to confirm the smaller value previously found. The value adopted for the spinnings of the third series before November 5 is

$$2668 \cdot 8 \text{ millims.}$$

and for November 5 and subsequent nights

$$2669 \cdot 4 \text{ millims.}$$

From these numbers we have to subtract 1·1 millim., as a correction for the thickness of the glass window; so that

$$D \text{ before November 5} = 2667 \cdot 7 \text{ millims.}$$

$$D \text{ November 5 and after} = 2668 \cdot 3 \text{ millims.}$$

These distances are expressed in terms of the divisions of the scale, whose exact agreement with millimetres is of no consequence.

Reduction of Results.

In order to give a clear idea of the results and of the manner in which they have been reduced, it will be advisable to quote from the note-book the details of one set of spinnings. I have chosen at random one of the third series made on October 31, 1881, with a speed of "45 teeth."

The first column gives the number of the spinning, the first six being made with wire circuit open, and the last six with the wire circuit closed. In spinnings I., III., V., VIII., X., XII., the rotation was in the direction reckoned negative, and in the remaining ones positive. The second column gives the time, the third the reading of the auxiliary magnetometer, the fourth the reading of the principal magnetometer, the fifth the result of correcting the latter by the former, the sixth and seventh the approximately constant sums and differences of consecutive pairs of numbers in the fifth column, and

TABLE I.

	Time* P.M.	Readings of auxiliary magneto-meter in millims	Readings of principal magneto-meter in millims	Readings corrected by auxiliary magneto-meter	Sum	Difference	Mean deflection	Standard minus copper, in bridge-wire divisions	Temperatures — Standard	Temperatures — Air	Temperatures — Fork	Beats per minute	Correction to middle of bridge	Correction to 13° of standard	Correction to 59 beats	Correction to 11° of fork	Corrected deflection
I.	8.16	81·3	593·38	593·38	1197·24	10·48											
II.	8.18	81·6	604·16	603·86	1197·27	10·45											
III.	8.20	82·2	594·31	593·41	1197·51	10·69											
IV.	··	83·0	605·80	604·10	1197·55	10·65	5·29										
V.	8.23	83·8	595·95	593·45	1197·50	10·60		+212	9·95	12·2	12·5	56					
VI.	8.25	84·2	606·95	604·05													
	8.36	··	··	··	··	··	··	Mean −52	Mean 10·02		Mean 13·05	Mean 56·5					
VII.	8.45	83·1	903·38	901·58	1197·69	605·47											
VIII.	8.47	82·5	297·31	296·11	1197·65	605·43											
IX.	8.50	81·4	901·64	901·54	1197·96	605·12	302·56										
X.	8.52	81·0	296·12	296·42	1197·75	604·91							+·03	−·27	−·10	−·12	302·10
XI.	8.55	81·0	901·03	901·33	1197·89	604·77											
XII.	8.58	81·1	296·36	296·56				−316·5									
	9.3	··	··	··	··	··	··		10·1	12·0	13·6	57					

Rows I.–VI. are grouped under **Contact open**; rows VII.–XII. under **Contact closed**.

* To prevent misapprehension it may be mentioned that the times given in this column are approximate only and were not relied upon to secure simultaneous readings of the two magnetometers. A signal was passed from one observer to the other at the beginning and end of the readings.

the eighth gives the mean deflection from zero, *i.e.*, 5·29 for the open contacts, and 302·56 for the closed contacts.

The ninth column shows the results of the resistance comparisons between the platinum-silver standard and the revolving copper coil before and after the closed contact spinnings. The first number (+ 212) means that at $8^h\ 36^m$ the resistance of the standard exceeded that of the copper by 212 bridge-wire divisions, each of which represents $\frac{1}{20000}$ of an ohm. It will be seen that during the spinnings the resistance of the copper increased, which accounts for the gradual fall observable in the seventh column. The mean of the comparisons before and after spinning is taken to correspond with the mean deflection 302·56. The three following columns show respectively the temperatures of the water in which the standard was immersed, of the air in the neighbourhood of the copper coil, and of the standard tuning-fork, while the thirteenth column gives the number of beats per minute between the electrically maintained and the standard fork.

For the sake of more convenient comparison of the results obtained at the same speed on different nights, small corrections are calculated to reduce the actually observed deflections in the eighth column to what they would have been in a standard condition of the resistance and of the speed. Under each of these heads we have two corrections to consider. In the first place the copper circuit differed in resistance from the standard coil by the outstanding (− 52) divisions of the bridge wire. The resistance of the whole being about 24 ohms, each division of the wire corresponds to one part in 480,000, so that in the present case the correction is additive and equal to 52 parts in 480,000, *i.e.*, is equal to + ·03 division of the scale. This is given in the fourteenth column. Secondly, the resistance of the standard itself depends upon a variable temperature. The mean temperature of the standard in this series was about 13°, to which all the observations are reduced. In the present case the temperature was below the normal, so that the resistances were too small and the deflections too large. Accordingly the correction is negative. To estimate its amount the change of resistance with temperature is taken at three parts in 10,000 per degree; so that in the present case we are to subtract 2·8 parts in 3000 of the whole deflection, *i.e.*, ·27, as entered in the fifteenth column. With use of these corrections we obtain the deflection as it would have been observed had the resistance of the revolving circuit (together with the long copper bars by which it was connected with the bridge) been on every occasion exactly that of the standard at 13°.

In like manner two other very small corrections have to be introduced to make the results correspond exactly to a normal speed of rotation. The standard number of beats is taken at 59, and the standard temperature of the fork at 17°. In the specimen set the number of beats is $2\frac{1}{2}$ per *minute* too small, which means that the octave of the electrically maintained fork made (relatively to the other fork) $2\frac{1}{2}$ complete vibrations per minute *too many*. The whole number of vibrations per minute being 60×127, the speed was too great by $2\frac{1}{2}$ parts in 60×127, by which fraction the observed deflection must be reduced. The correction is thus $- \cdot 10$. But besides this the standard fork at $13 \cdot 05°$ vibrated faster than its normal rate at $17°$, by about one part in 10,000 for each degree of difference. The correction for this is $- \cdot 12$.

In addition to the corrections already mentioned the observations of November 5 and after were subjected to another small correction for the observed change in D.

The accompanying Table (II.) exhibits the results of the second series in a manner which after what has been said will not require much explanation. Column VIII. gives in each case what the deflection would have been if the revolving circuit and the copper connecting bars had exactly balanced the platinum-silver standard at $16°$, the electric fork vibrating at such a speed as to give 59 beats per minute with the standard fork at $17°$, and thus allows us to test the agreement or otherwise of the results obtained on various occasions at the same speed. From this point onwards the means only need be considered; but as there is reason (as already explained) to distrust the observations of August 29, I have added a second mean from which the distrusted elements are excluded. The deflection (d) thus arrived at is equal to $D \tan 2\phi$, whereas what we require is $2D \tan \phi$. The connexion between the two quantities is obtained in a moment from the formula

$$2 \tan \phi = \tan 2\phi \, (1 - \tan^2 \phi),$$

by successive approximation. Thus

$$2 \tan \phi = \tan 2\phi \, \{1 - \tfrac{1}{4} \tan^2 2\phi + \tfrac{1}{8} \tan^4 2\phi\},$$

or

$$2D \tan \phi = d - \tfrac{1}{4} d^3/D^2 + \tfrac{1}{8} d^5/D^4.$$

Column X. gives the value of $2D \tan \phi$, XI. that of $2D \, (\tan \phi - \tan \phi_0)$ in the notation of p. [54], and XIII. that of $\log (\tan \phi - \tan \phi_0)$.

For the further calculation we require the value of ω. If f be the frequency of vibration of the electrically maintained fork, F that of the standard at $17°$, N the number of teeth,

$$\omega = 4\pi f/N,$$

and when the number of beats is 59 per minute,

$$f = \tfrac{1}{2} \, (F - \tfrac{59}{60}).$$

For F at $17°$ we take $128 \cdot 130$ (see p. [33] of former paper), so that

$$f = 63 \cdot 574.$$

TABLE II.—Second Series.

Date, August, 1881 (I.)	Open contact deflections (II.)	Closed contact deflections (III.)	Correction to middle of bridge wire (IV.)	Correction to 16° of standard (V.)	Correction to 59 beats (VI.)	Correction to 17° of fork (VII.)	Closed contact deflections, corrected to standard resistance and speed (VIII.)	Correction for scale reading $-\frac{d^3}{4D^3}+\frac{d^5}{8D^4}$ (IX.)	Closed contact deflections corrected for scale reading (X.)	X. minus II. (XI.)	(XII.)	Logarithm of $\tan\phi-\tan\phi_0$ (XIII.)
60 Teeth.												
15	3·97	228·60	− ·15	+ ·01	·00	·00	228·46	[− ·42			$D = 2669·0$ millims.	
23	3·87	228·68	− ·33	− ·01	·00	·00	228·34	+ ·00]	227·99	224·06		$\overline{2}·62298$
26	3·95	228·41	+ ·04	+ ·01	− ·03	·00	228·43	= − ·42				
Mean	3·93	··	··	··	··	··	228·41					
45 Teeth.												
19	5·25	302·22	− ·33	·00	·00	− ·01	301·88	[− ·97			$D = 2669·0$ millims.	
24	5·27	302·31	− ·34	+ ·01	− ·02	− ·01	301·95	+ ·01]	300·96	295·72		$\overline{2}·74850$
27	5·21	302·48	− ·50	+ ·00	− ·04	− ·01	301·93	= − ·96				
Mean	5·24	··	··	··	··	··	301·92					
35 Teeth.												
15	6·68	383·41	+ ·22	+ ·03	·00	·00	383·66	[− 1·98			$D = 2669·0$ millims.	
24	6·82	383·90	− ·18	+ ·02	+ ·02	− ·01	383·75	+ ·02]	381·60 or 381·74	374·83 or 374·97		$\overline{2}·84645$ or $\overline{2}·84662$
29	6·81	383·53	− ·18	− ·01	− ·05	− ·02	[383·27]	= − 1·96				
Mean	6·77	··	··	··	··	··	383·56 or 383·70	− 1·96				
30 Teeth.												
19	7·86	442·79	− ·01	+ ·02	·00	·00	442·80	[− 3·04			$D = 2669·0$ millims.	
26	7·80	442·73	− ·18	+ ·01	− ·03	·00	442·53	+ ·04]	439·45 or 439·66	431·63 or 431·84		$\overline{2}·90773$ or $\overline{2}·90794$
29	7·81	441·85	+ ·20	+ ·02	− ·03	− ·01	[442·03]	= − 3·00				
Mean	7·82	··	··	··	··	··	442·45 or 442·66	− 3·00				

Thus

$$\log (2\pi \cdot f \cdot \mathfrak{GK}) = 10{\cdot}77832 = \log (10^{10} \times 6{\cdot}0023),$$

and

$$R = \frac{10^{10} \times 6{\cdot}0023}{N\,(\tan \phi - \tan \phi_0)} \left[\tfrac{1}{2} (1 + {\cdot}00422 \sec \phi) \right.$$

$$\left. + \sqrt{\{\tfrac{1}{4} (1 + {\cdot}00422 \sec \phi)^2 - U (\tan \phi - \tan \phi_0)^2\}} \right],$$

in which

$$\log U = {\cdot}84325.$$

TABLE III.—Second Series.

	Number of teeth			
	60	45	35	30
R by preceding formula in ohms .	23·639	23·655	23·660 23·651	23·670 23·659
Resistance of standard at 16° . .	23·642	23·658	23·663 23·654	23·673 23·662
Resistance of standard at 13° . .	23·621	23·637	23·642 23·633	23·652 23·641

The immediate result of the formula is the resistance in absolute measure of the revolving circuit, on the supposition that with the connecting bars it exactly balances the standard at 16°. The resistance of the standard itself is therefore given by addition of the resistance of the bars, i.e., ·003 ohm. In the last line the results are reduced to the temperature of 13° for comparison with the third series.

TABLE IV.—Third Series.

I. Date, 1881	II. Open contact deflections	III. Closed contact deflections	IV. Correction to middle of bridge wire	V. Correction to 13° of standard	VI. Correction to 59 beats	VII. Correction to 17″ of fork	VIII. Correction to $D=2667\cdot7$	IX. Closed contact deflections corrected	X. Correction for scale reading as in Series II.	XI. Correction for curvature of scale	XII. Closed contact deflections corrected for scale	XIII. XII. *minus* II.	XIV.	XV. Logarithm of $\tan\phi - \tan\phi_0$
						60 TEETH.								
Oct. 22	3·92	228·47	+·19	−·06	−·06	−·07	··	228·47						
Oct. 29	3·95	228·74	−·08	−·10	−·06	−·08	··	228·42						
Nov. 7	3·92	228·21	+·27	+·13	−·06	−·01	−·05	228·49						
Mean	3·93	··	··	··	··	··	··	228·46	−·42	+·05	228·09	224·16	$D=2667\cdot7$ millims.	$\overline{2}\cdot62339$
						45 TEETH.								
Oct. 29	5·20	302·93	−·37	−·20	−·08	−·11	··	302·17						
Oct. 31	5·29	302·56	+·03	−·27	−·10	−·12	··	302·10						
Nov. 5	5·22	301·84	+·31	+·17	−·11	−·02	−·07	302·12						
Mean	5·24	··	··	··	··	··	··	302·13	−·96	+·08	301·25	296·01	$D=2667\cdot7$ millims.	$\overline{2}\cdot74414$
						35 TEETH.								
Oct. 22	6·77	383·49	+·64	−·16	−·05	−·12	··	383·80						
Oct. 31	6·77	383·92	+·39	−·33	−·10	−·15	··	383·73						
Nov. 7	6·73	383·48	+·29	+·18	−·10	−·03	−·09	383·73						
Mean	6·76	··	··	··	··	··	··	383·75	−1·96	+·11	381·90	375·14	$D=2667\cdot7$ millims.	$\overline{2}\cdot84702$
						30 TEETH.								
Oct. 29	7·92	443·64	−·25	−·26	−·12	−·17	−·10	442·84						
Nov. 5	··	442·31	+·54	+·22	−·17	−·03	··	442·77						
Nov. 7	7·90	442·74	+·16	+·19	−·11	−·04	−·10	442·84						
Nov.10	7·94	··	··	··	··	··	··							
Mean	7·92	··	··	··	··	··	··	442·82	−3·00	+·11	439·93	432·01	$D=2667\cdot7$ millims.	$\overline{2}\cdot90830$

TABLE V.—Third Series.

	Number of teeth			
	60	45	35	30
R by formula	23·616	23·618	23·627	23·635
Resistance of standard at 13° .	23·619	23·621	23·630	23·638

If we compare the results of the second and third series at the same speed, we find the agreement satisfactory (with a partial exception at the speed corresponding to 45 teeth), especially if we take the means from which the observations of August 29 in the second series are excluded. Adding together all the results of each series we should obtain from the second series 23·638, or with exclusion of August 29, 23·633, and from the third series 23·627, between which the extreme difference is less than one part in 2000. When, however, we compare the values obtained from observations at different speeds, we see from both series, but more especially from the third, evident signs of a tendency to rise with the speed, as if the self-induction of the revolving circuit had been underestimated. In view of the remarkable concordance of the results obtained at the same speed on different nights, it is impossible to attribute these discrepancies to errors of observation, and it is important to consider what cause of systematic disturbance can have remained unallowed for. The first question which presents itself is whether it is possible to admit an error in the adopted value of L sufficient to explain the progression. The proportional correction for self-induction is approximately $- U \tan^2 \phi$, or for the speed of 30 teeth ·0457. For the speed of 60 teeth the correction will be only one-fourth of this. To bring the results for the two speeds into agreement it would be necessary to increase the value of U by nearly 3 per cent., which would correspond to an increase of about one per cent. in L. It is difficult to believe that the value of L adopted for the wire circuit can be in error to this extent.

Another direction in which an explanation might be looked for would be the influence of air disturbance, or from tremor. The accompanying table, however, shows such an extraordinary agreement of the open contact deflections, both among themselves and with numbers proportional to the speeds of rotation, as to prevent us from supposing that this cause of disturbance can have operated.

On the whole, it would appear to be the most probable explanation that there were currents in the ring flowing in circuits not conjugate to the wire circuit, and therefore influencing the induction phenomena. But whatever view we may take on this matter, there is no reason to doubt that the true

TABLE VI.—Deflections with wire circuit open.

	Number of teeth			
	60	45	35	30
Mean of first series	3·89	5·22	6·74	7·85
„ second series	3·93	5·24	6·77	7·82
„ third series	3·93	5·24	6·76	7·92
After the wire had been removed, December 7	..	5·23	6·76	..
Numbers proportional to speed	3·94	5·25	6·76	7·88

value will be obtained by introducing such a correction proportional to the square of the speed as will harmonise the several results, a course equivalent to determining the coefficient of self-induction from the spinnings themselves. In this way the numbers corresponding to any two speeds may be made arbitrarily to agree, but the numbers for the two remaining speeds will afford a test of the admissibility of this procedure. The only hypothesis upon which the simple mean of the numbers already obtained for the various speeds should be taken as final would appear to be one that would attribute to the discrepancies an accidental character, and seems quite out of the question.

The simplest way to carry out the correction will be to determine the amount of the coefficient from the two extreme speeds. The squares of the speeds are as

$$1 : \tfrac{16}{9} : \tfrac{144}{49} : 4;$$

so that the difference of the numbers for the two extreme speeds, 23·638 − 23·619, i.e., ·019, is three times the quantity by which the lowest is to be reduced. We are accordingly to subtract respectively

$$·0063, \quad \tfrac{16}{9} \times ·0063, \quad \tfrac{144}{49} \times ·0063, \quad 4 \times ·0063,$$

with the following results.

TABLE VII.—Third series.

	Number of teeth				Mean
	60	45	35	30	
Resistance of standard at 13°, uncorrected	23·619	23·621	23·630	23·638	23·627
Correction proportional to square of speed	·006	·011	·018	·025	..
Resistance of standard at 13°, corrected .	23·613	23·610	23·612	23·613	23·612

It will be seen that the agreement is practically perfect, the coefficient given by the extreme speeds suiting also the requirements of the intermediate speeds. The maximum difference corresponds to about $\frac{3}{100}$ths of a millimetre only in the deflections of the principal magnetometer. The number $23 \cdot 612 \times 10^9$ is therefore to be regarded as the resistance in absolute C.G.S. measure of the platinum-silver standard at $13°$. If, however, the correction be rejected, the result will be different by decidedly less than one part in a thousand.

Although the experiments of the second series will not bear comparison with those of the third, it may be well to mention that they lead to substantially the same conclusion. The simple mean (taken with exclusion of August 29) of all the values is $23 \cdot 633$, and after introduction of the correction proportional to the square of the speed, $23 \cdot 618$.

The results of the first series of spinnings are given in Table VIII. They have been reduced by Dr Schuster, so as to show the value of the platinum-silver standard in absolute measure from the observations of each night at each speed. The mean radius of the coil was taken from the first measurements, and a somewhat higher value of U was employed than the subsequent calculation of the ring currents seemed to justify.

TABLE VIII.—First series.

Teeth.	Resistance in absolute measure of standard at 13°.				
80	$23 \cdot 651$,	$23 \cdot 632$,	$23 \cdot 628$	...	Mean $23 \cdot 637$
60	$23 \cdot 646$,	$23 \cdot 629$,	$23 \cdot 601$	...	Mean $23 \cdot 625$
45	$23 \cdot 678$,	$23 \cdot 691$,	$23 \cdot 686$	...	Mean $23 \cdot 685$
35	$23 \cdot 608$,	$23 \cdot 615$,	$23 \cdot 632$,	$23 \cdot 665$	Mean $23 \cdot 630$
30	$23 \cdot 644$,	$23 \cdot 639$,	$23 \cdot 628$	...	Mean $23 \cdot 637$

$23 \cdot 643$

Comparison with the standard B.A. units.

Four distinct sets of comparisons between the platinum-silver standard and the ultimate B.A. units have been effected in the course of these investigations, and two distinct methods have been followed. In the first method two coils of about five units, called for brevity [5]'s, were compared separately with five standard units combined in series with mercury cups. Secondly, the two [5]'s in series were compared with a [10]. Thirdly, the [10], the two [5]'s, and four singles were combined in series and compared with the platinum-silver standard [24]. The differences in every case were expressed in divisions of the wire of Fleming's bridge, whose value in terms of the

B.A. unit is known. This method is simple enough in principle, but the arrangement of so many mercury connexions is troublesome, and the calculation of the innumerable temperature corrections necessary is tedious. The labour would have been greater still had we not been able to avail ourselves of the previous work of Professor Fleming, who had carefully compared the various standard units, and had drawn up a chart on which is exhibited the comparative resistances of the coils over a considerable range of temperature. The mean B.A. unit, in terms of which our results are expressed, was defined by him, but the difference between the single standards is scarcely of importance for our purpose. In calculating the temperature corrections for the two [5]'s, the [10], and the [24], which were all of platinum-silver wire, the coefficient ·0003 per degree has been used. The temperatures were those of the water in which the coils were immersed. They never differed much from the temperature of the room, and were referred to a Kew standard. The results of three comparisons, executed by Mrs Sidgwick, are as follows:—

Resistance in mean B.A. units of platinum-silver standard at 13°.

July, 1881	23·9326
September, 1881	23·9341
November, 1881	23·9348

In February, 1882, a fourth determination was executed by myself, in which a different method was employed. Five coils approximately equal to each other and to five units were arranged in a closed case upon a tube of brass. The ten copper terminals emerged below from the ebonite bottom of the case, and rested in mercury cups upon an ebonite base-board, which was so arranged that by a single motion the terminals could be transferred from one set of cups which combined the coils in series to another set which combined them in multiple arc*. In this way resistances are obtained in the proportion of 25 : 1, independently of any exact equality between the single coils; for it is obvious that if the resistance in series is given, the resistance in multiple arc is a maximum in the case of equality, and therefore varies little, even if the equality be not exact†. By the aid of this apparatus the [24] was compared with a standard unit, without the assistance of other coils.

---

* I believe that Professor Rowland has used a contrivance of this sort.

† [1899. If there be n coils of mean resistance r and of actual resistances equal to $r(1+\alpha)$, $r(1+\beta)$, &c., so that

$$\alpha + \beta + \gamma + \ldots = 0,$$

the resistance of the coils combined in series is nr. Thus

$$\frac{\text{Resistance in series}}{\text{Resistance in parallel}} = \frac{n}{1+\alpha} + \frac{n}{1+\beta} + \ldots = n^2 \left\{ 1 + \frac{\alpha^2 + \beta^2 + \ldots}{n} \right\},$$

approximately.]

In the first place [24] + [1] was compared with the five coils in series, and in the second place the [1] was compared with the five coils in multiple arc. The only precaution necessary is to effect the second comparison so quickly after the first that the five coils have no time to change their temperature. Two determinations by this method on different days gave as the resistance of [24] at 13°

$$23\cdot9350, \ 23\cdot9358\text{—mean, } 23\cdot9354.$$

It would seem not impossible that the resistance of [24] has gradually increased, but the changes are unimportant. We will take as the resistance with which the absolute measurement is to be combined, that found in November, $23\cdot9348$; so that

$$23\cdot9348 \text{ B.A. units} = 23\cdot612 \times 10^9 \text{ C.G.S.} = 23\cdot612 \ \frac{\text{earth quadrant}}{\text{second}}.$$

Hence, as the result of the investigation, we conclude that

$$1 \text{ B.A. unit} = \cdot98651 \ \frac{\text{earth quadrant}}{\text{second}}.$$

[1899. Further experimental work by Mrs Sidgwick and myself upon this subject is recorded in *Phil. Trans.* 174, p. 295, 1882, Art. 94 below.]

81.

ON THE SPECIFIC RESISTANCE OF MERCURY.

By Lord RAYLEIGH and Mrs H. SIDGWICK.

[*Phil. Trans.* Vol. 174, pp. 173—185, 1882.]

OUR experiments on the determination of the British Association unit of electrical resistance in absolute measure are detailed in two memoirs communicated to the Society*. The conclusion to which they led us is that

$$1 \text{ B.A. unit} = \cdot 9865 \, \frac{\text{earth quadrant}}{\text{second}},$$

but this result differs considerably from that obtained by some other experimenters, the original Committee included. Although in the present state of the question it is not desirable that the B.A. unit should fall into disuse, there can be no question as to the importance of connecting it with the mercury unit introduced now more than twenty years ago by Siemens. It will then be possible, as recommended by the Paris Conference, to express our absolute measurements in terms of mercury, by stating what length of a column of mercury at 0° of 1 square millimetre section has a resistance of 1 ohm. Accordingly the experiments about to be described relate to the expression in terms of the B.A. unit of the resistances of known columns of mercury at 0°.

This investigation was the more necessary, as the principal authorities on the subject, Dr Werner Siemens and Dr Matthiessen, had obtained results differing by as much as ·8 per cent.

The earlier determinations of Siemens were vitiated by the assumption of an erroneous value (13·557) for the specific gravity of mercury, a constant

---

* *Proceedings*, April 12, 1881; *Phil. Trans.* 1882, Part II. [Arts. 79, 80].

which it is necessary to know in order to infer the mean section of a tube from the weight of contained mercury. The error, pointed out by Matthiessen, was afterwards * admitted by Siemens, who gives as the corrected expression of the relation between the two units,

$$1 \text{ mercury unit} = \cdot 9536 \text{ B.A. unit.}$$

On the other hand, the independent measurements of the resistance of mercury by Matthiessen and Hockin† gave

$$1 \text{ mercury unit} = \cdot 9619 \text{ B.A. unit,}$$

the mercury unit being defined as the resistance at 0° of a column of mercury 1 metre long and 1 square millimetre in section.

Our own experiments lead us to a value not differing much from that of Siemens. We find

$$1 \text{ mercury unit} = \cdot 95418 \text{ B.A. unit.}$$

If we assume that the B.A. unit is ·98651 ohm (in accordance with our determination), we find

$$1 \text{ mercury unit} = \cdot 94130 \text{ ohm,}$$

the ohm being 10^9 C.G.S. The same result may be expressed in another way by saying that the ohm is the resistance of a column of mercury at 0°, 1 square millimetre in section, and 1062·4 millims. in length.

Through the kindness of Dr C. W. Siemens we have had an opportunity of comparing with the B.A. units a standard mercury unit (No. 2513) issued by Messrs Siemens and Halske. At the proper temperature (16°·7) we find that its resistance is

$$\cdot 95365 \text{ B.A. unit,}$$

agreeing very closely with previous comparisons of Siemens' mercury measurements with the B.A. unit.

The determination of the specific resistance of mercury is simple enough in principle, though the execution is somewhat tedious, and the calculation of the results is complicated in practice by the necessity of introducing various temperature corrections. In a first sketch of the method it will be convenient to omit these corrections, which is tantamount to supposing that all the measurements are made at zero. If L be the length and s the section of the column of mercury, R its resistance, r the specific resistance of the metal,

$$R = \frac{rL}{s}, \quad \text{or} \quad r = R\frac{s}{L}.$$

* *Phil. Mag.* vol. xxxi. 1866.
† Reprint of British Association Reports, p. 114.

The length L can be measured directly, but s can only be found with the necessary accuracy from the contents. Thus if ρ be the specific gravity of mercury, and W the weight of the whole column in grammes, $\rho L s = W$, whence $s = W/\rho L$, and

$$r = \frac{RW}{\rho L^2}.$$

Apart from the temperature corrections already referred to, the simplicity of the formula is disturbed by the inevitable departure from the truly cylindrical form of the glass tubes used to contain the mercury. It is true indeed that to a first order of approximation the formula stands unaltered, as we may see if we understand by s the *mean* section of the tube. The volume is still truly expressed by sL, and the resistance is *approximately* expressed by rL/s. If, however, the squares of the variations of section cannot be neglected, the actual resistance is greater than the formula would lead us to suppose, as is evident if we imagine the section to become at one place very small.

In general we must regard s as a function of the position (x) along the tube at which it is taken. For the purposes of the present paper we may assume with sufficient approximation (see Lord Rayleigh's *Theory of Sound*, § 308)

$$R = r \int \frac{dx}{s}.$$

The necessary data with respect to s are obtained by a calibration of the tube. "If a small quantity of mercury is introduced into the tube and occupies a length λ of the tube, the middle point of which is distant x from one end of the tube, then the area s of the section near this point will be $s = C/\lambda$, where C is some constant. The weight of mercury which fills the whole tube is

$$W = \rho \int s\, dx = \rho C \Sigma \left(\frac{1}{\lambda}\right) \frac{L}{n},$$

where n is the number of points at equal distances along the tube, where λ has been measured, and ρ is the mass of unit of volume.

"The resistance of the whole tube is

$$R = \int \frac{r\, dx}{s} = \frac{r}{C} \Sigma\left(\lambda\right) \frac{L}{n}.$$

"Hence

$$WR = r\rho\, \Sigma\left(\lambda\right) \Sigma \left(\frac{1}{\lambda}\right) \frac{L^2}{n^2},$$

and

$$r = \frac{WR}{\rho L^2} \frac{n^2}{\Sigma\left(\lambda\right).\Sigma\left(\lambda^{-1}\right)}$$

gives the specific resistance of unit of volume" (Maxwell's *Electricity*, § 362).

In the sequel

$$\frac{\Sigma\,(\lambda)\,.\,\Sigma\,(\lambda^{-1})}{n^2}$$

is denoted by μ; it is a numerical quantity a little greater than unity.

Another correction is required in our method of working to take account of the resistance offered by that part of the mercury in the terminal cups, which is situated just beyond the ends of the tube. The question is identical with that of the correction necessary in calculations of pitch for the open ends of organ pipes (see *Theory of Sound*, § 307, and Appendix A), and it scarcely admits of absolutely definite solution. We cannot, however, be far wrong in adding to the actual length of the tube ·82 of its diameter, which corresponds to the supposition that the diameter of the mercury column suddenly becomes infinite. Since, in our experiments, the whole correction only amounts to about a thousandth part, even a ten per cent. error in our estimate would scarcely be material.

Let r = resistance of a column .of mercury 1 metre long and 1 square millimetre in section, at 0°, expressed in B.A. units.

R = resistance of the tube full of mercury at 0° in B.A. units.

L = length of the tube at t'° in centimetres as measured with brass rod.

l = length of a thread of mercury of nearly the length of the tube at t° as measured with brass rod.

W = weight of the same thread in grammes.

μ = coefficient correcting for conicality of tube.

δL = correction to L on account of the connecting rods not being close up to the ends of the tube = ·82 × diameter of tube.

ρ = specific gravity of mercury at 0° = 13·595.

γ = cubic expansion of mercury per degree = ·0001795.

g =   „   „   glass   „   = ·000025.

b = linear expansion of brass   „   = ·000018.

t_0 = temperature of brass measuring rod to which the lengths are corrected = 17°·2.

Then the volume of the thread at 0° = W/ρ.

$$\text{„} \qquad\qquad \text{„} \qquad\qquad t^{\circ} = \frac{W}{\rho}\,(1+\gamma t).$$

$$\text{Mean section of the tube at } t^{\circ} = \frac{W\,(1+\gamma t)}{\rho l\,\{1+b\,(t-t_0)\}}\,.$$

$$\text{Mean section at } 0^{\circ} \qquad = \frac{W\,(1+\gamma t)}{\rho l\,\{1+b\,(t-t_0)\}\,\{1+\tfrac{2}{3}gt\}}\,.$$

$$\text{Length of the tube at } 0^\circ = \frac{(L + \delta L)\left\{1 + b\left(t' - t_0\right)\right\}}{1 + \tfrac{1}{3}gt'}.$$

$$R = 10^{-4} \cdot r \cdot \mu \cdot \frac{(L + \delta L)\left\{1 + b\left(t' - t_0\right)\right\}}{1 + \tfrac{1}{3}gt'} \cdot \frac{\rho l\left\{1 + b\left(t - t_0\right)\right\}\left\{1 + \tfrac{2}{3}gt\right\}}{W\left(1 + \gamma t\right)}.$$

$$r = \frac{10^4 RW\left(1 + \gamma t\right)\left(1 + \tfrac{1}{3}gt'\right)}{\rho \mu\, lL\left(1 + \tfrac{2}{3}gt\right)}\left(1 - \frac{\delta L}{L}\right)\left\{1 - b\left(t + t' - 2t_0\right)\right\}.$$

The value of ρ is that used by the Committee of the British Association in reducing Dr Matthiessen's experiments (see reprint of *Reports on Electrical Standards*, p. 114), and stated to be the mean of the values given by Kopp, Regnault, and Balfour Stewart. The values of g, γ, and b are taken from Everett's *Units and Physical Constants*—γ being Regnault's value for the expansion of mercury. The measurements of the other quantities, which depend on the particular tube used, are given in the following table, together with the resulting value of r. The description of the means employed to obtain these data follows.

Number of observation	Date of observation, 1882	Number of tube	R	Temperature of coil F	Temperature of second coil	L in centimetres	μ
1	Feb. 23 & 24	I.	·79912	$\begin{cases}12\cdot2\\11\cdot5\end{cases}$	$\begin{cases}12\cdot1\\11\cdot5\end{cases}$	87·771	1·00314
2	,, 25 . .	I.	·79912	12·7	12·5	,,	,,
3	,, 21 . .	I.	,,	. .	. .	,,	,,
4	March 18 .	I.	·79920	13·7	13·75	,,	,,
5	Weighed Feb. 14	I.	·79912	. . .	. .	,,	,,
6	Feb. 24 . .	II.	·99088	12·0	. .	96·400	1·00007
7	,, 21 to 23	II.	·99081	13·2	. .	,,	,,
8	March 7 .	II.	·99081	11·5	. .	,,	,,
9	,, 8 .	II.	·99079	12·2	. .	,,	,,
10	,, 30 .	II.	·99085	13·25	. .	. .	. .
11	,, 6 .	III.	·99711	11·2	. .	123·566	1·00046
12	,, 10 .	III.	·99725	12·9	. .	,,	,,
13	,, 13 .	III.	·99720	12·7	. .	,,	,,
14	,, 14 .	III.	·99725	13·4	. .	,,	,,
15	,, 22 .	IV.	·50783	13·0	12·9	194·137	1·000838
16	,, 24 .	IV.	·50774	12·7	12·7	,,	,,

Number of observation	l in centimetres	W	t	t'	$\dfrac{\cdot 955 \times \delta L}{L}$	$rb(t + t' - 2t_0)$	r	Mean values of r from each tube
1	87·234	12·442	16·5	16·5	·00103	+·00002	·95386	
2	87·310	12·4545	17·2	17·2	,,	·00000	·95412	
3	87·035	12·4185	18·4	,,	,,	−·00002	·95424	·95416
4	87·558	12·486	20·6	,,	,,	−·00006	·95436	
5	87·771	12·523	16·5	16·5	,,	+·00002	·95421	
6	96·054	12·096	16·7	16·4	·000883	+·00002	·95389	
7	95·452	12·0245	16·4	,,	,,	+·00003	·95414	·95419
8	95·831	12·074	17·1	,,	,,	+·00002	·95437	
9	96·151	12·113	18·0	18·0	,,	−·00003	·95436	
10	..	..	..	..	..	..	..	
11	122·218	19·620	16·2	18·7	·000778	−·00001	·95424	
12	123·288	19·780	18·5	,,	,,	−·00005	·95418	·95416
13	123·221	19·7665	18·4	,,	,,	−·00005	·95399	
14	123·058	19·745	18·3	,,	,,	−·00005	·95425	
15	193·410	95·859	14·5	14·5	·000869	+·00009	·95440	·95427
16	192·576	95·402	16·8	..	,,	+·00005	·95415	

Mean of all the above values of r in B.A. units ·95418.

The mercury used for all the measurements except 10 and 14 was distilled *in vacuo* with an apparatus fitted up by Mr Shaw. In order to see whether a different result might not be obtained with other mercury, some was procured from the chemical laboratory for measurements 10 and 14. For the latter a portion of this mercury was treated with nitric acid and distilled at atmospheric pressure. For measurement 10 it was treated with nitric acid, but not distilled. An accident occurred in carrying out this measurement, so that only the resistance of the column was ascertained; but this agrees so well with the resistances found with the same tube for the other mercury, that there is no reason to suppose that any discrepancy would have appeared in proceeding with the measurement further.

The glass tubes used were supplied by Cassella, and were selected for uniformity of bore, so that the correction for conicality should be small. They were slender and easily broken, which made the manipulation of them difficult, and it was in fact owing to a breakage that the tube called No. I. was used so short. The measurements taken with it, at first intended to be preliminary, were, however, made with the same care as in the case of the other tubes, and the difference of length and resistance adds some variety to the data. Tubes II. and III. were cut so that their resistance should be as nearly as possible one B.A. unit. The section of tubes I., II., and III., was approximately 1 square millimetre.   Tube IV. was a much larger one, introduced with a view of varying the data as much as could conveniently be done.   The diameter of its bore was about 2 millims., and its length was nearly 2 metres.   It was cut so as to give a resistance of about half a B.A. unit.

The ends of the tubes were ground into a convex form with emery powder on a lathe, in order that the length (L) of the bore might be measured accurately.   This measurement was effected by setting two microscopes, which could be adjusted longitudinally to the exact position required by micrometer-screws graduated to $\frac{1}{10000}$ inch, so that their cross-wires should coincide with the ends of the tube.   Observations were made in three or four different positions as the tube was turned round its axis, and the mean taken.   After removal of the tube, a brass measuring rod belonging to the British Association was substituted for it, and the number of whole divisions corresponding most nearly to the distance between the cross-wires of the two microscopes was read off.   The outstanding fraction of a millimetre was then ascertained by screwing the microscope up to the whole division and reading the difference on the screw-head.   For the long tube the measuring rod was too short, and a third microscope had to be used to fix an intermediate point as a fresh departure for the scale.   A thermometer laid beside the tube during the measurement gave the temperature (t') at the moment.   The brass measuring rod was carefully examined, and its divisions were found to agree among themselves.

The tubes were cleaned by passing through them in succession, by means of a suction-pump, sulphuric acid, nitric acid, caustic potash, and distilled water, followed by air dried with chloride of calcium.   The process with omission of the acids was in general repeated between each refilling with mercury, but it was omitted in measurement 7, and there is no record of its having been done in 1, 3, and 6.

To calibrate the tubes a short thread of mercury was inserted and moved to the various positions required, by blowing through a chloride of calcium tube.   In the case of tubes I. and II., the length, λ, of the thread was measured by adjusting microscopes to its two ends, with subsequent substitution of an ivory scale divided in fiftieths of an inch.   But this method was troublesome; and with tubes III. and IV. the scale was simply placed against the thread and the length read off with a magnifying-glass, a procedure which was found to give sufficiently accurate results, notwithstanding the difficulty arising from parallax owing to the thickness of the glass.   The following table gives the different values of λ for each tube.

As a check upon the correction for conicality, two distinct values of μ were in some cases calculated from the alternate observations of λ, and were found to agree closely.   It may not be superfluous to mention that in carrying out the computations we must work to six or seven places, although the observed values of λ themselves may not be accurate beyond the third place.

The lengths are in fiftieths of an inch			
Tube I.	Tube II.	Tube III.	Tube IV.
80·8	104·5	135·0	171·0
80·0	104·1	134·0	172·0
77·0	104·5	133·0	171·5
75·8	105·0	132·0	170·5
76·0	104·5	131·5	171·5
76·4	105·2	130·5	174·5
75·0	104·3	128·0	175·0
74·0	104·0	127·5	174·5
73·4	104·7	126·5	175·5
73·0	104·0	126·5	176·5
72·7	103·0	126·5	177·0
72·3	101·8	126·0	180·0
72·5		125·0	180·5
71·9		125·5	180·7
71·1		126·0	182·2
70·1		126·0	183·7
69·7		126·0	183·5
68·0		126·5	182·5
67·9		127·0	184·0
67·6		127·0	186·0
65·9		128·5	186·5
65·3		128·0	
		128·5	
		128·0	

To find the mean section of the tubes we at first tried the method adopted by Messrs Matthiessen and Hockin in their experiments for the British Association. After aspirating the tube with dry air we placed it in a wooden trough full of mercury, and filled it by suction. It was then held down in the trough with iron weights till it was presumably of the same temperature as the mercury in the trough, which was taken at three places. It was then held by the fingers (previously cooled in other mercury), pressed against its two ends, and taken out of the trough, the mercury adhering to the outside was brushed off, and the contents of the tube were emptied into a small porcelain crucible and weighed. But there was no doubt that when the fingers holding the tube were bare they pressed a little way—how much it was difficult to determine—into the tube, and when they were covered with stiff leather, or other stiff material, it was difficult to get a sufficiently good hold. However, in one case (No. 5) r was calculated from the weight so obtained with leather on the fingers.

The method, followed by Siemens and Sabine, of screwing an iron plate up against the end of the tube, was attempted, but we did not succeed in closing the orifice sufficiently tightly in this way. Ultimately we came to the conclusion that the best results would be obtained by weighing a thread of mercury nearly as long as the tube, of which we could ascertain the actual length by direct measurement. We thought, also, that there might be some advantage in ascertaining the volume of the mercury from the same filling as that of which the resistance had been taken, as we could not be sure that the closeness of contact between the mercury and the glass was always the same, so that the same volume of mercury would always be contained in the same length of tube, nor that the tube itself was in no way altered by the action of the caustic potash used to clean it. The plan adopted was, therefore, after measuring the resistance, to keep the tube horizontal so as to retain in it most of the mercury while the terminals were removed, and then with microscopes and divided rod to measure the thread of mercury in the same way as the tubes were measured. The length so obtained is called in the table l. The greatest difference between l and L (that in measurement 11) is scarcely over 1 per cent., and in most cases the difference is considerably less, so that, considering how nearly cylindrical the tubes were, the error in the mean section introduced by using a thread of length l instead of L is quite inappreciable. It was another advantage of our method that it avoided the necessity of filling the tube under mercury, which it would have been difficult to do with a tube so long as IV.

The only difficulty in measuring the thread of mercury arose from the convexity of its ends. This was overcome by pressing them flat with little flat-ended vulcanite pins made to fit into the tube. The curvature of the ends when free was not always the same; but it was found that the length of the mercury held with pins varied little from the number calculated on the assumption that the ends were hemispherical, namely, the length of the portion of the column of mercury which was in contact with the glass added to two-thirds of the difference between this length and that between the convex extremities. In some cases, where, owing to the pins not fitting very well or other causes, there was a difficulty in flattening the ends properly, the calculated value was used. A thermometer lay beside the tube during the measurement, so as to give the temperature t. After the measurement, the mercury was blown out into a small crucible and weighed. Care had to be taken not to leave behind minute globules, which, owing probably to the small portion of the tube unoccupied by mercury during the measuring becoming damp from the air of the room or from the fingers, tended to adhere to the glass near the ends.

In three cases (No. 5 as above mentioned and Nos. 3 and 9) the mercury weighed and measured was not that of which the resistance was taken.

No. 3 was done before it occurred to us that there might be an advantage in carrying out both operations with the same filling, and in No. 9 about one-tenth of the mercury was spilt accidentally and had to be replaced.

The equality of the arms of the balance used for the weighing was tested. The weights were compared among themselves and found to be free from appreciable error.

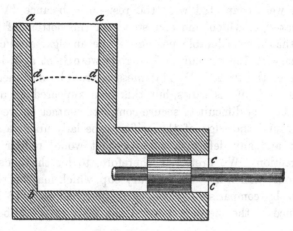

The terminals were composed of L-shaped pieces of ebonite, hollowed out in the manner shown (about full-size) in the figure. Each end of the tube was furnished with a short length of thick rubber tubing, by which the aperture between the glass and the ebonite was closed air-tight. As a further precaution, the space at cc beyond the rubber was filled up by pouring in melted paraffine wax.

After the terminals were fitted the tube was again aspirated with dry air through tubes in corks inserted at aa, and then filled with mercury, which was poured in to one terminal and allowed to run slowly through to the other till it stood at a considerable height, represented by dd, in both terminals. The tube was then placed in a wooden trough and covered with ice. Our reason for using vulcanite terminals rather than glass ones was the fear that under the influence of the ice moisture would collect on the portion of glass above the mercury and serve as a conductor. We certainly avoided all difficulty of this kind by using vulcanite. On the other hand, we probably increased a difficulty which would have existed in any case, namely, that of getting the temperature of the portion of the tube which was within the terminal down to 0°. This portion of the tube was about 2 centims. at each end, or about 5 per cent. of the length in the case of tube I., and about 2 per cent. in the case of tube IV. What the exact temperature of this part of the tube was it is impossible to say, but it was ascertained that the temperature of the mercury in the terminals with the copper connecting rods *in situ* was not higher than 5° or 6°, depending in some degree on the extent to which the ice was piled up round the cup. The mean temperature of the parts of the tubes not directly exposed to ice can hardly have been so high as 2°. Supposing it to have been 2°, and taking the case of tube I., where the largest proportion of the whole length was within the terminals, the effect would be an overestimate of r by about ·00008. In the case of tube IV. the error in r would be less than the half of this.

The tubes were connected with the resistance balance by copper rods, well amalgamated, of which one end stood on the bottom of the vulcanite terminals, so that a considerable portion of the amalgamated copper surface was in contact with the mercury. The rods were kept at a little distance from the ends of the tubes. Dr Matthiessen brought flattened copper rods up against the ends of his tubes, but this plan appeared open to objection, since it would be very difficult to secure complete contact between the copper and glass all round the edge of the orifice, especially under an opaque fluid like mercury; and any defect in such contact would render necessary an unknown correction. We preferred, therefore, to let the ends of the tube open without obstruction into the mercury cup, which may be regarded as of infinite extent by comparison. The correction necessary to take account of the resistance of the mercury beyond the ends of the tube has already been considered.

The resistance of the rods used to connect I., II., and III. with the bridge was about ·00215 B.A. unit. With tube IV. an additional rod had to be introduced to get the necessary length. This brought the resistance of the rods up to ·00291. The other end of the rods fitted into mercury cups on the resistance balance.

The balance used was one designed by Professor Fleming (*Phil. Mag.* IX. p. 109, 1880), in which Professor Carey Foster's method is employed of interchanging the resistances in the two arms of the balance containing the graduated wire, so that the difference between them is expressed in terms of the wire. One thousand divisions of the graduated wire are stated by Professor Fleming to equal ·0498 B.A. unit, and experiments of our own also showed it to be about ·05. The wire is of platinum-iridium, and as it has a high temperature coefficient compared with the platinum-silver of the standard coils, we thought it undesirable to use much over 100 divisions of it. In order to avoid this in the case of tubes I. and IV. it was necessary to introduce coils from a resistance box in multiple arc. The resistance box employed was one by Messrs Elliott Brothers. With tube I., 20 ohms from the box were used in multiple arc with the standards against which the tube was balanced, and in the case of tube IV. 24 ohms were used in multiple arc with the tube itself. Tubes II. and III. were balanced against the standard coil belonging to the British Association and deposited at the Cavendish Laboratory, called F. For tube IV. another of their unit coils, called the

Flat coil, was used in multiple arc with *F.* For tube I., *F* and a five-ohm coil were used in multiple arc. The standard coils belonging to the British Association have recently been carefully compared with each other by Professor Fleming, who has drawn out a chart in which is recorded their variation with temperature, together with their resistance in terms of the mean of their resistances at the temperatures at which they were originally considered to be correct. The values of *F* and of the *Flat coil*—both platinum-silver coils—were taken from this chart. The five-ohm coil had been compared with the British Association standards by ourselves. It was also of platinum-silver, and its temperature coefficient was assumed to be the same as that of the others.

The standard coils were immersed in water whose temperature was observed each time a resistance was measured. These temperatures are given in the table. It may be worth remarking that the resistances were taken in a different room from that in which the lengths were measured, which accounts for the difference between *t* and the temperature of the standards. The thermometer used to find all the temperatures was graduated to fifths, and was corrected by one which had been verified at Kew.

When one coil only was used to balance the tube, its terminals fitted directly into the mercury cups of the bridge, but when two were used in multiple arc their terminals were put into larger mercury cups, which were connected with the mercury cups of the bridge by short copper connecting pieces of about ·00017 ohm resistance.

All the measurements were repeated with reversed battery currents, in order to eliminate thermoelectric disturbance. The readings with battery current each way usually agreed very closely, and the mean of the two was adopted.

It will be observed that the values of *R* for tube IV. differ by nearly two parts in 10,000, and that there is a less proportional difference, but still an appreciable one, for the other tubes. The greatest actual difference between any two of the values in the table for the same tube is ·00014 ohm. Some small error is due to neglect of the change of resistance of the copper connecting rods and of the bridge wire with temperature. A change of 4° in the temperature of the rods would make a difference of about ·00003 ohm. There is further a probability of error in ascertaining the temperature of the standard coil. A difference of $\frac{1}{10}°$ in this also introduces a difference of ·00003 ohm in the resistance; and there is not only a probable error of perhaps $\frac{1}{10}$ in finding the temperature of the water in which the coil is immersed, but there is no certainty that the coil follows the water exactly. There is evidence, however, that the differences in *R* are partly due to a real difference in the resistance of different fillings of the tube—whether owing to microscopic bubbles or to a thin varying layer of air between the mercury and the glass, or to what cause, we were unable to determine*.

We found some reason for thinking that the resistance tended to diminish with time when the mercury remained long in the tube. To examine this we filled tube II. on April 3rd, and found its resistance to be ·99077. It was then left standing full of mercury till April 18th, when the resistance was ·99055. This difference can hardly be relied upon; and in any case the experiments we have tabulated cannot well be affected by any change of this kind, as the interval between the measurement of resistance and that of volume was very short, except in cases 1 and 7. In case 7 the tube stood full of mercury for two days after the resistance was taken. In case 1 the resistance was measured on two successive days, and the mean of the two values taken. The second was the lowest by ·00020, possibly owing to an error. The length was measured immediately after the last measurement of resistance.

The variations in the values of r are, as we should expect, greater than those in R, being affected by probable errors in the other data. The extreme difference amounts to less than 6 in 10,000, and the greatest divergence from the mean value is 3·3 in 10,000.

The mean value of r according to these experiments, ·95418, lies between that deduced from Dr Siemens' experiments for his 1864 standard, namely, ·9534, and Dr Matthiessen's value, namely, ·9619 (*Phil. Mag.* May, 1865), but the difference between our value and Dr Matthiessen's, namely, ·00772, is nearly ten times as great as that between ours and Dr Siemens'. We are unable to account satisfactorily for this large difference. One point, however, is worth noting. Dr Matthiessen measured the resistance of the mercury in his tubes, not at zero, but at temperatures between 18° and 19°·1 (Report of British Association Committee for 1864). To deduce the specific resistance at zero, therefore, he must have assumed the coefficient of variation with temperature, and presumably—though it is nowhere stated in the Report— he used that found from his own experiments (*Phil. Trans.* 1862), namely, ·074† per cent. per degree. Our own observations have led us to suspect that this value is too small. We made three comparisons of the resistance of tube III. in ice, and in water at approximately the temperature of the room, and one similar comparison with tube IV. The results are given in the following table. Our arrangements were not adapted for observing the resistance at other temperatures, as the open trough afforded no means of checking rapid change.

* A variation in the closeness of contact between mercury and glass amounting to less than one-fifth of a wave-length of mean light would account for the difference of resistances in the two fillings of tube IV.

† This is the value which results from the experiments made at 0° and at about 20°.

Date	No. of tube	Mean temperature of water in the trough	Resistance in water	Resistance at 0°	Difference for 1°÷resistance at 0°	Mean of the four values in the last column
March 13 .	III.	12·7	1·00814	·99720	·000863	
,, 14 .	III.	13·25	1·00874	·99725	·000870	
,, 28 .	III.	12·8	1·00810	·99720	·000854	·000861*
,, 24 .	IV.	12·5	·51318	·50774	·000857	

The above determined mean coincides with the value found by Schröder van der Kolk†, whose observations, however, related to a much greater range of temperature. An observation by Werner Siemens‡ between the temperature 18°·5 and 0° gives for the coefficient ·00090.

The difference between the coefficients ·00074 and ·00086, as applied to the reduction from 18°·7 (the mean temperature of the tubes in Dr Matthiessen's observations) to 0°, would account for about one quarter of the difference between his results and our own.

The remainder of the discrepancy may possibly be connected with the manner in which Dr Matthiessen's tubes were calibrated. Although in the description of the process a *small* column of mercury is spoken of (*Reprint*, p. 128), it is distinctly stated on the preceding page that the lengths of the columns of mercury were 383, 291, 245 millims. respectively, *i.e.*, nearly half the lengths of the tubes. It is possible that this may be a mistake; but if such lengths were really used, the correction for conicality would have been much underestimated, so that the specific resistance of mercury would come out too high. In the case of uniform conicality the true correction would be four times as great as that obtained by applying the formula applicable to short threads, to cases where the length is about half that of the tube.

[*January*, 1883.—The measuring rod and the weights used in the above investigation have been compared with standards verified by the Board of Trade, and the errors have been found to be negligible. But since the value of ρ employed relates to weighings *in vacuo*, a corresponding correction is called for here. On this account the final number, ·95418, should be reduced to

·95412.]

* It should be noticed that the resistances here compared are those of the contents of a certain glass tube at various temperatures, so that the accompanying temperature variations of length and section are determined by the properties of glass and not by the properties of mercury. The results are therefore not quite comparable with those obtained in similar experiments with solid metallic wires, which are free to determine for themselves their length and section.

† *Pogg. Ann.* vol. cx. 1860.

‡ *Ibid.* vol. cxiii. 1861.

82.

THE USE OF TELESCOPES ON DARK NIGHTS.

[*Proceedings of the Cambridge Philosophical Society*, IV. pp. 197, 198, 1882.]

IN *Silliman's Journal* for 1881 Mr E. S. Holden, after quoting observations to a like effect by Sir W. Herschel, gives details of some observations recently made with a large telescope at the Washburn Observatory, from which it appears that distant objects on a dark but clear night can be seen with the telescope long after they have ceased to be visible with the naked eye. He concludes, "It appears to me that this confirmation of Herschel's experiments is important, and worth the attention of physicists. So far as I know there is no satisfactory explanation of the action of the ordinary Night-glass, nor of the similar effect when large apertures are used."

It is a well-known principle that no optical combination can increase what is called the 'apparent brightness' of a distant object, and indeed that in consequence of the inevitable loss of light by absorption and reflection the 'apparent brightness' is necessarily *diminished* by every form of telescope. Having full confidence in this principle, I was precluded from seeking the explanation of the advantage in any peculiar action of the telescope, and was driven to the conclusion that the question was one of apparent magnitude only,—that a large area of given small 'apparent brightness' must be visible against a dark ground when a small area would not be visible. The experiment was tried in the simplest possible manner by cutting crosses of various sizes out of a piece of white paper and arranging them in a dark room against a black back-ground. A feeble light proceeded from a nearly turned-out gas-flame. The result proved that the visibility was a question of apparent magnitude to a greater extent than I had believed possible. A distance was readily found at which the larger crosses were plainly visible, while the smaller were quite indistinguishable. To bring the latter into view it was necessary either to increase the light considerably, to approach nearer, or lastly to use a telescope. With sufficient illumination the smallest crosses used were seen perfectly defined at the full distance.

There seems to be no doubt that the explanation is to be sought within the domain of Physiological Optics. It has occurred to me as possible that with the large aperture of the pupil called into play in a dark place, the focussing may be very defective on account of aberration. The illumination on the retina might then be really less in the image of a small than in the image of a large object of equal 'apparent brightness.'

[1899. See *Camb. Proc.* IV. p. 324, 1883; Art. 96 below.]

83.

ON A NEW FORM OF GAS BATTERY.

[Proceedings of the Cambridge Philosophical Society, IV. p. 198, 1882.]

IN Grove's well-known gas battery it would seem that the only efficient part of the platinum surface is where it meets both the gas and the liquid, or at any rate meets the liquid and is very near the gas. In order to render a larger area effective I have substituted for the usual platinum plates platinum gauze resting upon the surface of the liquid in a large trough in such a manner that the upper surface is damp but not immersed. One piece is exposed to the oxygen of the air; the other forms the bottom of an enclosed space into which hydrogen is caused to flow. The area of each piece is about 20 square inches.

To test the efficiency, the current was passed through an external resistance of about 6 ohms, including a galvanometer. Under these circumstances the permanent current was about one-fourth of that obtained when a large Daniell cell was substituted for the gas element. An inferior, but still considerable, current was observed when coal gas was used instead of hydrogen prepared from zinc.

84.

ACOUSTICAL OBSERVATIONS. IV.

[*Philosophical Magazine*, XIII. pp. 340—347, 1882.]

ON THE PITCH OF ORGAN-PIPES.—SLOW VERSUS QUICK BEATS FOR COM-
PARISON OF FREQUENCIES OF VIBRATION.—ESTIMATION OF THE
DIRECTION OF SOUNDS WITH ONE EAR.—A TELEPHONE-EXPERIMENT.—
VERY HIGH NOTES.—RAPID FATIGUE OF THE EAR.—SENSITIVE FLAMES.

On the Pitch of Organ-Pipes.

IN the *Philosophical Magazine* for June 1877 [Art. 46, vol. I. p. 320]
I described some observations which proved that the note of an open organ-
pipe, when blown in the normal manner, was *higher* in pitch than the natural
note of the pipe considered as a resonator. The note of maximum resonance
was determined by putting the ear into communication with the interior of
the pipe, and estimating the intensity of sounds of varying pitch produced
externally.

A more accurate result may be obtained with the method used by
Blaikley*, in which the external sound remains constant and the adjustment
is effected by tuning the resonator to it. About two inches were cut off
from the upper end of a two-foot metal organ-pipe, and replaced by an
adjustable paper slider. At a moderate distance from the lower end of the
pipe a tuning-fork was mounted, and was maintained in regular vibration by
the attraction of an electromagnet situated on the further side, into which
intermittent currents from an interrupter were passed. Neither the fork
nor the magnet were near enough to the end of the pipe to produce any
sensible obstruction. By comparison with a standard, the pitch of the fork

---

* *Phil. Mag.* May 1879.

thus vibrating was found to be 255 of König's scale. The resonance of the pipe was observed from a position not far from the upper end, where but little of the sound of the fork could be heard independently; and the paper slider was adjusted to the position of maximum effect. This observation was repeated many times, the distance between marks fixed on the pipe and on the slider respectively being recorded. The following numbers give the results, expressed in fiftieths of an inch [inch = 2·54 cm.]:—

31	33	30	25	31
25	32	31	34	29
35	28	29	30	

The extreme range being only one-fifth of an inch, shows that the observation is capable of considerable precision, corresponding as it does to only about 2 vibrations per second out of a total of 255. Finally, the slider was fixed at the mean of the above-determined positions, and the natural note of the pipe was then considered to be 255. The error in length was probably less than $\frac{1}{20}$ inch, and the error in pitch less than half a vibration per second.

The pipe was then blown from a well-regulated bellows; and the beats were counted between its note and that of the standard fork above referred to, the pressure being taken simultaneously with a water-manometer. Three observers were found to be necessary for accurate working—one to count the beats, rising to the rate of ten per second, one to keep the bellows uniformly supplied with wind, and one to observe the manometer. At pressures between 4·2 inches and 1·53 inch the pitch of the pipe was very well defined and considerably higher than the natural note. Below 1 inch the pitch became somewhat unsteady, and distinct fluctuations in the frequency of the beats were perceived, while no corresponding variation of pressure could be detected. At about ·8 inch the pitch of the pipe falls to unison with the natural note, and with further diminishing pressures becomes the graver of the two. Below ·7 inch the unsteadiness is such as to preclude accurate estimations of pitch.

The results are embodied in the accompanying table, which shows the correspondence of pitch and pressure. Instead of the actual number of beats counted, which involves a reference to the extraneous element of the pitch of the standard fork, the number (greater by unity) is given which expresses the excess in the frequency of vibration of the actual over that of the natural note of the pipe. It will be seen that at practical pressures the pitch is raised by the action of the wind, but that this rule is not universal.

Pressure, in inches	Difference of frequencies	Remarks
4·2	+11·0	
2·72	9·3	
2·26	8·4	
1·86	7·1	
1·53	5·6	
1·32	4·2	
1·06	2·1	
·88	1·5	
·82	+ ·1	
·75	− ·5	
·68	1·2	
·64	2·3	
·57	3·9*	About this point a discordant high note comes in alongside of the normal note.
·53	3·7	
·48	3·1*	Here the discordant note ceases.
·46		
·40		
·39 ·38	− 4·0	
·35	*	*About this point the octave of the normal note is heard, after which the normal note itself disappears.
·30	——	The normal note reappears, the octave continuing.
·26	——	The octave goes, and then the normal note, after which there is silence.
·20	+	Octave comes in again, and then the normal note, at a pitch which falls from considerably above to a little below the natural pitch. At the lowest pressures the normal note is unaccompanied by the octave.
·11	−	

Slow versus *quick Beats for comparison of Frequencies of Vibration.*

Most of those who have had experience in counting beats have expressed a preference for somewhat quick beats. Perhaps the favourite rapidity has been four beats per second. There is no doubt that in the case of insufficiently sustained sounds slow beats are embarrassing. The observer gets confused between the fall of sound which is periodic and that which is due to the dying away of the component vibrations, and loses his place, as it were, in the cycle. But it is also possible, I think, to trace an impression that, independently of the risk of confusion, quick beats can be counted with greater accuracy than slow ones. It is indeed true that the number of beats in a given time, such as a minute, can be determined with greater relative accuracy when there are many than when there are few; but it is also true, as a little consideration will show, that in the comparison of frequencies we are concerned not with the *relative*, but with the *absolute* number of beats executed in the given time. If we miscount the beats in a minute by one, it makes just the same error in the result whether the whole number of beats is 60 or 240.

When the sounds are pure tones and are well maintained, it is advisable to use beats much slower than four per second. By choosing a suitable position we may make the intensities at the ear equal; and then the phase of silence, corresponding to antagonism of equal and opposite vibrations, is extremely well marked. Taking advantage of this, we may determine slow beats with very great accuracy by observing the time which elapses between recurrences of silence. In favourable cases, the whole number of beats in the period of observation may be fixed to within one-tenth or one-twentieth of a single beat, a degree of accuracy which is of course out of the question when the beats are quick.

In some experiments, conducted by Dr Schuster and myself[*], to determine the absolute pitch of a König standard fork, I had occasion to observe some very slow beats. The beating sounds were of pitch 128. One of them was steady, proceeding from an electrically maintained fork; the other (from the standard fork) gradually died away. In order to be more independent of disturbing noises to which we were exposed, a resonator was used connected with the ear by an india-rubber tube. The standard fork was mounted at the end of a wooden stick, so that it might not be heated by the hand. As the vibrations became less powerful, the prongs of the fork were caused slightly to approach the mouth of the resonator, so as to maintain the equality of the two component sounds. In this way it was possible to obtain very definite silences, and to measure the interval of recurrence with accuracy. In one observation, extending over about two minutes, the beat occupied as much as twenty-four seconds, and there was no confusion. I have little doubt that even slower beats might be observed satisfactorily if both components were steadily maintained.

Estimation of the Direction of Sounds with one Ear.

In my former experiments (*Phil. Mag.* June 1877 [Art. 46, vol. I. p. 315]) I found it difficult to obtain satisfactory observations with one ear closed, although it was not doubtful that the power of estimating directions was greatly curtailed. My desire to experiment upon an observer deaf on one side has since been gratified by the kind assistance of Mr F. Galton. In January 1881 experiments were tried with him similar to those on normal hearers described in my former paper. It was found that Mr Galton made mistakes which would be impossible for normal ears, confusing the situation of voices and of clapping of hands when to his right or left, as well as when

* *Proc. Roy. Soc.* May 5, 1881, p. 137. [Art. 79.]

in front or behind him. Thus, when addressed loudly and at length by a little boy standing a few yards in front of him, he was under the impression that the voice was behind. In other cases, however, there seemed to be some clue, whose nature we could not detect. Bad mistakes were made; but the estimates were more often right than mere chance would explain.

After this experience it seemed unlikely that there could be any success in distinguishing whether pure tones came from right or left, and from in front or behind. The experiment was tried, however, with in the main the expected result. But when the sounds were close, there appeared to be some slight power of distinguishing right and left, which may perhaps have been due to incomplete deafness of the defective ear.

A Telephone-Experiment.

In Maxwell's *Electricity and Magnetism*, vol. II. § 655, it is shown that a perfectly conducting sheet acts as a barrier to the magnetic force:—"If the sheet forms a closed or infinite surface, no magnetic actions which may take place on one side of the sheet will produce any magnetic effect on the other side." In practice we cannot use a sheet of perfect conductivity; but the above-described state of things may be approximated to in the case of periodic magnetic changes, if the time-constants of the sheet circuits be large in comparison with the periods of the changes.

The experiment is made by connecting up into a primary circuit a battery, a microphone-clock, and a coil of insulated wire. The secondary circuit includes a parallel coil and a telephone. Under these circumstances the hissing sound is heard almost as well as if the telephone were inserted in the primary circuit itself. But if a large and stout plate of copper be interposed between the two coils, the sound is greatly enfeebled. By a proper choice of battery and of the distance between the coils, it is not difficult so to adjust the strength that the sound is conspicuous in the one case and inaudible in the other.

Very High Notes. Rapid Fatigue of the Ear.

In former experiments with bird-calls I had often been struck with what seemed to be the capricious behaviour of these sources of sound, but had omitted to follow up the observation. In the spring of last year the apparent caprice was traced to the ear, which very rapidly becomes deaf to sounds of high pitch and moderate intensity. A bird-call was mounted in connexion with a loaded gas-bag and a water-manometer, by which means the pressure could be maintained constant for a considerable time. When the ear is placed at a moderate distance from the instrument, a disagreeable sound

7—2

is heard at first, but after a short interval, usually not exceeding three or four seconds, fades away and disappears altogether. A very short intermission suffices for at any rate a partial recovery of the power of hearing. A pretty rapid passage of the hand, screening the ear for a fraction of a second, allows the sound to be heard again. During his visit to Cambridge in March 1881, I had the pleasure of showing this experiment to Prof. Helmholtz.

The uniformity of the sound in the physical sense may be demonstrated with a sensitive flame, which remains uniformly affected so long as the pressure indicated by the manometer does not vary. The sensitive flame may also be employed to determine the wave-length of the sound, in the manner described in the *Philosophical Magazine* for March 1879, p. 154 [Art. 61, vol. I. p. 406]. In the case of two bird-calls blown with a pressure of about $2\frac{1}{2}$ inches of water, the wave-lengths were found to be respectively 1·304 inches and 1·28 inches [one inch = 2·54 cm.]. The method was found to work easily and with considerable accuracy, almost identical results being obtained from observations of the *loops*, where the flame is *most* affected, and from the *nodes*, where it is *least* affected.

By modifying the pressures with pinch-cocks, the two notes could be brought into unison. Although both bird-calls were blown from the same gas-bag, it was not possible to keep the beats slow for more than a few seconds at a time; but that period was quite sufficient for the effects of the beats to manifest themselves in a striking manner by the behaviour of the flame. In repeating these experiments, it may be necessary to bear in mind that many people cannot hear these high notes at all, even at first. With a shorter wave-length of about $\frac{1}{2}$ inch, as determined by the flame, I was myself quite unable to hear any sound from the situation of the flame. A slight hissing was perceived when the ear was brought up close to the source; but it is probable that this was not the part of the sound that agitated the flame.

Sensitive Flames.

In the chapter devoted to this subject in Tyndall's *Sound* (third edition, p. 231) the accomplished author remarks :—" An essential condition to entire success in these experiments disclosed itself in the following manner. I was operating on two fishtail flames, one of which jumped to a whistle while the other did not. The gas of the non-sensitive flame was turned off, additional pressure being thereby thrown upon the other flame. It flared, and its cock was turned so as to lower the flame; but it now proved non-sensitive, however close it might be brought to the point of flaring. The narrow orifice of the half-turned cock interfered with the action of the sound. When the gas was fully turned on, the flame being lowered by opening the cock of the

other burner, it became again sensitive. Up to this time a great number
of burners had been tried, but with many of them the action was *nil*.
Acting, however, upon the hint conveyed by this observation, the cocks which
fed the flames were more widely opened, and our most refractory burners
thus rendered sensitive."   In the abstract of a Royal-Institution lecture
(*Phil. Mag.* Feb. 1867) a rather more definite view is expressed:—"Those
who wish to repeat these experiments would do well to bear in mind, as an
essential condition of complete success, that a free way should be open for
the transmission of the vibrations from the flame, *backwards*, through the
gas-pipe which feeds it.   The orifices of the stopcocks near the flame ought
to be as wide as possible."

During the preparation of some lectures on Sound in the spring of last
year, it occurred to me that light would probably be thrown upon these
interesting effects by introducing a manometer on a lateral branch near the
flame.   In the path of the gas there were inserted two stopcocks, one only
a little way behind the manometer-junction, the other separated from it by a
long length of india-rubber tubing.   When the first cock was fully open and
the flame was brought near the flaring-point by adjustment of the distant
cock, the sensitiveness to external sounds was great, and the manometer
indicated a pressure of ten inches of water.   But when the distant cock stood
fully open and the adjustment was effected at the other, high sensitiveness
could not be attained; and the reason was obvious, because the flame flared
without external excitation while the pressure was still an inch short of that
which had been borne without flinching in the former arrangement.   On
opening again the neighbouring cock to its full extent, and adjusting the
distant one until the pressure at the manometer measured nine inches, the
flame was found comparatively insensitive.

It appears, therefore, that the cause of the prejudicial action of partially
opened stopcocks in the neighbourhood of the flame is not so much that they
render the flame insensitive as that they induce premature flaring.   There
are two ways in which we may suppose this to happen.   It may be that, as
Prof. Barrett suggests (*Phil. Mag.* April 1867), the mischief is due to the
irregular flow and consequent ricochetting of the current of gas from side to
side of the pipe; or, again, the cause may lie in the actual production of
sonorous disturbance of the kind to which the flame is sensitive, afterwards
propagated forwards to the burner along the supply-pipe acting as a speaking-
tube.   The latter explanation was the one that suggested itself to my mind
at the time, in consequence of the observation that a hissing sound was
easily audible by the ear placed close to the half-open stopcock through
which gas was passing; and it was confirmed when I found that a screw
pinch-cock could be used for adjustment near the flame with impunity, in
which case no sound was perceptible.

Subsequently further experiments were tried with various nozzles inserted in the supply-tube. These included holes in thin metal plates and drawn-out glass tubes. Even though the rubber tubes were so bent that the streams issuing from the nozzles were directed against the sides, no sound was heard, and no loss of sensitiveness was apparent. It would seem that mere irregularity of flow produced no marked effect, and that, provided no sound attended it, the full pressure could be borne without flaring.

These observations in no way impair the value of the practical rule laid down by Tyndall. In some cases I have found a flame flare without external excitation when a neighbouring stopcock was partially closed, and in spite of the increase of pressure recover itself when the stopcock was completely opened. When the object is to investigate the conditions of flaring, the use of a manometer near the flame is decidedly to be recommended.

85.

FURTHER OBSERVATIONS UPON LIQUID JETS, IN CON-
TINUATION OF THOSE RECORDED IN THE ROYAL
SOCIETY'S 'PROCEEDINGS' FOR MARCH AND MAY, 1879.

[*Proceedings of the Royal Society*, XXXIV. pp. 130—145, 1882.]

THE experiments herein described were made in the spring and summer
of 1880, with the assistance of Mrs Sidgwick. Section 2 was indeed written
out as it now stands in August of that year. There were some other points
which I had hoped to submit to examination, but hitherto opportunity has
not been found.

*On some of the Circumstances which influence the Scattering of a nearly
Vertical Jet of Liquid.*

§ 1. It has been already shown [Art. 59, vol. I. p. 372] that the normal
scattering of a nearly vertical jet is due to the rebound of the drops when
they come into collision. If, by any means, the drops can be caused to
amalgamate at collision, the appearance of the jet is completely transformed.
This result occurs if a feebly electrified body be held near the place of
resolution into drops, and it was also observed to follow the addition of a
small quantity of soap to the water of which the jet was composed. In
trying to repeat the latter experiment in May, 1880, at Cambridge, I was
astonished to find that even large additions of soap failed to prevent the
scattering. Thinking that the difference might be connected with the
hardness of the Cambridge water—at home I had used rain water—I
repeated the observations with distilled water, but without finding any
explanation. The jet of distilled water scattered freely, both with and
without soap, and could only be prevented from doing so by electricity.
Eventually the anomalies were traced to differences in the character of the

soap. That used at Cambridge up to this point was a clarified specimen prepared for toilet use. On substitution for it of common yellow soap, the old effects were fully reproduced.

Further experiment seemed to prove that the real agent was not soluble soap at all. If water impregnated with the yellow soap was allowed to stand, a white deposit separated, after which the supernatant liquid was found to be inactive. But after shaking up the same effects were produced as at first. The addition of caustic potash to the unclarified soapy mixture destroyed its power. On the other hand, sulphuric acid rendered the clarified soap solution active.

The natural conclusion from these facts would be that the real agent is unsaponified greasy matter distributed through the liquid; and this view is confirmed by the striking results which follow the addition of small quantities of milk. The experiment may be made conveniently by connecting a Woulf's bottle with the water tap by a rubber tube fitted to one tubulure, while the vertical nozzle is in connexion with another tubulure. If a little milk be placed in the bottle, the jet of opalescent liquid apparently coheres, and passes the summit in one unbroken stream. After a time the milk is gradually washed out, and the scattering is re-established. About one drop of skimmed milk per ounce of water [say one part in 600] is sufficient to produce the effect.

I must not omit to mention that on several occasions distinct evidence was obtained that it is possible for soap to be in excess. With a large quantity the coherence of the jet was imperfect, and was improved by dilution. The complete elucidation of the subject probably requires more chemical knowledge and experience than is at my command.

Of the various other substances which have been tried, such as glycerine, sugar, gum arabic, alcohol, sulphuric acid, none have been found active.

Vertical fountains of mercury were found not to scatter. The head was about 15 inches [one inch = 2·54 cm.], and various glass nozzles were used from $\frac{1}{20}$ inch to $\frac{1}{50}$ inch in diameter. Also a nozzle terminating in an amalgamated brass plate, through which a hole of $\frac{1}{20}$ inch was pierced. In all these cases the drops of mercury coalesced at collision, behaving in the same way as drops of milky water issuing from the same nozzles. Fountains of clean water issuing from these nozzles under the same pressure scattered freely.

When the diameter of the nozzle from which a water jet issues is reduced to below $\frac{1}{100}$ inch, the scattering cannot be completely prevented by the presentation of an electrified body. One possible reason for this is evident. The mutual repulsion of the similarly electrified drops increases rapidly relatively to the masses as the size is reduced, and thus it may happen that before the *differential* electrification sufficient to rupture the separating envelope at contact is arrived at, the repulsion may be powerful enough to prevent most of the drops from coming into contact at all. In connexion with this it may be remarked that two perfectly equal and equally electrified spheres would repel one another at all distances; but that if there be the slightest difference in the size or electrification, the repulsion will be exchanged for attraction before actual contact is attained. This attraction will be local, and thus the opposed parts of the surfaces may come into contact with considerable violence, even when the relative motion of the centres of the masses is small. It is easily shown experimentally (see § 4) that violence of contact tends to promote coalescence, so that we have here a possible explanation of the action of electricity.

With respect to the persistent scattering of very fine jets, however, it would appear that the principal cause is simply that many of the fine drops fail to come into contact in any case. The capillary forces act with exaggerated power, and doubtless impress upon the minute drops irregular lateral velocities, which may easily reach a magnitude sufficient to cause them to clear one another as they pass. At any rate little difference is observable in this respect between a fine jet of clean water under feeble electrical influence, and one to which a little milk has been added, but without electrification.

With a suitable jet, say from a nozzle about $\frac{1}{20}$ inch diameter, and rising about 2 feet, the sensitiveness to electricity is wonderful, more especially when we remember that the effect is differential. I have often caused a jet to appear coherent, by holding near the place of resolution a brass ball about 1 inch in diameter, supported by a silk thread, and charged so feebly that a delicate gold-leaf electroscope would show nothing. Indeed, some care is necessary to avoid being misled by accidental electrifications. On one occasion the approach of a person, who had not purposely been doing anything electrical, invariably caused a transformation in the appearance of the jet.

The jets hitherto under discussion are such as resolve themselves naturally into drops soon after leaving the nozzle, or at any rate before approaching the summit of their path. If the diameter be increased, we may arrive at a condition of things in which the undisturbed jet passes the summit unbroken. In such a case the addition of milk, or the presentation of an electrified body, produces no special effect. One interesting observation, however, may be made. By the action of a vibrator of suitable pitch, e.g. a tuning-fork, resolution on the upward path may be effected. As the vibration gradually dies down, the place of resolution moves upwards, but it cannot pass a certain point. When the point is reached, resolution into actual drops ceases, the upper part of the jet exhibiting simple undulations, when viewed intermittently. The phenomenon is in perfect harmony

with theory. As it leaves the nozzle, the jet is unstable for the kind of
disturbance imposed upon it by the vibrator. The subsequent loss of velocity,
however, shortens the wave-lengths of disturbance, until at length they are
less than the circumference of the jet, after which the disturbance changes
its character from unstable to stable. The vibrator must evidently produce
its effect quickly, or not at all.

Influence of Regular Vibrations of Low Pitch.

§ 2. Towards the close of my former paper on the capillary phenomena
of jets [Art. 60, vol. I. p. 395], I hazarded the suggestion that the double
stream obtained when an obliquely ascending jet is subjected to the influence
of a vibration an octave graver than the natural note, is due to the *compound*
character of the vibration. At the time of Plateau's researches the fact that
most musical notes are physically composite was much less appreciated than
at present, and it is not surprising that this point escaped attention. I have
lately repeated Plateau's experiments under improved conditions, with results
confirmatory of the view that no adequate explanation of the phenomena can
be given which does not have regard to the possible presence of overtones;
and I have added some observations on the effects of the simultaneous action
of two notes forming a consonant chord.

In order to make a satisfactory examination of it, it is necessary to
employ some apparatus capable of affording an intermittent view of the jet
in its various stages of transformation. In the experiments formerly described
I used sparks from an induction coil, governed by the same tuning-fork which
determined the resolution of the jet. This has latterly been replaced by a
perforated disk of black cardboard, driven at a uniform speed by a small
water-motor. The diameter of the holes is one-fifth of an inch—about that
of the pupil of the eye, and the interval between the holes is about four
inches. Examined under these conditions the jet and resultant drops are
sufficiently well defined, and there is abundant illumination if the apparatus
is so arranged that the jet is seen projected against the sky. The speed of
the motor is regulated so that there is one view through the holes in about
one complete period of the phenomenon to be observed. If the power is a
little in excess, the application of a slight friction to the axle carrying the
disk renders the image steady, or, what is better, allows it to go forwards
through its phases with moderate slowness.

Although the multiple streams are better separated when the jet is
originally directed upwards at an angle of about 45°, I preferred to use a
horizontal direction as giving simpler conditions. The velocity and diameter
are then practically constant throughout the transformation, and may be
readily calculated from observations of the head and of the total quantity of
fluid discharged in a given time. The reservoir consisted of a large glass

bottle, provided with a tubulure near the bottom. Into this was fitted a 1-inch brass tube, closed at the end by a flat plate, in which a circular aperture was pierced of about $\frac{1}{12}$ of an inch [say, 2 mm.] in diameter.

If h = head, d = diameter of *jet*, v = velocity of issue, V = volume discharged in unit time, then

$$\tfrac{1}{4}\pi d^2 v = V, \quad v = \sqrt{(2gh)}.$$

Again, if N' be the frequency of the most rapid vibration which can influence the jet, we have by Plateau's theory—

$$N' = \frac{v}{\pi d} = \frac{v^{\frac{3}{2}}}{2\sqrt{(\pi V)}} = \frac{(2gh)^{\frac{3}{4}}}{2\sqrt{(\pi V)}}.$$

If N be the frequency of the principal note of the jet, then, as explained in my former paper,

$$N = \frac{3\cdot142}{4\cdot508} N'.$$

In the present experiment it was found that 1050 cub. centims. were discharged in four minutes, and the head was $7\frac{3}{4}$ inches, so that in c.g.s. measure—

$$V = \frac{1050}{240}; \quad h = 7\tfrac{3}{4} \times 2\cdot54; \quad g = 981;$$

whence

$$N' = 372, \quad N = 259.$$

As sources of sound tuning-forks, provided with adjustable sliding pieces, were employed. Except when it was important to eliminate the octave as far as possible, the vibration was communicated to the reservoir through the table on which it stood. The forks were either screwed to the table and vibrated with a bow, or mounted on stands (resting on the table) and maintained electrically. The former method was quite adequate when on'y one fork was wanted at a time.

With pitches ranging from 370 to about 180, the observed phenomena agreed perfectly with the unambiguous predictions of theory. From the point —decidedly below 370—at which a regular effect was first obtained, there was always one drop for each complete vibration of the fork, and a single stream, every drop breaking away under the same conditions as its predecessor. After passing 180 it becomes a question whether the octave of the fork's note may not produce an effect as well as the prime. If this effect be sufficient, the number of drops is doubled; and unless the prime be very subordinate indeed, there is a double stream, alternate drops taking sensibly different courses. In these experiments the influence of the prime was usually sufficient to determine the number of drops, even in the neighbourhood of pitch 128. Sometimes, however, the octave became predominant,

and doubled the number of drops.   It must be remembered that the relative
intensities with which the two vibrations reach the jet depend upon many
accidental circumstances.   The table has natural notes of its own, and even
the moving of a weight upon it may change the conditions very materially.
When the octave is not strong enough actually to double the drops, it often
produces an effect which is very apparent to an observer examining the
transformation through the revolving holes.   On one occasion a vigorous
bowing of the fork, which favours the octave, gave at first a double stream,
but this after a few seconds passed into a single one.   Near the point of
resolution those consecutive drops which ultimately coalesce as the fork dies
down, are connected by a ligament.   If the octave is strong enough, this
ligament breaks and the drops are separated; otherwise the ligament draws
the half-formed drops together, and the stream becomes single.   The transition from the one state of things to the other could be watched with facility.

In order to get rid entirely of the influence of the octave a different
arrangement is necessary.   It was found that the desired result could be
arrived at by holding a 128 fork in the hand over a resonator of the same
pitch resting on the table.   The transformation was now quite similar in
character to that effected by a fork of frequency 256, the only differences
being that the drops were bigger and twice as widely spaced, and that the
spherule, which results from the gathering together of the ligament, was
much larger.   We may conclude that the cause of the doubling of a jet by
the sub-octave of the note natural to it is to be found in the presence of the
second component, from which scarcely any musical notes are free.

When two forks of pitches 128 and 256 were sounded together, the single
or double stream could be obtained at pleasure by varying the relative
intensities.   Any imperfection in the tuning is rendered very evident by
the behaviour of the jet, which performs evolutions synchronous with the
audible beats.   This observation, which does not require the aid of the
revolving disk, suggests that the effect depends in some degree upon the
relative phases of the two tones, as might be expected *à priori*.   In some
cases the influence of the sub-octave is shown more in making the alternate
drops unequal in magnitude, than in projecting them into very different
paths.

Returning now to the case of a single fork screwed to the table, it was
found that as the pitch was lowered below 128, the double stream was
regularly established.   The action of the *Twelfth* below the principal note
(85⅓) demands special attention.   At this pitch we might in general expect
the first three components of a compound note to influence the result.   If
the third component were pretty strong it would determine the number of
drops, and the result would be a threefold stream.   In the case of a fork
screwed to the table the third component of the note must be extremely

weak, if not altogether missing; but the second (octave) component is fairly strong, and in fact determines the number of drops ($190\frac{2}{3}$). At the same time the influence of the prime ($85\frac{1}{3}$) is sufficient to cause the alternate drops to pursue different paths, so that a double stream is observed.

By the addition of a 256 fork there was no difficulty in obtaining the triple stream, but it was of more interest to examine whether it were possible to reduce the double stream to a single one with only $85\frac{1}{3}$ drops per second. In order to secure as strong and as pure a fundamental tone as possible and to cause it to act in the most favourable manner upon the jet, the air space over the water in the reservoir was tuned to the note of the fork by sliding a piece of glass over the neck so as partially to cover it. When the fork was held over the resonator thus formed, the pressure which expels the jet was rendered variable with a frequency of $85\frac{1}{3}$, and overtones were excluded as far as possible. To the unaided eye, however, the jet still appeared double, though on more attentive examination one set of drops was seen to be decidedly smaller than the other. With the revolving disk, giving about eighty-five views per second, the real state of the case was made clear. The smaller drops were the *spherules*, and the stream was single in the same sense as the streams given by pure tones of frequencies 128 and 256. The increased size of the spherule is of course to be attributed to the greater length of the ligament, the principal drops being now three times as widely spaced as when the jet is under the influence of the 256 fork.

With still graver forks screwed to the table the number of drops continued to correspond to the second component of the note. The double octave of the principal note (64) gave 128 drops per second, and the influence of the prime was so feeble that the duplicity of the stream was only just recognisable. Below 64 the observations were not carried. Attempts to get a single stream of 64 drops per second were unsuccessful, but it is probably quite possible to do so with vibrations of greater power than I could command.

In the case of a compound note of pitch 64 a considerable variety of effects might ensue, according to the relative strengths of the various components. Thus, the stream might be single (though this is unlikely), double, triple, four-fold, or even five-fold, with a corresponding number of drops.

Observations were next made on the effects of chords. For the chord of the Fifth the pitches taken were 256 and $\frac{2}{3} \times 256$. The two forks could be screwed to the table and bowed, or, as is preferable (especially in the case of the chords of the Fourth and Third to be spoken of presently), maintained in vibration electromagnetically by a periodic current from a break-fork of pitch $85\frac{1}{3}$, standing on another table. The revolving disk was driven at such a speed as to give about eighty-five views per second. As was to be expected, the number of drops was either 256 in a triple stream, or $\frac{2}{3} \times 256$ in a double stream, according to the relative intensities of the two vibrations. With the maintained forks the phenomenon is perfectly under control, and there is no difficulty in observing the transition from the one state of things to the other.

In like manner with forks 256 and $\frac{3}{4} \times 256$, driven by fork 64, and with sixty-four views per second, the stream is either triple or quadruple; and with forks 256 and $\frac{4}{5} \times 256$, we get at pleasure a four-fold or five-fold stream. To obtain a good result the intervals must be pretty accurately tuned. In the case of electrically maintained forks, the relative phase remains un-changed for any length of time, and the spectacle seen through the revolving holes is one of great beauty.

The actual results obtained experimentally by Plateau differ in some respects from mine, doubtless in virtue of the more composite character of the notes of the violoncello employed by him, but they are quite consistent with the views above expressed. The only point as to which I feel any difficulty relates to the single stream, which occasionally resulted from the action of the Twelfth below the principal note. It seems improbable that this could have been a single stream of the kind that I obtained with some difficulty from a pure tone; indeed the latter would have been pronounced to be a double stream by an observer unprovided with an apparatus for intermittent views. I should rather suppose that the number of drops really corresponded to an overtone, and that from some accidental cause the divergence of what would generally be separate streams failed to be sensible.

The Length of the Continuous Part.

§ 3. When a jet falls vertically downwards, the circumstances upon which its stability or instability depend are continually changing, more especially when the initial velocity is very small. The kind of disturbance to which the jet is most sensitive as it leaves the nozzle is one which impresses upon it undulations of length equal to about four and a half times the initial diameter. But as the jet falls its velocity increases (and consequently the undulations are lengthened), and its diameter diminishes, so that the degree of instability soon becomes small. On the other hand, the kind of disturb-ance which will be effective in a later stage is altogether ineffective in the earlier stages. The change of conditions during fall has thus a protective influence, and the continuous part tends to become longer than would be the case were the velocity constant, the initial disturbances being unaltered.

I have made many attempts to determine the origin of the disturbances which remain in operation when the jet is protected from ordinary tremors, but with little result. By suspending the reservoir with india-rubber straps, &c., from the top of a wooden tripod, itself resting upon the stone floor of one of the lower rooms of the Cavendish Laboratory, a considerable degree of isolation was attained. A stamp of the foot upon the floor, or the sounding of a note of suitable pitch of moderate intensity in the air, had no great effect. Without feeling much confidence I rather incline to the opinion that the residual disturbances are of internal origin. As the fluid flows up to the aperture along the inner surface of the plate which forms the bottom of the reservoir, eddying motions are almost certainly impressed upon it, and these may very possibly be the origin of the ultimate disintegration. With the view of testing this point, I arranged an experiment in which the velocity of the fluid over the solid walls should be as small as possible.

AB (fig. 1) represents a large brass tube, to which a smaller one is soldered at B, suitable for india-rubber connexion. The bottom of the large tube

Fig. 1.

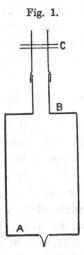

consists of a carefully worked plate in which is a circular hole of $\frac{1}{2}$ inch diameter. When the rubber tube is placed in connexion with the water supply, a jet drops from A, and may be made exceedingly fine by regulation of the pinch-cock C. By turning off the supply at C altogether, the jet at A may be stopped, without emptying the vessel. The stability, due to the capillary tension of the surface at A, preponderates over the instability due to gravity. By this device it is possible to obtain a jet whose velocity is acquired almost wholly *after* leaving the vessel from which it issues. In this form of the experiment, however, the jet is liable to disturbance depending upon the original velocity of the fluid as it passes through the comparatively narrow rubber tube, and when I attempted a remedy by suspending a closed reservoir (fig. 2), in which the water might be allowed first to come to rest, other difficulties presented themselves. The air confined over the surface of the water acts as a spring, and the flow of water below tends to become intermittent, when rendered sufficiently slow by limiting the admission of air. A definite cycle is often established, air flowing in and water flowing

Fig. 2.

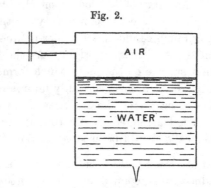

out alternatively at the lower aperture. The difficulty may be overcome by careful manipulation, but there is no easy means of making an adequate comparison with other jets, so that the question remains undecided whether the residual disturbances are principally of internal or of external origin.

Collision of Two Resolved Streams.

§ 4. In the case of a simple vertical fountain, when the scattering is prevented by electricity, there is every reason to believe that the action is differential, depending on a difference of potentials of colliding drops. The principal electrification, however, of the successive drops must be the same; and thus, sensitive as it is, this form of the phenomenon is not by any means the best calculated to render evident the smallest electrical forces. As was shown in my former paper [Art. 59, vol. I. p. 374], it is far surpassed by colliding *jets*, between which a difference of potential may be established, a subject to which we shall return in § 5. It is possible, however, to experiment upon the collision of two distinct streams of drops, which are differently— if we please, oppositely—electrified from the first. Apart from electrical influence, the collision of such streams presents points of interest which have been made the subject of examination.

Two similar brass nozzles, terminating in apertures about $\frac{3}{50}$ inch in diameter, were supplied from the same reservoir of water, and were held so that the jets rising obliquely from them were in the same plane and crossed each other at a moderate angle. The jets were resolved into regular series of drops by the action of a 256 fork screwed to the table and set in action by bowing. The periodic phenomenon thus established could be examined with facility by intermittent vision through a revolving perforated disk (§ 2), so arranged that about 256 holes passed the eye per second.

When the angle of collision is small, the disposition of the files of drops may be made such that they rebound without crossing, fig. 3; more often, however, the drops shoulder their way through after one or more collisions,

Fig. 3.

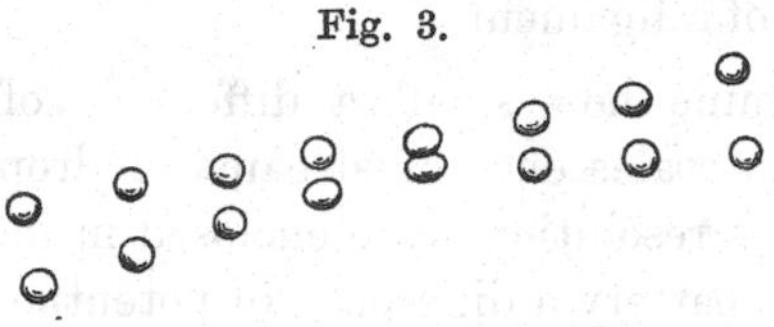

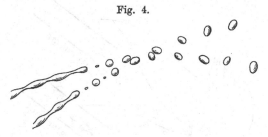

somewhat as in fig. 4. In both cases the presentation of an electrified body to one place of resolution will determine the amalgamation of colliding drops, with of course complete alteration of the subsequent behaviour. By judicious

Fig. 4.

management a feebly electrified body may be held in an intermediate position between the two points of resolution so as not to produce the effect, confirming the view that the action is differential.

At a somewhat higher angle of collision amalgamation will usually occur without the aid of electricity, but the fact may easily escape recognition when intermittent vision is not employed. The streams do not usually join into one, as we might perhaps expect, but appear to pass through one another, much as if no union of drops had occurred. With the aid of the revolving disk the course of things is rendered evident. The separating layer is indeed ruptured at contact, and for a short time the drops move as one mass. There is, however, in general, considerable outstanding relative

Fig. 5.

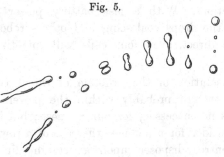

velocity, which is sufficient to bring about an ultimate separation, preceded by the formation of a ligament (fig. 5). In certain cases, although after contact a ligament is formed, the relative velocity is insufficient to overcome its tension, and the drops draw again together and ultimately cohere. If the impact is very direct, so that the relative velocity is almost entirely in the line of centres, the drops may flatten against one another and become united without the formation of a ligament.

In order to determine how small a difference of potential would be effective in causing the coalescence of streams of drops meeting at a small angle, the two places of resolution were enclosed in inductor-tubes, between which with the aid of a battery a difference of potential could be established. The arrangement is shown in fig. 6. One of the inductors is placed in

Fig. 6.

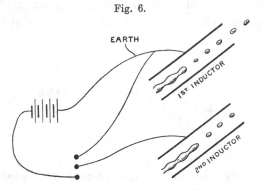

connexion with the earth, with the reservoir from which the water comes, and with one pole of the battery. By operating a key, the other inductor may be placed at pleasure in communication with the first inductor, or with the other pole of the battery. In the first case the battery is out of use, and in the second the difference of potential due to the battery is established between the two inductors.

Experiment showed that the effect depends a good deal upon the exact manner of collision. In almost all cases twenty cells of a De la Rue battery sufficed to produce amalgamation, with subsequent replacement of the original streams by a single one in a direction bisecting the angle between the original directions. With a less battery power the result may be irregular, some of the drops coalescing and others rebounding. When the collisions are very direct, even four cells will sometimes cause a marked transformation.

The complete solution of the problem of the direct collision of equal spheres of liquid, though probably within the powers of existing mathematical analysis, is not necessary for our purpose; but it may give precision to our ideas to consider for a moment the case of a row of equal spheres, or cylinders, with centres disposed upon a straight line, and so squeezed together that the distances between the centres must be less than the original diameters. By the symmetry, the common surfaces are planes, and the force between contiguous masses is found by multiplying the area of the common surface by the internal capillary pressure. When the amount of squeezing is small, the internal capillary pressure is approximately unaltered, and the force developed is simply proportional to the area of contact. In the case of the cylinder the problem admits of very simple solution, even when the squeezing is not small; for, as is easily seen, the free surfaces are necessarily semicircular, and thus the condition of unaltered volume is readily expressed. It will of course be noticed that as regards lateral displacements the equilibrium is unstable.

Collision of Streams before Resolution.

§ 5. The collision of unresolved streams was considered in my former paper. It appeared that the electromotive force of a single Grove cell, acting across the common surface, was sufficient to determine coalescence, and that the addition of a small quantity of soap made rebound impossible. Moreover, the " coalescence of the jets would sometimes occur in a capricious manner, without the action of electricity or other apparent cause."

As in many respects this form of the phenomenon is the most instructive, I was desirous of finding out the explanation of the apparent caprice, and many experiments have been made with this object in view. The observations on fountains recorded in § 1 having suggested the idea that the accidental presence of greasy matter, removable by caustic potash, might operate, this point was examined.

"*July* 8, 1880.*—*Colliding Jets.*—Two large glass bottles, with holes in the sides, close to the bottom, were fitted by means of corks with glass tubes, drawn out to nozzles of about $\frac{3}{40}$ of an inch in diameter. The bottles were well rinsed with caustic potash, to remove any possible traces of grease, and filled with tap water. The colliding jets coalesced in a manner apparently entirely capricious, the only principle observable being that they coalesced even more readily with high pressures (12 inches) than with low, and with lower pressures would stand collision at greater angles. The addition of caustic potash sufficient to give a very decided taste to the water produced no apparent effect." Subsequently the water used was boiled with caustic potash, but without success.

"July 27, 28, 29, 30.—On the theory that when the jets collide without uniting there is between them a thin film of air, which would be very liable to be sucked up by water not saturated with air, we tried jets of water through which a stream of atmospheric air had been passed for several hours. We tried it three times. The first time the jets seemed very decidedly less liable to unite capriciously. The second time they behaved even worse than ordinary tap water usually does. The third time we thought it rather better than tap water usually is, but not materially so."

* Mrs Sidgwick's *Note Book.*

Jets of hot water, and of mixtures of alcohol and water in various proportions, were also tried at this time, but without obtaining any clue as to the origin of the difficulty.

I had begun almost to despair of success, when a determined attempt to conjecture in what possible ways one part of the stirred liquid could differ from another part suggested the idea that the anomalies were due to dust.

"Aug. 1880.—We tried dropping dust on to the colliding jets just above the point of collision, and found that union was always produced. The following powders were tried—powdered cork, sand, lycopodium, plaster of Paris, flowers of sulphur, sugar, dust that had accumulated upon a shelf, and later emery and putty powder. The lycopodium was a little more uncertain in its action than the others, but apparently only because, owing to its lightness, it was difficult to ensure its falling upon the jets. Whenever we were sure it did so, union followed."

When mixed with the water, powders acted differently. Emery and putty powders were not effective, but sulphur caused immediate union. Much probably depends upon the extent to which the extraneous matter is wetted. A precipitate of chloride of silver, formed in the liquid itself, seemed to be without influence.

Acting upon this hint, Mrs Sidgwick made an extended series of observations upon the behaviour of jets composed of water which had been allowed to settle thoroughly, and which were protected from atmospheric dust. For this purpose the jets were enclosed in a beaker glass, the end of which was stopped by a plug of boxwood, fitted airtight. Through the plug passed horizontally the two inclined glass nozzles, and underneath a bent tube serving as a drain. The results, observed under these circumstances, were such as to render it almost certain that dust is the sole cause of the capricious unions. The protected jets of settled water were observed for a total period of 246 minutes, during which the unions were at the average rate of one in ten minutes. The longest intervals without unions were thirty-four minutes and twenty-nine minutes. Comparative experiments were made upon the behaviour of jets from the same nozzles under other conditions. Thus jets of unsettled water, but protected from atmospheric dust, united on an average twenty-four times in ten minutes. With unsettled water the protection from atmospheric dust is not of much use, as unprotected jets of the same water did not unite more than twenty-six times in ten minutes. On the other hand, jets of settled water, not protected from the atmosphere, united only twelve times in ten minutes. Although, no doubt, somewhat different numbers might be obtained on repetition of these experiments, they show clearly that the dust in the water is the more frequent cause of union under ordinary circumstances, but that when this is removed the atmospheric dust still exerts a powerful influence. The difficulty of getting water free from dust is well known from Tyndall's experiments, so that the residual tendency to unite under the most favourable conditions will not occasion surprise.

Although there is no reason to suppose that any other cause than dust was operative in the above experiments, it remains true that very little impurity of a greasy character will cause immediate union of colliding jets. For this purpose the addition of milk at the rate of one drop of milk to a pint of water [say one part in 10,000] is sufficient. It may be noticed too that the effect of milk is not readily neutralised by caustic potash.

With respect to the action of electricity, further experiments have been made to determine the minimum electromotive force competent to cause union. The current from a Daniell cell was led through a straight length of fine wire. One end of the wire was connected by platinum foil with the liquid in an insulated glass bottle, from which one of the jets was fed. The glass bottle supplying the second nozzle was similarly connected with a moveable point on the stretched wire. The electromotive force necessary to cause union, as measured by the distance between the two fine wire contacts, though definite at any one moment, was found to vary on different occasions, possibly in consequence of forces having their seat at the surfaces of the platinum oil. From one-half to three-quarters of the whole force of the Daniell was usually required.

With a view to further speculation upon this subject, an important question suggests itself as to whether or not there is electrical contact between colliding and rebounding jets. To solve this question it was only necessary to introduce a fine wire reflecting galvanometer into the arrangement just described, taking care that the electromotive forces employed fell short of what would be required to cause the union of the jets. Suitable keys were introduced for more convenient manipulation, and sulphuric acid was added to the water, in order to make sure that absence of strong galvanometer deflection could not be due merely to the high resistance of the thin columns of water composing the jets. Repeated trials under these conditions proved that so long as the jets rebounded their electrical insulation from one another was practically perfect.

As to the explanation of the action of electricity in promoting union, it would be possible to ascribe it to the additional pressure called into play by electrical attraction of the opposed water-surfaces, acting as plates of a condenser. But it appears much more natural to regard it as due rather to actual disruptive discharge, by which the separating skin is perforated and the equilibrium of the capillary forces is upset. A small electromotive force, incapable of overcoming the insulation of the thin separating layer, is without effect. [1900. See however *Phil. Mag.* XLVIII. p. 328, 1899.]

86.

ADDRESS TO THE MATHEMATICAL AND PHYSICAL SCIENCE SECTION OF THE BRITISH ASSOCIATION.

[*British Association Report*, pp. 437—441, 1882.]

In common with some of my predecessors in this chair, I recognise that probably the most useful form which a presidential address could take, would be a summary of the progress of physics, or of some important branch of physics, during recent years. But the difficulties of such a task are considerable, and I do not feel myself equal to grappling with them. The few remarks which I have to offer are of a general, I fear it may be thought of a commonplace character. All I can hope is that they may have the effect of leading us into a frame of mind suitable for the work that lies before us.

The diversity of the subjects which come under our notice in this section, as well as of the methods by which alone they can be adequately dealt with, although a sign of the importance of our work, is a source of considerable difficulty in the conduct of it. From the almost inevitable specialisation of modern science, it has come about that much that is familiar to one member of our section is unintelligible to another, and that details whose importance is obvious to the one fail altogether to rouse any interest in the mind of the other. I must appeal to the authors of papers to bear this difficulty in mind, and to confine within moderate limits their discussion of points of less general interest.

Even within the limits of those departments whose foundation is evidently experimental, there is room, and indeed necessity, for great variety of treatment. One class of investigators relies mainly upon reiterated appeals to experiment to resolve the questions which appear still to be open, while another prefers, with Thomas Young, to base its decisions as far as possible upon deductions from experiments already made by others. It is scarcely necessary to say that in the present state of science both methods are indispensable. Even where we may fairly suppose that the fundamental principles are well established, careful and often troublesome work is necessary to determine with accuracy the constants which enter into the expression of natural laws. In many cases the accuracy desirable, even from a practical point of view, is hard to attain. In many others, where the interest is mainly theoretical, we cannot afford to neglect the confirmations which our views may derive from the comparison of measurements made in different fields and in face of different experimental difficulties. Examples of the inter-dependence of measurements apparently distinct will occur to every physicist. I may mention the absolute determinations of electrical resistance, and of the amounts of heat developed from electrical and mechanical work, any two of which involve also the third, and the relation of the velocity of sound to the mechanical and thermal properties of air.

Where a measurement is isolated, and not likely to lead to the solution of any open question, it is doubtless possible to spend upon it time and attention that might with advantage be otherwise bestowed. In such a case we may properly be satisfied for a time with work of a less severe and accurate character, knowing that with the progress of knowledge the way is sure to be smoothed both by a better appreciation of the difficulties involved and by the invention of improved experimental appliances. I hope I shall not be misunderstood as underrating the importance of great accuracy in its proper place if I express the opinion that the desire for it has sometimes had a prejudicial effect. In cases where a rough result would have sufficed for all immediate purposes, no measurement at all has been attempted, because the circumstances rendered it unlikely that a high standard of precision could be attained. Whether our aim be more or less ambitious, it is important to recognise the limitations to which our methods are necessarily subject, and as far as possible to estimate the extent to which our results are uncertain. The comparison of estimates of uncertainty made before and after the execution of a set of measurements may sometimes be humiliating, but it is always instructive.

Even when our results show no greater discrepancies than we were originally prepared for, it is well to err on the side of modesty in estimating their trustworthiness. The history of science teaches only too plainly the lesson that no single method is absolutely to be relied upon, that sources of error lurk where they are least expected, and that they may escape the notice of the most experienced and conscientious worker. It is only by the concurrence of evidence of various kinds and from various sources that practical certainty may at last be attained, and complete confidence justified. Perhaps I may be allowed to illustrate my meaning by reference to a subject which has engaged a good deal of my attention for the last two years—the absolute measurement of electrical resistance. The unit commonly employed in this country is founded upon experiments made about twenty years ago by a distinguished committee of this Association, and was intended to represent an absolute resistance of 10^9. C.G.S., *i.e.* one ohm. The method employed by the committee at the recommendation of Sir W. Thomson (it had been originally proposed by Weber) consists in observing the deflection from the magnetic meridian of a needle suspended at the centre of a coil of insulated wire. This forms a closed circuit and is made to revolve with uniform and known speed about a vertical axis. From the speed and deflection, in combination with the mean radius of the coil and the number of its turns, the absolute resistance of the coil, and thence of any other standard, can be determined.

About ten years later Kohlrausch attacked the problem by another method, which it would take too long to explain, and arrived at the result that the B.A. unit was equal to 1·02 ohms—about two per cent. too large. Rowland, in America, by a comparison between the steady battery current flowing in a primary coil with the transient current developed in a secondary coil when the primary current is reversed, found that the B.A. unit was ·991 ohms. Lorenz, using a different method again, found ·980, while H. Weber, from distinct experiments, arrived at the conclusion that the B.A. unit was correct. It will be seen that the results obtained by these highly competent observers range over about four per cent. Two new determinations have lately been made in the Cavendish laboratory at Cambridge, one by myself with the method of the revolving coil, and another by Mr Glazebrook, who used a modification of the method followed by Rowland, with the result that the B.A. unit is ·986 ohms. I am now engaged upon a third determination, using a method which is a modification of that of Lorenz.

In another important part of the field of experimental science, where the subject-matter is ill understood, and the work is qualitative rather than quantitative, success depends more directly upon sagacity and genius. It must be admitted that much labour spent in this kind of work is ill-directed. Bulky records of crude and uninterpreted observations are not science, nor even in many cases the raw material out of which science will be constructed. The door of experiment stands always open; and when the question is ripe, and the man is found, he will nine times out of ten find it necessary to go through the work again. Observations made by the way, and under unfavourable conditions, may often give rise to valuable suggestions, but these must be tested by experiment, in which the conditions are simplified to the utmost, before they can lay claim to acceptance.

When an unexpected effect is observed, the question will arise whether or not an explanation can be found upon admitted principles. Sometimes the answer can be quickly given; but more often it will happen that an assertion of what *ought* to have been expected can only be made as the result of an elaborate discussion of the circumstances of the case, and this discussion must generally be mathematical in its spirit, if not in its form. In repeating, at the beginning of the century, the well-known experiment of the inaudibility of a bell rung *in vacuo*, Leslie made the interesting observation that the presence of hydrogen was inimical to the production of sound, so that not merely was the sound less in hydrogen than in air of equal pressure, but that the actual addition of hydrogen to rarefied air caused a diminution in the intensity of sound. How is this remarkable fact to be explained? Does it prove, as Herschel was inclined to think, that a mixture of gases of widely different densities differs in its acoustical properties from a single gas? These questions could scarcely be answered satisfactorily but by a mathematical investigation of the process by which vibrations are communicated from a vibrating solid body to the surrounding gas. Such an investigation, founded exclusively upon principles well established before the date of Leslie's observation, was undertaken years afterwards by Stokes, who proved that what Leslie observed was exactly what ought to have been expected. The addition of hydrogen to attenuated air increases the wave-length of vibrations of given pitch, and consequently the facility with which the gas can pass round the edge of the bell from the advancing to the retreating face, and thus escape those rarefactions and condensations which are essential to the formation of a complete sound wave. There remains no reason for supposing that the phenomenon depends upon any other elements than the density and pressure of the gaseous atmosphere, and a direct trial, *e.g.* a comparison between air and a mixture of carbonic anhydride and hydrogen of like density, is almost superfluous.

Examples such as this, which might be multiplied *ad libitum*, show how difficult it often is for an experimenter rightly to interpret his results without the aid of mathematics. It is eminently desirable that the experimenter himself should be in a position to make the calculations, to which his work gives occasion, and from which in return he would often receive valuable hints for further experiment. I should like to see a course of mathematical instruction arranged with especial reference to physics, within which those whose bent was plainly towards experiment might, more or less completely, confine themselves. Probably a year spent judiciously on such a course would do more to qualify the student for actual work than two or three years of the usual mathematical curriculum. On the other side, it must be remembered that the human mind is limited, and that few can carry the weight of a complete mathematical armament without some repression of their energies in other directions. With many of us difficulty of remembering, if not want of time for acquiring, would impose an early limit. Here, as elsewhere, the natural advantages of a division of labour will assert themselves. Innate dexterity and facility in contrivance, backed by unflinching perseverance, may often conduct to successful discovery or invention a man who has little taste for speculation; and on the other hand the mathematician, endowed with genius and insight, may find a sufficient field for his energies in interpreting and systematising the work of others.

The different habits of mind of the two schools of physicists sometimes lead them to the adoption of antagonistic views on doubtful and difficult questions. The tendency of the purely experimental school is to rely almost exclusively upon direct evidence, even when it is obviously imperfect, and to disregard arguments which they stigmatise as theoretical. The tendency of the mathematician is to overrate the solidity of his theoretical structures, and to forget the narrowness of the experimental foundation upon which many of them rest.

By direct observation, one of the most experienced and successful experimenters of the last generation convinced himself that light of definite refrangibility was capable of further analysis by absorption. It has happened to myself, in the course of measurements of the absorbing power of various media for the different rays of the spectrum, to come across appearances at first sight strongly confirmatory of Brewster's views, and I can therefore understand the persistency with which he retained his opinion. But the possibility of further analysis of light of definite refrangibility (except by polarisation) is almost irreconcilable with the wave theory, which on the strongest grounds had been already accepted by most of Brewster's contemporaries; and in consequence his results, though urgently pressed, failed to convince the scientific world. Further experiment has fully justified this scepticism, and in the hands of Airy, Helmholtz, and others, has shown that the phenomena by which Brewster was misled can be explained by the unrecognised intrusion of diffused light. The anomalies disappear when sufficient precaution is taken that the refrangibility of the light observed shall really be definite.

On similar grounds undulationists early arrived at the conviction that physically light and invisible radiant heat are both vibrations of the same kind, differing merely in wave-length; but this view appears to have been accepted slowly, and almost reluctantly, by the experimental school *.

* [1900. The reader may refer to a paper on "The History of the Doctrine of Radiant Energy," *Phil. Mag.* xxvii. p. 265, 1889.]

When the facts which appear to conflict with theory are well defined and lend themselves easily to experiment and repetition, there ought to be no great delay in arriving at a judgment. Either the theory is upset, or the observations, if not altogether faulty, are found susceptible of another interpretation. The difficulty is greatest when the necessary conditions are uncertain, and their fulfilment rare and uncontrollable. In many such cases an attitude of reserve, in expectation of further evidence, is the only wise one. Premature judgments err perhaps as much on one side as on the other. Certainly in the past many extraordinary observations have met with an excessive incredulity. I may instance the fireballs which sometimes occur during violent thunderstorms. When the telephone was first invented, the early reports of its performances were discredited by many on quite insufficient grounds.

It would be an interesting, but too difficult and delicate a task, to enumerate and examine the various important questions which remain still undecided from the opposition of direct and indirect evidence. Merely as illustrations I will mention one or two in which I happen to have been interested. It has been sought to remedy the inconvenience caused by excessive reverberation of sound in cathedrals and other large unfurnished buildings by stretching wires overhead from one wall to another. In some cases no difference has been perceived, but in others it is thought that advantage has been gained. From a theoretical point of view it is difficult to believe that the wires could be of service. It is known that the vibrations of a wire do not communicate themselves in any appreciable degree directly to the air, but require the intervention of a sounding-board, from which we may infer that vibrations in the air would not readily communicate themselves to stretched wires. It seems more likely that the advantage supposed to have been gained in a few cases is imaginary than that the wires should really have played the part attributed to them.

The other subject on which, though with diffidence, I should like to make a remark or two, is that of Prout's law, according to which the atomic weights of the elements, or at any rate of many of them, stand in simple relation to that of hydrogen. Some chemists have reprobated strongly the importation of à priori views into the consideration of the question, and maintain that the only numbers worthy of recognition are the immediate results of experiment. Others, more impressed by the argument that the close approximations to simple numbers cannot be merely fortuitous, and more alive to the inevitable imperfections of our measurements, consider that the experimental evidence against the simple numbers is of a very slender character, balanced, if not outweighed, by the à priori argument in favour of simplicity. The subject is eminently one for further experiment; and as it is now engaging the attention of chemists, we may look forward to the settlement of the question by the present generation. The time has perhaps come when a redetermination of the densities of the principal gases may be desirable—an undertaking for which I have made some preparations*.

If there is any truth in the views that I have been endeavouring to impress, our meetings in this section are amply justified. If the progress of science demands the comparison of evidence drawn from different sources, and fully appreciated only by minds of different order, what may we not gain from the opportunities here given for public discussion, and, perhaps more valuable still, private interchange of opinion? Let us endeavour, one and all, to turn them to the best account.

* [1899. See *Proc. Roy. Soc.* XLIII. p. 356, 1888; L. p. 449, 1892; LIII. p. 134, 1893.]

87.

ON THE TENSION OF MERCURY VAPOUR AT COMMON TEMPERATURES.

[*British Association Report*, p. 441, 1882.]

THE author called attention to the difficulty of reconciling the values of Regnault and Hagen with the phenomena observed by Crookes relating to the viscosity of gases at high exhaustions. The total gaseous pressure in the working chamber cannot be less than that of the mercury at the pump. If the penetration of mercury vapour be prevented by chemical means, some other gas must be present in equivalent quantity. If the value of Regnault and Hagen is substantially correct, it does not appear how the phenomena [of viscosity] could vary so much as they are observed to do at the highest degrees of exhaustion as measured by the McLeod gauge. The question then arises whether the value of mercury tension hitherto received may not be much in excess of the truth. In Hagen's researches it is assumed without reason that the pressure in a chamber of variable temperature is governed by the temperature of the coldest part, but this consideration tells in the wrong direction. It was suggested that possibly a change in the capillary constant, or currents in the fluid mercury at the chilled surface of the meniscus, might have had something to do with the minute changes of level which have been attributed to differences of pressure in the mercury vapour.

88.

ON THE ABSOLUTE MEASUREMENT OF ELECTRIC CURRENTS.

[*British Association Report*, pp. 445, 446, 1882.]

THE accurate absolute measurement of currents seems to be more difficult than that of resistance. The methods hitherto employed require either accurate measurements of the earth's horizontal intensity, or accurate measurements of coils of small radius and of many turns. If in the latter measurement we could trust to the inextensibility of the wire, as some experimenters have thought themselves able to do, the mean radius could be accurately deduced from the total length of wire and the number of turns; but actual trial has convinced me that fine wire stretches very appreciably under the tension necessary for winding a coil satisfactorily. Kohlrausch's method, in which the same current is passed through an absolute galvanometer and through a coil suspended bifilarly in the plane of the meridian, is free from the above difficulty; but it is not easy so to arrange the proportions that the suspended coil shall be sufficiently sensitive, and the galvanometer sufficiently insensitive. In this method, as in that of the dynamometer, the calculation of the forces requires a knowledge of the moment of inertia of the suspended parts.

When the electromagnetic action is a simple attraction or repulsion, it can be determined directly by balancing it against known weights. In Mascart's recent determination a long solenoid is suspended vertically in the balance, and is acted upon by a flat coaxial coil of much larger radius, whose plane includes the lower extremity of the solenoid. This arrangement, though simple to think about, does not appear to be the one best adapted to secure precise results. It is evident that a large part of the solenoid is really ineffective, those turns which lie nearly in the plane of the flat coil being but little attracted, as well as those which lie towards the further extremity. The result calculated from the total length of wire (even if this could be trusted), the length of the solenoid, and the number of turns, has an appearance of accuracy which is illusory unless it can be assumed that the distribution of the wire over the length is strictly uniform. It would appear that all the turns of the suspended coil should operate as much as possible, that is, that the suspended coil should be compact, and should be placed in the position of maximum effect.

There is a further incidental advantage in this arrangement which it is the principal object of the present note to point out. The expression for the attraction involves as factors the product of the numbers of turns, the square of the current, and a function of the mean radii of the two coils and of the distance between their mean planes. Now, as may be seen from the fact that the square of a current is already of the dimensions of a force, this function of three linear quantities is itself of no dimensions. In determining its actual value we should in general be subject to three errors; but when the position is such that the function (for two given coils) is a maximum, the result is practically dependent only upon the two mean radii, and being of no dimensions can involve them only in the form of a *ratio*. In order then to calculate the result, all that it is necessary to know with precision is the ratio of the mean radii of the two coils. This ratio can be obtained electrically, with full precision, and without any linear measurements. For, if the two coils considered as galvanometer coils are brought coaxially into the same plane, the ratio of their constants can be found by the known method of dividing a current between them in such a way that no effect is produced upon a small magnet suspended at their common centre. The ratio of the resistances in multiple arc gives the ratio of the currents, and this again (subject to small corrections for the finite size of the sections), gives the ratio of the mean radii.

It appears that in this way all that is necessary for the absolute determination of currents can be obtained without measurements of length, or of moments of inertia, or even of absolute angles of deflection. In practice it will be desirable to duplicate the fixed coil, placing the suspended coil midway between two similar fixed ones, through which the current passes in opposite directions. A rough approximation to the condition of things above described will be quite sufficient.

89.

ON THE DURATION OF FREE ELECTRIC CURRENTS IN AN INFINITE CONDUCTING CYLINDER.

[*British Association Report*, pp. 446, 447, 1882.]

TAKING the axis of the cylinder as that of z, we suppose that the currents are functions of $\sqrt{(x^2 + y^2)}$, or r, only, and flow in the circles $r = \text{constant}$. From the equations given in Maxwell's *Electricity*, vol. II. §§ 591, 598, 607, 610, 611, we may deduce for a conductor of constant μ

$$\left(\frac{d^2}{dx^2} + \frac{d^2}{dy^2} + \frac{d^2}{dz^2}\right) c = 4\pi\mu C \frac{dc}{dt},$$

with similar equations for b and a.

In the present case the magnetic forces b and a vanish, and c is a function of r only. Thus

$$\left(\frac{d^2}{dr^2} + \frac{1}{r}\frac{d}{dr}\right) c = 4\pi\mu C \frac{dc}{dt},$$

or, if c varies as e^{-nt},

$$\left(\frac{d^2}{dr^2} + \frac{1}{r}\frac{d}{dr} + 4\pi\mu nC\right) c = 0,$$

the solution of which, subject to the condition of finiteness at the centre, is

$$c = AJ_0\{\sqrt{(4\pi\mu nC)}.r\} = AJ_0(kr).$$

To determine the admissible values of n, we have only to form the condition which must be satisfied at the boundary of the cylinder $r = R$. It is evident that the magnetic force must here be zero, so that the condition is

$$J_0\{\sqrt{(4\pi\mu nC)}.R\} = 0.$$

The roots of the function are,

$$2\text{·}404,\ 5\text{·}520,\ 8\text{·}654,\ 11\text{·}792,\ \&\text{c}.$$

For the principal mode of longest duration

$$c = A J_0 (2\cdot404\, r/R),$$

and

$$n = \frac{2\cdot404^2}{4\pi\mu C R^2}.$$

If τ be the time in which the amplitude sinks in ratio $e : 1$,

$$\tau = \frac{1}{n} = \frac{4\pi\mu C R^2}{(2\cdot404)^2}.$$

For copper in C.G.S. measure $C = \dfrac{1}{1642}$, $\mu = 1$,
and thus

$$\tau = \frac{R^2}{800} \text{ nearly.}$$

In order that τ should be one second, the diameter of the cylinder would have to be about two feet.

[1900. In the case of iron subjected to small magnetic forces we may take (see *Phil. Mag.* vol. XXIII. p. 235, 1887)

$$C = \frac{1}{9611}, \qquad \mu = 100,$$

so that

$$\tau = \frac{R^2}{50} \text{ nearly.]}$$

90.

ON THE EQUILIBRIUM OF LIQUID CONDUCTING MASSES CHARGED WITH ELECTRICITY.

[*Philosophical Magazine*, XIV. pp. 184—186, 1882.]

IN consequence of electrical repulsion, a charged spherical mass of liquid, unacted upon by other forces, is in a condition of unstable equilibrium. If a_0 be the radius of the sphere, Q the charge of electricity, the original potential is given by

$$V = Q/a_0.$$

If, however, the mass be slightly deformed, so that the polar equation of its surface, expressed by Laplace's series, becomes

$$r = a\,(1 + F_1 + F_2 + \ldots + F_n + \ldots),$$

then

$$V = \frac{Q}{a_0}\left\{1 - \Sigma\,(n-1)\iint\frac{F_n{}^2 d\sigma}{4\pi}\right\};$$

and the potential energy of the system reckoned from the equilibrium position is

$$P' = -\frac{Q^2}{8\pi a_0}\Sigma\,(n-1)\iint F_n{}^2 d\sigma.$$

In actual liquids this instability, indicated by the negative value of P', is opposed by stability due to the capillary force. If T be the cohesive tension, the potential energy of cohesion is given by

$$P = \tfrac{1}{2}a_0{}^2 T\,\Sigma\,(n-1)(n+2)\iint F_n{}^2 d\sigma\,^*.$$

If $F_n \propto \cos\,(pt + \epsilon)$, we have for the motion under the operation of both sets of forces,

$$p^2 = \frac{n\,(n-1)}{\rho a_0{}^3}\left\{(n+2)\,T - \frac{1}{4\pi}\frac{Q^2}{a_0{}^3}\right\}.$$

* See *Proc. Roy. Soc.* May 15, 1879 [vol. I. p. 400].

If $T > Q^2/16\pi a_0^3$, the spherical form is stable for all displacements. When Q is great, the spherical form is unstable for all values of n below a certain limit, the maximum instability corresponding to a great, but still finite, value of n. Under these circumstances the liquid is thrown out in fine jets, whose fineness, however, has a limit.

The case of a cylinder, subject to displacement in two dimensions only, may be treated in like manner.

The equation of the contour being in Fourier's series

$$r = a (1 + F_1 + \dots + F_n + \dots),$$

we find as the expression for the potential energy of unit length

$$P' = -\frac{Q^2}{l^2} \Sigma (n-1) \int \frac{F_n^2 d\theta}{2\pi},$$

Q being the quantity of electricity resident on length l.

The potential energy due to capillarity is

$$P = \tfrac{1}{2}\pi a\, T \Sigma (n^2 - 1) \int \frac{F_n^2 d\theta}{2\pi},$$

and for the vibration of type n under the operation of both sets of forces,

$$p^2 = \frac{n^2 - n}{\rho a^3} \left\{ (n+1)\, T - \frac{2Q^2}{\pi l^2 a} \right\}.$$

The influence of electrical charge in diminishing the stability of a cylinder for transverse disturbances may be readily illustrated by causing a jet of water from an elliptical aperture to pass along the axis of an insulated inductor-tube, which is placed in connexion with an electrical machine. The jet is marked with a recurrent pattern, fixed in space, whose wave-length represents the distance travelled by the water in the time of one vibration of type $n = 2$ *. When the machine is worked, the pattern is thrust outwards along the jet, indicating a prolongation of the time of transverse vibration. The inductor should be placed no further from the nozzle than is necessary to prevent the passage of sparks, and must be short enough to allow the issue of the jet before its resolution into drops.

The value of T being known (81 C.G.S.), we may calculate what electrification is necessary to render a small rain-drop of, say, 1 millimetre diameter unstable. The potential, expressed in electrostatic measure, is given by

$$V = Q/a_0 = \sqrt{(16\pi a_0 T)} = 20.$$

The electromotive force of a Daniell cell is about ·004; so that an electrification of about 5000 cells would cause the division of the drop in question.

* [Art. 60, vol. I. p. 377.]

91.

ON AN INSTRUMENT CAPABLE OF MEASURING THE INTENSITY OF AERIAL VIBRATIONS.

[*Philosophical Magazine*, XIV. pp. 186, 187, 1882.]

THIS instrument arose out of an experiment described in the *Proceedings of the Cambridge Philosophical Society**, Nov. 1880, from which it appeared that a light disk, capable of rotation about a vertical diameter, tends with some decision to set itself at right angles to the direction of alternating aerial currents. In Fig. 1, A is a brass tube closed at one end with a glass

Fig. 1.

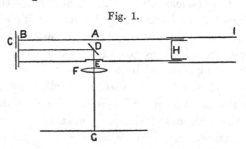

plate B, behind which is a slit C backed by a lamp. D is a light mirror with attached magnets, such as are used for reflecting-galvanometers, and is suspended by a silk fibre. The light from the slit is incident upon the mirror at an angle of 45°, and, after reflection, escapes from the tube through a glass window at E. It then falls upon a lens F, and throws an image of the slit upon a scale G. At a distance DH, equal to DC, the tube is closed by a diaphragm of tissue paper, beyond which it is acoustically prolonged by a sliding tube I.

When the instrument is exposed to sounds whose half wave-length is equal to CH, H becomes a node of the stationary vibrations, and the

---

* See also *Proc. Roy. Soc.* May 5, 1881, p. 110. [Art. 79, vol. II. p. 7.]

paper diaphragm offers but little impediment. Its office is to screen the suspended parts from accidental currents of air. At D there is a loop; and the mirror tends to set itself at right angles to the tube under the influence of the vibratory motion. This tendency is opposed by the magnetic forces; but the image upon the scale shifts its position through a distance proportional to the intensity of the action.

As in galvanometers, increased sensitiveness may be obtained by compensating the earth's magnetic force with an external magnet. Inasmuch, however, as the effect to be measured is not magnetic, it is better to obtain a small force of restitution by diminishing the moment of the suspended magnet rather than by diminishing the intensity of the field in which it works. In this way the zero will be less liable to be affected by accidental magnetic disturbances.

So far as I have tested it hitherto, the performance of the instrument is satisfactory. What strikes one most in its use is the enormous disproportion that it reveals between sounds which, when heard consecutively, appear to be of the same order of magnitude.

92.

COMPARISON OF METHODS FOR THE DETERMINATION OF RESISTANCES IN ABSOLUTE MEASURE.

[*Philosophical Magazine*, XIV. pp. 329—346, 1882.]

AT the present time, and in view of the projected conference at Paris, the subject of the present paper is engaging a large share of attention; and Prof. G. Wiedemann* has published an interesting discussion of some of the methods that have been employed. I have thought it might be of service if I also were to place upon record the views that I have been led to entertain, and which are the result of a good deal of experience.

Resistance being of the dimensions of velocity, its absolute measurement involves the absolute measurement of a length and of a time. The latter is usually the time of a vibration of a suspended magnet, and it can be determined without much difficulty. In the B.A. method it is the time of rotation of the revolving coil, and it can be obtained with all desirable accuracy. In this respect there is not much to choose between one method and another; but when we come to consider the manner in which the linear measurement enters, important differences reveal themselves. These will be discussed in detail presently; but for the moment it will be sufficient to say that the presumption is in favour of any method which requires only a single linear measurement. It is true that this question cannot be decided without regard to the subject of the measurement; but, with scarcely an exception, it is necessary to know the mean radius of a *coil* of several layers of insulated wire. This is apparently the measurement which fixes the limit of final accuracy; and, in comparison with it, determinations of the distances of mirrors and scales &c. are of secondary difficulty.

---

* "Ueber die bisherigen Methoden zur Feststellung des Ohm." Separatabdruck aus der *Electrotechnischen Zeitschrift*, July 1882. *Phil. Mag.* for October, p. 258.

It will be convenient now to enumerate the principal methods which have been proposed for determining absolute resistances. Minor details, which are not likely to influence the final value of the results, will in general be passed over.

I. Kirchhoff's *Method*, Maxwell's *Electricity and Magnetism*, § 759.

The magnitude of a continuous battery-current in a primary coil is compared with that of the transient current induced in a secondary coil when the primary circuit is removed. Rowland* effected an important improvement by simply reversing the battery-current without motion of the primary coil. The time of vibration of the ballistic galvanometer employed for the transient current is the principal time-measurement. In Rowland's investigation a second galvanometer was employed for the battery-current, and the ratio of constants had to be found by auxiliary experiments. In Glazebrook's† recent determination by this method only one galvanometer was used, the battery-current being reduced in a known manner by shunting. It is shown that the evaluation of the resistance-ratios presents no serious difficulty.

Let h denote the ratio in which the primary current is reduced when it produces a deflection α upon the galvanometer, θ the throw from rest due to the induction-current when the battery is reversed, T the time of vibration of the needle measured from rest to rest, M the coefficient of induction; then the resistance of the secondary circuit in absolute measure is given by

$$R = \frac{\pi M \tan \alpha}{T \cdot \sin \frac{1}{2}\theta} \div h.$$

Whenever, as in this method, the conductor whose resistance in absolute measure is first determined is composed of copper, frequent comparisons are necessary with standards of German silver or platinum-silver. Otherwise a variation of temperature of about $\frac{1}{4}$ of a degree Cent., which can hardly be detected with certainty by thermometers, would influence the result by as much as one part in a thousand.

If it be granted that the comparison of currents and the reference to the standard of resistance can be effected satisfactorily, we have only to consider the amount of error involved in the determination of M, the coefficient of mutual induction between the two circuits, which is the fundamental linear measurement. If the two coils are of very nearly the same size, it appears from symmetry that the result is practically a function of the mean of the mean radii only, and not of the two mean radii separately. It is also of course a function of the distance between the mean planes b. Leaving out of consideration the small corrections necessary for the finite size of the

* *American Journal*, vol. xv. 1878.
† *Proc. Roy. Soc.* June 1882.

sections, we consider M as equal to $4\pi\sqrt{(Aa)}$ multiplied by the function of γ given in tables appended to the second edition of Maxwell's *Electricity*, where

$$\sin\gamma = \frac{2\sqrt{(Aa)}}{\sqrt{\{(A+a)^2+b^2\}}},$$

or, if we identify A and a with their mean (A_0),

$$\tan\gamma = 2A_0/b.$$

The error in M will depend upon the errors committed in the estimates of A_0 and b. If we write

$$\frac{dM}{M} = \lambda\frac{dA_0}{A_0} + \mu\frac{db}{b},$$

then, since M is linear, $\lambda + \mu = +1.$

Thus, if b were great relatively to A_0, $\lambda = 4$, $\mu = -3$, a very unfavourable arrangement, even if it did not involve a great loss of sensitiveness. The object must be so to arrange matters that the errors in A_0 and b do not multiply themselves unnecessarily in M. But since μ is always negative, λ must inevitably be greater than unity.

The other extreme case, in which b is very small relatively to A_0, may also be considered independently of the general tables; for we may then take approximately (Maxwell's *Electricity*, § 705)

$$M = 4\pi A_0\left\{\log\frac{8A_0}{b} - 2\right\},$$

whence

$$\mu = -\frac{1}{\log(8A_0/b) - 2},$$

showing that as b diminishes μ approaches zero, and accordingly λ approaches unity, as is indeed otherwise evident. But when b is small, it is the absolute error db which we must regard as given rather than the relative error db/b; and thus we are directed to stop at a moderate value of b, even if the increased correction necessary for the size of the sections were not an argument in the same direction.

The following intermediate cases, calculated by the tables, will give an idea of the actual conditions suitable for a determination by this method:—

γ	$b/2A_0$	λ	μ	M
60°	·577	2·61	−1·61	·316
70	·364	2·18	−1·18	·597
75	·268	1·98	−0·98	·829
80	·176	1·76	−0·76	1·186

We may say that the error in the distance of mean planes will reproduce itself something like proportionally in the final result, and that the error of mean radius will be doubled.

Any uncertainty in the actual position of the mean planes relatively to the rings on which the wire is wound may be eliminated, as Glazebrook has shown, by reversing the rings relatively to the distance-pieces.

This method is subject to whatever uncertainty attaches to the use of a ballistic galvanometer*. In its favour it may be said that the apparatus and adjustments are simple, and that no measurement of distances between mirrors and scales is necessary for the principal elements. It should be noticed also that the error due to faulty determination of the distance of mean planes can be eliminated in great measure by varying this quantity, which can be done over a considerable range without much difficulty or expense.

With reference to the capabilities of the method for giving results of the highest accuracy when carried out in the most ambitious manner, it is important to consider the effect of increasing the size of the coils. The coils used by Glazebrook have a mean radius of about 26 centim.; the axial and radial breadths of the section are each about 2 centim. If we suppose the mean radius and the sides of the section to be doubled, the number of turns (about 800) remaining unaltered, the sensitiveness would be increased both by the doubling of M and by the diminished resistances of the coils, while at the same time the subjects of the linear measurements would be of more favourable magnitudes. To enhance the latter advantage, it would probably be an improvement to diminish the radial breadth of the section, on which much of the uncertainty of mean radius depends. In either case it is clear that the limit of accuracy obtainable by this method has not yet been reached.

II. Weber's *Method by Transient Currents*, Maxwell § 760.

"A coil of considerable size is mounted on an axle so as to be capable of revolving about a vertical diameter. The wire of this coil is connected with that of a tangent-galvanometer so as to form a single circuit. Let the resistance of this circuit be R. Let the large coil be placed with its positive face perpendicular to the magnetic meridian, and let it be quickly turned round half a revolution. There will be an induced current due to the earth's magnetic force ; and the total quantity of electricity in this current in electro-magnetic measure will be

$$Q = \frac{2g_1 H}{R},$$

* See *Phil. Trans.* 1882, p. 669. [Art. 80, vol. ii. p. 48.]

where g_1 is the magnetic moment of the coil for unit current, which in the case of a large coil may be determined directly by measuring the dimensions of the coil and calculating the sum of the areas of its windings; H is the horizontal component of terrestrial magnetism; and R is the resistance of the circuit formed by the coil and galvanometer together. This current sets the magnet of the galvanometer in motion."

"If the magnet is originally at rest, and if the motion of the coil occupies but a small fraction of the time of a vibration of the magnet, then, if we neglect the resistance to the motion of the magnet, we have, by § 748,

$$Q = \frac{H}{G} \frac{T}{\pi} 2 \sin \tfrac{1}{2}\theta,$$

where G is the constant of the galvanometer, T is the time of vibration of the magnet, and θ is the observed elongation. From these equations we obtain

$$R = \frac{\pi G g}{T \sin \tfrac{1}{2}\theta}.$$

The value of H does not appear in this result, provided it is the same at the position of the coil and at that of the galvanometer. This should not be assumed to be the case, but should be tested by comparing the time of vibration of the same magnet, first at one of these places, and then at the other."

If a be the mean radius of the coil of the inductor and A that of the galvanometer, we may write, neglecting the corrections for the finite sizes of the sections*,

$$g = \pi a^2, \qquad G = 2\pi/A ;$$

so that

$$gG = 2\pi^2 a^2/A.$$

This is the linear quantity of the method. With respect to the chances of error in determining it, we see that the error of the mean radius of the inductor enters doubly, and that of the mean radius of the galvanometer enters singly. Probably in this respect there is not much to choose between this method and the use in method I. of the same coils placed at a moderate distance apart.

A colossal apparatus for the use of the present method has been constructed and tested by MM. W. Weber and F. Zöllner†, the coils of which are as much as 1 metre in diameter. The principal difficulty arises in connexion with the galvanometer-magnet. Two magnets were used whose

---

* [1899. The factors expressive of the number of convolutions in the two coils are here omitted.]

† *Ber. d. Kön. Sächs. Ges. zu Leipzig*, 1880, vol. II. p. 77.

lengths were respectively 200 millim. and 100 millim.; and the results obtained in the two cases differed by as much as 2 per cent. The discrepancy is doubtless due to the influence of the finite length of the magnets causing the magnetic poles to be sensibly distant from the centre of the coil, for which point the effects are calculated; and the disturbance will be proportional to the square of the distance between the poles, or more properly to the "radius of gyration" of the ideal magnetic matter about the axis of rotation. But to assume that the disturbance from this source was exactly four times as great in the one case as in the other, and thence to deduce the result corresponding to an infinitely short magnet, appears to me to be a procedure scarcely consistent with the degree of accuracy aimed at. If this method is to give results capable of competing with those obtainable in other ways, it will be necessary to use a much shorter magnet; or, if that is not practicable, to devise some method by which the distance of the poles can be determined and a suitable correction calculated.

In carrying out the observations in the usual manner, it is necessary to measure the distance between a mirror and a scale. By using a double mirror with two scales and telescopes, MM. Weber and Zöllner avoid the principal cause of difficulty, *i.e.* the unsteadiness of the suspended mirror, all that is then necessary to know with accuracy being the distance between the two scales.

In using this and the three following methods great pains must be taken with the levelling of the earth inductor, since the deviation of the axis of rotation from the vertical (at least in the plane of the meridian) gives rise to an error of the first order with (in these latitudes) a high coefficient. In this respect it would be a decided advantage to carry out the experiments in a locality nearer to the magnetic equator (see "Account of Experiments to determine the Value of the B.A. Unit in Absolute Measure," *Phil. Trans.* 1882) [vol. II. p. 63]. It is to be hoped that the measurements commenced by Weber and Zöllner will be carried to a successful issue, as it is only by the coincidence of results obtained by various methods that the question can be satisfactorily settled. At present no value in absolute measure of the B.A. unit or of the Siemens unit has been published as the result of their work.

III.   *Method of Revolving Coil.*

This method, first, it would appear, suggested by Weber, was carried into execution by the celebrated Electrical Committee of the British Association*, and more recently by myself with the assistance of Dr Schuster and others†. The greater part of what I have to say upon this subject has been put

---

* *Brit. Assoc. Reports*, 1862–1867. Reprint, Spon, 1873.

† *Proc. Roy. Soc.* May 1881, Feb. 1882; *Phil. Trans.* 1882. [Arts. 79, 80.]

forward already in the papers referred to, from which alone the reader can form a complete opinion on the merits or demerits of the method as hitherto practised. On the present occasion I must take many of the conclusions there arrived at for granted, or at most give a mere indication of the nature of the arguments by which they may be supported.

Method III. differs from II. mainly in the fact that in III. the earth-inductor is, so to speak, its own galvanometer, the needle whose deflections measure the currents being suspended at the centre of the revolving coil itself instead of at the centre of another galvanometer-coil forming part of the same circuit. If, as in II., the inductor-coil were simply twisted through 180° when the needle passes its position of equilibrium, the disadvantages of the simplification would probably preponderate over the advantages. The diminution of effect due to the oblique position of the coil relatively to the needle (except at the moment of passing the magnetic meridian) would indeed be compensated by the diminished resistance of the complete circuit, and, as will presently appear, considerable advantage would arise in respect of errors in the measurement of the coil; but an almost fatal uncertainty would be introduced from the influence of self-induction.

The important advantage of III., obtained, as I believe, without any really important sacrifice, arises only when the inductor is set into uniform rotation. In II., if the connexions were maintained without a commutator, the current in the galvanometer-coil would be alternating, and therefore unsuitable for measurement with a magnetic needle; but in III., although the current in the coil itself alternates, the reversal of the coil relatively to the needle causes all the impulses to operate finally in the same direction. When, therefore, the coil is caused to revolve in a periodic time small relatively to that of the free vibration of the needle, a steady deflection is obtained which varies inversely with the absolute resistance of the coil.

If we omit for the moment all secondary considerations, although some of them may not be without importance, the formula by which the resistance (R) of the revolving circuit is given in terms of the mean radius (a), the number of turns (n), the angular velocity of rotation (ω), and the angle of deflection (ϕ), runs

$$R = \pi^2 n^2 a \omega \cot \phi;$$

from which it appears that, in respect of errors arising from the measurements of the coil, this method is much superior to those hitherto discussed. There is only one linear quantity concerned; and the error committed in its determination enters but singly into the final result. Indeed we may say that in this respect no improvement is possible, unless it be in the direction of substituting for the mean radius of a coil of several layers some other kind of linear quantity more easy to deal with.

In requiring the absolute measurement of angle, II. and III. stand precisely upon a level.

The time of vibration in the experiments of MM. Weber and Zöllner was 17 seconds or 30 seconds—none too long relatively to the time (2 seconds) occupied in turning the inductor. If we suppose the coil to be uniformly rotated at the rate of, say, 2 revolutions per second, there would be 68 or 120 impulses upon the needle in the time of 1 vibration. It would no doubt be a great exaggeration to represent the increase of sensitiveness as being in anything like this proportion, since by the method of recoil it is possible to make several observations of impulses during the time required for one observation of steady deflection. Nevertheless it cannot be doubted that the advantage of III. in respect of sensitiveness is very considerable.

Experience has shown that there is no difficulty in controlling and measuring the rotation of the coil; but of course some auxiliary apparatus is required for the purpose. Against this may be set the escape from observations of the time of vibration, and from any uncertainty which may attach to the ballistic use of a galvanometer-needle. The suspended magnet may easily be made of such dimensions that no appreciable error can arise from supposing it to be infinitely small.

On the other hand, some new complications enter in method III. which I desire to state in full. In the first place we have to take account of the fact that the inductor moves in a field of force due not only to the earth, but also to the suspended magnet itself. I do not think that the correction thus rendered necessary (about 4 parts per thousand in my experiments) adds in any appreciable degree to the uncertainty of the final result; but we may take note of the fact that an auxiliary determination must be made of the ratio of the magnetic moment of the suspended magnet to the earth's horizontal force.

If the metal ring on which the wire is wound be on a large scale and sufficiently massive for strength, currents may be developed in it, even although it is divided into two parts by ebonite insulation. In my experiments the effect of these currents was very sensible, and had to be allowed for by careful observations of the deflection produced when the ring was rotated with wire circuit open. In any future repetition it will be worthy of consideration whether the ring should not be formed of less conducting material. It does not appear, however, that the final result can be prejudicially influenced; and the effect produced by secondary closed circuits allows us to verify the insulation of contiguous layers or turns of the wire by comparing the deflections obtained before the wire is wound with those obtained after winding, but with main circuit open, any difference being due to leakage.

But the most serious complication in method III., and one which in the eyes of some good judges weighs strongly against it, is the disturbing influence of self-induction. With respect to this, the first point to be noticed is that the action is perfectly regular, and that the only question which arises is whether its magnitude can be determined with such accuracy that the final result does not suffer. Now the operation of self-induction is readily submitted to calculation if a certain coefficient (L) be known. We find

$$R = \pi^2 n^2 a \omega \cot \phi \{1 - U \tan^2 \phi - U^2 \tan^4 \phi\},$$

where U is a numerical quantity dependent upon L, so that the influence of self-induction is approximately proportional to the square of the speed of rotation. The same law applies also to any disturbances depending upon mutual induction between the wire circuit and subordinate circuits in the ring.

It will be seen that, if the law of squares may be depended upon, the influence of self-induction (and mutual induction) can be satisfactorily eliminated by combining observations taken at different speeds. In my experiments four speeds were used, of which the greatest and the least were in the ratio of $2 : 1$. The effect of self-induction was therefore four times as great at the high speed as at the low speed. In other words, the quantity (about 1 per cent.) by which the low-speed result is to be corrected in order to eliminate the influence of self-induction is only one-third of the discrepancy between the uncorrected results of the extreme speeds. If, therefore, the observations are good for anything at all, they are good enough to determine this correction with all desirable precision. If a check be considered necessary, it is supplied by the results of the intermediate speeds.

The above reasoning proceeds upon the supposition that we have no independent knowledge of the magnitude of the coefficient U. In point of fact, this coefficient can be calculated with considerable accuracy from the data of construction, so that the empirical correction is applied only to a small outstanding residue.

In considering the disadvantageous influence of self-induction as an argument in favour of II. as against III., we must remember that the magnitude of the influence can be greatly attenuated by simply diminishing the speed of rotation. At half the lowest speed above spoken of, for which the correction for self-induction would be reduced to $\frac{1}{4}$ per cent., the deflection (over 100 millim. at a distance of 2670 millim.) would probably correspond to a much greater sensitiveness than it is possible to obtain under II. If we prefer the higher speed, it is because we estimate the advantage of doubled sensitiveness as outweighing the disadvantage of a fourfold correction for self-induction.

The fourth objection which may be taken to this method, and it is one from which II. is free, lies in the necessary creation of mechanical disturbance in the neighbourhood of the suspended magnet.

How far these complications may be supposed to prejudice the result of carefully conducted experiments must be left to the estimation of the reader of my paper, in which very full data for a judgment are given. My own opinion is, that while in the aggregate they must be allowed to have some weight, they are far from preponderating over the advantages which the method possesses in comparison with II.

If we take the view that the method itself is trustworthy, the principal error will arise in connexion with the mean radius of the coil; and it becomes an interesting question to consider whether advantage may be expected from a further increase in the dimensions of the apparatus. For this purpose we may regard tan ϕ as given. The total resistance R will be proportional to n^2a/S, where S denotes the aggregate section of the copper, from which it follows that ωS may be regarded as given, while a is left undetermined by the consideration of sensitiveness. Thus, if we retain ω and S unaltered in a magnified apparatus, we shall have the same sensitiveness as before, while the increased diameter of the coil and the relatively decreased dimensions of the section will conduce to a more accurate determination of the mean radius.

The angular deflection being given, the correction for self-induction is nearly constant whatever may be the proportions of the coil.

If we are of opinion that there is danger in the operation of self-induction, the case becomes strong for the introduction of a second coil in a plane perpendicular to that of the first*. By this means the relative correction for self-induction would be reduced to one quarter, while the deflection remained unaltered. It scarcely needs to be remarked that this use of a second coil would not, as in II., increase the uncertainty depending upon the linear measurements, the two mean radii entering into the result as parts, and not as factors.

This combination would lend itself especially well to low speeds of rotation; for the deflecting force, being uniform in respect to time, would not give rise to forced vibrations of the needle. The latter would have nothing further to do than to indicate the direction of a constant field of force.

IV.

This method, which was proposed by Foster†, and more recently by Lippmann, and to a certain extent executed by the former, is a modification

* *Proc. Roy. Soc.* May 1881, p. 123 [vol. II. p. 19].
† *Brit. Assoc. Report*, 1881.

of III., in which the electromotive force generated during the rotation of the inductor is balanced by an external electromotive force, and thus not allowed to produce a current. The external electromotive force is due to the passage of a battery-current through certain resistance-coils; and the current is compared with the earth's horizontal intensity (H) by an absolute tangent-galvanometer. The difference of potential at the two points of derivation is thus known in terms of the included absolute resistance (R) and H. The circuit is continued through a sensitive galvanometer and the coil of the inductor, and is closed only when the latter coil is nearly in the plane of the meridian. When balance is obtained, the electromotive force of induction $n \cdot \pi a^2 \cdot H \cdot \omega$ is equal to $(RH/G) \tan \alpha$, where G is the constant of the tangent-galvanometer and α the angle of deflection. The result, from which H disappears, if it may be assumed to be the same in the two places, is thus

$$R = n\pi a^2 G \cdot \omega \cdot \cot \alpha,$$

or, if A be the mean radius of the galvanometer-coil,

$$R = 2n\pi^2 \omega \cot \alpha \cdot a^2/A,$$

from which the value of the resistance-coils is obtained in absolute measure. One advantage of this method, which it shares with VI. below, is that the resistance immediately expressed may be that of well-constructed coils of German silver or of platinum-silver at a known temperature.

This method is nearly free from the secondary objections to III. discussed above. The self-induction of the revolving wire-circuit does not enter, as no appreciable current is allowed to form itself; but there would appear to be a possibility of disturbance from mutual induction between the wire-circuit and secondary circuits in the ring. It would certainly be necessary to prevent the flow of currents round the ring by the insertion of an insulating layer; and even with this precaution some control in the way of a variation of speed would almost be necessary. Again, it is a question whether disturbance from thermo-electricity for instance, may not arise at the place where the contacts are made and broken.

It is to be hoped that a complete series of observations may be made by this method, which certainly possesses considerable merits; but at best it remains open to the objection mentioned under II., with which in this respect it stands upon a level, *i.e.* that errors may enter from the measurements of both coils, the error of A entering singly into the result, and that of a entering doubly.

In respect of requiring absolute measurements of angle, there is nothing to choose between II., III., IV., and V.

V.   Weber's *Method by Damping*.

This is the method followed by Kohlrausch* in his investigations upon this subject. It is founded upon II.; but in order to avoid the difficulty arising from the necessity of using a magnet small relatively to the coil in which it is suspended, no attempt is made to determine the constant from the data of construction. The inductor is connected with a sensitive galvanometer, and the constant of the latter is deduced from observations of the logarithmic decrement of the vibrations of the magnet when the circuit is closed (λ), and when it is open (λ_0). The result, however, involves H the horizontal intensity, K the moment of inertia of the needle, as well as the time of vibration T. Expressed roughly, in the notation previously employed, it is

$$ R = \frac{32a^4 H^2 T \lambda}{K} \cdot \frac{AB}{(A^2 + B^2)^2}, $$

where R is the resistance of the circuit composed of the inductor and galvanometer, A and B are the arcs of vibration in the method of recoil.

Interesting as this method is in some respects, I cannot but agree with Rowland in thinking that the final formula is enough to show that it cannot compete with others on equal terms, if the object be to obtain a result of high accuracy. The horizontal intensity itself is perhaps nearly as difficult to determine as absolute resistance; and the error thence arising doubles itself in the result. There is in addition the error of K. But even if H and K were not subject to error at all, I believe that the occurrence of the fourth power of the radius of the inductor is a fatal defect, and tends to explain the discrepant result obtained by Kohlrausch†. It is also worthy of note that the error of levelling enters twice as much as in II., III., and IV.

VI.   Lorenz's *Method*.

This method, which, with the introduction of certain modifications not affecting its essential character, I am disposed to consider the best of all, was proposed and executed by Lorenz, of Copenhagen, in 1873‡. A circular disk of metal, maintained in rotation about an axis passing through its

---

* *Pogg. Ann.* Ergänzungsband vi.; *Phil. Mag.* 1874, April and May.

† Oct. 1882.—It is very satisfactory to note that Kohlrausch (*Gött. Ges.* Sept. 1882) has recently detected an error in the value of the area of the windings of the inductor assumed in his previous calculations. Introducing the new value, obtained by an electrical process analogous to that described in Maxwell's *Electricity*, § 754, he finds

$$ 1 \text{ B.A. unit} = \cdot 990 \times 10^9. $$

‡ *Pogg. Ann.* vol. cxlix. p. 251.

centre at a uniform and known rate, is placed in the magnetic field due to a battery-current which circulates through a coaxal coil of many turns. The revolving disk is touched near its centre and circumference by two wires. If the circuit were simply closed through a galvanometer, the instrument would indicate the current due to the electromotive force of induction acting against the resistance of the circuit. The electromotive force corresponding to each revolution is the same as would be generated in a single turn of wire coincident with the circumference of the disk by the formation or cessation of the battery-current. If this be called γ, and M be the coefficient of induction between the coil and the circumference, m the number of revolutions per second, the electromotive force is $mM\gamma$. For the present purpose, however, the circuit is not simply closed, but its terminals are connected with the extremities of a resistance R through which the battery-current flows, and the variable quantities are so adjusted that the electromotive force $R\gamma$ exactly balances that of induction. When the galvanometer indicates no current, the following relation, independent, it will be observed, of the magnitude of the battery-current, must be satisfied,

$$R = mM;$$

and from this, M being known from the data of construction, the absolute resistance R of the conductor is determined.

It will be seen that this method has pretty close affinity to I. The secondary circuit is here, in a sense, reduced to a single turn, or rather to as many turns as the disk makes revolutions in a time comparable with the time of swing of the ballistic galvanometer; but the disadvantage of a reduced number of turns is probably more than compensated for by the continuous character of the induced current, which allows of its being brought into direct opposition to that of the battery. During the months from April to August of the present year I have been occupied in carrying out a determination by this method. Space will not permit of a detailed consideration of the various questions which presented themselves; and I must content myself with a brief statement of the procedure, and with such a discussion of the sources of error as will allow a comparison of this method with others. I hope shortly to communicate a detailed paper upon the subject to the Royal Society*.

One of the principal difficulties to be overcome arises from the exceeding smallness of the resistance R, less than $\frac{1}{200}$ B.A. in my experiments. Lorenz employed an actual column of mercury of known dimensions, so that the result is given at once in terms of mercury. I had intended to follow the same course, but, after some trials, came to the conclusion that there would be difficulties in the way of thus obtaining the degree of

* [See *Phil. Trans.* 1883 ; Art. 94 below.]

accuracy aimed at, and ultimately adopted a method of shunting. The main current from the battery was divided into two parts, the larger of which passed through a resistance of half a unit, formed by combining two singles in multiple arc. The resistance traversed by the other part of the main current was much larger (from 10 to 20); and it was to two points on this branch distant $\frac{1}{10}$ that the wires of the derived circuit were connected. With proper precautions this arrangement was found satisfactory, and the equivalent resistance R could be accurately expressed in terms of the standard B.A. units. The adjustment for obtaining the balance was effected by varying a large resistance placed in multiple arc with one of the others; or rather two effective resistances were used, one on either side of that required for balance, the latter being finally calculated by interpolation from the indications of the galvanometer.

By observing only the effect of reversing the battery-current the results are freed from the influence of terrestrial magnetism, and from the very sensible thermoelectric force having its seat at the sliding contact. These contacts were made by means of brushes of copper wire. One brush pressed against the cylindrical edge of the disk, which was about $\frac{1}{4}$ inch broad; and the other pressed against the shaft on which the whole turned. The area included by the secondary circuit was therefore not exactly that of the disk, but required a small correction, as to which, however, there is no difficulty.

The arrangements for driving the disk and for observing the speed were the same as for the revolving coil of method III. The results, which in the same arrangement have not differed by so much as $\frac{1}{1000}$ on different days, show that the sensitiveness was sufficient.

After these explanations I come to the main subject of the present remarks, viz. the degree of accuracy likely to be attained in the fundamental linear measurement. In the present case the quantity to be determined is M; and so far there is no difference between this method and I. But the fact that the secondary circuit is here represented by a disk whose diameter can be measured much more accurately than that of a coil introduces a certain modification. It is necessary also that the arrangements be symmetrical with respect to the middle plane of the disk, as, on account of the width of the brush, the place of contact cannot be considered as well defined. The necessary condition can be satisfied with a single coil by placing it so that its mean plane coincides with that of the disk. In this position slight errors of adjustment produce effects of the second order only, and everything depends upon the radii.

Preparatory to the design of the apparatus for my experiments, I made some calculations of the values of the induction-coefficient and of its rates of variation for various ratios of the radius of the coil (A) to that of the disk (a). The angle γ (see method I.) is here $(b = 0)$ determined by $\tan^2 \frac{1}{2}\gamma = a/A$. If we write

$$\frac{\delta M}{M} = \lambda \frac{\delta A}{A} + \nu \frac{\delta a}{a},$$

the sum of λ and ν will be unity. The following are the values found. Those under M are proportional only, and relate to the case in which A is constant.

a/A	λ	ν	M
·5	$-1\cdot2$	$+2\cdot2$	4·37
·6	$-1\cdot36$	$+2\cdot36$	6·65
·7	$-1\cdot5$	$+2\cdot5$	9·80
·8	$-2\cdot0$	$+3\cdot0$	14·4

In Lorenz's apparatus the value of a/A was even larger than the last in the table, and the radial dimension of the coil was no small fraction of $(A - a)$. On this account, as has already been pointed out by Rowland, no very accurate result could be expected.

In my experiments two similar coils were used [in series] whose radius $(A) =$ about 26 cm., and in two distinct arrangements. In the first arrangement the two cells were placed close together; so that the case corresponded pretty closely with that just spoken of. The radius of the disk is about 16 cm.; and thus the proportions are nearly those of the second example in the table. It will be seen that the circumstances are not unfavourable to accuracy, the error of mean radius of the coil entering into the result to a less extent than in any of the methods hitherto described, except III. and IV. The disk is so much more easily measured, that the larger coefficient 2·36, applicable to it, should not lead to much error in the result.

This arrangement was worked at two speeds of rotation in the proportion of 10 : 16, and gave with close accordance

1 B.A. unit = ·9867 × 10⁹ c.g.s.

In the other arrangement the two coils were separated to a considerable distance, and the induction-coefficient depended not only upon the mean radii of the coils (and of the disk), but also upon the distance of their mean planes. The peculiarity of this arrangement, to which I wish to draw special attention, is that it is possible so to proportion the quantities that *the error of mean radius of the coil does not affect the result*, which accordingly depends only upon the diameter of the disk and the distance of the coil's mean planes. How this may come about will be readily understood by considering the dependence of M upon A when a and b are given. It is clear that M vanishes, both when A is very small and when it is very large; from which it follows that there must be some value of A for which the effect is a maximum and therefore independent of small variations of A.

In carrying out this idea it is not necessary to approach the above-defined state of things very closely; for of course we have in reality a good approximate knowledge of the value of A. In my apparatus the distance of mean planes was about 30 cm., so that $b =$ about 15 cm. With the actual proportions a calculation of the effects of the various errors shows that

$$\frac{\delta M}{M} = \cdot 12 \, \frac{\delta A}{A} - \cdot 96 \, \frac{\delta b}{b} + 1 \cdot 8 \, \frac{\delta a}{a};$$

so that the error of A enters in quite a subordinate degree. The positive coefficient of δA shows that with the given coils and disk the separation was somewhat too great to secure the greatest independence of δA.

The success of this arrangement depends principally upon the degree of accuracy with which b can be determined. The two rings on which the wire is coiled are separated by distance-pieces; and, as in I., by reversing the rings relatively to the distance-pieces the result may be made to depend upon the mean length of these pieces and the mean thicknesses of the rings at the places of contact. The three distance-pieces were held together in one length and measured under microscopes; and the thicknesses of the rings were taken with verified callipers. There can hardly be a doubt but that this determination is much more accurate than that of the mean radius of a coil; and, what is also of some importance, it admits of repetition at pleasure with comparatively little trouble.

The value of the B.A. unit resulting from the measurement with this arrangement was $\cdot 9869 \times 10^9$ c.g.s.*

There seems no reason why a further increase of accuracy should not be obtainable by enlarging the scale of the apparatus. If we suppose the scale doubled, the number of turns in the coil and the angular speed of the disk being unaltered, the value of M would be doubled; and thus with the same battery-current the sensitiveness would be improved. Or, if we suppose the circumferential linear speed of the disk rather than its angular speed to be constant, the sensitiveness would be unchanged. If the larger coil were made of the same kind of wire as the smaller, its

---

* The reductions not being yet finally completed, these numbers are liable to a change of one or two units in the fourth place of decimals.

resistance would be augmented; but if the dimensions of the section were also doubled, so as to keep the proportions throughout, the advantage in this respect would lie with the larger apparatus.

On the whole, I am of opinion that if it is desirable at the present time to construct apparatus on the most favourable scale, so as to reach the highest attainable accuracy, the modification of Lorenz's method last described is the one which offers the best prospect of success. Before this is done, however, it appears to me important that the value now three times obtained in the Cavendish Laboratory by distinct methods should be approximately verified (or disproved) by other physicists. To distinguish between this value and those obtained, for instance, by Kohlrausch, by Lorenz, or by the first B.A. Committee, should not require the construction of unusually costly apparatus. Until the larger question is disposed of, it appears premature to discuss the details of arrangements from which the highest degree of precision is to be expected.

93.

ON THE DARK PLANE WHICH IS FORMED OVER A HEATED WIRE IN DUSTY AIR.

[*Proceedings of the Royal Society*, XXXIV. pp. 414—418, 1882.]

IN the course of his examination of atmospheric dust as rendered evident by a convergent beam from the electric arc, Professor Tyndall noticed the formation of streams of dust-free air rising from the summits of moderately heated solid bodies*. "To study this effect a platinum wire was stretched across the beam, the two ends of the wire being connected with the two poles of a galvanic battery. To regulate the strength of the current a rheostat was placed in the circuit. Beginning with a feeble current, the temperature of the wire was gradually augmented; but before it reached the heat of ignition, a flat stream of air rose from it, which, when looked at edgeways, appeared darker and sharper than one of the blackest lines of Fraunhofer in the solar spectrum. Right and left of this dark vertical band the floating matter rose upwards, bounding definitely the non-luminous stream of air."......

"When the wire is white hot, it sends up a band of intense darkness. This, I say, is due to the *destruction* of the floating matter. But even when its temperature does not exceed that of boiling water, the wire produces a dark ascending current. This, I say, is due to the *distribution* of the floating matter. Imagine the wire clasped by the mote-filled air. My idea is that it heats the air and lightens it, without in the same degree lightening the floating matter. The tendency, therefore, is to start a current of clean air through the mote-filled air. Figure the motion of the air all round the wire. Looking at its transverse section, we should see the air at the bottom of the wire bending round it right and left in two branch currents, ascending its sides, and turning to fill the partial vacuum created above the wire. Now

---

* *Proc. Roy. Inst.* vol. VI. p. 3, 1870.

as each new supply of air, filled with its motes, comes in contact with the hot wire, the clean air, as just stated, is first started through the inert motes. They are dragged after it, but there is a fringe of cleansed air in advance of the motes. The two purified fringes of the two branch currents unite above the wire, and, keeping the motes that once belonged to them right and left, they form by their union the dark band observed in the experiment. This process is incessant. Always the moment the mote-filled air touches the wire, the distribution is effected, a permanent dark band being thus produced. Could the air and the particles under the wire pass *through* its mass, we should have a vertical current of particles, but no dark band. For here, though the motes would be left behind at starting, they would hotly follow the ascending current, and thus abolish the darkness."

Professor Frankland*, on the other hand, considers that what is proved by the above described observations is that "a very large proportion of the suspended particles in the London atmosphere consists of water and other volatile liquid or solid matter."

Last summer (1881) I repeated and extended Tyndall's beautiful experiment, not feeling satisfied with the explanation of the dark plane given by the discoverer. Too much stress, it appeared to me, is placed upon the relative lightening of the air by heat. The original density is probably not more than about $\frac{1}{1000}$ part of that of the particles, and it is difficult to see how a slight further lightening could produce so much effect. In other respects, too, the explanation was not clear to me. At the same time I was not prepared to accept Professor Frankland's view that the foreign matter is volatilised.

The atmosphere of smoke was confined within a box (of about the size of a cigar-box), three of the vertical sides of which were composed of plates of glass. A beam of sunlight reflected into the darkened room from a heliostat was rendered convergent by a large lens of somewhat long focus, and made to pass in its concentrated condition through the box. The third glass side allowed the observer to see what was going on inside. It could be removed when desired so as to facilitate the introduction of smoke. The advantages of the box are twofold. With its aid much thicker smoke may be used than would be convenient in an open room, and it is more easy to avoid draughts which interfere greatly with the regularity of the phenomena to be observed. Smouldering brown paper was generally used to produce the smoke, but other substances, such as sulphur and phosphorus, have been tried. The experiment was not commenced until the smoke was completely formed, and had come nearly to rest. In some respects the most striking results were obtained from a copper blade, about $\frac{1}{4}$-inch broad, formed by hammering flat one end of a stout copper rod. The plane of the blade was horizontal, and

* *Proc. Roy. Soc.* vol. xxv. p. 542.

its length was in the line of sight. The unhammered end of the rod projected from the box, and could be warmed with a spirit-lamp. The dark plane was well developed. At a moderate distance above the blade it is narrow, sometimes so narrow as almost to render necessary a magnifying-glass; but below, where it attaches itself to the blade, it widens out to the full width, as shown in the figure. Whether the heated body be a thin blade or a cylindrical rod, the fluid passes round the obstacle according to the electrical law of flow, the stream-lines in the rear of the obstacle being of the same form as in front of it. This peculiarity of behaviour is due to the origin of the motion  being at the obstacle itself, especially at its hinder surface. If a stream be formed by other means and impinge upon the same obstacle without a difference of temperature, the motion is of a different character altogether, and eddies are formed in the shadow.

The difference of temperature necessary to initiate these motions with this dark plane accompaniment is insignificant. On July 20, 1881, a glass rod, about ¼-inch [6 mm.] in diameter, was employed. It was heated in a spirit-lamp, and then inserted in the smoke-box. The dark plane gradually became thinner as the rod cooled, but could be followed with a magnifier for a long time. While it was still quite distinct the experiment was stopped, and on opening the box the glass rod was found to be scarcely warmer than the fingers. It was almost impossible to believe that the smoky matter had been evaporated.

In order to test the matter more closely, smoke was slowly forced through a glass tube heated near the end pretty strongly by a spirit-lamp, and then allowed to emerge into the concentrated sunshine. No distinct attenuation of the smoke could be detected even under this treatment.

It is not necessary to dwell further upon these considerations, as the question may be regarded as settled by a decisive experiment tried a few days later. The glass rod before used was cooled in a mixture of salt and ice, and after wiping was placed in the box. In a short time a dark plane, extending *downwards* from the rod, clearly developed itself and persisted for a long while. This result not merely shows that the dark plane is not due to evaporation, but also excludes any explanation depending upon an augmentation in the difference of densities of fluid and foreign matter.

The experiment was varied by using a U-tube, through which cooled water could be made to flow. When the water was not very cold, the appearances were much the same as with the solid rod; but when, by means of salt and ice, the tube was cooled still further, a curious complication presented itself. Along the borders of the dark plane the smoke appeared considerably brighter than elsewhere. Sometimes when the flow was not very regular it looked at first as if the dark plane had been replaced by a bright one, but on closer examination the dark plane could be detected inside. There seems no doubt but that the effect is caused by condensation of moisture upon the smoke, due to the chilling which the damp air undergoes in passing close to the cold obstacle. Where the fog forms, more light is scattered; hence the increased brightness. That the fog should not form within the smoke-free plane itself is what we might expect from the interesting observations of Aitken.

With respect to the cause of the formation of the dark plane, the most natural view would seem to be that the relatively dense particles are thrown outwards by centrifugal force as the mixture flows in curved lines round the obstacle. Even when the fluid is at rest, a gradual subsidence must take place under the action of gravity; but this effect could at first only manifest itself at the top where the upper boundary of the gas prevents the entrance of more dust from above. It is known that air in a closed space will gradually free itself from dust, but the observation of a thin dust-free stratum at the top of the vessel is difficult. If we conceive a vessel full of dusty air to be set into rapid rotation, the dust might be expected to pass outwards in all directions from the axis, along which a dust-free line would form itself. I have tried this experiment, but looking along the axis through the glass top of the vessel I could see no sign of a dark line, so long as the rotation was uniform. When, however, the vessel was stopped, a column of comparatively smoke-free air developed itself along the axis. This I attributed to the formation of an inward flow along the top of the vessel, combined with a downward flow along the axis after the manner described and explained by Professor James Thomson, so that the purified air had been in intimate proximity with the solid cover. It would almost seem as if this kind of contact was sufficient to purify the air without the aid of centrifugal force.

The experiments made hitherto in order to elucidate this question have given no decisive result. If the thin convex blade already spoken of be held in the smoke-box in a vertical instead of in a horizontal plane, the lines of motion are much less curved, and we might expect to eliminate the influence of centrifugal force. I have not succeeded in this way in getting rid of the dark plane; but since under the magnifier the curvature of the motion was still quite apparent, no absolute conclusion can be drawn.

[1900. The reader is referred to interesting papers by Aitken (*Edin. Trans.* XXXII. p. 239, 1884) and by Lodge and Clark (*Phil. Mag.* XVII. p. 214, 1884) in which this question is further discussed. It seems clear that gravitation and a movement from hot to cold, somewhat as in Crookes' radiometer, are both concerned.]

94.

EXPERIMENTS, BY THE METHOD OF LORENZ, FOR THE
FURTHER DETERMINATION OF THE ABSOLUTE VALUE
OF THE BRITISH ASSOCIATION UNIT OF RESISTANCE,
WITH AN APPENDIX ON THE DETERMINATION OF THE
PITCH OF A STANDARD TUNING-FORK.

[*Phil. Trans.* CLXXIV. pp. 295—322, 1883.]

By Lord RAYLEIGH and Mrs H. SIDGWICK.

§ 1. IN this method, which was employed by Lorenz in 1873*, a circular disc of metal is maintained in rotation at a uniform and known rate about an axis passing through its centre, and is placed in the magnetic field due to a battery current which circulates through a coaxal coil of many turns. The revolving disc is touched at its centre and circumference by two wires. If the circuit were simply closed through a galvanometer, the instrument would indicate the current due to the electromotive force of induction acting against the resistance of the circuit. The electromotive force corresponding to each revolution is the same as would be generated in a single turn of wire coincident with the circumference of the disc by the formation or cessation of the battery current. If this be called γ, and M be the coefficient of induction between the coil and the circumference, m the number of revolutions per second, the electromotive force is $mM\gamma$. In the actual arrangement, however, the circuit is not simply closed, but its terminals are connected with the extremities of a resistance R, traversed by the battery current, and the variable quantities are so adjusted that the electromotive force $R\gamma$ exactly balances that of induction. When the galvanometer indicates no current, the

---

* *Pogg. Ann.* vol. CXLIX. p. 251.

following relation, independent, it will be observed, of the magnitude of the battery current, must be satisfied—

$$R = mM\,;$$

and from this, M being known from the data of construction, the absolute resistance R of the conductor is determined.

One of the principal difficulties to be overcome arises from the smallness of the resistance R, necessary for a balance, even when m and M are both increased as far as possible. Lorenz employed three resistances, ranging from ·0008 to ·002 of a mercury unit, and he evaded the necessity of comparing these small resistances with ordinary standards by constructing them of actual columns of mercury. His result was accordingly obtained directly in terms of mercury, and was to the effect that

$$1 \text{ mercury unit} = \cdot9337 \times 10^9 \text{ c.g.s.}$$

differing nearly 1 per cent. from the value (·941) obtained by ourselves.

§ 2. Under the conviction that this method offers in some respects important advantages, and influenced also by the fact that the arrangements for producing and measuring the uniform rotation necessary were ready to our hands, we determined to give it a trial, in the hope of obtaining confirmation of the results already arrived at by ourselves and by Glazebrook with other methods. At first the intention was to follow Lorenz in using for the resistance a glass tube full of mercury, with two points of which contact would be made by platinum wires passing through the glass. It appeared, however, that there would be difficulty in making the measurements with the degree of accuracy aimed at. If the wires were sealed into the glass, the section would probably be rendered irregular. An attempt was made to avoid this difficulty by using a tube from which the ends had been cut with the aid of heat. After small nicks had been filed sufficiently deep to receive the platinum wires, the ends were replaced in their original positions and secured with shellac. In this way a satisfactory uniformity of section near the points of derivation could be attained, but the measurement of the distance between these points, which is required to be known with full accuracy, was rendered difficult by the presence of the cement. It is possible that these difficulties might have been overcome, but at this point a method of shunting occurred to us, allowing the use of mercury to be dispensed with. Merely for the purpose of connecting the mercury unit with the B.A. unit or other standard of resistance, it would not be desirable to use tubes of such large bore*. This problem may more conveniently be taken by itself, and has already been treated by us in a former communication to the Society†.

---

* If the distance between the points of derivation were 1 metre, $R = \cdot002$ mercury unit would require a section equal to 500 square millims.

† *Phil. Trans.* 1883, p. 173 [vol. II. p. 78].

§ 3.   In the shunt method the greater part of the main current γ passes on one side through a relatively small resistance a (see fig. 1), and the difference of potentials at the points of derivation B, C, is due to the passage

Fig. 1.

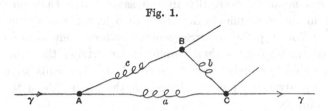

of a small fraction only of the total current, the resistance $(b + c)$ being great compared with a.   If at the same time b be small relatively to c, the difference of potentials is doubly attenuated.   Its value for a given main current γ is found at once from the consideration that the current divides itself between the two branches in the inverse ratio of the resistances.   The current through b is thus $\dfrac{a}{a + b + c}\,\gamma$, and the difference of potentials at the points of derivation is $\dfrac{ab}{a + b + c}\,\gamma$.   The quantity $\dfrac{ab}{a + b + c}$ thus takes the place of R in the simple formula, and is called the effective resistance.   By taking for instance $a = \tfrac{1}{2}$, $b = 1$, $c = 100$, we get an effective resistance of about $\tfrac{1}{200}$; and the resistances employed may be those of ordinary resistance coils, capable of accurate comparison with the standards.

§ 4.   In designing the apparatus we were influenced by the fact that we had at our disposal two very suitable coils of large radius, wound some years ago by Professor Chrystal, the same in fact as were used by Mr Glazebrook in his investigation by another method.   By bringing the two coils close to one another and to the plane of the disc, the inductive effect is rendered a maximum.   This arrangement accordingly was the one first experimented with, as being the most likely to prove successful.

The diameter of the disc is limited by two considerations.   If it be too small, the whole inductive effect, and with it the sensitiveness of the arrangement, suffers.   On the other hand if it be too large, the circumference enters the more intense region of magnetic force which lies near the wire, and the coefficient of induction changes its value rapidly when any alteration occurs in the mean radius of the coils, or in the diameter of the disc, and thus the final result becomes too sensitive to errors in the magnitude of these elements. In the *Phil. Mag.* for Nov., 1882, [Art. 92] the reader will find a calculation of the values of M for various cases, and a general comparison of the principal methods for determining absolute resistance, especially in respect of errors arising in connexion with the fundamental linear measurements.   For the experiments now to be described, the diameter of the disc was chosen so as to be somewhat more than half that of the coils (§§ 22, 23).

§ 5. The disc was of brass and turned upon a solid brass rod as axle. This axle was mounted vertically in the same frame that carried the revolving coil in the experiments described in a former communication to the Society* [see Vol. II. p. 39], an arrangement both economical and convenient, as it allowed the apparatus then employed for driving the disc and for observing the speed to remain almost undisturbed. The coils were supported horizontally upon wooden pieces screwed on the inner side of the three uprights of the frame.

During the earlier trials, extending over the month of May, 1882, the edge of the disc was bevelled, and contact was made with it by means of a brush of fine copper wires held in a nearly vertical position. No sufficiently regular results could be obtained until the sliding surfaces were amalgamated, and even then there were discrepancies between the work of one day and that of another, whose cause was not discovered until a later period. It soon became manifest, however, that the bevelled edge would not answer the purpose, for it cut its way by degrees into the wires of the brush in such a manner as to render the effective radius uncertain. The substitution of a cylindrical for a bevelled edge promised better results. The width of the edge (equal to the thickness of the disc) was $4\frac{1}{2}$ millims. and allowed sufficient room for the contact of the brush though placed tangentially. In this way broader bearing surfaces were available, and the small extension of the contact in the direction of the axis is unobjectionable, provided everything be arranged symmetrically with respect to the middle plane of the disc.

As will presently appear, the success of the method is independent of any constant thermo-electric force at the sliding contact, but it is evident that good readings cannot be taken if the thermo-electric force changes its magnitude often and suddenly. It was found advisable to renew the amalgamation of the edge at the commencement of each day's work. The excess of mercury, if any, attaches itself to the brush, and does not appear to render the diameter of the disc uncertain.

The inner contact was made in a similar manner by a brush pressing against the shaft itself at a place a little below that at which the disc was attached. The coefficient of induction to be employed in the calculation is the difference between the coefficients for the coil and the outer and inner circles of sliding contact respectively, but the latter is quite subordinate (§ 25).

§ 6. The disc was driven by the same water-engine that was employed for the revolving coil of former determinations†, the connexion being made

---

* _Phil. Trans._ Part II. 1882 [Art. 80].

† _Proc. Roy. Soc._ May 5, 1881 [Art. 79]; _Phil. Trans._ Part II. 1882 [Art. 80].

by a long cord passing round a wooden pulley attached to the lower part of the shaft.   To the upper face of the disc was cemented a circle of paper on which were marked a series of circles of alternately black and white teeth. One observer looking through the prongs of an electro-magnetically maintained fork regulated the speed of the disc by application of the necessary friction to the driving-cord, which passed through his fingers.   When one of the series of circles is seen to be stationary, a simple and easily expressed relation is established between the frequency of the fork and that of revolution.   At intervals the number of beats per minute is counted between the notes of a standard fork, and (the octave of) the electric fork.   There is no difficulty in thus determining the speed of rotation to within one part in 10,000.   With respect to the absolute pitch of the standard fork itself, see the Appendix to this Memoir.

§ 7.   When the disc is caused to rotate, and the galvanometer circuit is closed, a deflexion is observed, although the battery which generates the main current is not in action.   This deflexion is due to two causes—thermo-electric force at the sliding contact, and induction dependent upon the vertical component of the earth's magnetism.   Although not a direct source of error, this deflexion is better avoided, both for convenience in reading the galvano-meter and because it implies the actual passage of a not insensible current through the sliding contacts and thus brings into consideration the *resistance* of these contacts.   The compensation was effected by the introduction of an opposing electromotive force; for which purpose two terminals of the galvano-meter circuit J, K, fig. 2, instead of being connected directly, were attached

Fig. 2.

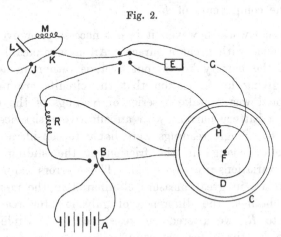

by binding screws to two points on a stout copper wire forming part of a circuit which included a sawdust Daniell (L) and a resistance coil of 100 ohms (M).   By shifting one of the binding screws, the galvanometer reading, in the absence of the main battery current, and after attainment of the proper speed, was made to be nearly the same as when the galvanometer contact was broken.

§ 8. The general plan of the connexions and the *modus operandi* will now be intelligible from fig. 2. The poles of the battery A, consisting of 20 Daniell cells, were connected with a mercury reversing key B, the two positions of which were distinguished by the letters E and W (east and west). From thence the current passed through the induction coils C and the equivalent resistance R, of which the details are reserved for the moment. The reflecting galvanometer, G, is placed at a considerable distance in order to avoid the direct influence of the coils, and is connected with the inner sliding contact, F. Its resistance is about $\frac{1}{2}$ ohm; and by the aid of the compensating magnet the vibrations of the needle were made slow enough to be readily observed. The terminals of the galvanometer branch, which includes also a commutator, I, are connected to the extremities of the resistance, R.

If, while the disc is maintained in uniform rotation, the reading of the galvanometer is the same whichever way the battery key may stand (correction being made, if necessary, for a direct effect upon the needle), it is a proof the contemplated balance is actually attained. In this way all disturbance from the earth's magnetism, and from thermo-electric forces whether situated at the sliding contacts, or within the resistance coils of which R is composed, or at any other part of the galvanometer circuit, is eliminated from the result. The adjustment is effected by varying a comparatively large resistance, taken from a box, and placed in multiple arc with one of the components of R.

§ 9. In actual work, however, it is not necessary, or even desirable, to hit off the balance with great accuracy. An unmistakeable difference of readings when the battery key is put over, is rather an advantage than otherwise, as giving an indication that the circuits are properly closed. The plan adopted was to take a series of readings of the effect ($E-W$) of reversing the battery current with an effective resistance R_1, not very different from R. Single readings were liable to considerable irregularity in consequence of change in the friction at the sliding contacts, and of momentary variations in the speed. These errors cannot possibly be systematic, and are in great measure eliminated in the mean of a series. Having thus obtained the difference of galvanometer readings ($E-W$) corresponding to R_1, we altered the resistance in multiple arc so as to change R_1 into R_2, the difference being some such fraction as $\frac{1}{100}$ of the whole, and in such a direction that the sign of $E-W$ is changed. The two series give by simple interpolation (after correction for the direct effect) the true value of R, that is the effective resistance corresponding to the balance. In order to get the best result relatively to the time occupied, the number of observations of $E-W$ in each set was taken roughly in inverse proportion to the values. To diminish the influence of a progressive change in the strength of the battery current, the observations with R_2 were interspersed between those with R_1 as effective resistance. The readings were usually taken continuously, with no more delay than was necessary to allow the vibrations of the needle to become of moderate extent after each change. When they were completed, the driving cord was reversed, as well as the commutator, I, and a similar set of observations was taken with rotation in the opposite direction.

§ 10. In the earlier experiments the resistance coils composing the effective resistance were arranged as in fig. 1, in which A, B, C may be supposed to represent mercury cups, the bottoms of which were formed of amalgamated copper discs. On these discs rested the amalgamated terminals of the various resistance coils and connecting wires. The shunt a consisted of two unit coils in multiple arc, between which the greater part of the main current was equally divided. The magnitude of the main current was less than $\frac{1}{10}$ ampère. The resistance b between the points of derivation was a unit, while the third resistance c was alternately 105 and 106.

In reckoning the resistance of the galvanometer circuit we have to include b. The remainder scarcely exceeds the $\frac{1}{2}$ ohm due to the galvanometer itself. It appears therefore that the deflections obtained with the arrangement described are only one-third part as great as they would be if a quite small resistance were substituted for the unit in b. As the sensitiveness appeared likely to be inadequate, we afterwards replaced the unit by $\frac{1}{10}$, using for c a coil of ten units. As in this case the addition or subtraction of a whole ohm in c would make too great a difference, the adjustment was obtained by varying a comparatively large resistance placed in multiple arc with a.

In the light of subsequent experience it is doubtful whether this change was an improvement. The increase of galvanometer deflection was not really of much advantage, since the difficulty of getting sharp results arose from electromotive disturbances, and these were magnified in the same proportion. It would probably have been better to have retained the unit in b, and to have replaced the galvanometer by one of higher resistance.

§ 11. Preliminary trials having given apparently satisfactory results, we proceeded to make regular series of observations in the manner already described. We had not gone far before anomalies revealed themselves of such a character as to prove that we were not yet masters of the method. It usually happened that each day's observations agreed well together, showing that the sensitiveness was sufficient; but when we came to compare the results obtained on different days unaccountable discrepancies became apparent. The first result of the more severe criticism to which the arrangements were then subjected was to show that sufficient thought had not been given to the question of insulation. The wire composing the induction coils, or rather one extremity of it, is necessarily at a high potential, and a very moderate leakage from the coils to the frame, and thence to the disc, might cause great disturbance. Some such leakage was in fact detected on application of appropriate tests. Ebonite insulation was accordingly introduced into the supports of the coils. The battery was carefully insulated from the ground, as was also the frame carrying the revolving disc, and other precautions were taken which it is unnecessary here to detail. For the sake of definiteness one point of the galvanometer commutator was connected to earth. With these improvements tests were satisfied more severe than that of actual use, and these tests were renewed at intervals during the spinnings.

The results however still showed that some defect existed which we had not yet succeeded in detecting. It made no appreciable difference which way the disc rotated, but the means of different days' work failed to exhibit the desired accordance. Two months' work had already been spent upon the experiments, and we had begun to despair of a satisfactory issue, when it occurred to us that the connexion of the coils for compounding the effective resistance was faulty.

§ 12. By reference to fig. 1 it will be seen that the main current traverses part of the cup C, and that part of the same cup is also included in b. Now, although for all ordinary purposes the resistance of the parts of the cup might be neglected, in the present case it is the small effective resistance R with which it comes into comparison. If we aim at an accuracy of $\frac{1}{10000}$, we cannot afford to overlook a resistance entering in this manner, even though it may not exceed $\frac{1}{2000000}$ ohm. The discrepancies were doubtless due to small differences in the position of the wires and coils in cup C, moved as they were from day to day in order to verify the soundness of the contacts.

In order to avoid the difficulty we have only to take care that no part of b can possibly be traversed by the main current, and this is easily done by the introduction of another mercury cup. Fig. 3 shows the arrangement adopted. The main current enters at the cups A and D, and the greater part is taken by the two unit coils in multiple arc whose terminals rest in these cups. The galvanometer terminals are led into two other cups B and C. The ends of these are beaten flat and the legs of the $\frac{1}{10}$ rest upon them. The connexion between C and D was through a stout copper rod, which may be regarded as part of c. For the first series the connexion between A and B was through a single coil of 10 units' resistance, replaced in subsequent series by other coils giving altogether 16 and 20 units' resistance respectively.

Fig. 3.

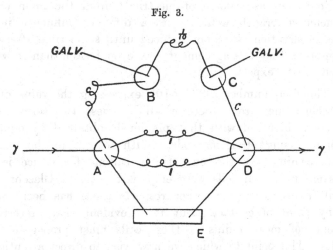

To make the necessary adjustment and variation of resistance, a box, E, was placed in multiple arc with the two unit coils. The resistances taken from the box were afterwards carefully determined, but they enter into the final results in quite a subordinate manner.

§ 13. Further trials now led to the satisfactory conclusion that the defect was remedied, for the means obtained on different days agreed well together, even although the resistance coils were taken down and remounted in the interval. As we had now every reason to suppose that our experiments would have a successful issue, we proceeded to make the final adjustments preparatory to a complete series of observations.

In the first and second series the two [induction-] coils were near one another, separated only by three slips of glass, and held firmly together by wooden clamps. The adjustments presented no particular difficulty. By means of an iron finger clamped to the disc and carried gradually round, it could be verified that the coils and disc were concentric and in parallel planes. The coils were gradually wedged into their places, and secured when their mean planes occupied the desired symmetrical positions relatively to the disc. It is evident that errors of maladjustment influence the result only in the second order.

§ 14. Experience in this series having shown that the arrangement was satisfactory, and that the sensitiveness was fully sufficient, we proceeded to make a second series of observations without displacement of the induction coils, but at a speed of rotation lower than before in about the ratio of 16 : 10. This, of course, entailed a corresponding change in R, which was effected by increasing the component c. An agreement between the final results of the two series would give an important confirmation, inasmuch as leakage of electricity from the main circuit into the galvanometer branch would exert a different influence in the two cases. The observations were not reduced until some time afterwards, and it then appeared that the agreement was even better than it would have been reasonable to expect.

§ 15. The final number, $\cdot9867 \times 10^9$, expressing the value of the B.A. unit in absolute measure as determined by these two series of observations, is almost identical with that previously obtained by ourselves, and by Glazebrook using other methods. With respect to the independence of these determinations, the only thing calling for notice is the fact that the same induction coils were employed both by Glazebrook and in the present investigation. In other respects there has been, we believe, scarcely any point of contact. But it is evident that an error in the measurements of mean radius of these coils must propagate itself into both results. The point to which we now wish to direct attention, is that the error of mean radius will influence the final number in *opposite directions*. In the method employed by Glazebrook, an under-estimate of the mean radius would lead to an under-estimate of the induction coefficient, whereas with us it would lead to an over-estimate of that quantity. So far, therefore, as the error of mean radius is concerned, it would appear that the use of the same coils is far from impairing the value of the results. Even with respect to the number of turns, an error, if that be supposed possible, would affect the results in a different manner, for Glazebrook was concerned with the *product* of the numbers for the two coils, while we evidently are concerned with the *sum*.

§ 16. In researches of this kind it is proper to calculate the influence upon the result of errors in the fundamental measurements. The value of M depends upon three linear quantities: the radius of the disc (a), the mean radius of the two coils (A), and the distance between their mean planes ($2b$). In the present case, however, the latter element enters in a very subordinate degree. From § 25 it appears that

$$\frac{dM}{M} = -1\cdot4\,\frac{dA}{A} + 2\cdot4\,\frac{da}{a}.$$

It has been shown* that these conditions compare favourably with those of most of the other methods that have been employed. From its nature a is much more easily measured than the diameter of a coil.

§ 17. The results deduced from the several days' observations, when corrected for slight variations of temperature of the resistance coils, &c.,

* *Phil. Mag.* Nov. 1882 [Art. 92].

exhibit a remarkable accordance. By reference to the tables (§ 27) the reader will see that the maximum divergence from the mean in Series I. is only about one part in 4000, while in Series II. it is even less. We were thus encouraged to carry out a modification of the method which we had had in view all along, and the results of which would be in great measure independent of those of Series I. and II.

§ 18. The modification referred to relates to the position of the induction coils relatively to the disc. In the arrangement with which we have been dealing hitherto, the mean planes of the coils are nearly coincident with that of the disc, and the accuracy of the final number depends upon an exact knowledge of the mean radius of the coils. It has, on the other hand, the advantages of being practically independent of measurements parallel to the axis, and of giving the maximum coefficient of induction. In the new arrangement the coils are separated to such a distance that the *result is nearly independent of a knowledge of the mean radius.* How this may come about will be readily understood by considering the dependence of the coefficient of induction M upon A, when a and b are given. It is clear that M vanishes, both when A is very small, and also when it is very large; from which it follows that there must be some value of A for which the effect is a maximum, and therefore independent of small variations of A.

In carrying out this idea, it is not necessary to approach the above defined state of things very closely; for of course we have in reality a good approximate knowledge of the value of A. In our apparatus the distance of mean planes was about 30 centims., so that $b =$ about 15 centims. ($A = 26$, $a = 16$). From the calculations in § 25 it appears that with the actual proportions

$$\frac{dM}{M} = + \cdot 12 \, \frac{dA}{A} - \cdot 96 \, \frac{db}{b} + 1 \cdot 8 \, \frac{da}{a} \, ;$$

so that the error of A enters in quite a subordinate degree. The positive coefficient of dA shows that with the given coils and the given disc the separation was somewhat too great to secure the utmost independence of dA.

§ 19. The success of this arrangement depends principally upon the degree of accuracy with which b can be determined. The two rings upon which the coils are wound were held apart by three equal distance-pieces, against which they were firmly pressed by wooden clamps. The distance-pieces were hollow, of massive brass, and the terminal faces were carefully turned. Central marks upon them facilitated the adjustment of the coils into the symmetrical positions. The distance of mean planes does not however depend solely upon the distance-pieces. Even if we could assume that the mean planes are symmetrically situated relatively to the grooves in which the wire is wound, we should still have to take account of the thicknesses of the flanges. All uncertainty in this matter is eliminated by following the plan adopted by Glazebrook of reversing the rings (without interchange), and then repeating the measurements. Whatever may be the situation of the mean planes and the thicknesses of the flanges, the mean result thus obtained corresponds to a distance equal to the length of the pieces *plus* half the total outside thicknesses of the rings. These quantities can all be measured with great precision, and as easily after the coils are wound as before. Full particulars are given in § 24. There can hardly be a doubt but that the determination is much more accurate than that of the mean radius of a coil; and, what is also of some importance, it admits of repetition at pleasure with comparatively little trouble.

§ 20. The sensitiveness of this arrangement was about the same as in Series II., and the table shows a good agreement among the results obtained on different days. The final number from this series is ·9868, almost the same as from Series I. and II.

The small difference of effective resistances required for balance in the two positions of the induction coils, amounting to about one part per thousand, is almost exactly accounted for by the small difference of distances of mean planes in the two cases, as deduced from Professor Chrystal's measurements of the thicknesses of the flanges. In the first position (see § 24) the coils are nearer together by almost exactly one part per thousand, a difference which, according to the formula given above (§ 18), should be reproduced almost without change in M and therefore in R, the greater values of M and R corresponding to the smaller distance.

§ 21. If we combine all the results of the present investigation, giving equal weights to the two arrangements of the induction coils, we have

$$1 \text{ B.A. unit} = ·98677 \times 10^9 \text{ C.G.S.}$$

With use of the ratio between the mercury unit and the B.A. unit found by us (*Proc. Roy. Soc.*, May, 1882 [Art. 81]), this gives

$$1 \text{ mercury unit} = ·94150 \times 10^9 \text{ C.G.S. ;}$$

or, which is the same thing, the ohm is the resistance of a column of mercury at 0° centigrade whose section is 1 square millim., and whose length is

$$1062·14 \text{ millims.}$$

We now pass on to the details of the measurements.

DETAILS OF MEASUREMENTS.

Diameter of disc.

§ 22. Preliminary measurements of the disc while still mounted were made on August 11, 1882, with callipers by Messrs Elliott. Read by the vernier of the instrument itself the mean diameter was

$$2a = 310\cdot76 \text{ millims.}$$

The opening of the callipers was also determined independently by reference with the aid of microscopes to a verified scale of millimetres. In this way

$$2a = 310\cdot77 \text{ millims.}$$

The circumference was also measured by a steel tape, afterwards compared with the millimetre scale. Correction being made for the thickness of the tape, the result was

$$2a = 310\cdot84 \text{ millims.}$$

After the disc had been dismounted, the diameter could be determined more advantageously by direct observation through microscopes focussed upon its edge with subsequent reference to the standard scale. It was found (August 19, 1882) that a very appreciable difference existed between the diameter of the upper and lower faces, showing that the edge was somewhat conical. At the upper edge the diameter was 310·80, and at the lower edge 310·58. These were the extremes. At the middle of the thickness the diameter was 310·75. This departure from the truly cylindrical form was undoubtedly a defect in the apparatus, which could easily have been avoided if detected in time. When the apparatus was first set up, the success of the experiment was problematical, and a minute examination of the disc seemed premature. The diameter to be adopted is an average taken with reference to the conductivity of brush contact. The whole width of the brush being decidedly less than the thickness of the disc, and the pressure being greatest at the central parts, we decided (of course without knowing to what precise final result the estimate would lead) to take the mean of 310·75 and $\frac{1}{2}(310\cdot58 + 310\cdot80)$. Thus

$$2a = 310\cdot72 \text{ millims.}$$

The error due to the conicality of the edge cannot exceed one part in 5000 at the worst, and thus it appeared scarcely worth while to correct the defect and repeat the spinnings.

The diameter of the shaft at the place where the other brush contact was made, was found to be ·825 inch, or 20·96 millims.

The induction coils.

§ 23.   These are the same as were used in Mr Glazebrook's measurement, and were wound by Professor Chrystal in 1878.   The following are the dimensions; for further particulars reference may be made to Mr Glazebrook's Memoir*.

	A	B	Mean
Mean radius in centims. (A)	25·753	25·766	25·760
Radial width of section (2h)	1·92	1·90	1·91
Axial width of section (2k)	1·896	1·899	1·897
Number of windings	797	791	½ × 1588
Resistance (approximate) in B.A. units .	84	83	½ × 167

Since the coils are so nearly similar and were used symmetrically, it is sufficient to use the numbers in the last column.   The section of the ring is shown in fig. 4 full size.

To find the distance of mean planes the following measurements of the thicknesses of the rims are required.   They are given in centimetres.

	A	B
Rim (marked side)	·478	·446
Channel	1·896	1·899
Rim (unmarked side) . . .	·488	·465
Total thickness of ring . .	2·862	2·810

Now that the rings are wound it is difficult to verify these numbers. However, the total thickness of the rings at the places touched by the distance-pieces in the arrangement used for Series III. was taken, with the result

	A	B
Mean of three places . . .	2·8625	2·8067

These latter values of the thicknesses will be used in the calculation of Series III.

* *Phil. Trans.* 1883, p. 223.

In Series I. and II. the rings were not reversed, and we must use the numbers above given for the thicknesses of rims which were contiguous to the slips of glass; but in this case the result is not at all sensitive to changes in the distance of mean planes. The rims contiguous to the glass were for both coils the *marked* rims, of which the aggregate thickness is ·924. If we add to this the thickness of the glass strips ·454, we obtain 1·378 as the distance between the wire sections. Again, adding the mean axial width of section 1·897, we find as the distance of the mean planes

Fig. 4.

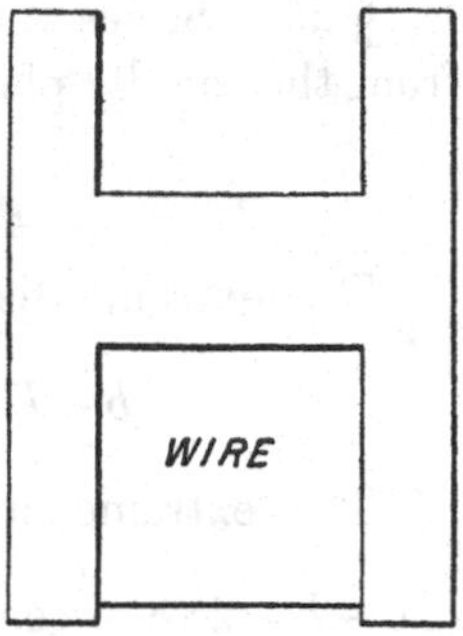

WIRE

$$2b = 3·275 \text{ centims.}$$

The distance-pieces.

§ 24. The measurement of the distance-pieces used for the third series was made with great care. As only the mean is required, the three pieces were held under the microscopes in one length by a nut and a long bolt running through. Readings were taken in several positions, as the pieces were turned round, and reference was finally made to the standard scale. Two independent measurements gave 83·580 and 83·579, mean 83·5795 centims., as the aggregate length. This was further verified by measuring each piece separately with callipers, the sum of the lengths thus found being 83·582. For the mean length of these distance-pieces we take

$$27·8598 \text{ centims.}$$

As has been already explained, the rings were used in two positions relatively to the distance-pieces, with the view of eliminating any uncertainty as to the situation of the mean planes, and of rendering the final result independent of all measurements of thickness except that of the total thicknesses of the rings. Thus the mean distance of mean planes in the two positions is

$$27·8598 + \tfrac{1}{2}(2·8625 + 2·8067) = 30·6944 \text{ centims.}$$

To compare the partial results for the two positions separately, we must use the thicknesses of the rims which were in contact with the distance-pieces. In the first position these were the marked rims, and thus the distance of mean planes

$$= 27·860 + ·478 + ·446 + 1·897 = 30·681 \text{ centims.}$$

In like manner for the second position we find

$$27·860 + ·488 + ·465 + 1·897 = 30·710 \text{ centims.}$$

The induction-coefficients.

§ 25.   Series I. and II.   The distance (*b*) of the mean planes of the coils from the middle plane of the disc is

$$b = 1\text{·}637 \text{ centim.}$$

The extreme distances, required to be known for the quadrature, are

$$b + k = 2\text{·}585 \text{ centims.,} \qquad b - k = \text{·}689 \text{ centim.}$$

The extreme and mean radii are

$$A - h = 24\text{·}805 \text{ centims.,} \qquad A = 25\text{·}760 \text{ centims.,} \qquad A + h = 26\text{·}715 \text{ centims.}$$

while

$$a = 15\text{·}536 \text{ centims.}$$

The coefficient of induction between the disc and the middle turn of the coil, denoted by $M(A, a, b)$, is equal to $4\pi \sqrt{(Aa)} . f(\gamma)$, where $f(\gamma)$ is a function of γ given by tables*.   The angle γ itself is defined by

$$\sin \gamma = \frac{2 \sqrt{(Aa)}}{\sqrt{\{(A + a)^2 + b^2\}}} .$$

It is not necessary to give the details of the calculations, which have been carefully checked.   The tabular interval being 6′, it was found desirable in many cases to proceed beyond the simple interpolation by first differences. The results are

$$M(A, a, b) = 215\text{·}4674$$
$$M(A + h, a, b) = 205\text{·}1917$$
$$M(A - h, a, b) = 226\text{·}9835$$
$$M(A, a, b + k) = 211\text{·}7246$$
$$M(A, a, b - k) = 217\text{·}5972.$$

The mean coefficient for the area of the section is found by doubling the first of these values, adding in the others, and then dividing by 6.

Thus

$$M = 215\text{·}405\dagger.$$

The separate values allow us to form an estimate of the effect of errors in the fundamental data.   If we write

$$\frac{dM}{M} = \lambda \frac{dA}{A} + \mu \frac{db}{b} + \nu \frac{da}{a} ,$$

---

* Maxwell's *Electricity and Magnetism*, 2nd edition, § 706.
† The factor expressing the number of windings is omitted.

we may take approximately

$$\lambda = \frac{M(A+h,a,b) - M(A-h,a,b)}{2h} \div \frac{M}{A} = -1\cdot36.$$

In like manner, $\mu = -\cdot02$, whence, since $\lambda + \mu + \nu = 1$, $\nu = +2\cdot38$.

Series III. In this case the data remain precisely as before, except that we now have $b = 15\cdot3472$.

We find

$$M(A,a,b) = 110\cdot9240$$
$$M(A+h,a,b) = 111\cdot2573$$
$$M(A-h,a,b) = 110\cdot2442$$
$$M(A,a,b+k) = 104\cdot5571$$
$$M(A,a,b-k) = 117\cdot6519,$$

whence

$$M = 110\cdot926.$$

Determining λ, μ, ν, as in the former case, we find

$$\frac{dM}{M} = +\cdot123\,\frac{dA}{A} - \cdot956\,\frac{db}{b} + 1\cdot833\,\frac{da}{a}.$$

From these values, calculated for the circumference of the disc, we have to subtract the value (M_0) applicable to the small circuit touched by the inner brush. The area of this is $\frac{1}{4}\pi(2\cdot096)^2$. For the first and second series we have

$$M_0 = \frac{2\pi A^2}{(A^2+b^2)^{\frac{3}{2}}}\,\frac{1}{4}\pi(2\cdot096)^2 = \cdot836.$$

For the third series in like manner

$$M_0 = \cdot534.$$

Thus finally for the first and second series

$$M - M_0 = 214\cdot569,$$

and for the third series

$$M - M_0 = 110\cdot392.$$

The resistance-coils.

§ 26. In all three series the resistance b, fig. 3, was a German-silver coil of about $\frac{1}{10}$, referred to for brevity as the $[\frac{1}{10}]$; and the resistance a was composed of three resistances in multiple arc, the first two being standard singles, and the third a resistance such as 7 B.A. units taken from a box. To make the necessary change, according to the plan already explained in § 9, the 7 would be replaced by 8. The value of a is of course determined principally by the unit resistance-coils, and only secondarily by the resistance taken from the box.

The third element of the system of resistances was varied in the different series. In the first series c was a [10], in the second series it was [10] + [5] + [1], and in the third series [10] + [5] + [5']. Besides the standard singles, whose values at various temperatures was already known in terms of the mean B.A. unit, we had to determine accurately the values of the [$\frac{1}{10}$], the [10], the [5], and the [5'], as well as the small resistances of the various connecting pieces employed.

The [10] has been determined in various ways, but principally by means of the device referred to in the former paper*. Three German-silver wires of about 3 units each are wound on the same tube, and their terminals are so arranged that by means of a base board containing mercury cups they can be combined either in multiple arc or in series. In the former combination they are compared with a standard single, and the resistance is found to be (say) $1 + \alpha$, where α is small. The coils are now without loss of time combined in series, a change which can be effected in a moment. The resistance in series is very approximately $9 + 9\alpha$; by the addition of the standard single it becomes $10 + 9\alpha$, and can now be compared with the [10]. If the difference observed be β we have [10] $= 10 + 9\alpha + \beta$. By this method it is easy to obtain an accuracy of at least $\frac{1}{10000}$.

The [5]'s were determined in two ways. Five singles were combined in series and compared with one of the [5]'s; afterwards the two [5]'s were compared with one another. In the second method, which is probably preferable, the sum of the two [5]'s was found by comparison with the [10]. From the sum and difference the separate values can of course be deduced.

The measurement of the [$\frac{1}{10}$] demanded some precaution on account of its smallness. Two standard singles, the [10], and the [$\frac{1}{10}$], were combined with four insulated mercury cups, and without the use of connecting pieces,

Fig. 5.

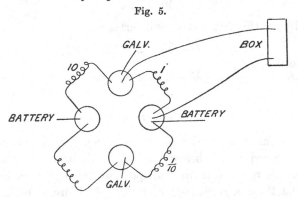

so as to form a Wheatstone's balance (fig. 5), care being taken to bring the associated battery and galvanometer terminals into immediate contact with the legs of the [$\frac{1}{10}$] (see § 12). To get the means of adjustment, a box,

* *Phil. Trans.* Part II. 1882, p. 697 [vol. II. p. 75.].

giving resistances up to 10,000, was placed in multiple arc with one of the singles. If, as was the case, the four coils be so nearly in proportion that a resistance of several hundreds from the box is needed for balance, the delicacy of the arrangement is all that can be desired. Readings are taken also with battery reversed, to eliminate thermo-electric disturbances. Especial pains were taken with the measurement of the $[\frac{1}{10}]$, and of the [10], errors of which would be propagated into the results of all three series.

§ 27. The various temperatures of the coils at the time of use, and the fluctuations from day to day, complicate the calculation of the effective resistances R_1 and R_2, which in principle is simple enough. The results are given in column II. of the Tables. Thus in the first series on July 14, when the effective resistance was ·0044076 B.A., as calculated from the values of $a, b, c,$ for the observed temperatures of the coils, the effect $(E-W)$ of reversing the battery key (corrected for direct effect) was − 30 divisions of the galvanometer scale, the direction of rotation being positive. When the effective resistance was altered to ·0044430, the difference $E-W$ became + 10 divisions. From these results we infer that $E-W$ would vanish for the effective resistance ·0044341, as given in column V. The corresponding result with negative rotation is given in column VI. These resistances relate to the actual speed of rotation determined by the frequency of vibration of the electric fork (§ 6). To render the results of different days fairly comparable, two small corrections have to be introduced, the first relating to small alterations in the relative frequencies of the two forks, as shown by the number of beats per minute (column VII.), the second to variations in the frequency of the standard fork itself, dependent upon change of temperature. The temperatures were read by a thermometer which stood between the prongs of the standard, and are given in column IX. The corrections necessary for reduction to a standard number of beats (16 per minute) and to a standard temperature (16°) are tabulated in columns VIII. and X., and the corrected results themselves in XI. and XII. In all cases the electric fork vibrated *more* quickly than the standard.

The degree of accordance in the numbers entered in these columns shows the success of the observations, so far as relates to errors of a casual character. In column XIII. the results of the positive and negative rotations are combined, so as to exhibit the total result of the day's work.

The Table, showing the results of the third series, is divided into two parts, corresponding to the two positions of the induction coils, before and after reversal (§ 19). In each position, it will be seen that two sets of observations were taken upon one of the days. Both sets, however, were complete, and in the interval between them the resistance-coils were all dismounted. A similar precaution was taken at least once in each of Series I. and II.

FIRST SERIES.

Coils near together.

Speed of disc about 12·8 revolutions per second.

Approximate resistances $a = \frac{1}{2}$, $b = \frac{1}{10}$, $c = 10$.

Date	Effective resistance (in B.A. units) used	Difference of reading of galvanometer on reversal of current		Effective resistance (in B.A. units) corresponding to zero difference in galvanometer		Beats between forks	Correction to 16 beats	Temperature of standard fork	Correction to 16°	Effective resistance (in B.A. units) as finally corrected		Means of resistances with both directions of rotation
		Rotation +	Rotation −	Rotation +	Rotation −					Rotation +	Rotation −	
July, 1882								°				
14th	·0044076	−30·0	+32·0	·0044341	·0044371	17	−·0000006	16·7	+·0000003	·0044338	·0044368	·00443530
„	·0044430	+10·0	−6·3									
15th	·0044084	−24·6	+25·8	·0044333	·0044346	16	0	17·1	+·0000005	·0044338	·0044351	·00443445
„	·0044438	+10·5	−9·2									
17th	·0044090	−25·4	+27·0	·0044322	·0044315	15	+·0000006	17·3	+·0000006	·0044334	·0044327	·00443305
„	·0044444	+13·4	−15·5									
18th	·0044095	−27·4	+25·9	·0044331	·0044332	16	0	17·6	+·0000008	·0044339	·0044340	·00443395
„	·0044449	+18·7	−12·8									
19th	·0044100	−23·9	+27·4	·0044324	·0044337	16	0	18·0	+·0000010	·0044334	·0044347	·00443405
„	·0044454	+13·9	−13·5									
20th	·0044100	−24·4	+25·4	·0044326	·0044328	16	0	17·9	+·0000009	·0044335	·0044337	·00443360
„	·0044454	+13·8	−14·0									
Means . .										·00443363	·00443450	·00443407

SECOND SERIES.

Coils near together.

Speed of disc about 8 revolutions per second.

Approximate resistances $a = \frac{1}{2}$, $b = \frac{1}{10}$, $c = 16$.

Date	Effective resistance (in B.A. units) used	Difference of reading of galvanometer on reversal of current		Effective resistance (in B.A. units) corresponding to zero difference in galvanometer		Correction for change of speed of fork				Effective resistance (in B.A. units) as finally corrected		Means of effective resistances with both directions of rotation
		Rotation +	Rotation −	Rotation +	Rotation −	Beats between forks	Correction to 72 beats	Temperature of standard fork	Correction to 16°	Rotation +	Rotation −	
August, 1882												
7th	·0027827	− 8·2	+ 9·4	·0027914	·0027918	73	− ·0000004	17·6°	+ ·0000005	·0027915	·0027919	·00279170
,,	·0028126	+ 20·1	− 21·5									
8th	·0027821	− 8·6	+ 9·5	·0027908	·0027920	73	− ·0000004	17·2	+ ·0000004	·0027908	·0027920	·00279140
,,	·0028120	+ 21·0	− 19·1									
9th	·0027826	− 7·9	+ 7·5	·0027912	·0027910	72	0	17·5	+ ·0000005	·0027917	·0027915	·00279160
,,	·0028125	+ 19·5	− 19·1									
									Means · ·	·00279133	·00279180	·00279157

THIRD SERIES.

Coils separated.
Speed of disc about 12·8 revolutions per second.
Approximate resistances $a = \frac{1}{2}$, $b = \frac{1}{10}$, $c = 20$.

Date	Effective resistance (in B.A. units) used	Difference of reading of galvanometer on reversal of current Rotation +	Rotation −	Effective resistance (in B.A. units) corresponding to zero difference in galvanometer Rotation +	Rotation −	Beats	Correction to 72 beats	Temperature of standard fork	Correction to 16°	Effective resistance (in B.A. units) as finally corrected Rotation +	Rotation −	Means for each day of observations with both directions of rotation	Means of all the observations with direction of rotation Rotation +	Rotation −	Means of all observations with both directions of rotation
August, 1882.															
14th	·0022981	+ ·1	− ·3	·0022982	·0022985	73	− ·0000003	18·4	+ ·0000006	·0022985	·0022988	·00229865			
,,	·0023100	− 11·0	+ 9·1												
15th	·0022976	− ·1	− 1·9	·0022975	·0022995	72	0	18·4	+ ·0000006	·0022981	·0023001	·00229910	·00229853	·00229910	·00229881
,,	·0023095	− 11·7	+ 10·0												
15th	·0022973	+ 1·1	− ·6	·0022985	·0022979	72	0	18·1	+ ·0000005	·0022990	·0022984	·00229870			
,,	·0023092	− 9·9	+ 11·1												
Second position of induction coils.															
16th	·0022969	− ·3	0	·0022966	·0022969	72	0	17·6	+ ·0000004	·0022970	·0022973	·00229715			
,,	·0023088	− 11·9	+ 7·8												
17th	·0022965	− ·8	+ ·7	·0022956	·0022958	72	0	17·4	+ ·0000004	·0022960	·0022962	·00229610	·00229633	·00229650	·00229642
,,	·0023084	− 11·2	+ 12·1												
17th	·0022966	− ·8	+ 1·0	·0022956	·0022956	72	0	17·4	+ ·0000004	·0022960	·0022960	·00229600			
,,	·0023085	− 10·2	+ 12·4												
												Means .	·00229743	·00229780	·00229762

§ 28. The results given in these tables are the effective resistances required to obtain a balance, expressed in terms of the B.A. unit. To reduce them to absolute measure we must multiply by 10^9, and by a factor, which we may call x, expressing the absolute value of the B.A. unit in terms of 10^9, and which it is our object to determine.

The actual value of the same quantities in absolute measure is found by multiplying the coefficients of induction $(M - M_0)$ already given (§ 25) by the number of turns in the coils 1588, and by the number of revolutions per second.

In the first series the frequency of vibration (f) of the electric tuning-fork was in the standard case (see Appendix)

$$f = \tfrac{1}{2}(128 \cdot 140 + \tfrac{16}{60}) = \tfrac{1}{2} \times 128 \cdot 407$$

and the number of revolutions per second is equal to $2f \div 10$. In the second and third series $2f = 129 \cdot 340$, a number which in the second series is to be divided by 16, and in the third series by 10, in order to obtain the number of revolutions per second.

The equation to determine x is thus for the first series of observations

$$214 \cdot 569 \times 1588 \times 12 \cdot 8407 = x \times \cdot 00443407 \times 10^9,$$

whence

$$x = \cdot 98674.$$

From the second series

$$x = \frac{214 \cdot 569 \times 1588 \times 129 \cdot 340}{16 \times 10^9 \times \cdot 00279157} = \cdot 98669.$$

From the third series

$$x = \frac{110 \cdot 392 \times 1588 \times 129 \cdot 340}{10 \times 10^9 \times \cdot 00229762} = \cdot 98683.$$

These are the final results already considered in § 21.

APPENDIX.

Frequency of Vibration of Standard Fork.

All our measurements, both by this method and by that of the revolving coil, being dependent upon the pitch of a standard tuning-fork, we have considered it advisable to determine this element afresh. As in the first determination*, a fork vibrating about 32 times per second rendered intermittent an electric current, which, passing through the coils of small

* Proc. Roy. Soc. May 1881, p. 137 [vol. II. p. 33].

electromagnets, maintained in vibration not only the interrupter fork itself, but also a second fork of pitch about 128. After the apparatus has been a short time in operation, the vibrations of the second fork are exactly four times as quick as those of the first, independently of any precise tuning; and they give rise to audible beats when the standard fork is simultaneously excited. In the presence of extraneous noises the observation of the beats is much facilitated by the use of resonators, with one of which the ear may be connected by an indiarubber tube. The object to be aimed at is to make the intensities of the two sounds (as they reach the ear) very nearly equal. The moment of antagonism is then marked by a well-defined silence, whose occurrence can be timed to within a second, although the whole duration of the beat may be 20 seconds or more. Without fresh bowing of the standard, the silences can be observed satisfactorily for at least a minute.

In the first determination the comparison between the fork of frequency 32 and the pendulum of the clock was made directly. The observer, looking over a plate carried by the upper prong of the fork, obtained 32 views per second, i.e., 64 views of the pendulum in one complete vibration. The immediate subject of observation is a silvered bead attached to the bottom of the pendulum, upon which as it passes the position of equilibrium the light of a paraffin lamp is concentrated. Close in front of the pendulum is placed a screen perforated by a somewhat narrow vertical slit. If the period of the pendulum were a precise multiple of that of the fork, the flash of light which to ordinary observation would be visible at each passage, would either be visible, or be obscured, in a permanent manner. If, as in practice, the coincidence be not perfect, the flashes appear and disappear in a regular cycle, whose period is the time in which the fork gains (or loses) one complete vibration. This period can be determined with any degree of precision by a sufficient prolongation of the observations.

On account of the large number of views per second, the interval between successive visible positions of the bead, even when it is moving with maximum velocity, is rather small; and thus the adjustment of the apparatus is somewhat delicate*. In order to meet this objection, a modification has been introduced, which must now be explained†.

* In the earliest use of this method (*Nature*, vol. XVII. p. 12, 1877) [vol. I. p. 333] the break-fork had a frequency of about 13, and no difficulty of this kind was experienced.

† July, 1883.—It should be stated, however, that the wheel may easily be dispensed with, if proper care be taken in the illumination of the bead and in the management of the fork. The vibration should be vigorous, and the screens so arranged that the view past the fork at the moment of greatest elongation should be of short duration. Determinations by this method (without the wheel) have often been made successfully by students in the Cavendish Laboratory.

A few years ago it was shown almost simultaneously by La Cour and by Lord Rayleigh [Art. 56, vol. I. p. 355], that an electromagnetic engine could be accurately governed by an interrupter-fork. The construction (fig. 6) which has been found most suitable is similar to that of Froment's engine. A horizontal shaft revolving upon steel points carries a number of parallel soft iron armatures, disposed symmetrically round the circumference. In the course of the revolution these armatures pass in succession between the poles of a vertical horse-shoe electromagnet, so as almost to complete the magnetic circuit. It is much better that the armatures should pass *between* the poles than *over* them, as in the most usual arrangement, for in the latter case the bearings are subjected to an unnecessary and prejudicial strain. The wheel may be used either with or without an independent driving power. In the former case the power should be very steady, and adjusted so as to give by itself nearly the speed intended. The currents from the interrupter-fork are passed also through the electromagnet of the engine, and give the force required to accelerate or retard the motion so that it may exactly synchronise with the fork, one armature passing for each complete vibration. If the independent power is in excess, the phase of the motion is such that the electromagnet is excited principally after the armatures have passed through the electromagnet; if the independent power is in defect, the electromagnet is excited principally while the armatures are approaching it. Within certain limits any necessary acceleration or retardation is obtained by suitable self-acting adjustment of phase.

Fig. 6.

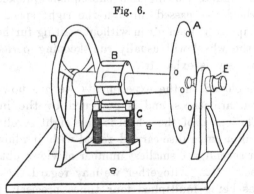

If when the wheel is moving steadily under the influence of the intermittent currents, a slight disturbance is communicated to it, oscillations will set in, the wheel being alternately in advance and in the rear of its proper position. In some cases these oscillations are very persistent, and interfere seriously with the utility of the instrument. To check them, a hollow ring filled with water is attached to the shaft and revolves with it. When the rotation is perfectly regular, the water behaves as if it were a rigid body and offers no impediment to the motion, but it tends to check variations of speed of moderate period. The oscillations, when they exist, are usually audible; and in any case the behaviour of the wheel in this and other respects may be examined by looking at the interrupter-fork through a paper disc carried by the wheel and perforated symmetrically along a circle with holes equally numerous with the armatures. When all is regular, the prongs of the fork are seen in one phase only, so long as the eye retains a position fixed in space.

When the wheel runs lightly, independent driving power may be dispensed with, a sufficient amount of work being obtainable from the intermittent governing current. In the present case the whole apparatus, consisting of the two forks and the wheel, was driven by one current supplied from three Grove cells. The only difficulty experienced is in starting the wheel. By means of string passed once round the shaft, alternately tightened for the advance and slackened for the return, it is easy to cause the wheel to achieve a speed in excess of the necessary eight revolutions per second. But it will not usually happen, every time the speed falls through the proper value, that the wheel will engage with the fork. For this purpose it is necessary that at the moment in question the phase of the wheel should be correct, within limits, which may be narrow when there is no great margin of power; and this can only happen by chance. Several attempts may be necessary before success is reached. With a little practice, however, there is no great loss of time, the ear learning to recognise, by the gradual slowing and subsequent quickening of a sort of beat, when the wheel has passed through the right speed without engagement. A fresh impulse is then given without waiting further. After a start is once effected, the wheel will usually run, keeping perfect time with the fork, until the battery is exhausted.

The wheel employed in the experiments we are now concerned with, has *four* soft iron armatures, and is governed by the interrupter-fork of frequency 32. The speed of the wheel is thus eight revolutions per second; and a single hole in a paper disc carried round with it allows eight views of the pendulum per second, the smallest number of views obtainable by direct use of the fork being 32. Altogether we may regard the frequency of the interrupter-fork as being multiplied four times precisely in the frequency of the auxiliary fork, and as divided four times precisely in the frequency of the wheel. The former is directly comparable with the standard fork, and the latter with the clock. The standard fork was screwed to the table precisely as during the electrical measurements. A thermometer placed between the prongs gave the temperature with fair accuracy.

The calculation of the results is very simple. Supposing in the first instance that the clock is correct, let a be the number of cycles per second (perhaps $\frac{1}{40}$) between the wheel and the clock. Since the period of a cycle is the time required for the wheel to gain, or to lose, one revolution upon the clock, the frequency of revolution is $8 \pm a$. The frequency of the auxiliary fork is precisely 16 times as great, *i.e.*, $128 \pm 16a$. If b be, the number of beats per second between the two forks, the frequency of the standard is

$$128 \pm 16a \pm b.$$

To give an idea of the magnitudes of the numbers concerned, it will be advisable to quote in detail the results of one day's observations. On October 19, with a certain loading of the interrupter-fork, the cycle of the pendulum occupied about 78 seconds, and the beats were at the rate of about six per minute. The interrupter was then *sharpened*, after which several observations were taken of the duration of five cycles of the pendulum, and of 16 beats between the forks. For the former the times found were 210, 210, 212 seconds; for the latter by simultaneous observation 58, 58, 59, 59, 59, 60, 60 seconds. The temperature, as given by the thermometer, ranged from $17°\cdot2$ to $17°\cdot4$. After the sharpening of the interrupter, the frequency both of the wheel and of the auxiliary fork was increased, so that the sign of $16a$ in the expression written above is determined to be $+$ and that of b to be $-$. Using the mean values we find

$$16a = \cdot3797, \qquad b = \cdot2712,$$

whence

$$128 + 16a - b = 128\cdot108.$$

To this we must add $\cdot009$, making altogether $128\cdot117$, to allow for the gaining rate of the clock, which was $6\frac{1}{2}$ seconds per diem. This corresponds to a mean temperature $17°\cdot3$.

The procedure adopted was quite good enough for our purpose; but if it were desired to push the power of the method to its limit, the work should be undertaken at an astronomical observatory, and extended over the whole time required to rate the clock by observations of the stars. In this way the comparison of the period of vibration of the standard fork with the mean solar second could be effected with the same degree of accuracy as that to which the former quantity is capable of definition. Without this precaution we cannot be quite sure that the rate of the clock at the time of the observations is identical with the mean rate employed in the calculation. It is scarcely necessary to say that the uncertainty which arises under this head is common to every method by which absolute pitch could be determined.

The results obtained, including those recorded previously*, are given in the accompanying table. They are well represented by the formula

$$128\cdot140 \times \{1 - (t - 16)° \times \cdot00011\},$$

* *Proc. Roy. Soc.* May 1881, p. 138 [vol. II. p. 33].

in which the temperature coefficient used (·00011) is that found by M'Leod and Clarke*. The numbers in the fourth column are calculated from the formula.

Date	Temperature	Frequency by observation	Frequency by calculation
1881	13°	128·180	128·182
1881	14°·6	128·161	128·160
October, 1882 . .	15°·98	128·141	128·140
October, 1882 . .	17°·45	128·122	128·120
October, 1882 . .	17°·6	128·119	128·118
October, 1882 . .	17°·3	128·117	128·122

Of the small discrepancies which the table exhibits it is probable that the larger part is due to imperfect knowledge of the actual temperatures of the standard fork. The use of screens to cut off radiation from the observers would probably have effected an improvement. For the highest accuracy some sort of jacket, or chamber, would have to be contrived.

SECOND APPENDIX.

(Added July, 1883.)

On the Effect of the Imperfect Insulation of Coils.

In a former paper (*Phil. Trans.* 1882 [vol. II. p. 51]), it was pointed out that the method of the revolving coil, employed by the first B.A. Committee, possesses the important advantage that it is possible to detect the existence of leakage from turn to turn, or from layer to layer, of the coil of wire. The general influence of such leakage, if undetected, upon the final number x expressing the ratio of the resistance of the coil when measured (R) in absolute units to its resistance $r \times 10^9$ as referred to B.A. units, is easily seen by supposing that one turn of the coil is simply short-circuited. The formula in C.G.S. measure is

$$x = \frac{R}{r \times 10^9} = \frac{\pi^2 n^2 a \, \omega \cot \phi}{r \times 10^9}. \dots\dots\dots\dots(1)$$

During the revolutions the short-circuited turn produces its full effect in deflecting the magnet, and error arises only in the comparison with the standard of resistance. The quantity r will evidently be under-estimated

* *Phil. Trans.* Part I. 1880.

by $1/n$, and this will lead to an *over*-estimate of x, also by $1/n$. This result, however, is modified, if as in practice we take only the *difference* of effects observed when the wire contact is open and closed. The short-circuited turn will produce its effects in *both* cases, and its influence will therefore disappear from the result. For all purposes it will be virtually non-existent, and the error produced is the same as if n had simply been miscounted. The final number x will thus be over-estimated by the fraction $2/n$.

In Lorenz's method the effect of a short circuit in the induction coil is in the same direction. M, and therefore R and x, will be over-estimated by $1/n$.

If we examine the formulæ applicable to determinations by other methods, we shall see that a similar conclusion holds good, so that in every case leakage leads to an over-valuation of x, at least whenever the result is calculated from the number of turns of wire in a coil*. Even without such an examination, it is pretty evident from consideration of the magnitudes involved that the large factor 10^9 in the denominator of the formula corresponding to (1) can only be compensated by one or more large factors expressive of the number of windings in a coil or coils. An over-valuation of these factors, due to leakage, will therefore lead to an over-valuation of x.

In carefully constructed coils serious leakage is, perhaps, not likely to occur, but its presence in a smaller degree is more probable, and is usually difficult of detection. So far as this argument applies, we may say that the smaller values of the number expressive of the B.A. unit, or of the mercury unit, in absolute measure are to be preferred to the larger.

---

* The case is different when the constants of a coil of many turns are determined by electrical comparison, as for instance in Kohlrausch's recent correction of the constant of his earth-inductor.

95.

ON THE MEAN RADIUS OF COILS OF INSULATED WIRE.

[*Proceedings of the Cambridge Philosophical Society*, IV. pp. 321—324, 1883.]

In electrical work it is often necessary to use coils of such proportions that their constants cannot well be obtained from the data of construction, but must be determined by electrical comparison with other coils whose proportions are more favourable. A method for comparing the galvanometer-constants of two coils, *i.e.* of finding the ratio of magnetic forces at their centres when they are traversed by the same current, is given in Maxwell's *Treatise*, vol. II. § 753.

I have used a slight modification of Maxwell's arrangement which is perhaps an improvement, when the coils to be compared are of copper and therefore liable to change their resistance pretty quickly in sympathy with variations of temperature. The coils are placed as usual approximately in the plane of the meridian so that their centres and axes coincide, and a very short magnet with attached mirror is delicately suspended at the common centre. If the current from a battery be divided between the coils, connected in such a manner that the magnetic effects are opposed, it will be possible by adding resistance to one or other of the branches in multiple arc to annul the magnetic force at the centre, so that the same reading is obtained whichever way the battery current may circulate. The ratio of the galvanometer constants is then simply the ratio of the resistances in multiple arc.

To obtain this ratio in an accurate manner, the two branches already spoken of are combined with two other resistances of german silver, so as to form a Wheatstone's balance. Of these resistances both must be accurately known, and one at least must be adjustable. The electromagnetic balance is first secured by variation of the resistance associated with one of the given coils, which resistance does not require to be known. During this operation the galvanometer of the Wheatstone's bridge is short-circuited. Afterwards the galvanometer is brought into action and the resistance-balance is adjusted. The ratio of the galvanometer-constants is thus equal to the ratio of the german silver resistances. The two adjustments may be so rapidly alternated as to eliminate any error due to changes of temperature in the copper wires. Indeed, if desired, the final tests of the electromagnetic and resistance-balances might be made simultaneously.

If the ratio of galvanometer-constants be the final object of the measurement, there is nothing more to be done; but if we desire to know the ratio of the mean radii of the coils we must introduce certain small corrections for the finite dimensions of the sections. In the first place, however, it will be desirable to consider a little more closely what should be understood by the *mean radius* of a coil.

In Maxwell's treatment of the subject (§ 700) the mean radius of a coil is considered to correspond with the geometrical centre of its rectangular section, that is to say, the windings are assumed to be uniformly distributed over the section. In practice absolute uniformity is not attainable, and it is therefore proper to take into account the effect of a small imperfection in this respect. The *density* of the windings, *i.e.* the number of windings per unit area, may be denoted by ρ, and is to be regarded as approximately constant.

The introduction of the factor ρ makes but little difference in the investigation of § 700. If we take the origin of co-ordinates x and y, no longer at the geometrical centre, but at what may be called the centre of density of the section, we shall have (as in the ordinary theory of the centre of gravity)

$$\iint \rho \, x \, dx \, dy = 0, \qquad \iint \rho \, y \, dx \, dy = 0,$$

the integrations being extended over the area of the section. If P be any function of x and y, $\overline{P}$ the mean value of the function (with reference to ρ), P_0 the value at the origin, we have

$$\overline{P} \iint \rho \, dx \, dy = \iint P \rho \, dx \, dy$$

$$= P_0 \iint \rho \, dx \, dy + \tfrac{1}{2} \frac{d^2 P}{dx_0^2} \iint \rho x^2 dx \, dy + \frac{d^2 P}{dx_0 \, dy_0} \iint \rho \, xy \, dx \, dy + \tfrac{1}{2} \frac{d^2 P}{dy_0^2} \iint \rho y^2 \, dx \, dy,$$

the terms of the first order disappearing in consequence of the choice of origin. In the terms of the second order we may neglect the effect of variable density, and write

$$\iint \rho x^2 dx \, dy = \tfrac{1}{12} \xi^2 \iint \rho \, dx \, dy, \quad \iint \rho y^2 dx \, dy = \tfrac{1}{12} \eta^2 \iint \rho \, dx \, dy, \quad \iint \rho \, xy \, dx \, dy = 0,$$

ξ, η being the breadths in the directions of x and y of the rectangular section. Thus

$$\overline{P} = P_0 + \tfrac{1}{24} \left(\xi^2 \frac{d^2 P}{dx_0^2} + \eta^2 \frac{d^2 P}{dy_0^2} \right).$$

The *form* of this expression is the same as when the windings are supposed to be distributed with absolute uniformity, but the mean radius and mean plane are to be reckoned with reference to the density of the windings.

In the application to the galvanometer-constant of a coil, we have, if A be the mean radius, ξ the radial and η the axial dimension of the section,

$$G_1 = \frac{2\pi}{A}\left(1 + \tfrac{1}{12}\frac{\xi^2}{A^2} - \tfrac{1}{8}\frac{\eta^2}{A^2}\right),$$

by means of which, ξ and η being approximately known, G_1 may be inferred from A, *or conversely* A may be inferred from G_1. If the ratio of galvanometer-constants of two coils has been determined by the electrical process, the ratio of mean radii can be accurately deduced by use of the above formula.

When the mean radius of a coil has been determined in this manner by comparison with another of proportions more favourable for calculation from the data of construction, other quantities relating to the coil may be deduced by mere calculation. For instance, the important constant g_1, denoting the mean area included by the windings, is connected with the mean radius A by the equation

$$g_1 = \pi A^2 + \tfrac{1}{12}\pi\xi^2.$$

A more direct process for determining g_1 electrically is given by Maxwell § 754, and has recently received an important application in the hands of Kohlrausch. In this method the quantity sought is proportional to the cube of a distance not very easy of precise measurement; and it is possible that the less direct method explained above may be the more accurate in practice.

96.

ON THE INVISIBILITY OF SMALL OBJECTS IN A BAD LIGHT.

[Proceedings of the Cambridge Philosophical Society, IV. p. 4, 1883.]

In a former communication to the Society (March 6, 1882) [Art. 82, vol. II. p. 92] I made some remarks upon the extraordinary influence of apparent magnitude upon the visibility of objects whose 'apparent brightness' was given, and I hazarded the suggestion that in consequence of aberration (attending the large aperture of the pupil called into operation in a bad light) the focussing might be defective. Further experiment has proved that in my own case at any rate much of the effect is attributable to an even simpler cause. I have found that in a nearly dark room I am distinctly short-sighted. With concave spectacles of 36 inches negative focus my vision is rendered much sharper, and is attended with increased binocular effect. On a dark night small stars are much more evident with the aid of the spectacles than without them.

In a moderately good light I can detect no signs of short-sightedness. In trying to read large print at a distance I succeeded rather better without the glasses than with them*. It seems therefore that the effect is not to be regarded as merely an aggravation of permanent short-sightedness by increase of aperture.

The use of spectacles does not however put the small and the large objects on a level of brightness when seen in a bad light, and the outstanding difference may still be plausibly attributed to aberration.

---

* [1899. It may be worthy of record that sixteen years later, having now lost nearly all power of accommodation, I find lenses of about 36 inches negative focus necessary in order to see distant objects perfectly.]

97.

ON MAINTAINED VIBRATIONS.

[*Philosophical Magazine*, xv. pp. 229—235, 1883.]

WHEN a vibrating system is subject to dissipative forces, the vibrations cannot be permanent, since they are dependent upon an initial store of energy which suffers gradual exhaustion. In the usual equation

$$\frac{d^2\theta}{dt^2} + \kappa \frac{d\theta}{dt} + n^2\theta = 0 \quad \dots\dots\dots\dots\dots\dots\dots(1)$$

κ is positive, and the solution indicates the progressive decay of the vibrations in accordance with the exponential law. In order that the vibrations may be maintained, the vibrating body must be in connexion with a source of energy. This condition being satisfied, two principal classes of maintained vibrations may be distinguished. In the first class the magnitude of the force acting upon the body in virtue of its connexion with the source of energy is proportional to the amplitude, and its phase depends in an approximately constant manner upon the phase of the vibration itself; in the second class the body is subject to influences whose phase is independently determined.

The first class is by far the more extensive, and includes vibrations maintained by wind (organ-pipes, harmonium-reeds, æolian harps, &c.), by heat (singing flames, Rijke's tubes, &c.), by friction (violin-strings, finger-glasses, &c.), as well as the slower vibrations of clock-pendulums and of electromagnetic tuning-forks. When the amplitude is small, the force acting upon the body may be divided into two parts, one proportional to the displacement θ (or to the acceleration), the second proportional to the velocity $d\theta/dt$. The inclusion of these forces does not alter the *form* of (1). By the first part (proportional to θ) the pitch is modified, and by the second the coefficient of decay*. If the altered κ be still positive, vibrations

---

* For more detailed application of this principle to certain cases of maintained vibrations, see *Proceedings of the Royal Institution*, March 15, 1878. [Art. 55, vol. I. p. 348.]

gradually die down; but if the effect of the included forces be to render the complete value of κ negative, vibrations tend on the contrary to increase. The only case in which according to (1) a steady vibration is possible, is when the complete value of κ is zero. If this condition be satisfied, a vibration of any amplitude is permanently maintained.

When κ is negative, so that small vibrations tend to increase, a point is of course soon reached after which the approximate equations cease to be applicable. We may form an idea of the state of things which then arises by adding to equation (1) a term proportional to a higher power of the velocity. Let us take

$$\frac{d^2\theta}{dt^2} + \kappa \frac{d\theta}{dt} + \kappa' \left(\frac{d\theta}{dt}\right)^3 + n^2\theta = 0, \quad \dots\dots\dots\dots\dots(2)$$

in which κ and κ' are supposed to be small. The approximate solution of (2) is

$$\theta = A \sin nt + \frac{\kappa' n A^3}{32} \cos 3nt, \quad \dots\dots\dots\dots\dots(3)$$

in which A is given by

$$\kappa + \tfrac{3}{4}\kappa' n^2 A^2 = 0. \quad \dots\dots \dots\dots\dots\dots\dots(4)$$

From (4) we see that no steady vibration is possible unless κ and κ' have different signs. If κ and κ' be both positive, the vibration in all cases dies down; while if κ and κ' be both negative, the vibration (according to (2)) increases without limit. If κ be negative and κ' positive, the vibration becomes steady and assumes the amplitude determined by (4). A smaller vibration increases up to this point, and a larger vibration falls down to it. If, on the other hand, κ be positive, while κ' is negative, the steady vibration abstractedly possible is unstable, a departure in either direction from the amplitude given by (4) tending always to increase.

Of the second class the vibrations commonly known as *forced* have the first claim upon our attention. The theory of these vibrations has long been well understood, and depends upon the solution of the differential equation formed by writing as the right-hand member of (1) $P \cos pt$ in place of zero. The period of steady vibration is coincident with that of the force, and independent of the natural period of vibration; but the amplitude of vibration is greatly increased by a near agreement between the two periods. In all cases the amplitude is definite and is proportional to the magnitude of the impressed force. When the force, though strictly periodic, is not of the simple harmonic type, vibrations may be maintained by its operation whose period is a submultiple of the principal period.

There is also another kind of maintained vibration which from one point of view may be considered to be forced, inasmuch as the period is imposed from without, but which differs from the kind just referred to in that the imposed periodic variations do not tend directly to displace the body from its configuration of equilibrium. Probably the best-known example of this kind of action is that form of Melde's experiment in which a fine string is maintained in transverse vibration by connecting one of its extremities with the vibrating prong of a massive tuning-fork, *the direction of motion of the point of attachment being parallel to the length of the string**. The effect of the motion is to render the tension of the string periodically variable; and at first sight there is nothing to cause the string to depart from its equilibrium condition of straightness. It is known, however, that under these circumstances the equilibrium position may become unstable, and that the string may settle down into a state of permanent and vigorous vibration, *whose period is the double of that of the point of attachment†*.

The theory of vibrations of this kind presents some points of difficulty, and does not appear to have been treated hitherto. In the present investigation we shall start from the assumption that a steady vibration is in progress, and inquire under what circumstances the assumed state of things is possible.

If the force of restitution, or 'spring,' of a body susceptible of vibration be subject to an imposed periodic variation, the differential equation becomes

$$\frac{d^2\theta}{dt^2} + \kappa \frac{d\theta}{dt} + (n^2 - 2\alpha \sin 2pt)\,\theta = 0, \quad \ldots\ldots\ldots\ldots\ldots(5)$$

in which κ and α are supposed to be small. A similar equation would apply approximately in the case of a periodic variation in the effective mass of the body. The motion expressed by the solution of (5) can only be regular when it keeps perfect time with the imposed variations. It will appear that the necessary conditions cannot be satisfied rigorously by any simple harmonic vibration; but we may assume

$$\theta = A_1 \sin pt + B_1 \cos pt + A_3 \sin 3pt + B_3 \cos 3pt + A_5 \sin 5pt + \ldots, \ldots(6)$$

in which it is not necessary to provide for sines and cosines of even multiples of pt. . If the assumption is justifiable, the series in (6) must be convergent. Substituting in the differential equation, and equating to zero the coefficients of $\sin pt$, $\cos pt$, &c., we find

$$A_1 (n^2 - p^2) - \kappa p B_1 - \alpha B_1 + \alpha B_3 = 0,$$
$$B_1 (n^2 - p^2) + \kappa p A_1 - \alpha A_1 - \alpha A_3 = 0,$$
$$A_3 (n^2 - 9p^2) - 3\kappa p B_3 - \alpha B_1 + \alpha B_5 = 0,$$
$$B_3 (n^2 - 9p^2) + 3\kappa p A_3 + \alpha A_1 - \alpha A_5 = 0,$$
$$A_5 (n^2 - 25p^2) - 5\kappa p B_5 - \alpha B_3 + \alpha B_7 = 0,$$
$$B_5 (n^2 - 25p^2) + 5\kappa p A_5 + \alpha A_3 - \alpha A_7 = 0,$$

$$\ldots\ldots\ldots\ldots\ldots\ldots\ldots\ldots\ldots\ldots\ldots\ldots\ldots\ldots\ldots\ldots$$

---

* When the direction of motion is transverse, the case falls under the head of ordinary forced vibrations.

† See Tyndall's *Sound*, 3rd ed. ch. III. § 7, where will also be found a general explanation of the mode of action.

These equations show that relatively to A_1, B_1, A_3, B_3 are of the order α; that relatively to A_3, B_3, A_5, B_5 are of the order α, and so on. If we omit A_3, B_3 in the first pair of equations, we find as a first approximation,

$$A_1 (n^2 - p^2) - (\kappa p + \alpha) B_1 = 0,$$

$$A_1 (\kappa p - \alpha) + (n^2 - p^2) B_1 = 0;$$

whence

$$\frac{B_1}{A_1} = \frac{n^2 - p^2}{\kappa p + \alpha} = \frac{\alpha - \kappa p}{n^2 - p^2} = \frac{\sqrt{(\alpha - \kappa p)}}{\sqrt{(\alpha + \kappa p)}}, \quad \ldots\ldots\ldots\ldots\ldots(7)$$

and

$$(n^2 - p^2)^2 = \alpha^2 - \kappa^2 p^2. \quad \ldots\ldots\ldots\ldots\ldots\ldots(8)$$

Thus, if α be given, the value of p necessary for a regular motion is definite; and p having this value, the regular motion is

$$\theta = P \sin (pt + \epsilon),$$

in which ϵ, being equal to $\tan^{-1} (B_1/A_1)$, is also definite. On the other hand, as is evident at once from the linearity of the original equation, there is nothing to limit the amplitude of vibration.

These characteristics are preserved however far it may be necessary to pursue the approximation. If A_{2m+1}, B_{2m+1}, may be neglected, the first m pairs of equations determine the *ratios* of all the coefficients, leaving the absolute magnitude open; and they provide further an equation connecting p and α, by which the pitch is determined.

For the second approximation the second pair of equations gives

$$A_3 = \frac{\alpha B_1}{n^2 - 9p^2}, \qquad B_3 = -\frac{\alpha A_1}{n^2 - 9p^2},$$

whence

$$\theta = P \sin (pt + \epsilon) + \frac{\alpha P}{9p^2 - n^2} \cos (3pt + \epsilon); \quad \ldots\ldots\ldots\ldots(9)$$

and from the first pair

$$\tan \epsilon = \left\{ n^2 - p^2 - \frac{\alpha^2}{n^2 - 9p^2} \right\} \div (\alpha + \kappa p), \quad \ldots\ldots\ldots\ldots(10)$$

while p is determined by

$$\left[(n^2 - p^2) - \frac{\alpha^2}{(n^2 - 9p^2)^2} \right]^2 = \alpha^2 - \kappa^2 p^2. \quad \ldots\ldots\ldots\ldots(11)$$

Returning to the first approximation, we see from (8) that the solution is only possible under the condition that $\alpha > \kappa p$. If $\alpha = \kappa p$, then $p = n$; *i.e.* the imposed variation in the 'spring' must be exactly twice as quick as the natural vibration of the body would be in the absence of friction. From (7) it appears that in this case $\epsilon = 0$, which indicates that the spring is a minimum one-eighth of a period *after* the body has passed its position of equilibrium, and a maximum one-eighth of a period *before* such passage. Under these circumstances the greatest possible amount of energy is communicated to the system; and in the case contemplated it is just sufficient to balance the loss by dissipation, the adjustment being evidently independent of the amplitude.

If $\alpha < \kappa p$, sufficient energy cannot pass to maintain the motion, whatever may be the phase-relation; but if $\alpha > \kappa p$, the equality between energy supplied and energy dissipated may be attained by such an alteration of phase as shall diminish the former quantity to the required amount. The alteration of phase may for this purpose be indifferently in either direction; but if ϵ be positive, we must have

$$p^2 = n^2 - \sqrt{(\alpha^2 - \kappa^2 p^2)};$$

while if ϵ be negative,

$$p^2 = n^2 + \sqrt{(\alpha^2 - \kappa^2 p^2)}.$$

If α be very much greater than κp, $\epsilon = \pm \frac{1}{4}\pi$, which indicates that when the system passes through its position of equilibrium the spring is at its maximum or at its minimum.

The inference from the equations that the adjustment of pitch must be absolutely rigorous for steady vibration will be subject to some modification in practice; otherwise the experiment could not succeed. In most cases n^2 is to a certain extent a function of amplitude; so that if n^2 have very nearly the required value, complete coincidence is attainable, without other alteration in the conditions of the system, by the assumption of an amplitude of large and determinate amount.

When a particular solution of (5) has been found, it may be generalized by a known method. Thus, if $\theta = A\theta_1$, we have as the complete solution

$$\theta = A\theta_1 + B\theta_1 \int_0^t \theta_1^{-2} e^{-\kappa t}\, dt,$$

which may be put into the form

$$\theta = P\theta_1 - B\theta_1 \int_t^\infty \theta_1^{-2} e^{-\kappa t}\, dt. \quad \ldots\ldots\ldots\ldots\ldots(12)$$

When t is great, the second term diminishes rapidly, and the solution tends to assume the original form $\theta = P\theta_1$.

The number of cases falling under the present head which have been discovered and examined hitherto is not great. The mysterious *son rauque* of Savart, which sometimes accompanies the longitudinal vibrations of bars, and which is attributed by Terquem to an associated transverse vibration, is doubtless of this character. Just as in Melde's experiment already spoken of, the periodic variations of tension accompanying the longitudinal vibrations will throw the bar into lateral vibration, if there happen to be a mode of such vibration whose pitch is nearly enough coincident with the *suboctave* of the principal note.

For a lecture illustration we may take a pendulum formed of a bar of soft iron and vibrating on knife-edges. Underneath the pendulum is placed symmetrically a vertical bar electromagnet, through which is caused to pass an electric current rendered intermittent by an interrupter whose frequency is twice that of the pendulum. The magnetic force does not tend to displace the pendulum from its equilibrium position, but produces the same sort of effect as if gravity were subject to a periodic variation.

A similar result is obtained by causing the point of support of the pendulum to vibrate in a *vertical* path. If we denote this motion by $\eta = \beta \sin 2pt$, the effect is as if gravity were variable by the term $4p^2\beta \sin 2pt$. Of the same nature are the crispations observed by Faraday and others on the surface of water which oscillates vertically. Faraday arrived experimentally at the conclusion that there were two complete vibrations of the support for each complete vibration of the liquid. This view has been contested by Matthiessen*, who maintains that the vibrations are isoperiodic. By observations, which I hope to find another opportunity of detailing†, I have convinced myself that in this matter Faraday was perfectly correct. The vibrations of water standing upon a horizontal glass plate, which was attached to the centre of a vibrating iron bar, were at the rate of 15 per second when the vibrations of the bar were at the rate of 30 per second. The only difference of importance between this case and that of the pendulum is that, whatever may be the rate of vibration of the plate, there is always possible a free water-vibration of nearly the same frequency, and that consequently no special tuning is called for.

* *Pogg. Ann.* vol. CXLI. 1870.

† [1899. See Art. 102 below. It should be remarked that corrections have been introduced in equations (10), (11) above.]

98.

THE SOARING OF BIRDS.

[*Nature*, XXVII. pp. 534, 535, 1883.]

THE recent correspondence in *Nature* upon this subject ought not to close without some reference to a possible explanation of *soaring* which does not appear to have been yet suggested.

I premise that if we know anything about mechanics it is certain that a bird *without working his wings* cannot, either in still air or in a uniform horizontal wind, maintain his level indefinitely. For a short time such maintenance is possible at the expense of an initial relative velocity, but this must soon be exhausted. Whenever therefore a bird pursues his course for some time without working his wings we must conclude either (1) that the course is not horizontal, (2) that the wind is not horizontal, or (3) that the wind is not uniform. It is probable that the truth is usually represented by (1) or (2); but the question I wish to raise is whether the cause suggested by (3) may not sometimes come into operation*.

In *Nature*, Vol. XXIII. p. 10, Mr S. E. Peal makes very distinct statements as to the soaring of pelicans and other large birds in Assam. The course is in large and nearly circular sweeps, and at each lap some 10 or 20 feet of elevation is gained. *When there is a wind*, the birds may in this way "without once flapping the wings" rise from a height of 200 to a height of 8000 feet.

That birds do not soar when there is no wind is what we might suppose, but it is not evident how the existence of a wind helps the matter. If the wind were horizontal and uniform, it certainly could not do so. As it does not seem probable that at a moderate distance from the ground there could

---

* [1899. Under this head reference may be made to Langley's Memoir on the Internal Work of the Wind, *Smithsonian Contributions to Knowledge*, 1893.]

be a sufficient vertical motion of the air to maintain the birds, we are led to inquire whether anything can be made of the difference of horizontal velocities which we know to exist at different levels.

In a uniform wind the available energy at the disposal of the bird depends upon his velocity *relatively* to the air about him. With only a moderate waste this energy can at any moment be applied to gain elevation, the gain of elevation being proportional to the loss of relative velocity squared. It will be convenient for the moment to ignore the waste referred to, and to suppose that the whole energy available remains constant, so that however the bird may ascend or descend, the relative velocity is that due to a fall from a certain level to the actual position, the certain level being of course that to which the bird might just rise by the complete sacrifice of relative velocity.

For distinctness of conception let us now suppose that above and below a certain plane there is a uniform horizontal wind, but that in ascending through this plane the velocity increases, and let us consider how a bird sailing somewhat above the plane of separation, and endowed with an initial relative velocity, might take advantage of the position in which he finds himself.

The first step is, if necessary, to turn round until the relative motion is to leeward, and then to drop gradually down through the plane of separation. In falling down to the level of the plane there is a gain of relative velocity, but this is of no significance for the present purpose, as it is purchased by the loss of elevation; but in passing through the plane there is a really effective gain. In entering the lower stratum the actual velocity is indeed unaltered, but the velocity relatively to the surrounding air is *increased*. The bird must now wheel round in the lower stratum until the direction of motion is to windward, and then return to the upper stratum, in entering which there is a second increment of relative velocity. This process may evidently be repeated indefinitely; and if the successive increments of relative velocity squared are large enough to outweigh the inevitable waste which is in progress all the while, the bird may maintain his level, and even increase his available energy, without doing a stroke of work.

In nature there is of course no such abrupt transition as we have just now supposed, but there is usually a continuous increase of velocity with height. If this be sufficient, the bird may still take advantage of it to maintain or improve his position without doing work, on the principle that has been explained. For this purpose it is only necessary for him to descend while moving to leeward, and to ascend while moving to windward, the simplest mode of doing which is to describe circles on a plane which inclines downwards to leeward. If in a complete lap the advantage thus obtained compensates the waste, the mean level will be maintained without expenditure of work; if there be a margin, there will be an outstanding gain of level susceptible of indefinite repetition.

A priori, I should not have supposed the variation of velocity with height to be adequate for the purpose; but if the facts are correct, some explanation is badly wanted*. Mr Peal makes no mention of the circular sweeps being inclined to the horizon, a feature which is essential to the view suggested. It is just possible, however, that the point might escape attention not specially directed to it.

However the feat may be accomplished, if it be true that large birds can maintain and improve their levels without doing work, the prospect for human flight becomes less discouraging. Experimenters upon this subject would do well to limit their efforts for the present to the problem of gliding or sailing through the air. When a man can launch himself from an elevation and glide long distances before reaching the ground, an important step will have been gained, and until this can be done, it is very improbable that any attempt to maintain the level by expenditure of work can be successful. Large birds cannot maintain their levels in still air without a rapid horizontal motion, and it is easy to show that the utmost muscular work of a man is utterly inadequate with any possible wings to allow of his maintenance in a fixéd position relatively to surrounding air. With a rapid horizontal motion, the thing may perhaps be possible, but for further information bearing upon this subject, I must refer to a paper on the resistance of fluids published in the *Philosophical Magazine* for December, 1876.

[1899. The maintenance of a fixed position in still air, whether by a bird or by a man or by an engine, can only be secured by the generation of a downward current of air, *e.g.* by a screw, whose momentum shall balance the weights to be supported. If v denote the velocity, S the section of the stream, ρ the density of air, the momentum generated in unit time is $S\rho v^2$; and the work done in the same time is $\frac{1}{2}S\rho v^3$. Thus if gM be the whole weight sustained,

$$gM = S\rho v^2 \dots\dots\dots\dots\dots\dots\dots\dots\dots\dots\dots(1)$$

---

* [1899. A good deal depends upon the velocity of flight. If this reckoned relatively to the surrounding air be called v, and if it become v', whether owing to a passage of the bird into another stratum or to a freshening of the wind in the same stratum, the gain (h) of potential elevation is given by

$$v'^2 - v^2 = 2gh,$$

from which we see that the effect of a given change ($v' - v$) increases with v.

If we suppose that $v=30$ miles per hour, and that $h=10$ feet, we find $v'=34\cdot7$ miles per hour; so that at this speed a gain of 10 feet requires a freshening of the wind amounting to $4\cdot7$ miles per hour.

See further a letter on the Sailing Flight of the Albatross, *Nature*, xl. p. 34, 1889.]

Again, if V denote the rate at which the weight must be lifted in order to represent the work done by the driving engine,

$$gMV = \tfrac{1}{2} S\rho v^3 \dots\dots\dots\dots\dots\dots\dots\dots\dots(2)$$

Thus $v = 2V$, and

$$S = \frac{gM}{4\rho V^2}. \quad\dots\dots\dots\dots\dots\dots\dots\dots\dots(3)$$

So far as these equations are concerned, any weight can be sustained by a limited expenditure of work, but the smaller the power available the larger must be the section of the stream of air and consequently of the mechanism by which the air is set in motion.   Again, from (3)

$$gMV = \frac{(gM)^{\frac{3}{2}}}{(4\rho S)^{\frac{1}{2}}}, \quad\dots\dots\dots\dots\dots\dots\dots\dots(4)$$

so that, if S be given, the whole power required varies as $(gM)^{\frac{3}{2}}$.

To obtain numbers applicable to the case of a man supporting himself by his own muscular power, we take in C.G.S. measure,

$$M = 68000, \quad V = 15, \quad \rho = \tfrac{1}{800}, \quad g = 981,$$

thus finding

$$S = 6\cdot 0 \times 10^7 \text{ sq. cm.}$$

This represents the cross-section of the descending column of air. If we equate S to $\tfrac{1}{4}\pi d^2$, d will be the diameter of the screw required, and we get $d = 90$ metres. It is to be observed that the assumed nature of V corresponds to the power which a man may exercise when working for 8 hours a day. But even if he could do ten times as much for a few minutes, d would still amount to 9 metres, and in this estimate nothing has been allowed for the weight of the mechanism, or for frictional losses.

The present subject is further discussed in the Wilde Lecture on the Mechanical Principles of Flight (*Manchester Proceedings* 1900).]

99.

DISTRIBUTION OF ENERGY IN THE SPECTRUM.

[*Nature*, XXVII. pp. 559, 560, 1883.]

In the reaction against the arbitrariness of prismatic spectra there seems to be danger that the claim to ascendancy of the so-called diffraction spectrum may be overrated. On this system the rays are spaced so that equal intervals correspond to equal differences of wave-length, and the arrangement possesses indisputably the advantage that it is independent of the properties of any kind of matter. This advantage, however, would not be lost, if instead of the simple wave-length we substituted any function thereof; and the question presents itself whether there is any reason for preferring one form of the function to another.

On behalf of the simple wave-length, it may be said that this is the quantity with which measurements by a grating are immediately concerned, and that a spectrum drawn upon this plan represents the results of experiment in the simplest and most direct manner. But it does not follow that this arrangement is the most instructive.

Some years ago Mr Stoney proposed that spectra should be drawn so that equal intervals correspond to equal differences in the *frequency of vibration*. On the supposition that the velocity of light in vacuum is the same for all rays, this is equivalent to taking as abscissa the *reciprocal* of the wave-length instead of the wave-length itself. A spectrum drawn upon this plan has as much (if not more) claim to the title of *normal*, as the usual diffraction spectrum.

The choice that we make in this matter has an important influence upon the curve which represents the distribution of energy in the spectrum. In all cases the intensity of the radiation belonging to a given range of the spectrum is represented by the area included between the ordinates which correspond to the limiting rays, but the form of the curve depends upon what function of the ray we elect to take as abscissa. Thus in the ordinary prismatic spectrum of the sun, the curve culminates in the ultra-red, but in the diffraction spectrum the maximum is in the yellow, or even in the green, according to the recent important observations of Prof. Langley. If we wish to change the function of the ray represented by the abscissa, we can of course deduce by calculation the transformed curve of energy without fresh experiments. To pass from the curve with abscissæ proportional to wavelength to one with abscissæ proportional to reciprocals of wave-length, we must magnify the ordinates of the former in the ratio of the square of the wave-length, and this will give us an energy curve more like that obtained with a prismatic spectrum.

There is another method of representation intermediate between these two, which is not without advantage. In the diffraction spectrum the space devoted to a lower octave (if we may borrow the language of acoustics) is greater than that devoted to a higher octave. In Mr Stoney's map the opposite is the case. If we take the *logarithm* of the wave-length (or of the frequency) as abscissa, we shall obtain a map in which every octave occupies the same space, and this perhaps gives a fairer representation than either of the others. To deduce the curve of energy from that appropriate to the diffraction spectrum, we should have to magnify the ordinates in the ratio of the first power of the wave-length.

My object, however, is not so much to advocate any particular method of representation, as to point out that the curve of energy of the diffraction spectrum has no special claim to the title of "normal."

100.

INVESTIGATION OF THE CHARACTER OF THE EQUILIBRIUM OF AN INCOMPRESSIBLE HEAVY FLUID OF VARIABLE DENSITY*.

[*Proceedings of the London Mathematical Society*, XIV. pp. 170—177, 1883.]

THE well-known condition of equilibrium requires that the fluid be arranged in horizontal strata, so that its density σ is a function of the vertical coordinate z only. If this state of things be slightly departed from, we may regard the actual density at any point x, y, z as equal to $\sigma + \rho$, where ρ is a function of x, y, z, and the time t, which always remains small during the period contemplated. The component velocities u, v, w are equally to be regarded as small; they are connected by the equation of continuity

$$\frac{du}{dx} + \frac{dv}{dy} + \frac{dw}{dz} = 0. \quad \dots\dots\dots\dots\dots\dots\dots\dots(1)$$

The equilibrium pressure p is a function of z only. If the actual pressure be called $p + \delta p$, the dynamical equations become, with omission of the squares of small quantities,

$$\frac{d\delta p}{dx} = -\sigma \frac{du}{dt}, \quad \frac{d\delta p}{dy} = -\sigma \frac{dv}{dt}, \quad \frac{d\delta p}{dz} = -g\rho - \sigma \frac{dw}{dt}. \quad \dots\dots(2)$$

One further equation is supplied by the condition that the density of every particle remains unchanged.

Thus

$$\frac{d\rho}{dt} + w \frac{d\sigma}{dz} = 0. \quad \dots\dots\dots\dots\dots\dots\dots\dots(3)$$

---

* These calculations were written out in 1880, in order to illustrate the theory of cirrous clouds propounded by the late Prof. Jevons (*Phil. Mag.* XIV. p. 22, 1857). Pressure of other work has prevented me hitherto from pursuing the subject.

By Fourier's theorem and the general theory of disturbed equilibrium, we know that the complete solution of the present problem may be decomposed into partial solutions, for any one of which the variable quantities considered as functions of x vary as e^{ikx}, as functions of y vary as $e^{ik'y}$, and as functions of t vary as e^{int}. The *wave-lengths* of the disturbances parallel to x and y are λ, λ', where $\lambda = 2\pi/k$, $\lambda' = 2\pi/k'$.

The introduction of these suppositions into (1), (2), and (3) leads to

$$iku + ik'v + dw/dz = 0, \quad \ldots\ldots\ldots\ldots\ldots\ldots\ldots(4)$$

$$k\delta p = -n\sigma u, \quad k'\delta p = -n\sigma v, \quad d\delta p/dz = -g\rho - in\sigma w, \ldots\ldots\ldots(5)$$

$$in\rho + w\, d\sigma/dz = 0. \quad \ldots\ldots\ldots\ldots\ldots\ldots(6)$$

Eliminating u and v between (4) and the two first of equations (5), we get

$$i(k^2 + k'^2)\,\delta p - n\sigma\, dw/dz = 0. \quad \ldots\ldots\ldots\ldots\ldots(7)$$

Next eliminating δp between (7) and the last of equations (5), we find

$$i(k^2 + k'^2)(g\rho + in\sigma w) + n\frac{d}{dz}\left(\sigma\frac{dw}{dz}\right) = 0. \quad \ldots\ldots\ldots\ldots(8)$$

Finally between (6) and (8) we eliminate ρ, and thus obtain

$$\frac{d}{dz}\left(\sigma\frac{dw}{dz}\right) - (k^2 + k'^2)\left(\frac{g}{n^2}\frac{d\sigma}{dz} + \sigma\right)w = 0; \quad \ldots\ldots\ldots\ldots(9)$$

or, as it may be also written,

$$\frac{d^2w}{dz^2} - (k^2 + k'^2)\,w + \frac{d\sigma}{\sigma\, dz}\left\{\frac{dw}{dz} - \frac{g(k^2 + k'^2)}{n^2}w\right\} = 0. \ldots\ldots\ldots(10)$$

We will first apply this equation to the well-known case of two uniform fluids of densities σ_1, σ_2, separated by a horizontal boundary ($z = 0$), and for brevity we will omit to write k'. For both regions of fluid, the general equation (10) reduces to

$$\frac{d^2w}{dz^2} - k^2w = 0, \quad \ldots\ldots\ldots\ldots\ldots\ldots\ldots(11)$$

of which the solution is

$$w = A\,e^{kz} + B\,e^{-kz}. \ldots\ldots\ldots\ldots\ldots\ldots(12)$$

By the condition at infinity, we are to take for the upper fluid $A = 0$, and for the lower $B = 0$. Moreover by continuity the value of w must be the same for both fluids at the separating surface. Thus we may write for the upper fluid $w = B\,e^{-kz}$, and for the lower $w = B\,e^{kz}$. The second boundary condition is obtained by integrating equation (9) across the surface of transition. Thus

$$\left[\sigma\frac{dw}{dz}\right]_2 - \left[\sigma\frac{dw}{dz}\right]_1 - \frac{gk^2}{n^2}(\sigma_2 - \sigma_1) = 0;$$

whence

$$n^2 = gk \, \frac{\sigma_1 - \sigma_2}{\sigma_1 + \sigma_2}, \dots\dots\dots\dots\dots\dots\dots\dots(13)$$

the known solution.

If the upper fluid be the lighter, $\sigma_2 < \sigma_1$, and n^2 is positive. This indicates stability with harmonic oscillations, whose frequency increases without limit with k; that is, as the wave-length diminishes. If, on the other hand, $\sigma_2 > \sigma_1$, the equilibrium is unstable, and the instability (measured by the rate at which a small disturbance is multiplied in a given time) is greater the smaller the wave-length. If the disturbance be not limited to two dimensions, we have simply to replace k by $\sqrt{(k^2 + k'^2)}$.

We know from the general theory that only real values of n^2 are admissible in (9), and that if $d\sigma/dz$ be negative throughout, all the values of n^2 are positive, but if $d\sigma/dz$ be positive throughout, all the values of n^2 are negative. In order to prove this from the equation, suppose that w and w' are two solutions corresponding to different values of n^2, say n^2 and n'^2. Then

$$\int w' \frac{d}{dz}\left(\sigma \frac{dw}{dz}\right) dz = k^2 \int \left(\frac{g}{n^2} \frac{d\sigma}{dy} + \sigma\right) ww' \, dz,$$

or, on integration by parts between two finite or infinite limits for which w, w' vanish,

$$\int \sigma \frac{dw}{dz} \frac{dw'}{dz} dz + k^2 \int \sigma ww' dz + k^2 \frac{g}{n^2} \int \frac{d\sigma}{dz} ww' dz = 0. \dots\dots(14)$$

In this equation w and w' may be interchanged if n'^2 be written for n^2. Hence

$$\int \sigma \frac{dw}{dz} \frac{dw'}{dz} dz + k^2 \int \sigma ww' dz = 0, \dots\dots\dots\dots(15)$$

$$\int \frac{d\sigma}{dz} ww' dz = 0. \dots\dots\dots\dots\dots\dots\dots(16)$$

If now n^2 could be complex, there would be two solutions of the form $w = \alpha + i\beta$, $w' = \alpha - i\beta$, and equation (15) would become

$$\int \sigma \left\{ \left(\frac{d\alpha}{dz}\right)^2 + \left(\frac{d\beta}{dz}\right)^2 \right\} dz + k^2 \int \sigma \, (\alpha^2 + \beta^2) \, dz = 0,$$

which cannot be true if, as we suppose, σ is everywhere positive.

Again, suppose in (14) that $w' = w$. Thus

$$\int \sigma \left\{ \left(\frac{dw}{dz}\right)^2 + k^2 w^2 \right\} dz + \frac{k^2 g}{n^2} \int \frac{d\sigma}{dz} w^2 dz = 0, \dots\dots\dots(17)$$

from which it is evident that, if $d\sigma/dz$ be of one sign throughout, n^2 can only be of the opposite sign.

These conclusions are limited to the cases for which every mode of disturbance is stable, or every mode unstable; but we know that if $d\sigma/dz$ be *anywhere* positive, instability must ensue. To see this from equation (9), we may regard it as the condition (according to the methods of the Calculus of Variations) that $\int(d\sigma/dz)\,w^2dz$ is a maximum or minimum, while $\int\sigma\{(dw/dz)^2 + k^2w^2\}\,dz$ is given, $-n^2/gk^2$ being the then value of the ratio of the integrals. If $d\sigma/dz$ be anywhere positive, the first integral admits of a positive value, and therefore of a positive maximum, so that one value at least of n^2 is negative, and one mode of disturbance is unstable.

The simplest case of a variable density which we can consider is that obtained by supposing $\sigma^{-1}d\sigma/dz$ to be constant, equal say to β, or, on integration,

$$\sigma = \sigma_0 e^{\beta z} ; \dots\dots(18)$$

so that all strata of equal thickness are similarly constituted, differing only in absolute density. In this case, with omission of k' as before, (10) becomes

$$\frac{d^2w}{dz^2} + \beta\frac{dw}{dz} - k^2\left(1 + \frac{g\beta}{n^2}\right)w = 0. \dots\dots(19)$$

If m_1, m_2 be the roots of

$$m^2 + \beta m - k^2(1 + g\beta n^{-2}) = 0, \dots\dots(20)$$

the general solution of (19) is

$$w = Ae^{m_1z} + Be^{m_2z}, \dots\dots(21)$$

A and B being arbitrary constants.

Let us now suppose that the fluid is bounded by impenetrable horizontal planes at $z = 0$ and at $z = l$. Since w vanishes with z, $B = -A$, so that (21) becomes

$$w = A(e^{m_1z} - e^{m_2z}). \dots\dots(22)$$

Again, since w vanishes when $z = l$, $e^{m_1l} - e^{m_2l} = 0$, or $e^{(m_1-m_2)l} = 1$, whence

$$(m_1 - m_2)l = 2\alpha i\pi, \dots\dots(23)$$

α being an integer. Thus (22) may be written

$$w = Ae^{\frac{1}{2}(m_1+m_2)z}\{e^{\frac{1}{2}(m_1-m_2)z} - e^{-\frac{1}{2}(m_1-m_2)z}\} = A'e^{-\frac{1}{2}\beta z}\sin(\alpha\pi z/l), \dots\dots(24)$$

by (20), (23), A' being a new arbitrary constant. The values of n corresponding to the various values of α are obtained by comparison of (20) and (23). From the former

$$(m_1 - m_2)^2 l^2 = \beta^2l^2 + 4k^2l^2(1 + g\beta n^{-2}), \dots\dots(25)$$

so that

$$k^2l^2(1 + g\beta n^{-2}) = -\tfrac{1}{4}\beta^2l^2 - \alpha^2\pi^2, \dots\dots(26)$$

or

$$\frac{g\beta}{n^2} = -\frac{\frac{1}{4}\beta^2 l^2 + k^2 l^2 + \alpha^2 \pi^2}{k^2 l^2}. \quad \dots\dots\dots\dots\dots(27)$$

From (27) we see that the disturbances are all stable if β is negative, that is, if the density diminishes upwards, and that in the contrary case they are all unstable. The smallest admissible value of α is unity, and this corresponds to the greatest numerical value of n^2. Contrary to what is met with in most vibrating systems, there is (in the case of stability) a limit on the side of rapidity of vibration, but none on the side of slowness. In the case of instability we are principally interested in the mode for which the instability is greatest, and this also corresponds to the unit value of α. When α is greater than unity, there are internal nodal planes, as appears from (24).

If l, k, and α are given, n^2 is numerically greatest when β is such that

$$\tfrac{1}{4}\beta^2 l^2 = k^2 l^2 + \alpha^2 \pi^2.$$

If l, α, β be regarded as given, n^2 increases numerically from zero when k is zero, up to a finite limit when k is infinite; or, in the case of stability, as the wave-length diminishes from ∞ to 0, the frequency of vibration rises from 0 to a finite value, given by

$$n^2 = -g\beta, \quad \dots\dots\dots\dots\dots\dots\dots\dots(28)$$

which is independent both of α and of l. These vibrations are isochronous with the vibrations of a pendulum whose length is equal to the distance between two strata of which the densities are as $e : 1$.

If the disturbance be not limited to two dimensions, we must write $\sqrt{(k^2 + k'^2)}$ for k^2. The completely expressed value of w, corresponding to one normal mode of disturbance, is then

$$w = A e^{-\frac{1}{2}\beta z} \sin \frac{\alpha \pi z}{l} \cos \frac{2\pi (x - x_0)}{\lambda} \cos \frac{2\pi (y - y_0)}{\lambda'} \cos n (t - t_0). \quad \dots\dots(29)$$

We will now apply the solution to the investigation of the case in which the density for all values of z less than 0 is σ_1, and for all values of z greater than l is σ_2, the transition from the one density to the other being in accordance with the law $\sigma = \sigma_1 e^{\beta z}$, so that

$$\sigma_2 = \sigma_1 e^{\beta l}. \quad \dots\dots\dots\dots\dots\dots\dots(30)$$

When $z > l$, $w \propto e^{-kz}$, so that for $z = l$, $dw/w dz = -k$; similarly for the lower fluid, when $z = 0$, $dw/w dz = +k$. Thus, by (21), the boundary conditions are

$$m_1 A e^{m_1 l} + m_2 B e^{m_2 l} = -k (A e^{m_1 l} + B e^{m_2 l}),$$

$$m_1 A + m_2 B = +k (A + B),$$

whence, by elimination of $A : B$,

$$e^{m_1 l} (m_1 + k) (m_2 - k) = e^{m_2 l} (m_2 + k) (m_1 - k).$$

This, in connection with (20), determines the admissible values of n. It may be written

$$\frac{m_1 m_2 + k(m_2 - m_1) - k^2}{m_1 m_2 - k(m_2 - m_1) - k^2} = e^{(m_2 - m_1)l},$$

or

$$\frac{k(m_2 - m_1)}{m_1 m_2 - k^2} = \tanh \tfrac{1}{2}(m_2 - m_1)\, l.$$

By (20) this may be put into the form

$$\frac{k(m_2 - m_1)}{\tfrac{1}{4}\beta^2 - k^2 - \tfrac{1}{4}(m_2 - m_1)^2} = \tanh \tfrac{1}{2}(m_2 - m_1)\, l,$$

or, if for brevity we write θ for $(m_2 - m_1)\, l$,

$$\frac{kl \cdot \theta}{\tfrac{1}{4}\beta^2 l^2 - k^2 l^2 - \tfrac{1}{4}\theta^2} = \tanh \tfrac{1}{2}\theta. \quad\dots\dots\dots\dots\dots(31)$$

This equation determines θ; and then, by (20),

$$\theta^2 = l^2 (m_2 - m_1)^2 = l^2 \{(m_2 + m_1)^2 - 4m_1 m_2\} = \beta^2 l^2 + 4k^2 l^2 (1 + g\beta n^{-2}),$$
$$\dots\dots\dots(32)$$

giving n in terms of θ.

Before going farther we may verify these results by applying them to the case of a sudden transition, for which l vanishes, while βl remains finite. The principal solution of (31) gives $\theta^2 = \beta^2 l^2$ approximately, so that

$$\tfrac{1}{4}\beta^2 l^2 - \tfrac{1}{4}\theta^2 = kl \cdot \beta l \cdot \coth \tfrac{1}{2}\beta l.$$

Using this in (32), we get

$$g\beta n^{-2} = - k^{-1}\beta \coth \tfrac{1}{2}\beta l,$$

whence

$$n^2 = - gk \tanh \tfrac{1}{2}\beta l = - gk \frac{\sigma_2 - \sigma_1}{\sigma_2 + \sigma_1},$$

as before.

Other solutions of (31) are obtained by supposing $\theta^{-1} \tanh \tfrac{1}{2}\theta$ to vanish, whence $\theta = i \cdot \alpha \cdot 2\pi$, α being an integer other than zero. These are of no importance, as the corresponding values of n vanish.

When the layer of transition is of finite thickness, the general solution expressed by (31), (32) is rather complicated. A simplification, which does not involve much loss of interest, may be effected by supposing that the whole change of density is small, so that (31), (32) become

$$\frac{kl \cdot \theta}{k^2 l^2 + \tfrac{1}{4}\theta^2} = - \tanh \tfrac{1}{2}\theta, \dots\dots\dots\dots\dots\dots(33)$$

$$g\beta n^{-2} = \frac{\theta^2}{4k^2 l^2} - 1 \dots\dots\dots\dots\dots\dots(34)$$

From (33),

$$\pm \frac{kl - \frac{1}{2}\theta}{kl + \frac{1}{2}\theta} = e^{\frac{1}{2}\theta},$$

whence

$$- \tanh \tfrac{1}{4}\theta = \frac{\theta}{2kl}, \quad \text{or} \quad \frac{2kl}{\theta}. \quad\quad\quad\quad (35)$$

Equation (35) cannot be satisfied by any real value of θ. If we write $\theta = i\phi$, we get in place of it,

$$\tan \tfrac{1}{4}\phi = - \frac{\phi}{2kl}, \quad \text{or} \quad \frac{2kl}{\phi}; \quad\quad\quad\quad (36)$$

and in place of (34),

$$g\beta n^{-2} = - \frac{\phi^2}{4k^2l^2} - 1. \quad\quad\quad\quad (37)$$

The series of admissible values of ϕ, given by (36), extends to infinity, but the higher roots correspond to small values of n^2, which are of little interest. Whether the equilibrium be stable or unstable, the most important root is the smallest. It lies in the first quadrant, and is given by the second alternative of (36). The progress of n^2 as a function of kl is easily traced. When kl is small, $\phi^2 = 8kl$, and $g\beta/n^2 = - 2/kl$, which leads to $n^2 = - gk(\sigma_2 - \sigma_1)/2\sigma$, the known result for a rapid transition. As kl increases, $\tfrac{1}{4}\phi$ ranges from 0 to $\tfrac{1}{2}\pi$, and $\phi^2/4k^2l^2$ or $\cot^2 \tfrac{1}{4}\phi$ ranges from infinity to zero. Thus the numerical value of n^2 continually increases, until for an infinitely small wave-length it approaches the finite limit $- g\beta$, beyond which it cannot pass. The principal result of the substitution of a gradual for an abrupt transition is to arrest the further increase of n^2, after the wave-length has diminished so far as to become comparable in magnitude with the thickness of the layer of transition. In the case of the limiting value of n^2, the length of the equivalent pendulum is

$$l \div (\log \sigma_2 - \log \sigma_1).$$

If, for example, the extreme difference of densities amounted to one per cent., the length of the equivalent pendulum would be 100 times the thickness of the layer of transition.

For actual calculation (36), (37) may advantageously be written

$$\tfrac{1}{2}kl = \tfrac{1}{4}\phi \times \tan \tfrac{1}{4}\phi, \quad\quad\quad\quad (38)$$

$$- \tfrac{1}{2}kg \cdot \beta l \cdot n^{-2} = \tfrac{1}{2}kl \div \sin^2 \tfrac{1}{4}\phi, \quad\quad\quad\quad (39)$$

the right-hand member of (39) being equal to unity, when kl is small. Ascribing arbitrary values to $\tfrac{1}{4}\phi$, we can readily calculate corresponding values of kl and $\tfrac{1}{2}kl/\sin^2 \tfrac{1}{4}\phi$, and thus exhibit the effect upon the equilibrium of a gradual increase in the thickness of the layer of transition, the extreme densities (determined by βl) and the wave-length being given.

$\frac{1}{4}\phi$	kl	$\frac{1}{2}kl/\sin^2\frac{1}{4}\phi$
0°	kl	1·000
10°	·06155	1·021
20°	·2541	1·086
30°	·6046	1·209
40°	1·172	1·418
50°	2·080	1·772
60°	3·628	2·418
70°	6·713	3·801
80°	15·838	8·165
90°	kl	$\frac{1}{2}kl$

101.

ON THE VIBRATIONS OF A CYLINDRICAL VESSEL CONTAINING LIQUID.

[*Philosophical Magazine*, xv. pp. 385—389, 1883.]

THE problem of a uniform cylinder vibrating in two dimensions is considered in my book on the *Theory of Sound*, § 233. If the displacements at any point a, θ of the circumference be δr, $a\,\delta\theta$, then for a single component

$$\delta r = aA_n \cos n\theta, \qquad \delta\theta = -n^{-1}A_n \sin n\theta. \dots\dots\dots\dots(1)$$

If d be the thickness and σ the volume-density of the material, the kinetic energy of the motion for a length z measured parallel to the axis is

$$T = \tfrac{1}{2}\pi z d\sigma a^3 (1 + n^{-2}) \left(\frac{dA_n}{dt}\right)^2 .\dots\dots\dots\dots\dots(2)$$

The corresponding potential energy is

$$V = \frac{\pi B z}{2a} (n^2 - 1)^2 A_n^2,$$

in which B is a constant depending upon the material and upon the thickness. As a function of thickness $B \propto d^3$; so that we may write $B = B_0 d^3$, in which B_0 depends upon the material only. Thus

$$V = \frac{\pi B_0 d^3 z}{2a} (n^2 - 1)^2 A_n^2 .\dots\dots\dots\dots\dots\dots(3)$$

If the cylinder be empty, these expressions suffice to determine the periods of vibration. Thus, if $A_n \propto \cos p_0 t$,

$$p_0^2 = \frac{B_0 d^2}{\sigma a^4} \frac{(n^2 - 1)^2}{1 + n^{-2}} ,\dots\dots\dots\dots\dots\dots(4)$$

showing that for a given material the frequency is proportional to the thickness and inversely as the square of the radius.

If the cylinder contain frictionless fluid, the motion of the fluid will depend upon a velocity-potential ϕ which satisfies the equation

$$\left(\frac{d^2}{dr^2} + \frac{1}{r}\frac{d}{dr} + \frac{1}{r^2}\frac{d^2}{d\theta^2} + \frac{d^2}{dz^2} + k^2\right)\phi = 0,\ldots\ldots\ldots(5)$$

in which

$$k = p/a', \ldots\ldots\ldots\ldots\ldots(6)$$

a' being the velocity of propagation of sound within the fluid. If the fluid can be treated as incompressible, we may put $k = 0$. For the present purpose we will retain k, but we will assume that the motion is strictly in two dimensions. Introducing the further assumption that $\phi \propto \cos n\theta$, we get in place of (5),

$$\left(\frac{d^2}{dr^2} + \frac{1}{r}\frac{d}{dr} - \frac{n^2}{r^2} + k^2\right)\phi = 0, \ldots\ldots\ldots\ldots(7)$$

of which the solution is

$$\phi = a_n \cos pt \cos n\theta\, J_n(kr).\ldots\ldots\ldots\ldots(8)$$

The relation between a_n and A_n of (1) is readily found by equating the value of $d\phi/dr$, when $r = a$, to $d\delta r/dt$, both of which represent the normal velocity at the circumference. We get

$$a_n \cos pt = \frac{a}{kJ_n'(ka)}\frac{dA_n}{dt}. \ldots\ldots\ldots\ldots(9)$$

The kinetic energy of the fluid motion is given by

$$T = \tfrac{1}{2}\rho \iiint \left\{\left(\frac{d\phi}{dx}\right)^2 + \left(\frac{d\phi}{dy}\right)^2\right\} dx\,dy\,dz$$

$$= \tfrac{1}{2}\rho \left[z\int \phi\,\frac{d\phi}{dr}\,a\,d\theta + k^2 \iiint \phi^2\,dx\,dy\,dz \right]$$

$$= \tfrac{1}{2}\pi\rho z a_n^2 \cos^2 pt\,\{ka\,.\,J_n(ka)\,J_n'(ka) + k^2 \int_0^a J_n^2(kr)\,r\,dr\}. \ldots(10)$$

For the potential energy of the liquid, if compressible, we have

$$V = \tfrac{1}{2}\rho a'^{-2} \iiint \left(\frac{d\phi}{dt}\right)^2 dx\,dy\,dz = \tfrac{1}{2}\pi\rho z a_n^2 \sin^2 pt\,.\,k^2 \int_0^a J_n^2(kr)\,r\,dr. \ldots(11)$$

The sum of the potential and kinetic energies for the solid and liquid together must be independent of the time. The unintegrated terms in (10) and (11) cancel, and we find

$$\frac{B_0 d^3\,(n^2-1)^2}{a^3 p^2} = a\sigma d\,(1 + n^{-2}) + \frac{\rho a}{k}\frac{J_n(ka)}{J_n'(ka)}. \ldots\ldots\ldots\ldots(12)$$

In the application of (12) ka is a small quantity. From the ascending series for $J_n(ka)$ we find

$$\frac{a}{k}\frac{J_n(ka)}{J_n'(ka)} = \frac{a^2}{n}\left(1 + \frac{k^2 a^2}{n\,.\,2n+2} + \ldots\ldots\right), \ldots\ldots\ldots\ldots(13)$$

so that approximately

$$\frac{B_0 d^3 (n^2-1)^2}{a^4 p^2} = \sigma d \left(1+n^{-2}\right) + n^{-1} \rho a \left(1 + \frac{k^2 a^2}{n \cdot 2n+2}\right). \quad \ldots\ldots\ldots (14)$$

If p_0 be the value of p when $\rho = 0$,

$$\frac{p_0^2}{p^2} = 1 + \frac{n}{n^2+1} \frac{\rho a}{\sigma d} \left(1 + \frac{k^2 a^2}{n \cdot 2n+2}\right). \quad \ldots\ldots\ldots\ldots (15)$$

From (14) or (15) we see that the effect of a finite as compared with an infinitely small compressibility is to *increase* the depression of pitch due to the fluid. As the velocity of sound is greater in liquids than in air, it would seem that $\frac{1}{12} k^2 a^2$ would generally be negligible. In this case, for the principal mode of vibration corresponding to $n = 2$, (15) becomes simply

$$\frac{p_0^2}{p^2} = 1 + \frac{2}{5} \frac{\rho a}{\sigma d}. \quad \ldots\ldots\ldots\ldots\ldots\ldots (16)$$

In Auerbach's recent paper upon this subject* various observations upon the depression of pitch due to the action of liquid are given. In his notation $p_0/p = G$. From (15) we see that if G_0 be the value of G for water, the same vessel being used in both cases,

$$\frac{G^2-1}{G_0^2-1} = s, \quad \ldots\ldots\ldots\ldots\ldots\ldots\ldots\ldots (17)$$

if s denote the specific gravity of the liquid, referred as usual to water as a standard. Auerbach's observations are fairly accordant with (17); and there seems to be scarcely sufficient warrant for attributing the discrepancies to the influence of compressibility.

In observations with different vessels of the same material and filled with the same fluid, difficulty was experienced in obtaining by direct measurement a sufficiently accurate value of d. To meet this, d was determined indirectly from the pitch. By (4) we have

$$\frac{p_0^2}{p^2} = 1 + \frac{n^4-n^2}{(n^2+1)^{\frac{3}{2}}} \frac{\rho B_0^{\frac{1}{2}}}{\sigma^{\frac{3}{2}} p_0 a}, \quad \ldots\ldots\ldots\ldots\ldots (18)$$

from which it appears that $G^2 - 1$ is inversely proportional to the pitch (before filling), as well as inversely proportional to the radius of the cylinder. In Auerbach's notation a constant C is employed, whose value for the case $n = 2$ would be by (18)

$$C = \frac{6}{5^{\frac{3}{2}}\pi} \frac{\rho B_0^{\frac{1}{2}}}{\sigma^{\frac{3}{2}} a}. \quad \ldots\ldots\ldots\ldots\ldots\ldots (19)$$

In actual experiment the two-dimensional character of the fluid motion is disturbed by the existence of a free surface at which a special condition

---

* *Wied. Ann.* Bd. XVII. p. 964.

must be satisfied. Hence arises a vertical motion of the surface, which is the proximate cause of the "crispations" usually to be observed under these circumstances. In considering this question we may leave the force of gravity out of account, inasmuch as the period of free waves of length comparable with the diameter of the cylinder is much greater than that of the actual motion.

In accordance with (5), if the fluid be treated as incompressible, we may take

$$\phi = a_n \cos pt \cos n\theta \, r^n + \Sigma A_k \cos pt \cos n\theta \, e^{-kz} J_n(kr), \quad \ldots\ldots(20)$$

in which z is measured downwards from the surface, and k denotes a root of

$$J_n'(ka) = 0. \quad \ldots\ldots\ldots\ldots\ldots\ldots\ldots\ldots\ldots (21)$$

The coefficients A_k are to be determined by the condition at the surface which is simply $\phi = 0$. Thus for each value of k

$$a_n \int_0^a r^{n+1} J_n(kr)\, dr + A_k \int_0^a J_n^2(kr)\, r\, dr = 0. \quad \ldots\ldots\ldots(22)$$

Now (see *Theory of Sound*, §§ 203, 332)

$$\int_0^a r^{n+1} J_n(kr)\, dr = \frac{na^n}{k^2} J_n(ka), \quad \int_0^a J_n^2(kr)\, r\, dr = \tfrac{1}{2}a^2 \left(1 - \frac{n^2}{k^2 a^2}\right) J_n^2(ka),$$

so that

$$\phi = a_n \cos pt \cos n\theta \left\{ r^n - 2na^n \Sigma \frac{e^{-kz} J_n(kr)}{(k^2 a^2 - n^2) J_n(ka)} \right\}. \quad \ldots\ldots(23)$$

To calculate the kinetic energy we have to integrate $\phi \, d\phi/dn$ over the whole boundary of the fluid. Now at the free surface $\phi = 0$, and at a great depth the motion becomes two-dimensional. We have therefore only to consider the cylindrical surface. By supposition $J_n'(ka) = 0$, and thus

$$\frac{d\phi}{dn} = \frac{d\phi}{dr} = n a_n a^{n-1} \cos pt \cos n\theta.$$

We get therefore

$$T = \tfrac{1}{2}\rho a_n^2 n a^{2n} \cos^2 pt \iint \left\{ 1 - 2n \Sigma \frac{e^{-kz}}{k^2 a^2 - n^2} \right\} \cos^2 n\theta \, d\theta \, dz$$

$$= \tfrac{1}{2}\pi\rho a_n^2 n a^{2n} \cos^2 pt \left\{ z - 2n \Sigma \frac{k^{-1}}{k^2 a^2 - n^2} \right\}. \quad \ldots\ldots\ldots\ldots(24)$$

The value of T is less than if the motion were strictly two-dimensional by a quantity corresponding to the length

$$2na \, \Sigma \frac{k^{-1} a^{-1}}{k^2 a^2 - n^2}. \quad \ldots\ldots\ldots\ldots\ldots\ldots\ldots(25)$$

For $n = 2$, the values of ka from (21) are 3·054, 6·705, 9·965, 13·1, 16·3, &c.; and thus (25) becomes ·2674 a.

102.

ON THE CRISPATIONS OF FLUID RESTING UPON A VIBRATING SUPPORT.

[*Philosophical Magazine*, XVI. pp. 50—58, 1883.]

IF a glass plate, held horizontally and made to vibrate as for the production of Chladni's figures, be covered with a thin layer of water or other mobile liquid, the phenomena in question may be readily observed. Over those parts of the plate which vibrate sensibly the surface of the liquid is ruffled by minute waves, the degree of fineness increasing with the frequency of vibration. Similar crispations are observed on the surface of liquid in a large wine-glass or finger-glass which is caused to vibrate in the usual manner by carrying the moistened finger round the circumference. All that is essential to the production of crispations is that a body of liquid with a free surface be constrained to execute a vertical vibration. It is indifferent whether the origin of the motion be at the bottom, as in the first case, or, as in the second, be due to the alternate advance and retreat of a lateral boundary, to accommodate itself to which the neighbouring surface must rise and fall.

More than fifty years ago the nature of these vibrations was examined by Faraday with great ingenuity and success. His results are recorded in an Appendix to a paper on a Peculiar Class of Acoustical Figures*, headed " On the Forms and States assumed by Fluids in Contact with Vibrating Elastic Surfaces." In more recent times Dr L. Matthiessen has travelled over the same ground†, and on one very important point has recorded an opinion in opposition to that of Faraday. In order more completely to satisfy myself, I have lately repeated most of Faraday's experiments, in some cases with improved appliances, and have been able to add some further observations in support of the views adopted.

---

* *Phil. Trans.* 1831.

† *Pogg. Ann.* t. CXXXIV. 1868; t. CXLI. 1870.

The phenomenon to be examined is evidently presented in its simplest form when the motion of the vibrating horizontal plate on which the liquid is spread is a simple up-and-down motion without rotation.   To secure this, Faraday attached the plate to the centre of a strip of glass or lath of deal, supported at the nodes, and caused to vibrate by friction.   In my experiments an iron bar was used about 1 metre long and ·0064 metre thick (in the plane of vibration).   The bar was supported horizontally at the nodes; and to its centre a glass plate was attached by gutta-percha and carefully levelled.   The vibrations of the bar were maintained electromagnetically, as in tuning-fork interrupters, with the aid of an electromagnet placed under the centre, the circuit being made and broken at a mercury-cup by a dipper carried at one end of the bar.   By calculation from the dimensions*, and without allowance for the load at the centre, the frequency of (complete) vibration is 33.   Comparisons with a standard tuning-fork gave more accurately for the actually loaded bar a frequency of 31.

The vibrating liquid standing upon the plate presents appearances which at first are rather difficult to interpret, and which vary a good deal with the nature of the liquid in respect of transparency or opacity, and with the incidence of the light.   The vibrations of the liquid, whether at the rate of 31 per second, or, as in fact, at the rate of $15\frac{1}{2}$ per second, are too quick to be followed by the eye; and thus the effect observed is an average, due to the superposition of an indefinite number of components corresponding to the various phases of vibration.

The motion of the liquid consists of two sets of stationary vibrations superposed, the ridges and furrows of the two sets being perpendicular to one another, and usually parallel to the edges of the (rectangular) plate. Confining our attention for the moment to one set of stationary waves, let us consider what appearance it may be expected to present.   At one moment the ridges form a set of parallel and equidistant lines, the interval being the wave-length.   Midway between these are the lines which represent at that moment the position of the furrows.   After the lapse of $\frac{1}{4}$ period, the surface is flat; after another $\frac{1}{4}$ period, the ridges and furrows are again at their maximum development, but the *positions are exchanged.*   Now, since only an average effect can be perceived, it is clear that no distinction can be recognized between the ridges and the furrows, and that the observed effect must be periodic within a distance equal to *half* a wave-length of the real motion. If the liquid on the plate be rendered moderately opaque by addition of aniline blue, and be seen by diffused transmitted light, the lines of ridge and furrow will appear bright in comparison with the intermediate nodal lines where the normal depth is preserved throughout the vibration.   The gain of light when the thickness is small will, in accordance with the law of absorp-

* *Theory of Sound,* § 171.

tion, outweigh the loss of light which occurs half a period later when the furrow is replaced by a ridge.

The actual phenomenon is more complicated in consequence of the co-existence of the two sets of ridges and furrows in perpendicular directions (x, y). In the adjoining figure the thick lines represent the ridges, and the

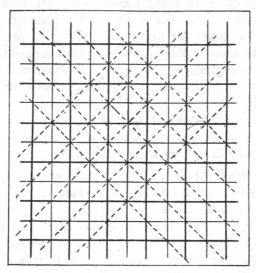

thin lines the furrows, of the two systems at a moment of maximum excur-sion. One quarter period later the surface is flat, and one half a period later the ridges and furrows are interchanged. The places of maximum elevation and depression are the intersections of the thick lines and of the thin lines, not distinguishable by ordinary vision; and these regions will appear like holes in the sheet of colour. The nodal lines, where the normal depth of colour is preserved, are shown dotted; they are inclined at 45°, and pass through the intersections of the thick lines with the thin lines. The pattern is recurrent in the directions of both x and y, and in each case with an interval equal to the real wave-length (λ). The distance between the bright spots measured parallel to x or y is thus λ; but the shortest distance between these spots is in directions inclined at 45°, and is equal to $\frac{1}{2}\sqrt{2} \cdot \lambda$.

In order to determine the relation of the frequency of the liquid vibrations to that of the bar, an apparatus was fitted up capable of giving an inter-mittent view of the vibrating system. This consisted of a blackened paper disk pierced with three sets of holes, mounted upon an axle, and maintained in rotation by a small electromagnetic engine of Apps's construction. The whole was fastened to one base-board, and could be moved about freely, the leading wires from the battery being flexible. The current was somewhat in excess; so that the desired speed could be attained by the application of moderate friction. At a certain speed of rotation the appearances were as follows.   Through the set of four holes (giving four views for each rotation of the disk) the bar was seen double.   Through the set of two holes the bar was seen single, and the water-waves were seen double.   Through the single hole the bar was seen single, and the waves also were seen single.   From this it follows that the water vibrations are not, as Matthiessen contends, synchronous with those of the bar, but that there are two complete vibrations of the support for each complete vibration of the water, in accordance with Faraday's original statement.

An attempt was made to calculate the frequency of liquid vibration from measurements of the wave-length and of the depth.   The depth (h), deduced from the area of the plate and the whole quantity of liquid, was ·0681 centim.; and by direct measurement $\lambda = \cdot 848$ centim.   Sir W. Thomson's formula connecting the velocity of propagation with the wave-length, when the effect of surface-tension is included, is

$$v^2 = \frac{\lambda^2}{\tau^2} = 982 \left(\frac{\lambda}{2\pi} + \frac{\cdot 074 \times 2\pi}{\lambda} \right) \times \frac{e^\alpha - e^{-\alpha}}{e^\alpha + e^{-\alpha}}, \dots\dots\dots\dots\dots\text{(A)}$$

where $\alpha = 2\pi h/\lambda$.   With the above data we find for the frequency (τ^{-1}) of vibration 20·8.   This should have been 15·5; and the discrepancy is probably to be attributed to friction, whose influence must be to diminish the efficient depth, and may easily rise to importance when the total depth is so small.

Another method by which I succeeded in determining the frequency of these waves requires a little preliminary explanation.   If $n = 2\pi/\tau$, and $k = 2\pi/\lambda$, the stationary waves parallel to y may be expressed as the resultant of opposite progressive waves in the form

$$\cos (kx + nt) + \cos (kx - nt) = 2 \cos kx \cos nt. \dots\dots\dots\dots\text{(1)}$$

This represents the state of things referred to an origin fixed in space. But now let us refer it to an origin moving forward with the velocity (n/k) of the progressive waves, so as to obtain the appearance that would be presented to the eye, or to the photographic camera, carried forward in this manner. Writing $kx' + nt$ for kx, we get

$$\cos (kx' + 2nt) + \cos kx'. \dots\dots\dots\dots\dots\dots\text{(2)}$$

Now the average effect of the first term is independent of x', so that what is seen is simply that set of progressive waves which moves with the eye.   In this way a kind of resolution of the stationary wave into its progressive components may be effected.

In the actual experiment two sets of stationary waves are combined; and the analytical expression is

$$\cos (kx + nt) + \cos (kx - nt) + \cos (ky + nt) + \cos (ky - nt), \dots\dots\text{(3)}$$

which is equal to

$$2 \cos kx \cos nt + 2 \cos ky \cos nt, \dots\dots\dots\dots\text{(4)}$$

or to

$$4 \cos \{\tfrac{1}{2} k (x + y)\} \cos \{\tfrac{1}{2} k (x - y)\} \cos nt. \quad\dots\dots\dots\dots(5)$$

If, as before, we write $kx' + nt$ for kx, we get

$$\cos (kx' + 2nt) + \cos kx' + 2 \cos ky \cos nt. \quad\dots\dots\dots\dots(6)$$

The eye, travelling forward with the velocity n/k, sees mainly the corresponding progressive waves, whose appearance, however, usually varies with y, i.e. along the length of a ridge or furrow. If the effect could be supposed to depend upon the *mean* elevation only, this complication would disappear, as we should be left with the term $\cos kx'$ standing alone. With the semiopaque coloured water the variation along y is evident enough; but the experiment may be modified in such a manner that the ridges and furrows appear sensibly uniform. For this purpose the coloured water may be replaced by milk, lighted from above, but very obliquely. The appearance of a set of (uniform) ridges and furrows varies greatly with the direction of the light. If the light fall upon the plate in a direction nearly parallel to the ridges, the disturbance of the surface becomes almost invisible; but if, on the other hand, the incidence be perpendicular to the line of ridges, the disturbance is brought into strong relief. The application of this principle to the case before us shows that, when the eye is travelling parallel to x, the ridges and furrows will look nearly uniform if the incidence of the light be also nearly parallel to x; but if the incidence of the light be nearly parallel to y, the ridges will show marked variations along their length, and in fact be resolved into a series of detached humps. The former condition of things is the simplest, and the most suitable as the subject of measurement.

In order to see the progressive waves it is not necessary to move the head as a whole, but only to turn the eye as when we look at an ordinary object in motion. To do this without assistance is not at first very easy, especially if the area of the plate be somewhat small. By moving a pointer at various speeds until the right one is found, the eye may be guided to do what is required of it; and after a few successes repetition becomes easy. If we wish not merely to see the progressive waves, but to measure the velocity of propagation with some approach to accuracy, further assistance is required. In my experiments an endless string, passing over pulleys and driven by a small water-engine, travelled at a small distance above the plate so that its length was in the direction of wave-propagation. A piece of wire was held at one end by the fingers, and at the other rested upon the travelling string and was carried forward with it. In this way, by adjusting the water supply, the speed of the string could be made equal to that of wave-propagation; and the former could easily be determined from the whole length of the string, and from the time required by a knot upon it to make a complete circuit. Thus (on February 7) the velocity of propagation was found to be 5·4 inches per second. At the same time, by measurement of the pattern as seen by ordinary vision, $14\lambda = 4\frac{7}{8}$ inches. Hence frequency $= 5\cdot4/\lambda = 15\cdot5$ per second; exactly one half the observed frequency of the bar, viz. 31.

In addition to the phantoms which may be considered to represent the four component progressive waves, others may be observed travelling in directions inclined at 45°. If we take coordinates ξ, η in these directions, (5) may be written

$$4 \cos(k\xi/\sqrt{2}) \ \cos(k\eta/\sqrt{2}) \ \cos nt; \ \dots\dots\dots\dots\dots(7)$$

in which if we put

$$k\xi/\sqrt{2} = nt + k\xi'/\sqrt{2}$$

(*i.e.* if we suppose the eye to travel with velocity $\sqrt{2}\,.\,n/k$), we get

$$2 \cos(k\eta/\sqrt{2}) \ \cos(k\xi'/\sqrt{2}) + \text{terms in } 2nt.$$

The non-periodic part may be supposed roughly to represent the phenomenon.

In order if possible to settle the question beyond dispute, I made yet another comparison of the frequencies of vibration of the fluid and of the support, using a plan not very different from that originally employed by Faraday. A long plank was supported on trestles at the nodes, and could be tuned within pretty wide limits by shifting weights which rested upon it near the middle and ends. At the centre was placed a beaker $4\frac{1}{4}$ inches in diameter, and containing a little mercury. The plank was set into vibration by properly timed impulses with the hand, and the weights were adjusted until the period corresponded to one mode of free vibration of the pool of mercury. When the adjustment is complete, a very small vibration of the plank throws the mercury into great commotion, and unless the vessel is deep there is risk of the fluid being thrown out. The question now to be decided is whether, or not, the vibrations of the mercury are executed in the same time as those of the plank.

On March 18 the plank was adjusted so as to excite that mode of vibration of the mercury in which there are two nodal diameters. Two other diameters bisecting the angles between these give the places of maximum vertical motion. At one moment the mercury is elevated at *both* ends of one diameter and depressed at both ends of the perpendicular diameter; half a period later the case is reversed. The frequency of the fluid vibrations could be counted by inspection, and was found to be 30 (complete) vibrations in 15 seconds, or exactly two vibrations per second. The vibrations of the plank were counted by allowing it to tap slightly against a pencil held in the hand. In five seconds there were 21 complete vibrations, *i.e.* $4\frac{1}{5}$ vibrations per second, almost exactly twice as many as was found for the mercury. The measurements were repeated several times; and the general result is beyond question.

On another occasion the mode of fluid vibration was that in which there is but one nodal diameter, the fluid being most raised at one end of the perpendicular diameter and most depressed at the other end. The frequency of fluid vibration was $30/22 = 1\cdot36$; while that of the plank was $27/10 = 2\cdot7$. Here again the fluid vibrations are proved to be only half as quick as those of the support.

The mechanics of the question are considered in a communication to the *Philosophical Magazine* for April, 1883*, to which reference must be made. Merely to observe the phenomenon, it is sufficient to take a porcelain evaporating-dish containing a shallow pool of mercury 2 or 3 inches in diameter, and, holding it firmly with both hands, to impose upon it a vertical vibratory motion. After a few trials of various speeds it is possible to excite various modes of vibration, including those referred to in connexion with the plank. The first (with two nodal diameters) is more interesting in itself, and is more certainly due to a vertical as opposed to a horizontal vibration of the support. The gradually shelving bank presented by the dish adds to the beauty of the experiment by its tendency to prevent splashing.

Dr Matthiessen, in the papers referred to, records a long series of measurements of the wave-lengths of crispations corresponding to various frequencies of vibration, not only in the case of water, but also of mercury, alcohol, and other liquids. He remarks that the nature of the liquid affects the relation in a marked manner, contrary to the theoretical ideas of the time, which recognized gravity only as a "motive" for the vibrations. In the following year Sir W. Thomson gave the complete theory of wave-propagation†, in which it is shown that in the case of wave-lengths so short as most of those experimented upon by Matthiessen, the influence of cohesion, or capillary tension, far outweighs that of gravity. In general, if T be the tension, $k = 2\pi/\lambda$, the velocity of propagation (v) is given by

$$v = \sqrt{(Tk + g/k)}; \quad \dots\dots\dots\dots\dots(8)$$

or, when λ is small enough,

$$v = \sqrt{(Tk)}. \quad \dots\dots\dots\dots\dots(9)$$

Since $\lambda = v\tau$, the relation between τ and λ is, by (9),

$$2\pi T\tau^2 = \lambda^3; \quad \dots\dots\dots\dots\dots(10)$$

or, if N be the frequency of vibration,

$$N^{\frac{2}{3}}\lambda = \text{constant}. \quad \dots\dots\dots\dots\dots(11)$$

Dr Matthiessen's results agree pretty well with (11), much better in fact than with the formula proposed by himself.

There is another point of some interest on which the views expressed by Matthiessen call for correction. It was observed by Lissajous some years

* [Art. 97, vol. II. p. 190.]
† *Phil. Mag.* Nov. 1871.

ago, that if two vibrating tuning-forks of slightly different pitch are made to touch the surface of water, the nearly stationary waves formed midway between the sources of disturbance travel slowly towards the graver. We may take as the expression for the two progressive waves

$$\cos(kx - nt) + \cos(k'x + n't),$$

or, which is the same,

$$2\cos\{\tfrac{1}{2}(k + k')x + \tfrac{1}{2}(n' - n)t\} \times \cos\{\tfrac{1}{2}(k' - k)x + \tfrac{1}{2}(n' + n)t\}.$$

The position at any time of the crests of the nearly stationary waves is given by

$$\tfrac{1}{2}(k + k')x + \tfrac{1}{2}(n' - n)t = 2m\pi,$$

where m is an integer. The velocity of displacement V is thus

$$V = \frac{n - n'}{k + k'}; \quad\dots\dots\dots\dots\dots\dots(12)$$

from which it appears that in every case the shifting is in the direction of propagation of waves of higher pitch, or towards the source of graver pitch.

According to Matthiessen, the shifting takes place with a velocity equal to half the difference of velocities of the component trains, i.e.

$$2V = \frac{n}{k} - \frac{n'}{k'}, \dots\dots\dots\dots\dots\dots(13)$$

and in the direction of that component train which moves with greatest velocity. So far as regards the direction merely, the two rules come to the same thing for the range of pitch used by Lissajous and Matthiessen, since over this range the velocity increases with pitch. If, however, we have to deal with waves longer than the critical value (1·7 centim. for water), the two rules are at issue, since now the velocity increases as the pitch diminishes. The following are a few corresponding values, in C.G.S. measure, of wave-length, velocity, and frequency of vibration calculated by Thomson's formula (A).

Wave-length ...	·5	1·0	1·7	2·5	3·0	5·0
Velocity	31·48	24·75	23·11	23·94	24·92	29·54
Frequency	62·97	24·75	13·60	9·579	8·306	5·908

I have examined the matter experimentally with the aid of vibrators making from 12 to 7 complete vibrations per second, and therefore well below the critical point, with the result that the transference is towards the source of graver pitch, although this is the direction of propagation of the component which travels with the smaller velocity. I reserve for the present a more detailed description of the apparatus, as I propose to apply it to the general verification of Thomson's law of velocities.

103.

ON POROUS BODIES IN RELATION TO SOUND.

[*Philosophical Magazine*, XVI. pp. 181—186, 1883.]

IN Acoustics we have sometimes to consider the incidence of aerial waves upon porous bodies, in whose interstices some sort of aerial continuity is preserved. Tyndall has shown that in many cases sound penetrates such bodies, *e.g.* thick pieces of felt, more freely than would have been expected, though it is reflected from quite thin layers of continuous solid matter. On the other hand, a hay-stack seems to form a very perfect obstacle. It is probable that porous walls give a diminished reflection, so that within a building so bounded resonance is less prolonged than would otherwise be the case.

When we inquire into the matter mechanically, it is evident that sound is not destroyed by obstacles as such. In the absence of dissipative forces, what is not transmitted must be reflected. Destruction depends upon viscosity and upon conduction of heat; but the influence of these is enormously augmented by the contact of solid matter exposing a large surface. At such a surface the tangential as well as the normal motion is hindered, and a passage of heat to and fro takes place, as the neighbouring air is heated and cooled during its condensations and rarefactions. With such rapidity of alternations as we are concerned with in the case of audible sounds, these influences extend to only a very thin layer of the air and of the solid, and are thus greatly favoured by attenuation of the masses.

I have thought that it might be interesting to consider a little more definitely a problem sufficiently representative of that of a porous wall, in order to get a better idea of the magnitudes of the effects to be expected. We may conceive an otherwise continuous wall, presenting a flat face, to be perforated by a great number of similar small channels, uniformly distributed, and bounded by surfaces everywhere perpendicular to the face. If the channels be sufficiently numerous, the transition from simple plane waves outside to the state of aerial vibration corresponding to the interior of a channel of infinite length, occupies a space which is small relative to the wave-length of the vibration, and then the connexion between the condition of things inside and outside admits of simple expression.

Considering first the interior of one of the channels, and taking the axis of x parallel to the axis of the channel, we suppose that as functions of x the velocity-components u, v, w, and the condensation s are proportional to e^{ikx}, while as functions of t everything is proportional to e^{int}, n being real. The relationship between k and n depends on the nature of the gas and upon the size and form of the channel, and must be found in each case by a special investigation. Supposing it known for the present, we will go on to show how the problem of reflection is to be dealt with.

For this purpose consider the equation of continuity as integrated over the cross section of the channel σ. Since the walls are impenetrable,

$$\frac{d}{dt}\iint s\,d\sigma + \frac{d}{dx}\iint u\,d\sigma = 0,$$

so that

$$n\iint s\,d\sigma + k\iint u\,d\sigma = 0. \quad\dots\dots\dots\dots\dots\dots(1)$$

This result is applicable at points distant from the open end more than several diameters of the channel.

Taking now the origin of x at the face of the wall, we have to form corresponding expressions for the waves outside; and we may here neglect the effects of friction and heat-conduction. If a be the velocity of sound in the open, and $k_0 = n/a$, we may write

$$s = (+ e^{ik_0 x} + B e^{-ik_0 x})\, e^{int}, \quad\dots\dots\dots\dots(2)$$

$$u = a\,(- e^{ik_0 x} + B e^{-ik_0 x})\, e^{int}; \quad\dots\dots\dots\dots(3)$$

so that the incident wave is

$$s = e^{i\,(nt + k_0 x)}, \quad\dots\dots\dots\dots\dots\dots(4)$$

or, on throwing away the imaginary part,

$$s = \cos(nt + k_0 x). \quad\dots\dots\dots\dots\dots\dots(5)$$

These expressions are applicable when x exceeds a moderate multiple of the distance between the channels. Close up to the face the motion will be more complicated; but we have no need to investigate it in detail. The ratio of u and s at a place near the wall is given with sufficient accuracy by putting $x = 0$ in (2) and (3),

$$\frac{u}{s} = \frac{a\,(-1 + B)}{1 + B}. \quad\dots\dots\dots\dots\dots\dots(6)$$

We now assume that a region about $x = 0$, on one side of which (6) is applicable and on the other side of which (1) is applicable, may be taken so small relatively to the wave-length that the mean pressures are sensibly the same at the two boundaries, and that the flow into the region at the one boundary is sensibly equal to the flow out of the region at the other boundary. The equality of flow does not imply an equality of mean velocities, since the areas concerned are different. The mean velocities will be inversely proportional to the corresponding areas—that is, in the ratio $\sigma : \sigma + \sigma'$, if σ' denote the area of the unperforated part of the wall corresponding to each channel. By (1) and (6) the connexion between the inside and outside motion is expressed by

$$- \frac{n}{k}\sigma = \frac{(B-1)a}{B+1}(\sigma + \sigma').$$

We will denote the ratio of the unperforated to the perforated parts of the wall by g, so that $g = \sigma'/\sigma$. Thus,

$$\frac{1-B}{1+B} = \frac{k_0}{k(1+g)}. \quad\ldots\ldots\ldots\ldots\ldots\ldots\ldots\ldots(7)$$

If $g = 0$, $k = k_0$, there is no reflection; if there are no perforations, $g = \infty$, and then $B = 1$, signifying a complete reflection. In place of (7) we may write

$$B = \frac{k(1+g)-k_0}{k(1+g)+k_0}, \quad\ldots\ldots\ldots\ldots\ldots\ldots\ldots\ldots(8)$$

which is the solution of the problem proposed. It is understood that waves which have once entered the wall do not return. When dissipative forces act, this condition may always be satisfied by supposing the channels long enough. The necessary length of channel, or thickness of wall, will depend upon the properties of the gas and upon the size and shape of the channels.

Even in the absence of dissipative forces there must be reflection, except in the extreme case $g = 0$. Putting $k = k_0$ in (8), we have

$$B = \frac{g}{2+g}. \quad\ldots\ldots\ldots\ldots\ldots\ldots\ldots\ldots(9)$$

If $g = 1$ (that is, if half the wall be cut away), $B = \frac{1}{3}$, $B^2 = \frac{1}{9}$, so that the reflection is but small. If the channels be circular, and arranged in square order as close as possible to each other, $g = (4-\pi)/\pi$, whence $B = \cdot121$, $B^2 = \cdot015$, nearly all the motion being transmitted.

It remains to consider the value of k. The problem of the propagation of sound in a circular tube, having regard to the influence of viscosity and heat-conduction, has been solved analytically by Kirchhoff*, on the suppo-

---

* *Pogg. Ann.* cxxxiv. 1868.

sitions that the tangential velocity and the temperature-variation vanish at the walls. In discussing the solution, Kirchhoff takes the case in which the dimensions of the tube are such that the immediate effects of the dissipative forces are confined to a relatively thin stratum in the neighbourhood of the walls. In the present application interest attaches rather to the opposite extreme, viz. when the diameter is so small that the frictional layer pretty well. fills the tube. Nothing practically is lost by another simplification which it is convenient to make (following Kirchhoff)— that the velocity of propagation of viscous and thermal effects is negligible in comparison with that of sound.

One result of the investigation may be foreseen. When the diameter of the tube is very small, the conduction of heat from the centre to the circumference of the column of air becomes more and more free. In the limit the temperature of the solid walls controls that of the included gas, and the expansions and rarefactions take place isothermally. Under these circumstances there is no dissipation due to conduction, and everything is the same as if no heat were developed at all. Consequently the coefficient of heat-conduction will not appear in the result, which will involve, moreover, the Newtonian value of the velocity of sound (b) and not that of Laplace (a).

Starting from Kirchhoff's formulæ, we find as the value of k^2 applicable when the diameter ($2r$) is very small,

$$k^2 = -\frac{8in\mu'}{b^2r^2}, \quad \dots\dots\dots\dots\dots\dots\dots\dots(10)$$

μ' being the kinematic coefficient of viscosity. The wave propagated into the channels is thus proportional to

$$e^{px}\cos(nt + px + \epsilon), \quad \dots\dots\dots\dots\dots\dots\dots(11)$$

where

$$p = \frac{k}{1-i} = \frac{2\sqrt{(n\mu')}}{br} = \frac{2\sqrt{(n\gamma\mu')}}{ar}, \quad \dots\dots\dots\dots(12)$$

γ being the ratio of the specific heats, equal to 1·41. In the derivation of (10), $nr^2/(8\nu)$, ν being the thermometric coefficient of conductivity, is assumed to be small.

To take a numerical example, suppose that the pitch is 256 (middle c of the scale), so that $n = 2\pi \times 256$. The value of μ' for air is ·16 C.G.S. (Maxwell), and that of ν is ·256. If we take $r = \frac{1}{1000}$ centim., we find $nr^2/8\nu$ equal to about $\frac{1}{1000}$. If r were 10 times as great, the approximation would perhaps still be sufficient.

From (12), if $n = 2\pi \times 256$,

$$p = \frac{1 \cdot 15 \times 10^{-3}}{r};\dots\dots\dots\dots\dots\dots(13)$$

so that if $r = \frac{1}{1000}$, $p = 1 \cdot 15$. In this case the amplitude is reduced in ratio $e : 1$ in passing over the distance p^{-1}—that is, about one centimetre. The distance penetrated is proportional to the radius of the channel.

The amplitude of the reflected wave is, by (8),

$$B = \frac{p(1+g)(1-i) - k_0}{p(1+g)(1-i) + k_0},$$

or, as we may write it,

$$B = \frac{p'(1-i)-1}{p'(1-i)+1} = \frac{p'-1-ip'}{p'+1-ip'}, \quad\dots\dots\dots\dots(14)$$

where

$$p' = (1+g)\, p/k_0. \dots\dots\dots\dots\dots\dots\dots(15)$$

If I be the intensity of the reflected sound, that of the incident sound being unity,

$$I = \frac{2p'^2 - 2p' + 1}{2p'^2 + 2p' + 1}. \quad\dots\dots\dots\dots\dots(16)$$

The intensity of the intromitted sound is given by

$$I' = 1 - I = \frac{4p'}{2p'^2 + 2p' + 1}. \quad\dots\dots\dots\dots(17)$$

By (12), (15),

$$p' = \frac{2(1+g)\sqrt{(\mu'\gamma)}}{r\sqrt{n}}. \quad\dots\dots\dots\dots\dots(18)$$

If we suppose $r = \frac{1}{1000}$ centim., and $g = 1$, we shall have a wall of pretty close texture. In this case, by (18), $p' = 47 \cdot 4$, and $I' = \cdot 0412$. A four-per-cent. loss may not appear to be much; but we must remember that in prolonged resonance we are concerned with the accumulated effects of a large number of reflections, so that rather a small loss in a single reflection may well be material. The thickness of the porous layer necessary to produce this effect is less than one centimetre.

Again, suppose $r = \frac{1}{100}$ centim., $g = 1$. We find $p' = 4 \cdot 74$, $I' = \cdot 342$, and the necessary thickness would be less than 10 centimetres.

If r be much greater than $\frac{1}{100}$ centim., the exchange of heat between the air and the walls of the channels is no longer sufficiently free for the expansions to be treated as isothermal. When r is so great that the thermal and viscous effects extend only through a small fraction of it, we have the case discussed by Kirchhoff. If we suppose for simplicity $g = 0$

(a state of things, it is true, not strictly consistent with channels of circular section*), we have

$$I = \frac{\gamma'^2}{4nr^2}, \dots\dots\dots\dots\dots\dots\dots\dots(19)$$

in which

$$\gamma' = \sqrt{\mu'} + \left(\frac{a}{b} - \frac{b}{a}\right)\sqrt{\nu}. \dots\dots\dots\dots(20)$$

The incident sound is absorbed more and more completely as the diameter of the channels increases; but at the same time a greater thickness becomes necessary in order to prevent a return from the further side. If $g = 0$, there is no theoretical limit to the absorption; and, as we have seen, a moderate value of g does not by itself entail more than a comparatively small reflection. A loosely compacted hay- or straw-stack would seem to be as effective an absorbent of sound as anything likely to be met with.

In large spaces bounded by non-porous walls, roof, and floor, and with few windows, a prolonged resonance seems inevitable. The mitigating influence of thick carpets in such cases is well known. The application of similar material to the walls, or to the roof, appears to offer the best chance of further improvement.

* The problem in two dimensions is somewhat simpler than that treated by Kirchhoff. Although it would allow us without violence to suppose $g = 0$, it seems scarcely worth while to enter upon it here, as the results are of precisely the same character. The principal difference is that the hyperbolic functions cosh &c. replace that of Bessel. [1900. The analysis for this case, as well as those treated by Kirchhoff, will be found in *Theory of Sound*, vol. II. §§ 348—351.]

104.

SUGGESTIONS FOR FACILITATING THE USE OF A DELICATE BALANCE.

[*British Association Report*, pp. 401, 402, 1883.]

In some experiments with which I have lately been occupied a coil of insulated wire, traversed by an electric current, was suspended in the balance, and it was a matter of necessity to be able quickly to check the oscillation of the beam, so as to bring the coil into a standard position corresponding to the zero of the pointer. A very simple addition to the apparatus allowed this to be done. The current from a Leclanché cell is led into an auxiliary coil of wire, coaxal with the other, and is controlled by a key. When the contact is made, a vertical force acts upon the suspended coil, but ceases as soon as the contact is broken. After a little practice the beam may be brought to rest at zero at the first or second application of the retarding force.

This control over the oscillations has been found so convenient that I have applied a similar contrivance in the case of ordinary weighings, and my object in the present note is to induce chemists and others experienced in such operations to give it a trial. Two magnets of steel wire, three or four inches long, are attached vertically to the scale-pans, and underneath one of them is fixed a coil of insulated wire of perhaps 50 or 100 turns, and of 4 or 5 inches in diameter. The best place for the coil is immediately underneath the bottom of the balance-case. It is then pretty near the lower pole of the magnet, and is yet out of the way. The circuit is completed through a Leclanché cell and a common spring contact-key, placed in any convenient position. The only precaution required is not to bring other magnets into the neighbourhood of the balance, or at any rate not to move them during a set of weighings.

The other point as to which I wish to make a suggestion relates to the time of vibration of the beam. I think that, with the view of obtaining a high degree of sensitiveness, the vibrations are often made too slow. Now the limit of accuracy depends more upon the smallness of the force which can be relied upon to displace the beam in a definite manner than upon the magnitude of the displacement so produced. As in other instruments whose operation depends upon similar principles, *e.g.* galvanometers, it is useless to endeavour to increase the sensitiveness by too near an approach to instability, because the effect of casual disturbances is augmented in the same proportion as that of the forces to be estimated. If the time of vibration be halved, the displacement due to a small excess of weight is indeed reduced in the ratio of four to one, but it is not necessarily rendered any more uncertain. The mere diminution in the amount of displacement may be compensated by lengthening the pointer, or by optical magnification of its motions. By the method of mirror-reading such magnification may be pushed to almost any extent, but I am dealing at present only with an arrangement adapted for ordinary use.

In the balance (by Oertling) that I am now using, the scale-divisions are finer than usual, and the motion of the pointer is magnified four or five times without the slightest inconvenience by a lens fixed in the proper position. The pointer being in the same plane as the scale-divisions, there is no sensible parallax. In this way the advantage of quick vibrations is combined with easy visibility of the motion due to the smallest weights appreciable by the balance.

To illuminate the scale the image of a small and distant gas-flame is thrown upon it by means of a large plate-glass lens. This artificial illumination is found to be very convenient, as the instrument stands at some distance from a window, but it is not at all called for in consequence of the use of the magnifying lens.

105.

ON THE IMPERFECTION OF THE GALVANOMETER AS A TEST OF THE EVANESCENCE OF A TRANSIENT CURRENT.

[*British Association Report*, pp. 444, 445, 1883.]

In certain electrical measurements a galvanometer is used to indicate whether or not the integral value of a current of short duration is zero. For example, in the method given in Maxwell's *Electricity*, § 755, for comparing the coefficients of mutual induction, M, of two pairs of coils, the evanescence of the integral current through the galvanometer is made the test of the fulfilment of a certain relation between the coefficients of induction and the resistances. The two primary coils are joined up in simple circuit with a battery. The two secondaries are also connected together in such a way that the inductive electro-motive forces conspire, and two points P, Q, one on each connector, are brought into contact with the galvanometer terminals. In special cases, as for instance when the two pairs of coils are similar, there is no current through the galvanometer, whatever may happen in the primary circuit; but in general the establishment or interruption of the primary current will cause a deflection of the galvanometer indicative of the integral value of the current passing. The method consists in adding inductionless resistance coils to one or other of the secondaries until this current vanishes.

The required conditions are most readily obtained by supposing the galvanometer circuit broken, and inquiring into the value of the electro-motive force E between the points P and Q. The same current y flows in both secondaries, and if x be the primary current, the equations are :—

$$N_1 \frac{dy}{dt} + M_1 \frac{dx}{dt} + Ry = E, \qquad N_2 \frac{dy}{dt} + M_2 \frac{dx}{dt} + Sy = -E.$$

M_1, M_2, are the induction coefficients to be compared; R, S, the resistances of the two secondaries (with associated resistance coils); N_1, N_2, their coefficients of self-induction. Thus—

$$(M_1 + M_2) E = (M_2 N_1 - M_1 N_2) \frac{dy}{dt} + (M_2 R - M_1 S) y.$$

Since y begins from 0 and ends at 0, the integral electro-motive force vanishes if

$$M_2 R - M_1 S = 0.$$

If this condition is satisfied, there is no integral current through the galvanometer, and then the ratio of induction coefficients is known by the ratio of resistances.

In general, however, the evanescence of the integral current is obtained by the opposition of consecutive positive and negative parts, and even although the whole duration of the effect be but a small fraction of the time of vibration, the needle of the galvanometer will be disturbed in such a manner as to make it difficult to say whether or not the whole impulse acting upon it be zero. To obtain a satisfactory measurement it is necessary to secure at least an approximate fulfilment of the second condition required in order that the current may be zero throughout, viz.—

$$M_2 N_1 - M_1 N_2 = 0.$$

In this there is no difficulty, as we can easily increase the defective self-induction by the addition of other coils, placed at a sufficient distance. The most convenient plan is to include two coils by the variation of the relative situation of which the self-induction can be adjusted. With moderate care the initial impulsive electro-motive force, caused by a sudden variation of the primary current, and dependent only upon the induction coefficients, may be made so small that the needle shows no uneasiness when the other adjustment relative to the resistances is complete.

In March 1881 I attempted, in conjunction with Messrs Glazebrook and Dodds, to carry out the plan above suggested for the comparison of two coefficients of mutual induction. No satisfactory result could be obtained in the ordinary method of working, the needle showing uneasiness whatever resistances were employed, so that it was impossible to fix upon any particular value as corresponding to a zero integral current. The addition of other coils to increase the self-induction of one of the secondaries was so far successful that the needle could be reduced to quietness, but calculation showed that the additional self-induction found to be necessary in experiment was much in excess of what the above theory would indicate. The explanation which afterwards suggested itself to me was that the anomalous effect was due to the conducting rings upon which some of the coils were wound, and whose presence complicates the otherwise simple theory. We verified this view by bringing a coil of wire into the neighbourhood of one of the principal coils, the behaviour of the galvanometer needle being very sensibly different according as the auxiliary coil was open or closed.

The kind of embarrassment to which measurements of this kind are subject is well illustrated by placing the galvanometer in a tertiary circuit, not directly influenced at all by the battery current in the primary. A pair of coils with double wires, such as are often used for large electro-magnets, is suitable for the experiment. One wire of the first coil is connected with the battery, and forms the primary circuit. The second wire of the first coil and the first wire of the second coil are connected, and constitute together the secondary circuit. The second wire of the second coil and the galvanometer form the tertiary circuit. The apparatus must be so adjusted that no effect is perceived at the galvanometer when the secondary is broken, whatever may happen in the primary. When this adjustment is complete the secondary is closed, and the effect is observed of opening or closing the primary. If the contacts are properly made, the integral current through the galvanometer at each operation is rigorously zero, but in the experiments that I have made no one could infer the fact from the behaviour of the galvanometer needle. The effect may be exaggerated by the insertion of a few iron wires into the induction coils.

106.

ON LAPLACE'S THEORY OF CAPILLARITY.

[*Philosophical Magazine*, XVI. pp. 309—315, 1883.]

FROM the hypothesis of forces sensible only at insensible distances Laplace *, it is well known, arrived at the conclusion that the pressure within a sphere of liquid of radius b may be expressed by

$$K + \frac{H}{b} . \quad\dots\dots\dots\dots\dots\dots\dots\dots\dots\dots\dots(1)$$

H is the constant on which capillary phenomena depend, and the effect of the second term may be represented by the fiction of a constant tension in the superficial layer. According to Laplace's theory, however, the first term K is enormously the greater; only, being the same at all points in the interior of the fluid, whatever may be the form of the boundary, it necessarily escapes direct observation.

When two liquids are in contact the difference of pressures within them will still be of the form (1), but the values of K and H will depend upon the properties of both kinds of matter.

The existence of an intense molecular pressure K is a necessary part of Laplace's, and probably of any similar, theory of these phenomena; but it has not met with universal acceptance†. The difficulty which has been felt appears to depend upon an omission in the theory as hitherto presented. Before we can speak of K as a molecular pressure proper to the liquid, it is necessary to show that the change, which we may denote by K_{13}, experienced in passing the surface dividing liquid I. from liquid III. is identical with the sum of the changes denoted by K_{12} and K_{23}; so that it makes no difference whether we pass from I. to III. directly or by way of II. That this should be the case upon Laplace's principles will be shown further on. The point,

---

* *Mécanique Céleste*, Supplement to Tenth Book.
† Quincke, *Pogg. Ann.* 1870. Also Riley, *Phil. Mag.* March 1883.

however, is so important that I propose to give in addition a proof of much wider generality, by which the relation is placed upon a sound basis. The existence of an intense internal pressure is probable for many reasons; and it is hoped that no further difficulty need be felt in admitting it as a legitimate hypothesis.

Let us imagine different kinds of liquids, varying continuously or discontinuously, to be arranged in plane strata, and let us examine the difference of pressure, due to the attracting forces, at two points A and B, round each of which the fluid is uniform to a distance exceeding the range of the forces. The difference of pressure in crossing any infinitely thin stratum at P is due to the forces operative between P and all the other strata. The force between one of the interior strata Q and P will depend upon the thicknesses of the strata, upon the nature and condition of the fluids composing them, and upon the distance PQ. But whatever may be the law of the action in these respects, the force exerted by Q upon P must be absolutely the same as the force exerted by P upon Q. Now, as we pass downwards from A to B, every pair of elements between A and B comes into consideration twice. In passing through P we find an increase of pressure due to the action of Q upon P, but in passing through Q we have an equal diminution of pressure due to the action of P upon Q. Along the whole path from A to B the only elements which can contribute to a final difference of pressure are those which lie outside, i.e. in the fluid above A and below B. By hypothesis the action of the fluid above A on the strata traversed in going towards B ceases within the limits of the uniform fluid about A; and consequently the whole difference of pressure due, according to this way of treating the matter, to the fluid above A depends only upon the properties of A. In like manner the difference due to the fluid below B depends only upon the properties of B; and we conclude that the whole difference of pressure due to the action of the forces along the path AB depends upon the properties of the fluids at A and B, and not upon the manner in which the transition between the two is made. In particular the difference is the same whether we pass direct from one to the other, or through an intermediate fluid of any properties whatsoever.

It is evident that the enormous pressure which Laplace's theory indicates as prevalent in the interior of liquids cannot be submitted to any direct test. Capillary observations can neither prove nor disprove it. But it seems to have been thought that the relation

$$K_{13} = K_{12} + K_{23} \qquad \dots\dots\dots\dots\dots\dots\dots(2)$$

implies a corresponding relation between the capillary constants

$$H_{13} = H_{12} + H_{23}; \qquad \dots\dots\dots\dots\dots\dots\dots(3)$$

and the fact that (3) is inconsistent with observation is supposed to throw doubts upon (2). Indeed Mr Riley*, in his interesting remarks upon Capillary Phenomena, goes the length of asserting that, according to Laplace, K is a function of H. It is thus important to show that Laplace's principles, even in their most restricted form, are consistent with the violation of (3).

In attempting calculations of this kind we must make some assumption as to the forces in operation when more than one kind of fluid is concerned. The simplest supposition is that the law of force between any two elements is always the same, $\phi(r)$, as a function of the distance, and that the difference between one fluid and another shows itself only in the intensity of the action. The coefficient proper to each fluid may be called the "density," without meaning to imply that it has any relation to inertia or weight. The force between two elements (of unit volume) of fluids I. and II. may thus be denoted by $\rho_1\rho_2\,\phi(r)$; that between two elements of the same fluid by $\rho_1^2\,\phi(r)$, or $\rho_2^2\,\phi(r)$, as the case may be.

We will first examine the forces operative in a fluid whose density varies slowly, that is to say undergoes only a small change in distances of the order of the range of the forces, supposing, for simplicity, that the strata are surfaces of revolution round the axis of z. The first step will be to form an expression for the force at any point O on the axis.

The direction of this force is evidently along z, and its magnitude depends upon the variation of density in the neighbourhood of O. If the density were constant, there would be no force. We may write

$$\delta\rho = \frac{d\rho}{dz}\,z + \frac{d^2\rho}{dz^2}\frac{z^2}{2} + \frac{d^2\rho}{dx^2}\frac{x^2}{2} + \dots,$$

or in polar coordinates,

$$\delta\rho = \frac{d\rho}{dz}\,r\cos\theta + \frac{d^2\rho}{dz^2}\frac{r^2\cos^2\theta}{2} + \frac{d^2\rho}{dx^2}\frac{r^2\sin^2\theta}{2} + \text{terms in } r^3. \dots\dots(4)$$

For the attraction of the shell of radius r and thickness dr we have

$$2\pi r^2\,dr\,\phi(r)\!\int_0^\pi \delta\rho\,\cos\theta\,\sin\theta\,d\theta = \frac{4\pi}{3}\,r^3\,\phi(r)\,dr\,\frac{d\rho}{dz} + \dots;$$

and for the complete attraction,

$$\frac{4\pi}{3}\frac{d\rho}{dz}\int_0^\infty r^3\,\phi(r)\,dr + \text{terms in }\int_0^\infty r^5\,\phi(r)\,dr.$$

The difference of pressure corresponding to a displacement dz is found by multiplying this by $\rho\,dz$. Thus

$$dp = \frac{2\pi}{3}\frac{d\rho^2}{dz}\int_0^\infty r^3\,\phi(r)\,dr + \dots,$$

* *Loc. cit.* p. 193.

and

$$p_1 - p_2 = \frac{2\pi}{3}\,(\rho_1{}^2 - \rho_2{}^2)\int_0^\infty r^3\,\phi\,(r)\,dr + \text{terms in} \int_0^\infty r^5\,\phi\,(r)\,dr. \ldots\ldots(5)$$

Laplace employs a function ψ, such that

$$\tfrac{1}{3}\int_0^\infty r^3\,\phi\,(r)\,dr = \int_0^\infty \psi\,(r)\,dr\,; \ \ldots\ldots\ldots\ldots\ldots\ldots(6)$$

and he finds that in the case of a uniform fluid in contact with air the principal term, K, depends upon $\int \psi\,(r)\,dr$, and the second, H, upon $\int r\,\psi\,(r)\,dr$. For the continuously varying fluid here considered, we see from (5) that

$$p_1 - p_2 = 2\pi\,(\rho_1{}^2 - \rho_2{}^2)\int_0^\infty \psi\,(r)\,dr, \ \ldots\ldots\ldots\ldots\ldots(7)$$

and that there is no term of the order of the capillary force. Equation (7) agrees with our general result that the difference of pressures required to equilibrate the forces operating between two points depends only upon the nature of the fluid at the final points; and it shows further that, under the more special suppositions upon which the present calculation proceeds, the molecular pressure at any point is to be regarded as proportional to the square of the density.

But what is more particularly to be noticed is that, in spite of the curvature of the strata, there is no variation of pressure of the nature of the capillary force; from which we may infer that the existence of a capillary force is connected with suddenness of transition from one medium to another, and that it may disappear altogether when the transition is sufficiently gradual.

For the further elucidation of this question we will now consider the problem of an abrupt transition. It does not appear that Laplace has anywhere investigated the forces operative at the common surface of two fluids of finite density, but the results given by him for a single fluid are easily extended.

Let OA (equal to a) be the radius of a spherical mass of liquid of "density" ρ_2, surrounded by an indefinite quantity of other fluid of density ρ_1, and let us consider the variation of pressure along a line from a point (say O) removed from the surface on one side to a point B also removed from the surface on the other side.

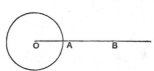

The difference of pressure corresponding to each element of the path OB is found by multiplying the length of the element by the local density of the fluid and by the resultant attraction at the point.

The attraction of the whole mass of fluid may be regarded as due to an uninterrupted mass of infinite extent of density ρ_1, and to a spherical mass

OA of density $(\rho_2 - \rho_1)$. Since the first part can produce no effect at any part of OB, we have to deal merely with the attraction of the sphere OA.

Laplace has shown that if OA were of unit density, its action along the line OA would be

$$K + \frac{H}{a},$$

where

$$K = 2\pi \int_0^\infty \psi(r)\, dr, \qquad H = 2\pi \int_0^\infty r\, \psi(r)\, dr; \quad\ldots\ldots\ldots(8)$$

while along AB its action would be

$$K - \frac{H}{a}.$$

The loss of pressure in going outwards from O to A is thus

$$(\rho_2 - \rho_1)\rho_1 \left(K + \frac{H}{a}\right);$$

and from A to B,

$$(\rho_2 - \rho_1)\rho_2 \left(K - \frac{H}{a}\right)$$

Accordingly the whole difference of pressure between O and B is

$$K(\rho_2^2 - \rho_1^2) + \frac{H}{a}(\rho_2 - \rho_1)^2. \quad\ldots\ldots\ldots\ldots\ldots\ldots(9)$$

Thus, in addition to the former result that the difference of pressure independent of curvature varies as $(\rho_2^2 - \rho_1^2)$, we see that the capillary pressure, proportional to the curvature, varies as $(\rho_2 - \rho_1)^2$.

The reasoning just given is in fact little more than an expansion of that of Young*. If the effect depends only upon the difference of densities, it cannot fail to be proportional to $(\rho_2 - \rho_1)^2$.

Writing $H_{12} = H(\rho_1 - \rho_2)^2$, we see that there is no reason whatever for supposing that the capillary constants of three liquids should be subject to the relation

$$H_{13} = H_{12} + H_{23}.$$

On the contrary, the relation to be expected, if the suppositions at the basis of the present calculations agree with reality, is

$$\sqrt{H_{13}} = \sqrt{H_{12}} + \sqrt{H_{23}}. \quad\ldots\ldots\ldots\ldots\ldots\ldots(10)$$

In (10) the three radicals are supposed to be positive, and H_{13} is the greatest.

If we suppose that the third fluid is air, and put $\rho_3 = 0$, we have

$$\sqrt{H_{12}} = \sqrt{H_1} - \sqrt{H_2}, \quad\ldots\ldots\ldots\ldots\ldots\ldots(11)$$

* *Encyc. Brit.* 1816. Young's *Works*, vol. I. p. 463.

in which $H_1 > H_2$.   From (11)

$$H_{12} = H_1 - H_2 - 2\sqrt{H_1}\sqrt{H_2},$$

so that

$$H_{12} < (H_1 - H_2). \dots\dots\dots\dots\dots\dots\dots\dots(12)$$

The reason why the capillary force should disappear when the transition between two liquids is sufficiently gradual will now be evident. Suppose that the transition from 0 to ρ is made in two equal steps, the thickness of the intermediate layer of density $\frac{1}{2}\rho$ being large compared to the range of the molecular forces, but small in comparison with the radius of curvature. At each step the difference of capillary pressure is only $\frac{1}{4}$ of that due to the sudden transition from 0 to ρ, and thus altogether half the effect is lost by the interposition of the layer. If there were three equal steps, the effect would be reduced to one-third, and so on. When the number of steps is infinite, the capillary pressure disappears altogether.

Although the relation (12) is given by Quincke* as the result of experiment, the numerical values found by him do not agree with (11). In most cases the tension at the common surface of two liquids exceeds that calculated from the separate tensions in contact with air. This result, which must be considered to disprove the applicability of our special hypotheses, need not much surprise us. There was really no ground for the assumption that the law of force is always the same with the exception of a constant multiplier. The action of one fluid upon another might follow an altogether different law from its action upon itself. Besides this we are not entitled to assume that a fluid retains its properties close to the surface of contact with another fluid. Even if the hypothesis, which would refer everything to a difference of "densities," were correct, its application would be rendered uncertain by any modifications which the contiguous layers of different liquids might impose upon one another. As we have seen, if this modification were of the nature of making the transition less abrupt, the capillary forces would be thereby diminished.

[1899.   Reference may be made to further papers "On the Theory of Surface Forces," *Phil. Mag.* XXX. pp. 285, 456, 1890; XXXIII. p. 209, 1892; XXXIII. p. 468, 1892.]

---

* *Loc. cit.* pp. 27, 87.

107.

ON THE MEASUREMENT OF ELECTRIC CURRENTS.

[*Proceedings of the Cambridge Philosophical Society*, v. pp. 50—52, 1883.]

PERHAPS the simplest way of measuring a current of moderate intensity, when once the electro-chemical equivalent of silver is known, is to determine the quantity of metal thrown down by the current in a given time in a silver voltameter. According to Kohlrausch the electro-chemical equivalent of silver is in C.G.S. measure $1\cdot136 \times 10^{-2}$, and according to Mascart $1\cdot124 \times 10^{-2}$. Experiments conducted in the Cavendish Laboratory during the past year by a method of current weighing described in the British Association Report for 1882* have led to a lower number, viz. $1\cdot119 \times 10^{-2}$. At this rate the silver deposited per ampere per *hour* is $4\cdot028$ grams, and the method of measurement founded upon this number may be used with good effect when the strength of the current ranges from $\frac{1}{20}$ ampere to perhaps 4 amperes. It requires however a pretty good balance, and some experience in chemical manipulation. [See Art. 112.]

Another method which gives good results and requires only apparatus familiar to the electrician, depends upon the use of a standard galvanic cell. The current from this cell is passed through a high resistance, such as 10,000 ohms, and a known fraction of the electro-motive force is taken by touching this circuit at definite points. The current to be measured is caused to flow along a strip of sheet German silver, from which two tongues project. The difference of potential at these tongues is the product of the resistance included between them and of the current to be measured, and it is balanced by a fraction of the known electro-motive force of the standard cell (fig. 1). With a sensitive galvanometer the balance may be adjusted to about $\frac{1}{4000}$. The German silver strip must be large enough to avoid heating. The resistance between the tongues may be $\frac{1}{200}$ ohm, and may be determined by a method similar to that of Matthiessen and Hockin (Maxwell's *Electricity*,

* [Art. 88, vol. II. p. 126.]

§ 352).   The proportions above mentioned are suitable for the measurement of such currents as 10 amperes.

Fig. 1.

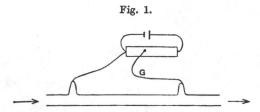

Another method, available with the strong currents which are now common, depends upon Faraday's discovery of the rotation of the plane of polarization by magnetic force.   Gordon found 15°* as the rotation due to the reversal of a current of 4 amperes circulating about 1000 times round a column of bisulphide of carbon.   With heavy glass, which is more convenient in ordinary use, the rotation is somewhat greater.   With a coil of 100 windings we should obtain 15° with a current of 40 amperes; and this rotation may easily be tripled by causing the light to traverse the column three times, or what is desirable with so strong a current, the thickness of the wire may be increased and the number of windings reduced. With the best optical arrangements the rotation can be determined to one or two minutes, but in an instrument intended for practical use such a degree of delicacy is not available.   One difficulty arises from the depolarizing properties of most specimens of heavy glass.   Arrangements are in progress for a redetermination of the rotation in bisulphide of carbon. [See Art. 118.]

---

* [Jan. 1884. In a note recently communicated to the Royal Society (*Proceedings*, Nov. 15, 1883) Mr Gordon points out that, owing to an error in reduction, the number given by him for the value of Verdet's constant is twice as great as it should be. The rotations above mentioned must therefore be halved, a correction which diminishes materially the prospect of constructing a useful instrument upon this principle.]

108.

ON THE CIRCULATION OF AIR OBSERVED IN KUNDT'S TUBES, AND ON SOME ALLIED ACOUSTICAL PROBLEMS.

[*Philosophical Transactions*, CLXXV. pp. 1—21, 1883.]

EXPERIMENTERS in Acoustics have discovered more than one set of phenomena, apparently depending for their explanation upon the existence of regular currents of air resulting from vibratory motion, of which theory has as yet rendered no account. This is not, perhaps, a matter for surprise, when we consider that such currents, involving as they do *circulation* of the fluid, could not arise in the absence of friction, however great the extent of vibration. And even when we are prepared to include in our investigations the influence of friction, by which the motion of fluid in the neighbourhood of solid bodies may be greatly modified, we have no chance of reaching an explanation, if, as is usual, we limit ourselves to the supposition of infinitely small motion and neglect the squares and higher powers of the mathematical symbols by which it is expressed.

In the present paper three problems of this kind are considered, two of which are illustrative of phenomena observed by Faraday*. In these problems the fluid may be treated as incompressible. The more important of them relates to the currents generated over a vibrating plate, arranged as in Chladni's experiments. It was discovered by Savart that very fine powder does not collect itself at the nodal lines, as does sand in the production of Chladni's figures, but gathers itself into a cloud which, after hovering for a time, settles itself over the places of maximum vibration. This was traced by Faraday to the action of currents of air, rising from the plate at the places of maximum vibration, and falling back to it at the nodes. In a

---

* "On a Peculiar Class of Acoustical Figures; and on certain Forms assumed by groups of particles upon Vibrating Elastic Surfaces," *Phil. Trans.* 1831, p. 299.

vacuum the phenomena observed by Savart do not take place, all kinds of powder collecting at the nodes. In the investigation of this, as of the other problems, the motion is supposed to take place in two dimensions.

It is probable that the colour phenomena observed by Sedley Taylor* on liquid films under the action of sonorous vibrations are to be referred to the operation of the aerial vortices here investigated. In a memoir on the colours of the soap-bubble†, Brewster has described the peculiar arrangements of colour, accompanied by whirling motions, caused by the impact of a gentle current of air. In Mr Taylor's experiments the film probably divides itself into vibrating sections, associated with which will be aerial vortices reacting laterally upon the film.

The third problem relates to the air currents observed by Dvorak in a Kundt's tube, to which is apparently due the formation of the dust figures. In this case we are obliged to take into account the compressibility of the fluid.

My best thanks are due to Mr W. M. Hicks, who has been good enough to examine the mathematical work of the paper. The results are thus put forward with greater confidence than I could otherwise have felt.

§ 1.   In the usual notation the equations of motion in two dimensions are

$$\left.\begin{aligned}\frac{1}{\rho}\frac{dp}{dx} &= -\frac{du}{dt} + \nu \nabla^2 u - u\frac{du}{dx} - v\frac{du}{dy}\\ \frac{1}{\rho}\frac{dp}{dx} &= -\frac{dv}{dt} + \nu \nabla^2 v - u\frac{dv}{dx} - v\frac{dv}{dy}\end{aligned}\right\}, \dots\dots\dots\dots(1)$$

and since the fluid is incompressible,

$$\frac{du}{dx} + \frac{dv}{dy} = 0. \quad \dots\dots\dots\dots\dots\dots(2)$$

In virtue of (2) we may write

$$u = d\psi/dy, \qquad v = -d\psi/dx. \quad \dots\dots\dots\dots(3)$$

Eliminating p between equations (1), we get

$$\nu \nabla^2 \left(\frac{du}{dy} - \frac{dv}{dx}\right) - \frac{d}{dt}\left(\frac{du}{dy} - \frac{dv}{dx}\right) = \frac{d}{dy}\left(u\frac{du}{dx} + v\frac{du}{dy}\right) - \frac{d}{dx}\left(u\frac{dv}{dx} + v\frac{dv}{dy}\right).$$

Now

$$u\frac{du}{dx} + v\frac{du}{dy} = \tfrac{1}{2}\frac{d(u^2+v^2)}{dx} + v\left(\frac{du}{dy} - \frac{dv}{dx}\right),$$

$$u\frac{dv}{dx} + v\frac{dv}{dy} = \tfrac{1}{2}\frac{d(u^2+v^2)}{dy} - u\left(\frac{du}{dy} - \frac{dv}{dx}\right),$$

* Proc. Roy. Soc. 1878.
* Proc. Roy. Soc. 1878.
† Edinburgh Transactions, 1866—67.

and

$$\frac{du}{dx} - \frac{dv}{dx} = \nabla^2 \psi,$$

so that

$$\nabla^4 \psi - \frac{1}{\nu} \frac{d}{dt} \nabla^2 \psi = \frac{u}{\nu} \frac{d\nabla^2 \psi}{dx} + \frac{v}{\nu} \frac{d\nabla^2 \psi}{dy} \quad \dots\dots\dots\dots\dots(4)$$

For the first approximation we neglect the right-hand member of (4), as being of the second order in the velocities, and take simply

$$\nabla^2 \left(\nabla^2 - \frac{1}{\nu} \frac{d}{dt} \right) \psi = 0. \quad \dots\dots\dots\dots\dots(5)$$

The solution of (5) may be written*

$$\psi = \psi_1 + \psi_2, \quad \dots\dots\dots\dots\dots(6)$$

where

$$\nabla^2 \psi_1 = 0, \qquad \left(\nabla^2 - \frac{1}{\nu} \frac{d}{dt} \right) \psi_2 = 0. \quad \dots\dots\dots\dots(7)$$

We will now introduce the suppositions that the motion is periodic with respect to x, and also (to a first approximation) with respect to t. We thus assume that ψ_1 and ψ_2 are proportional to $\cos kx$, and also to e^{int}. The wave-length (λ) along x is $2\pi/k$, and the period τ is $2\pi/n$. The equations (7) now become

$$\left(\frac{d^2}{dy^2} - k^2 \right) \psi_1 = 0, \qquad \left(\frac{d^2}{dy^2} - k^2 - \frac{in}{\nu} \right) \psi_2 = 0, \quad \dots\dots\dots(8)$$

by which ψ_1 and ψ_2 are to be determined as functions of y. If we write

$$k'^2 = k^2 + in/\nu, \quad \dots\dots\dots\dots\dots(9)$$

we have as the most general solutions of (8)

$$\psi_1 = A e^{-ky} + B e^{+ky}, \quad \dots\dots\dots\dots\dots(10)$$

$$\psi_2 = C e^{-k'y} + D e^{+k'y}. \quad \dots\dots\dots\dots(11)$$

With respect to the value of k', we see from (9) that it is complex. If we write

$$k^2 = P^2 \cos 2\alpha, \qquad n/\nu = P^2 \sin 2\alpha,$$

then

$$k' = P \cos \alpha + i P \sin \alpha.$$

In all the applications that we shall have occasion to make, an approximate value of k' is admissible. On account of the smallness of ν, n/ν is very large in comparison with k^2, that is to say, the thickness of the stratum through which the tangential motion can be propagated in time τ is very small relatively to the wave-length λ. We may therefore neglect k^4 in the equation

$$P^4 = k^4 + n^2/\nu^2,$$

and take simply

$$P^2 = n/\nu.$$

---

* Stokes, "On Pendulums," *Camb. Phil. Trans.* vol. IX. 1850.

Again $\qquad$ $(\sin\alpha - \cos\alpha)^2 = 1 - \sin 2\alpha = \frac{1}{2}k^4\nu^2/n^2,$

so that the difference between $\cos\alpha$ and $\sin\alpha$ may be neglected. We will therefore write

$$k' = \beta(1 + i), \dots\dots\dots\dots\dots\dots\dots\dots(12)$$

where

$$\beta = \sqrt{(n/2\nu)}. \dots\dots\dots\dots\dots\dots\dots(13)$$

We must now distinguish the cases which we have to investigate. In the first we suppose that a wave motion is in progress in a vessel whose horizontal bottom occupies a fixed plane $y = 0$. We may conceive the fluid to be water vibrating in stationary waves under the action of gravity, the question being to examine the influence of the bottom upon the motion. If there are no other solids in the neighbourhood of the bottom, we may put $D = 0$, y being measured upwards, and β being taken positive.

The conditions to be satisfied at $y = 0$ are that u and v should there vanish. Thus

$$A + B + C = 0, \qquad -kA + kB - k'C = 0,$$

so that $\qquad \psi = C\{-\cosh ky + (k'/k)\sinh ky + e^{-k'y}\},$

and $\qquad u = C\{-k\sinh ky + k'\cosh ky - k'e^{-k'y}\}.$

At a short distance from the bottom, $u = k'C$. If we denote by u_0 the maximum value of u near the bottom, we have

$$k'C = u_0 e^{int}\cos kx,$$

and then

$$\psi = u_0 e^{int}\cos kx\left\{-\frac{\cosh ky}{k'} + \frac{\sinh ky}{k} + \frac{e^{-k'y}}{k'}\right\}, \dots\dots\dots(14)$$

$$u = u_0 e^{int}\cos kx\left\{-\frac{k}{k'}\sinh ky + \cosh ky - e^{-k'y}\right\}, \dots\dots(15)$$

$$v = u_0 e^{int}\sin kx\left\{-\frac{k}{k'}\cosh ky + \sinh ky + \frac{k}{k'}e^{-k'y}\right\}\dots\dots(16)$$

These are the symbolical values. If we throw away the imaginary parts, we have as the solution in real quantities by (12),

$$\psi = u_0\cos kx\left\{-\frac{\cosh ky}{\beta\sqrt{2}}\cos(nt - \tfrac{1}{4}\pi) + \frac{\sinh ky}{k}\cos nt\right.$$
$$\left. + \frac{e^{-\beta y}}{\beta\sqrt{2}}\cos(nt - \tfrac{1}{4}\pi - \beta y)\right\}, \dots\dots(17)$$

$$u = u_0\cos kx\left\{-\frac{k\sinh ky}{\beta\sqrt{2}}\cos(nt - \tfrac{1}{4}\pi) + \cosh ky\cos nt\right.$$
$$\left. - e^{-\beta y}\cos(nt - \beta y)\right\}, \dots\dots(18)$$

$$v = u_0\sin kx\left\{-\frac{k\cosh ky}{\beta\sqrt{2}}\cos(nt - \tfrac{1}{4}\pi) + \sinh ky\cos nt\right.$$
$$\left. + \frac{ke^{-\beta y}}{\beta\sqrt{2}}\cos(nt - \tfrac{1}{4}\pi - \beta y)\right\}\dots\dots(19)$$

This is the solution to a first approximation. At a very small distance from the bottom the terms in $e^{-\beta y}$ become insensible.

Although the values of u and v in (18) and (19) are strictly periodic, it is proper to notice that the same property does not attach to the motions thereby defined of the particles of the fluid. In our notation u is not the velocity of any particular particle of the fluid, but of the particle, whichever it may be, that *at the moment under consideration* occupies the point x, y. If $x + \xi$, $y + \eta$ be the actual position at time t of the particle whose mean position during several vibrations is x, y, then the real velocities of the particle at time t are not u, v, but

$$u + \frac{du}{dx}\xi + \frac{du}{dy}\eta, \qquad v + \frac{dv}{dx}\xi + \frac{dv}{dy}\eta;$$

and thus the mean velocity parallel to x is not necessarily zero, but is equal to the mean value of

$$\frac{du}{dx}\xi + \frac{du}{dy}\eta,$$

in which again

$$\xi = \int u\, dt, \qquad \eta = \int v\, dt.$$

From the general form of u, viz., $\cos kx\, F(y, t)$, it follows readily that

$$\int \frac{du}{dx}\xi\, dt = 0.$$

For the second term we must calculate from the actual values as given in (18), (19). Thus

$$\eta = \frac{u_0 \sin kx}{n}\left\{ - \frac{k \cosh ky}{\beta\sqrt{2}} \sin (nt - \tfrac{1}{4}\pi) + \sinh ky\ \sin nt \right.$$

$$\left. + \frac{k\, e^{-\beta y}}{\beta\sqrt{2}} \sin (nt - \tfrac{1}{4}\pi - \beta y) \right\},$$

$$\frac{du}{dy} = u_0 \cos kx \left\{ - \frac{k^2 \cosh ky}{\beta\sqrt{2}} \cos (nt - \tfrac{1}{4}\pi) + k \sinh ky\ \cos nt \right.$$

$$\left. + \sqrt{2}\, . \, \beta e^{-\beta y} \cos (nt + \tfrac{1}{4}\pi - \beta y) \right\},$$

of which the two first terms may be neglected relatively to the third (containing the large factor β). The product of η and du/dy will consist of two parts, the first independent of t, and the second harmonic functions of $2nt$. It is with the first only that we are here concerned. The mean value of the velocity parallel to x is thus

$$\frac{u_0^2 \sin 2kx\, e^{-\beta y}}{4n}\left\{ k \cosh ky \cos \beta y + \sqrt{2}\, . \, \beta \sinh ky \sin (\beta y - \tfrac{1}{4}\pi) - k\, e^{-\beta y} \right\}.$$

On account of the factor $e^{-\beta y}$, this quantity is insensible except when ky is extremely small.   We may therefore write it

$$\frac{u_0{}^2 \sin 2kx \, e^{-\beta y}}{4V} \left\{ \cos \beta y + \beta y \, (\sin \beta y - \cos \beta y) - e^{-\beta y} \right\}, \ldots\ldots(20)$$

V (equal to k/n) being the velocity of propagation of waves corresponding to k and n.

The only approximation employed in the derivation of (15) and (16) is the neglect of the right-hand member of (4), and the corresponding real values of u and v could, if necessary, be readily exhibited without the use of a merely approximate value of k'.   To proceed further we must calculate the value of

$$\frac{u}{\nu}\frac{d\nabla^2\psi}{dx} + \frac{v}{\nu}\frac{d\nabla^2\psi}{dy} \quad \ldots\ldots\ldots\ldots\ldots\ldots(21)$$

in (4), for which it will be sufficient to take the values given by the first approximation.   Thus

$$\nabla^2\psi = \nabla^2\psi_2 = \nu^{-1} d\psi_2/dt,$$

and by (17)

$$\frac{d\psi_2}{dt} = - \frac{n u_0 \cos kx \, e^{-\beta y}}{\beta\sqrt{2}} \sin (nt - \tfrac{1}{4}\pi - \beta y),$$

from which we find as the value of (21),

$$\frac{n k u_0{}^2 \sin 2kx \, e^{-\beta y}}{4\nu^2\beta\sqrt{2}} \left\{ \left(\frac{k}{\beta\sqrt{2}} - \frac{\beta\sqrt{2}}{k} \right) \sinh ky \sin \beta y - \sqrt{2} \cosh ky \cos \beta y + \sqrt{2} \, e^{-\beta y} \right\}$$

$$+ \text{ terms in } 2nt.$$

On account of the factor $e^{-\beta y}$ this quantity is sensible only when y is very small.   We may write it with sufficient approximation

$$\frac{n k u_0{}^2 \sin 2kx \, e^{-\beta y}}{4\nu^2\beta} \left\{ - \beta y \sin \beta y - \cos \beta y + e^{-\beta y} \right\} . \ldots\ldots\ldots(22)$$

The terms in $2nt$, corresponding to motions of half the original period, are not required for our purpose, which is to investigate the non-periodic motion of the second order.   The equation with which we have to proceed is found by equating (22) to $\nabla^4\psi$.   The solution will consist of two parts, one resulting from the direct integration of (22) and involving the factor $e^{-\beta y}$, the second a complementary function with arbitrary coefficients satisfying $\nabla^4\psi = 0$.   In the calculation of the first part we may identify ∇^4 with d^4/dy^4, on account of the smallness of k relatively to β.   In this way our equation becomes

$$\frac{d^4\psi}{d(\beta y)^4} = \frac{n k u_0{}^2 \sin 2kx \, e^{-\beta y}}{4\nu^2\beta^5} \left\{ - \beta y \sin \beta y - \cos \beta y + e^{-\beta y} \right\}, \ldots(23)$$

of which the solution is

$$\psi = \frac{n k u_0{}^2 \sin 2kx \, e^{-\beta y}}{4\nu^2\beta^5} \left\{ \tfrac{3}{4} \cos \beta y + \tfrac{1}{2} \sin \beta y + \tfrac{1}{4}\beta y \sin \beta y + \tfrac{1}{16} e^{-\beta y} \right\} . \ldots(24)$$

The complementary function, being proportional to $\sin 2kx$, may be written

$$\frac{nku_0^2 \sin 2kx}{4\nu^2\beta^5} \{(A + By)\, e^{-2ky} + (A' + B'y)\, e^{+2ky}\}.$$

If the fluid be uninterrupted by a free surface, or otherwise, within distances for which ky is sensible, we must suppose $(A' + B'y) = 0$, so that by (13) the complementary function may be written

$$\frac{u_0^2 \sin 2kx}{\beta V} (A + By)\, e^{-2ky}.$$

The condition that v (equal to $-d\psi/dx$) must vanish when $y = 0$, gives $A = -\frac{13}{16}$. For the velocity parallel to x we have

$$u = \frac{u_0^2 \sin 2kx}{V} [e^{-\beta y} \{-\sin \beta y - \tfrac{1}{4} \cos \beta y + \tfrac{1}{4}\beta y \cos \beta y - \tfrac{1}{4}\beta y \sin \beta y - \tfrac{1}{8}e^{-\beta y}\}$$
$$+ \beta^{-1} e^{-2ky} \{B - 2k\,(A + By)\}].$$

In order that u should vanish when $y = 0$, we must have

$$B = 2kA + \tfrac{3}{8}\beta = \tfrac{3}{8}\beta - \tfrac{13}{8}k = \tfrac{3}{8}\beta,$$

approximately. Thus

$$u = \frac{u_0^2 \sin 2kx}{V} [e^{-\beta y} \{-\sin \beta y - \tfrac{1}{4} \cos \beta y + \tfrac{1}{4}\beta y \cos \beta y - \tfrac{1}{4}\beta y \sin \beta y - \tfrac{1}{8}e^{-\beta y}\}$$
$$+ \tfrac{3}{8}e^{-2ky} \{1 - 2ky\}], \quad \dots(25)$$

and

$$v = -\frac{2ku_0^2 \cos 2kx}{\beta V} [e^{-\beta y} \{\tfrac{3}{4} \cos \beta y + \tfrac{1}{2} \sin \beta y + \tfrac{1}{4}\beta y \sin \beta y + \tfrac{1}{16}e^{-\beta y}\}$$
$$+ e^{-2ky} \{-\tfrac{13}{16} + \tfrac{3}{8}\beta y\}]. \quad \dots(26)$$

To obtain the mean velocity parallel to x of a particle, we must add to (25), the terms previously investigated and expressed by (20). If we call the total u', we have

$$u' = \frac{u_0^2 \sin 2kx}{V} [e^{-\beta y} \{-\sin \beta y - \tfrac{3}{8}e^{-\beta y}\} + \tfrac{3}{8}e^{-2ky} \{1 - 2ky\}]. \quad \dots\dots(27)$$

At a short distance from the bottom $e^{-\beta y}$ becomes insensible, and we have simply

$$u' = \tfrac{3}{8} \frac{u_0^2 \sin 2kx}{V} e^{-2ky} (1 - 2ky), \quad \dots\dots\dots\dots\dots\dots(28)$$

$$v' = -\frac{2ku_0^2 \cos 2kx}{\beta V} e^{-2ky} (-\tfrac{13}{16} + \tfrac{3}{8}\beta y). \quad \dots\dots\dots\dots(29)$$

The steady motion expressed by (28) and (29) is of a very simple character. It consists of a series of vortices periodic with respect to x in a distance $\tfrac{1}{2}\lambda$. For a given x the horizontal motion is of one sign near the bottom, and of the opposite sign at a distance from it, the place of transition being at $y = (2k)^{-1} = \lambda/4\pi$. The horizontal motion of the first order near the bottom being by (18) $u = u_0 \cos kx \cos nt$, we see that it is a maximum when $kx = 0, \pi, 2\pi, \ldots$ If we call these places loops, and the places of minimum velocity nodes, (29) shows that v' is negative and a maximum at the loops, positive and a maximum at the nodes. The fluid therefore rises from the bottom over the nodes and falls back again over the loops, the horizontal motion near the bottom being thus directed towards the nodes and from the loops. The maximum horizontal motion is simply $\tfrac{3}{4} u_0^2 / V$, and is *independent of the value of v.* We cannot, therefore, avoid considering this motion by supposing the coefficient of viscosity to be very small, the maintenance of the vortices becoming easier in the same proportion as the forces tending to produce the vortical motion diminish.

To ascertain the character of the motion quite close to the bottom, we must include the terms in $e^{-\beta y}$. When y is extremely small

$$u' = u_0^2 V^{-1} \sin 2kx \left\{ - \tfrac{1}{4}\beta y + \ldots \right\}, \ldots\ldots\ldots\ldots\ldots\ldots(30)$$

so that the motion is here in the opposite direction to that which prevails when $e^{-\beta y}$ can be neglected.

A few corresponding values of βy and of $-(\sin \beta y + \tfrac{3}{2} e^{-\beta y}) e^{-\beta y} + \tfrac{3}{2}$ are annexed, in order to show the distribution of velocities within the thin frictional layer.

βy		βy	
$\dfrac{\pi}{16}$	$- \cdot 038$	$\dfrac{3\pi}{8}$	$+ \cdot 055$
$\dfrac{\pi}{8}$	$- \cdot 054$	$\dfrac{\pi}{2}$	$+ \cdot 151$
$\dfrac{3\pi}{16}$	$- \cdot 049$	π	$+ \cdot 374$
$\dfrac{\pi}{4}$	$- \cdot 025$	$\dfrac{3\pi}{2}$	$+ \cdot 384$

It appears that ($\sin 2kx$ being positive) the velocity is negative from the plate outwards until βy somewhat exceeds $\tfrac{1}{4}\pi$, after which it is positive, until reversed by the factor $(1 - 2ky)$. The greatest negative velocity in the layer is about $\tfrac{1}{7}$ of that which is found at a little distance outside the layer.

Faraday found that fine sand, scattered over the bottom, tends to collect at the loops. This is in agreement with what the present calculation would lead us to expect, provided that we can suppose that the sand is controlled by the layer at the bottom whose motion is negative. The exceeding thinness of the layer, however, presents itself as a difficulty. The subject requires further experimental investigation; but in the meantime the following data may be worth notice, though in some respects, *e.g.*, the shallowness of the liquid in relation to the wave-length, the circumstances differed materially from those assumed in the theoretical investigation.

The liquid was water ($\nu = \cdot 014$ C.G.S.), and the period of vibration was $\frac{1}{15}$, so that $n = 2\pi \times 15$. The thickness of the layer

$$= \tfrac{1}{4}\pi \sqrt{(2\nu/n)} = \cdot 0135 \text{ centim.}$$

Measurements of the diameters of the particles of sand gave about $\cdot 02$ centim., so that the grains would be almost wholly immersed in the negative layer, even if isolated. It seems therefore that the observed motion to the loops gives rise in this case to no difficulty. But it is possible that the behaviour of the sand is materially influenced by the vertical motion of the vessel by which in these experiments the liquid vibrations are maintained*.

§ 2.   In the problem to which we now proceed the motion will be supposed to have its origin in the assumed motion of a flexible plate situated when in equilibrium at $y = 0$. Thus for a first approximation we take $u = 0$, $v = v_0 \sin kx \, e^{int}$, when $y = 0$, and the question is to investigate the resulting motion of the fluid in contact with the plate.

The solution to a first approximation is readily obtained. As in (10), (11), we have

$$\psi = \psi_1 + \psi_2 = e^{int} \cos kx \, (A \, e^{-ky} + C e^{-k'y}), \quad \text{.............(31)}$$

in which we may take as before

$$k' = \sqrt{(n/2\nu)} . (1 + i) = \beta \, (1 + i). \quad \text{....................(32)}$$

By the condition at $y = 0$,

$$A = -\frac{k'}{k} C, \qquad\qquad C = \frac{v_0}{k' - k},$$

so that

$$\psi = \frac{v_0 e^{int} \cos kx}{k - k'} \left\{ -\frac{k'}{k} e^{-ky} + e^{-k'y} \right\}, \quad \text{.............(33)}$$

$$u = \frac{v_0 e^{int} \cos kx}{k - k'} \left\{ k' e^{-ky} - k' e^{-k'y} \right\}. \quad \text{.................(34)}$$

In passing to real quantities it will be convenient to write

$$\frac{v_0}{k - k'} = H e^{i\epsilon}. \quad \text{.........................(35)}$$

Thus throwing away the imaginary parts of (33), (34), we get

$$\psi = \cos kx \left\{ -\frac{\beta \sqrt{2}}{k} e^{-ky} \cos (nt + \epsilon + \tfrac{1}{4}\pi) + e^{-\beta y} \cos (nt + \epsilon - \beta y) \right\}, \quad \text{.......(36)}$$

$$u = \sqrt{2} . \beta H \cos kx \left\{ e^{-ky} \cos (nt + \epsilon + \tfrac{1}{4}\pi) - e^{-\beta y} \cos (nt + \epsilon + \tfrac{1}{4}\pi - \beta y) \right\}, \quad \text{...(37)}$$

$$v = H \sin kx \left\{ -\beta \sqrt{2} e^{-ky} \cos (nt + \epsilon + \tfrac{1}{4}\pi) + k e^{-\beta y} \cos (nt + \epsilon - \beta y) \right\}. \quad \text{......(38)}$$

* See a paper " On the Crispations of Fluid resting upon a Vibrating Support," *Phil. Mag.* July, 1883.   [Art. 102, vol. ii. p. 212.]

From (32), (35), the approximate value of H is $-v_0/\beta\sqrt{2}$, and that of ϵ is $-\frac{1}{4}\pi$. More exact values will however be required later. We find

$$H = -\frac{v_0}{\sqrt{\{(\beta-k)^2+\beta^2\}}} = -\frac{v_0}{\beta\sqrt{2}}\left(1+\frac{k}{2\beta}\right), \quad \ldots\ldots\ldots\ldots(39)$$

$$\cos\epsilon = \frac{\beta-k}{\sqrt{\{(\beta-k)^2+\beta^2\}}} = \frac{1}{\sqrt{2}}\left(1-\frac{k}{2\beta}\right). \quad \ldots\ldots\ldots\ldots\ldots(40)$$

The values of u and v above expressed give $u=0$, $v=v_0\sin kx\cos nt$, when $y=0$. This is sufficient for a first approximation, but in proceeding further we must remember that these prescribed velocities apply in strictness not to $y=0$, but to

$$y = \frac{v_0}{n}\sin kx \sin nt.$$

Substituting the latter value of y in the expressions (37), and (38), we find

$$u = \sqrt{2}.\beta H \cos kx \{-ky\cos(nt+\epsilon+\tfrac{1}{4}\pi)+\sqrt{2}.\beta y\cos(nt+\epsilon+\tfrac{1}{2}\pi)\}$$

$$= \frac{\beta^2 v_0 H}{n}\sin 2kx \sin nt \left\{-\frac{k}{\beta\sqrt{2}}\cos(nt+\epsilon+\tfrac{1}{4}\pi)+\cos(nt+\epsilon+\tfrac{1}{2}\pi)\right\}$$

$$= \frac{\beta^2 v_0 H}{2n}\sin 2kx \left\{\frac{k}{\beta\sqrt{2}}\sin(\epsilon+\tfrac{1}{4}\pi)-\sin(\epsilon+\tfrac{1}{2}\pi)\right\} + \text{terms in } 2nt.$$

The first term within the bracket is of the *second* order in k/β relatively to the latter term, and may be omitted. Thus

$$u = -\frac{\beta^2 v_0 H}{2n}\sin 2kx \cos \epsilon.$$

The terms in $2nt$ we need not further examine. From (39), (40), $H\cos\epsilon = -v_0/2\beta$ very approximately, so that we may write

$$u = \frac{\beta v_0^2}{4n}\sin 2kx. \quad \ldots\ldots\ldots\ldots\ldots\ldots\ldots(41)$$

To the same degree of approximation, $v=v_0\sin kx\cos nt$, simply.

We have next, as in the first problem, to consider the complete equation

$$\nabla^4\psi = \frac{u}{\nu^2}\frac{d^2\psi_2}{dx\,dt} + \frac{v}{\nu^2}\frac{d^2\psi_2}{dy\,dt} \quad \ldots\ldots\ldots\ldots\ldots\ldots(42)$$

in the right-hand member of which we use the approximate values given by (36), (37), (38). Thus

$$\frac{d\psi_2}{dt} = -nH\cos kx\, e^{-\beta y}\sin(nt+\epsilon-\beta y),$$

and (42) becomes

$$\nabla^4\psi = \frac{nk\beta H^2\sin 2kx\, e^{-\beta y}}{4\nu^2}\left\{e^{-ky}\left(\frac{2\beta}{k}\sin\beta y-\sin\beta y-\cos\beta y\right)+2e^{-\beta y}\right\}. \quad \ldots(43)$$

It will be found presently that the term divided by k disappears from the final result, and thus we have to pursue the approximation further than might at first appear necessary. We may however neglect terms of order k^2/β^2, in comparison with the principal term. Thus ∇^4 may be identified with d^4/dy^4, and the equation becomes

$$\frac{d^4\psi}{d(\beta y)^4} = \frac{nkH^2 \sin 2kx \, e^{-\beta y}}{4v^2\beta^3} \left\{\left(\frac{2\beta}{k} - 1\right)\sin\beta y - \cos\beta y - 2\beta y \sin\beta y + 2e^{-\beta y}\right\}, \quad (44)$$

whence

$$\psi = \frac{nkH^2 \sin 2kx \, e^{-\beta y}}{4v^2\beta^3} \left\{\left(-\frac{\beta}{2k} + \frac{5}{4}\right)\sin\beta y + \frac{5}{4}\cos\beta y + \frac{1}{2}\beta y \sin\beta y + \frac{1}{8}e^{-\beta y}\right\}. \quad (45)$$

And

$$u = \frac{d\psi}{dy} = \frac{nkH^2 \sin 2kx \, e^{-\beta y}}{4v^2\beta^2}\left\{\left(\frac{\beta}{2k} - 2\right)\sin\beta y - \frac{\beta}{2k}\cos\beta y - \frac{1}{2}\beta y \sin\beta y\right.$$

$$\left. + \frac{1}{2}\beta y \cos\beta y - \frac{1}{4}e^{-\beta y}\right\}. \quad\dots\dots\dots(46)$$

To obtain the value of u at the surface of the plate it will be sufficient to put $y = 0$ in (46). Thus

$$u = \frac{nkH^2 \sin 2kx}{4v^2\beta^2}\left\{-\frac{\beta}{2k} - \frac{1}{4}\right\}. \quad\dots\dots\dots\dots\dots(47)$$

By (32), (39)

$$\frac{nkH^2}{4v^2\beta^2} = \frac{kv_0^2}{2n}\left(1 + \frac{k}{\beta}\right) = \frac{v_0^2}{2V}\left(1 + \frac{k}{\beta}\right),$$

if as before we put V for k/n. Thus in (47)

$$u = \frac{v_0^2}{4V}\left(-\frac{\beta}{k} - \frac{3}{2}\right)\sin 2kx. \quad\dots\dots\dots\dots\dots(48)$$

To obtain the complete value of u at the surface of the plate, corresponding to (37), (46), we have to add to (48) that given in (41). The term of lowest order disappears, and we are left simply with

$$u = -\frac{3v_0^2}{8V}\sin 2kx. \quad\dots\dots\dots\dots\dots(49)$$

In like manner we find for the complete value of v at the surface of the plate corresponding to (38), (45),

$$v = v_0 \sin kx \cos nt - \frac{11v_0^2 k \cos 2kx}{8\beta V}. \quad\dots\dots\dots (50)$$

The values of u and v expressed in (49) and the second part of (50) must be cancelled by a suitable choice of the complementary function, satisfying $\nabla^4\psi = 0$, so that to the second order of approximation the fluid in contact with the plate may have no relative motion.

The complementary function is

$$\psi = (A + By) e^{-2ky} \sin 2kx,$$

whence

$$u = \{B - 2k(A + By)\} e^{-2ky} \sin 2kx,$$

$$v = -2k(A + By) e^{-2ky} \cos 2kx.$$

Determining the constants as indicated above, we get

$$u = \frac{3v_0^2}{8V} (1 - 2ky) e^{-2ky} \sin 2kx, \quad \dots\dots\dots\dots\dots(51)$$

$$v = -\frac{kv_0^2}{8\beta V} (11 + 6\beta y) e^{-2ky} \cos 2kx. \quad \dots\dots\dots\dots(52)$$

The velocities given by (51), (52) are the only part of the motion of the second order which is sensible beyond a very small distance from the vibrating plate. The nodes of the plate (where sand would collect) are at the points given by $kx = 0, \pi, 2\pi\dots$, and the loops at the points $kx = \frac{1}{2}\pi, \frac{3}{2}\pi\dots$ At the former points v is negative, and at the latter positive. For $kx = \frac{1}{4}\pi$, u is positive, and for $kx = \frac{3}{4}\pi$, u is negative.

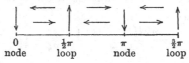

The magnitude of the vortical motion is independent of the coefficient of friction.

The complete value of u to the second order of approximation (except the terms in $2nt$) is obtained by adding together (37), (46), and (51), and it will contain the term divided by k in (46), whose appearance, however, is misleading. The objectionable term will be got rid of, if we express the mean velocity of a particle, instead of as in (46), the mean velocity *at a point*. For this purpose we are to add to (46), (51), the mean value of

$$\xi \frac{du}{dx} + \eta \frac{du}{dy},$$

as calculated from the first approximation, where

$$\xi = \int u\, dt, \qquad \eta = \int v\, dt.$$

As in the former problem the mean value of $\xi\, du/dx$ is zero.

Multiplying together du/dy, and $\int v\, dt$ as found from (37), (38), and rejecting the terms in $2nt$, we get with omission of k^2,

$$-\frac{k\beta^2 H^2 e^{-\beta y} \sin 2kx}{n} \left\{\frac{\beta}{2k} (1 - ky)(\sin \beta y - \cos \beta y) + \frac{1}{2} e^{-\beta y}\right\}, \quad \dots\dots(53)$$

in which we may write

$$-\frac{k\beta^2 H^2}{n} = -\frac{k\beta^4 H^2}{n\beta^2} = -\frac{nkH^2}{4\nu^2\beta^2}.$$

Combining (53), (46), and (51), we get finally

$$u' = \frac{nkH^2\, e^{-\beta y} \sin 2kx}{4\nu^2\beta^2} \left\{ -2\sin\beta y - \tfrac{3}{4}e^{-\beta y} \right\} + \frac{3v_0^2}{8V}(1-2ky)e^{-2ky}\sin 2kx$$

$$= \frac{v_0^2 \sin 2kx}{2V}\left\{ -e^{-\beta y}(2\sin\beta y + \tfrac{3}{4}e^{-\beta y}) + \tfrac{3}{4}(1-2ky)e^{-2ky} \right\}, \ldots\ldots(54)$$

which expresses the mean particle velocity.

When βy is very small, (54) gives

$$u' = \frac{v_0^2 \sin 2kx}{2V}(-\tfrac{1}{2}\beta y + \ldots)\ldots\ldots\ldots\ldots(55)$$

from which it appears that quite close to the plate the mean velocity is in the opposite direction to that which is found outside the frictional layer.

§ 3. In the third problem, relating to Kundt's tubes, the fluid must be treated as compressible, as the motion is supposed to be approximately in one dimension, parallel (say) to x. The solution to a first approximation is merely an adaptation to two dimensions of the corresponding solution for a tube of revolution by Kirchhoff*, simplified by the neglect of the terms relating to the development and conduction of heat. It is probable that the solution to the second order would be practicable also for a tube of revolution, but for the sake of simplicity I have adhered to the case of two dimensions. The most important point in which the two problems are likely to differ can be investigated very simply, without a complete solution.

If we suppose $p = a^2\rho$, and write σ for $\log\rho - \log\rho_0$, the fundamental equations are

$$a^2\frac{d\sigma}{dx} = -\frac{du}{dt} - u\frac{du}{dx} - v\frac{du}{dy} + \nu\nabla^2 u + \nu'\frac{d}{dx}\left(\frac{du}{dx}+\frac{dv}{dy}\right), \ldots\ldots(56)$$

with a corresponding equation for v, and the equation of continuity,

$$\frac{du}{dx} + \frac{dv}{dy} + \frac{d\sigma}{dt} + u\frac{d\sigma}{dx} + v\frac{d\sigma}{dy} = 0. \ldots\ldots\ldots\ldots(57)$$

Whatever may be the actual values of u and v, we may write

$$u = \frac{d\phi}{dx} + \frac{d\psi}{dy}, \qquad v = \frac{d\phi}{dy} - \frac{d\psi}{dx}, \ldots\ldots\ldots\ldots(58)$$

in which

$$\nabla^2\phi = \frac{du}{dx} + \frac{dv}{dy}, \qquad \nabla^2\psi = \frac{du}{dy} - \frac{dv}{dx}. \ldots\ldots\ldots\ldots(59)$$

From (56), (57),

$$\left(a^2 + \nu'\frac{d}{dt}\right)\frac{d\sigma}{dx} = -\frac{du}{dt} + \nu\nabla^2 u - u\frac{du}{dx} - v\frac{du}{dy} - \nu'\frac{d}{dx}\left(u\frac{d\sigma}{dx}+v\frac{d\sigma}{dy}\right), \ldots\ldots(60)$$

$$\left(a^2 + \nu'\frac{d}{dt}\right)\frac{d\sigma}{dy} = -\frac{dv}{dt} + \nu\nabla^2 v - u\frac{dv}{dx} - v\frac{dv}{dy} - \nu'\frac{d}{dy}\left(u\frac{d\sigma}{dx}+v\frac{d\sigma}{dy}\right). \ldots\ldots(61)$$

* *Pogg. Ann. t. cxxxiv. 1868.*

Again, from (60), (61),

$$\left(a^2 + v\frac{d}{dt} + v'\frac{d}{dt}\right)\nabla^2\sigma - \frac{d^2\sigma}{dt^2} = \frac{d}{dt}\left(u\frac{d\sigma}{dx} + v\frac{d\sigma}{dy}\right) - (v+v')\nabla^2\left(u\frac{d\sigma}{dx} + v\frac{d\sigma}{dy}\right)$$

$$- \frac{d}{dx}\left(u\frac{du}{dx} + v\frac{du}{dy}\right) - \frac{d}{dy}\left(u\frac{dv}{dx} + v\frac{dv}{dy}\right). \quad\ldots(62)$$

For the first approximation the terms of the second order in u, v, and σ are to be omitted. If we assume that as functions of t, all the periodic quantities are proportional to e^{int}, and write q for $a^2 + inv + inv'$, (62) becomes

$$q\nabla^2\sigma + n^2\sigma = 0. \quad\ldots\ldots\ldots\ldots\ldots\ldots\ldots\ldots(63)$$

Now by (57), (59),

$$\nabla^2\phi = -in\sigma = iqn^{-1}\nabla^2\sigma,$$

so that

$$\phi = iqn^{-1}\sigma,*$$

and

$$u = \frac{iq}{n}\frac{d\sigma}{dx} + \frac{d\psi}{dy}, \qquad v = \frac{iq}{n}\frac{d\sigma}{dy} - \frac{d\psi}{dx}. \quad\ldots\ldots\ldots\ldots(64)$$

Substituting in (60), (61), with omission of terms of the second order, we get in view of (63),

$$(\nu\nabla^2 - in)\frac{d\psi}{dy} = 0, \qquad (\nu\nabla^2 - in)\frac{d\psi}{dx} = 0,$$

whence

$$(\nu\nabla^2 - in)\psi = 0. \quad\ldots\ldots\ldots\ldots\ldots\ldots(65)$$

If we eliminate σ directly from the fundamental equations (56), we get

$$\left(\nu\nabla^4 - \frac{d}{dt}\nabla^2\right)\psi = \frac{d}{dy}\left(u\frac{du}{dx} + v\frac{du}{dy}\right) - \frac{d}{dx}\left(u\frac{dv}{dx} + v\frac{dv}{dy}\right) = \frac{d}{dy}(v\nabla^2\psi) + \frac{d}{dx}(u\nabla^2\psi)$$

$$= \left(\frac{du}{dx} + \frac{dv}{dy}\right)\nabla^2\psi + u\frac{d\nabla^2\psi}{dx} + v\frac{d\nabla^2\psi}{dy}. \quad\ldots\ldots\ldots\ldots\ldots(66)$$

If we now assume that as functions of x the quantities σ, ψ, &c., are proportional to e^{ikx}, equations (63), (65) may be written

$$(d^2/dy^2 - k''^2)\sigma = 0, \qquad \text{where } k''^2 = k^2 - n^2/q, \quad\ldots\ldots(67)$$
$$(d^2/dy^2 - k'^2)\psi = 0, \qquad \text{where } k'^2 = k^2 + in/\nu. \quad\ldots\ldots(68)$$

If the origin for y be in the middle between the two parallel boundaries, σ must be an even function of y, and ψ must be an odd function. Thus we may write

$$\sigma = A\cosh k''y \cdot e^{int}e^{ikx}, \qquad \psi = B\sinh k'y \cdot e^{int}e^{ikx}, \quad\ldots(69)$$

$$u = \left(-\frac{kq}{n}A\cosh k''y + k'B\sinh k'y\right)e^{int}e^{ikx} \Bigg\}$$

$$v = \left(\frac{ik''q}{n}A\sinh k''y - ikB\sinh k'y\right)e^{int}e^{ikx} \Bigg\} \quad\ldots\ldots\ldots(70)$$

---

* It is unnecessary to add a complementary function ϕ', satisfying $\nabla^2\phi' = 0$, as the motion corresponding thereto may be regarded as covered by ψ.

If the fixed walls are situated at $y = \pm y_1$, u and v must vanish for these values of y. Eliminating from (70) the ratio of A to B, we get as the equation for determining k,

$$k^2 \tanh k'y_1 = k'k'' \tanh k''y_1, \quad \ldots\ldots\ldots\ldots\ldots\ldots(71)$$

in which k', k'' are given as functions of k by (67), (68). We now introduce further approximations dependent upon the assumption that the direct influence of friction extends through a layer whose thickness is a small fraction only of y_1. On this supposition k' is large, and k'' is small, so that we may put $\tanh k'y_1 = \pm 1$, $\tanh k''y_1 = \pm k''y_1$. Equation (71) then becomes

$$k^2 = k'k''^2 y_1, \quad \ldots\ldots\ldots\ldots\ldots\ldots\ldots\ldots(72)$$

or if we introduce the values of k', k'' from (67), (68),

$$k^2 = (k^2 - n^2/q)\, y_1\, \sqrt{(k^2 + in/\nu)}.$$

Since in/ν is great, $k^2 = n^2/q = n^2/a^2$ approximately.

Thus

$$k^2 = \frac{n^2}{q} + \frac{k^2}{y_1 \sqrt{(k^2 + in/\nu)}} = \frac{n^2}{a^2}\left\{1 + \frac{1}{y_1 \sqrt{(in/\nu)}}\right\},$$

and

$$k = \pm \frac{n}{a}\left\{1 + \frac{1-i}{2y_1 \sqrt{(2n/\nu)}}\right\}. \quad \ldots\ldots\ldots\ldots\ldots(73)$$

If we write $k = k_1 + ik_2$,

$$k_1 = \pm \frac{n}{a}\left\{1 + \frac{\sqrt{(\nu/2n)}}{2y_1}\right\}, \qquad k_2 = \mp \frac{n}{a}\frac{\sqrt{(\nu/2n)}}{2y_1}, \quad \ldots\ldots\ldots(74)$$

which agrees with the result given in § 347 (11) of my book on the Theory of Sound.

In taking approximate forms for (70), we must distinguish which half of the symmetrical motion we contemplate. If we choose that for which y is *negative*, we replace $\cosh k'y$ and $\sinh k'y$ by $\tfrac{1}{2}e^{-k'y}$. For $\cosh k''y$ we may write unity, and for $\sinh k''y$ simply $k''y$. If we change the arbitrary multiplier so that the maximum value of u is unity, we have

$$u = (-1 + e^{-k'(y+y_1)})\, e^{ikx}\, e^{int}, \qquad v = \frac{ik}{k'}\left(\frac{y}{y_1} + e^{-k(y+y_1)}\right) e^{ikx}\, e^{int}, \quad \ldots\ldots(75)$$

in which, of course, u and v vanish when $y = -y_1$.

If in (75) we change k into $-k$, and then take the mean, we obtain

$$u = (-1 + e^{-k'(y+y_1)})\cos kx\, e^{int}, \qquad v = -\frac{k}{k'}\left(\frac{y}{y_1} + e^{-k'(y+y_1)}\right)\sin kx\, e^{int}. \quad (76)$$

Although k is not absolutely a real quantity, we may consider it to be so with sufficient approximation for our purpose. If we write as before

$$k' = \sqrt{(n/2\nu)}\,.\,(1+i) = \beta(1+i),$$

we get from (76) in terms of real quantities

$$u = \cos kx \left[-\cos nt + e^{-\beta(y+y_1)} \cos \{nt - \beta (y + y_1)\} \right]$$

$$v = -\frac{k}{\beta\sqrt{2}} \sin kx \left[\frac{y}{y_1} \cos (nt - \tfrac{1}{4}\pi) + e^{-\beta(y+y_1)} \cos \{nt - \tfrac{1}{4}\pi - \beta(y+y_1)\} \right] \Bigg\} \; . \quad (77)$$

It will shorten the expressions with which we have to deal if we measure y from the wall (on the negative side) instead of as hitherto from the plane of symmetry, for which purpose we must write y for $y + y_1$. Thus

$$u = \cos kx \{ -\cos nt + e^{-\beta y} \cos (nt - \beta y)\}$$

$$v = \frac{k \sin kx}{\beta\sqrt{2}} \left\{ \frac{y_1 - y}{y_1} \cos (nt - \tfrac{1}{4}\pi) - e^{-\beta y} \cos (nt - \tfrac{1}{4}\pi - \beta y) \right\} \Bigg\} \; \dots\dots(78)$$

From (78) approximately

$$\nabla^2 \psi = \beta\sqrt{2} \, . \, \cos kx \, e^{-\beta y} \sin (nt - \tfrac{1}{4}\pi - \beta y), \dots\dots\dots\dots(79)$$

$$\frac{du}{dx} + \frac{dv}{dy} = k \sin kx \cos nt, \dots\dots\dots\dots\dots\dots(80)$$

$$u \frac{d\nabla^2\psi}{dx} + v \frac{d\nabla^2\psi}{dy} = \tfrac{1}{2}k\beta \sin 2kx \, e^{-\beta y} (-\cos \beta y + e^{-\beta y}) + \text{terms in } 2nt, \quad (81)$$

$$\left(\frac{du}{dx} + \frac{dv}{dy} \right) \nabla^2\psi = -\tfrac{1}{4}k\beta \sin 2kx \, e^{-\beta y} (\sin \beta y + \cos \beta y) + \text{terms in } 2nt. \quad (82)$$

As in former problems the periodic terms in $2nt$ will be omitted. For the non-periodic part of ψ of the second order, we have from (66)

$$\nabla^4\psi = -\frac{k\beta}{4\nu} \sin 2kx \, e^{-\beta y} \{\sin \beta y + 3 \cos \beta y - 2 e^{-\beta y}\}. \quad \dots\dots(83)$$

In this we identify ∇^4 with d^4/dy^4, so that

$$\psi = \frac{k \sin 2kx \, e^{-\beta y}}{16\nu\beta^3} \{\sin \beta y + 3 \cos \beta y + \tfrac{1}{2}e^{-\beta y}\}, \quad \dots\dots\dots(84)$$

to which must be added a complementary function, satisfying $\nabla^4\psi = 0$, of the form

$$\psi = \frac{\sin 2kx}{16\nu\beta^3} \{A \sinh 2k (y_1 - y) + B (y_1 - y) \cosh 2k (y_1 - y)\}, \dots(85)$$

or as we may take it approximately, if y_1 be small compared with the wavelength λ,

$$\psi = \frac{k \sin 2kx}{16\nu\beta^3} \{A' (y_1 - y) + B' (y_1 - y)^3\}. \quad \dots\dots\dots(86)$$

The value of σ to a second approximation would have to be investigated by means of (62). It will be composed of two parts, the first independent of t, the second a harmonic function of $2nt$. In calculating the part of $d\phi/dx$ independent of t from

$$\nabla^2\phi = -\frac{d\sigma}{dt} - u \frac{d\sigma}{dx} - v \frac{d\sigma}{dy},$$

we shall obtain nothing from $d\sigma/dt$. In the remaining terms on the right-hand side it will be sufficient to employ the values of u, v, σ of the first approximation. From

$$\frac{d\sigma}{dt} = -\frac{du}{dx} - \frac{dv}{dy},$$

in conjunction with (80), we get

$$\sigma = -(u_0/a) \sin kx \sin nt,$$

whence

$$\frac{d^2\phi}{d(\beta y)^2} = \frac{ku_0^2}{2a\beta^2} \cos^2 kx \, e^{-\beta y} \sin \beta y.$$

It is easily seen from this that the part of u resulting from $d\phi/dx$ is of order $k^2\beta^2$ in comparison with the part (87) resulting from ψ, and may be omitted.

Accordingly by (84), with introduction of the value of β and (in order to restore homogeneity) of u_0^2

$$u = -\frac{u_0^2 \sin 2kx \, e^{-\beta y}}{8a} \{4 \sin \beta y + 2 \cos \beta y + e^{-\beta y}\}, \quad \ldots\ldots\ldots(87)$$

$$v = -\frac{2ku_0^2 \cos 2kx \, e^{-\beta y}}{8\beta a} \{\sin \beta y + 3 \cos \beta y + \tfrac{1}{2}e^{-\beta y}\}; \quad \ldots\ldots(88)$$

and from (86)

$$u = -\frac{u_0^2 \sin 2kx}{8\beta a} \{A' + 3B'(y_1 - y)^2\}, \quad \ldots\ldots\ldots\ldots\ldots\ldots(89)$$

$$v = -\frac{2ku_0^2 \cos 2kx}{8\beta a} \{A'(y_1 - y) + B'(y_1 - y)^3\}. \quad \ldots\ldots\ldots\ldots(90)$$

When $y = 0$, the complete values of u and v, as given by the four last equations, must vanish. Determining in this way the arbitrary constants A' and B', we get as the complete values at any point,

$$u = -\frac{u_0^2 \sin 2kx}{8a} \left\{ e^{-\beta y}(4 \sin \beta y + 2 \cos \beta y + e^{-\beta y}) + \tfrac{3}{2} - \tfrac{9}{2}\frac{(y_1 - y)^2}{y_1^2} \right\}, \quad \ldots\ldots(91)$$

$$v = -\frac{2ku_0^2 \cos 2kx}{8\beta a} \left\{ e^{-\beta y}(\sin \beta y + 3 \cos \beta y + \tfrac{1}{2}e^{-\beta y}) \right.$$

$$\left. + \tfrac{3}{2}\beta(y_1 - y) - \tfrac{3}{2}\beta\frac{(y_1 - y)^3}{y_1^2} \right\}. \quad \ldots\ldots\ldots(92)$$

Outside the thin film of air immediately influenced by the friction we may put $e^{-\beta y} = 0$, and then

$$u = -\frac{3u_0^2 \sin 2kx}{16a} \left\{ 1 - 3\frac{(y_1 - y)^2}{y_1^2} \right\}, \quad \ldots\ldots\ldots\ldots\ldots(93)$$

$$v = -\frac{3u_0^2 \, 2k \cos 2kx}{16a} \left\{ y_1 - y - \frac{(y_1 - y)^3}{y_1^2} \right\}. \quad \ldots\ldots\ldots(94)$$

From (93) we see that u changes sign as we pass from the boundary $y = 0$ to the plane of symmetry $y = y_1$, the critical value of y being $y_1 (1 - \sqrt{\tfrac{1}{3}})$, or $\cdot 423\, y_1$.

The principal motion being $u = -\, u_0 \cos kx \cos nt$, the loops correspond to $kx = 0$, π, $2\pi, \ldots$, and the nodes correspond to $\tfrac{1}{2}\pi$, $\tfrac{3}{2}\pi, \ldots$. Thus v is positive at the nodes and negative at the loops, vanishing of course in either case both at the wall $y = 0$, and at the plane of symmetry $y = y_1$.

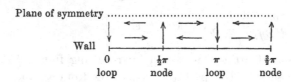

To obtain the mean velocities of the *particles* parallel to x, we must make an addition to u, as in the former problems.

In the present case the mean value of

$$\frac{du}{dx}\xi + \frac{du}{dy}\eta = -\frac{u_0{}^2 \sin 2kx\, e^{-\beta y}}{4a}\left\{ e^{-\beta y} - \cos \beta y \right\},$$

so that

$$u' = -\frac{u_0{}^2 \sin 2kx}{8a}\left\{ e^{-\beta y}(4 \sin \beta y + 3e^{-\beta y}) + \tfrac{3}{2} - \tfrac{9}{2}\frac{(y_1 - y)^2}{y_1{}^2} \right\}. \quad \ldots(95)$$

When βy is small,

$$u' = -\frac{u_0{}^2 \sin 2kx}{8a}\left\{ -2\beta y + \ldots \right\}. \quad \ldots\ldots\ldots\ldots(96)$$

Inside the frictional layer the motion is in the same direction as just beyond it.

We have seen that the width of the direct current along the wall is $\cdot 423\, y_1$, and that of the return current (measured up to the plane of symmetry) is $\cdot 577\, y_1$, so that the direct current is distinctly narrower than the return current. This will be still more the case in a tube of circular section. The point under consideration depends only upon a complementary function analogous to (86), and is so simple that it may be worth while to investigate it.

The equation for ψ is

$$\left(\frac{d^2}{dr^2} - \frac{1}{r}\frac{d}{dr} - 4k^2 \right)^2 \psi = 0, \quad \ldots\ldots\ldots\ldots\ldots(97)$$

but if we suppose that the radius of the tube is small in comparison with λ, k^2 may be omitted. The general solution is

$$\psi = \{A + Br^2 + B'r^2 \log r + Cr^4\} \sin 2kx, \quad \ldots\ldots\ldots\ldots(98)$$

so that

$$u = \frac{1}{r}\frac{d\psi}{dr} = \{2B + B'\,(2\log r + 1) + 4Cr^2\}\sin 2kx,$$

whence $B' = 0$, by the condition at $r = 0$. Again,

$$v = -\frac{1}{r}\frac{d\psi}{dx} = -2k\,\{Ar^{-1} + Br + Cr^3\}\cos 2kx,$$

whence $A = 0$.

We may take therefore

$$u = \{2B + 4Cr^2\}\sin 2kx, \qquad v = -2k\,\{Br + Cr^3\}\cos 2kx. \quad \ldots(99)$$

If $v = 0$, when $r = R$, $B + CR^2 = 0$, and

$$u = 2C\,(2r^2 - R^2)\sin 2kx. \quad \ldots\ldots\ldots\ldots\ldots\ldots\ldots(100)$$

Thus u vanishes, when

$$r = \frac{R}{\sqrt{2}} = \cdot707\,R, \qquad R - r = \cdot293\,R.$$

The direct current is thus limited to an annulus of thickness $\cdot293\,R$, the return current occupying the whole interior, and having therefore a diameter of

$$2 \times \cdot707\,R = 1\cdot414\,R.$$

109.

THE FORM OF STANDING WAVES ON THE SURFACE OF RUNNING WATER.

[*Proceedings of the London Mathematical Society*, xv. pp. 69—78, 1883.]

THE present investigation had its origin in an attempt to explain more fully some interesting phenomena described by Scott Russell[*] and Thomson[†], and figured by the former. When a small obstacle, such as a fishing line, is moved forward slowly through still water, or (which of course comes to the same thing) is held stationary in moving water, the surface is covered with a beautiful wave-pattern, fixed relatively to the obstacle. On the up-stream side the wave-length is short, and, as Thomson has shown, the force governing the vibrations is principally cohesion. On the down-stream side the waves are longer, and are governed principally by gravity. Both sets of waves move with the same velocity relatively to the water; namely, that required in order that they may maintain a fixed position relatively to the obstacle. The same condition governs the velocity, and therefore the wave-length, of those parts of the wave-pattern where the fronts are oblique to the direction of motion. If the angle between this direction and the normal to the wave-front be called θ, the velocity of propagation of the waves must be equal to $v_0 \cos \theta$, where v_0 represents the velocity of the water relatively to the (fixed) obstacle.

Thomson has shown that, whatever the wave-length may be, the velocity of propagation of waves on the surface of water cannot be less than about 23 centims. per second. The water must run somewhat faster than this in order that the wave-pattern may be formed. Even then the angle θ is subject to a limit defined by $v_0 \cos \theta = 23$, and the curved wave-front has a corresponding asymptote.

The immersed portion of the obstacle disturbs the flow of the liquid

---

[*] *Brit. Assoc. Report* for 1844.
[†] *Phil. Mag.* Nov. 1871.

independently of the deformation of the surface, and renders the problem in its original form one of great difficulty. We may, however, without altering the essence of the matter, suppose that the disturbance is produced by the application to one point of the surface of a slightly abnormal pressure, such as might be produced by electrical attraction, or by the impact of a small jet of air. Indeed, either of these methods—the latter especially—gives very beautiful wave-patterns.

Even with this simplification, the difficulties remain considerable. It would appear. to be a necessary first step to solve the problem in two dimensions; that is, to find the standing wave-form produced in running water by the impact of a sheet of wind, which strikes the surface along a straight line. Of this I have succeeded in obtaining the solution, and it accounts satisfactorily for one of the leading features of the phenomenon,— the existence of the waves of small wave-length only on the up-stream side, and of the waves of greater wave-length only on the down-stream side of the place of disturbance. In terms of this solution, that of the original problem is analytically expressible, since we may imagine the pressure localised round a point to be the result of the superposition of an infinite system of linear pressures, whose lines of action pass through the point, and are distributed equally in every direction. But the expression in terms of an integral is not readily interpretable, and it is even doubtful—see (23)— whether it has a definite limit when the viscosity of the liquid is supposed to be infinitely small. In fact, that element of the integral which represents a system of parallel waves, travelling (perpendicularly to their own fronts) with the minimum velocity, has an infinite coefficient, as might perhaps have been expected from the corresponding problem for sound, where all waves travel with the same velocity. The prominence of this part of the system is a marked feature of the observed wave-pattern.

But, without an exact solution, it is possible to determine the form of the curved wave-fronts, considered as the envelope of a system of straight lines, and thus to obtain from theory a pretty good general idea of the phenomenon as a whole. In fig. 3 this construction is carried out for the particular case in which the asymptotes include a right angle.

Let us suppose that deep water, originally in motion with uniform velocity c parallel to the horizontal coordinate x, is disturbed slightly in two dimensions. If ϕ and ψ be the potential and stream functions, we may take

$$\phi = cx + \Sigma \alpha e^{-kz} \sin (kx + \epsilon), \quad \psi = cz - \Sigma \alpha e^{-kz} \cos (kx + \epsilon). \ldots \ldots (1)$$

In (1) z is measured downwards from the undisturbed surface, the wave-length is $2\pi/k$, and, for each value of k, α and ϵ are arbitrary. For the velocity at any point, we have, from (1),

$$\tfrac{1}{2} U^2 = \tfrac{1}{2} c^2 + c \Sigma k\alpha e^{-kz} \cos (kx + \epsilon). \ldots \ldots \ldots \ldots \ldots (2)$$

17—2

In calculating the pressure, we will suppose that the motion of each element is opposed by a retarding force proportional to the velocity*, of which therefore the components parallel to the axes may be denoted by $-hu$, $-hv$, h being positive. This (*Theory of Sound*, § 239) is not inconsistent with the existence of a velocity potential, but we must imagine a bodily force to act throughout the fluid sufficient to maintain the velocity c. The only other force acting within the fluid is gravity. Hence, on the supposition that the motion is steady, the equation for the pressure takes the form

$$p/\rho = \text{const.} + gz - h(\phi - cx) - \tfrac{1}{2}U^2$$
$$= \text{const.} + gz - h\Sigma\alpha e^{-kz}\sin(kx + \epsilon) - c\Sigma k\alpha e^{-kz}\cos(kx + \epsilon). \quad\ldots\ldots\ldots(3)$$

The equation of the surface, found from (1) by putting $\psi = 0$, is

$$cz = \Sigma\alpha\cos(kx + \epsilon). \quad\ldots\ldots\ldots\ldots\ldots\ldots\ldots(4)$$

Thus, for the variable part of the pressure just below the surface, we get

$$\delta p/\rho = \Sigma\alpha(gc^{-1} - kc)\cos(kx + \epsilon) - h\Sigma\alpha\sin(kx + \epsilon). \quad\ldots\ldots\ldots(5)$$

In passing from (5) to the expression for the pressure ϖ which must act externally upon the surface, we must include the effect of the capillary tension T. The curvature of the surface is

$$c^{-1}\Sigma\alpha k^2\cos(kx + \epsilon),$$

and thus

$$c\varpi/\rho = \Sigma\alpha(g + T'k^2 - kc^2)\cos(kx + \epsilon) - hc\Sigma\alpha\sin(kx + \epsilon), \quad\ldots\ldots(6)$$

in which T' is written for T/ρ.

If we introduce a new angle ϵ', defined by

$$\tan\epsilon' = \frac{hc}{g + T'k^2 - kc^2}, \quad\ldots\ldots\ldots\ldots\ldots\ldots(7)$$

(6) may be written

$$c\varpi/\rho = \Sigma\alpha\sqrt{\{(g + T'k^2 - kc^2)^2 + h^2c^2\}}\cos(kx + \epsilon + \epsilon'). \quad\ldots\ldots\ldots(8)$$

In the problem before us, we are to regard ϖ as given, and thence determine the form of the surface. If we suppose

$$\varpi/\rho = \Sigma\beta\cos(kx + e), \quad\ldots\ldots\ldots\ldots\ldots\ldots\ldots(9)$$

where β and e are given for each value of k, then $\epsilon + \epsilon' = e$, and

$$c\beta = \alpha\sqrt{\{(g + T'k^2 - kc^2)^2 + h^2c^2\}}.$$

Accordingly, by (4),

$$z = \Sigma\frac{\beta\cos(kx + e - \epsilon')}{\sqrt{\{(g + T'k^2 - kc^2)^2 + h^2c^2\}}}$$

$$= \Sigma\frac{\beta(g + T'k^2 - kc^2)\cos(kx + e) + \beta hc\sin(kx + e)}{(g + T'k^2 - kc^2)^2 + h^2c^2}, \quad\ldots\ldots(10)$$

gives the equation to the surface corresponding to the applied pressures (9).

* January, 1884. The dissipative forces here introduced are ultimately supposed to vanish, but without them it did not seem easy to interpret the analytical expressions to which we are led.

If we suppose that h is small, and limit ourselves to the case of a single train of waves, i.e., to a single value of k, we see that the phases of ϖ and z are in general coincident or opposite, according as $(g + T'k^2 - kc^2)$ is positive or negative. The first case arises when the wave-length is either very great or very small, and then the pressure is in excess over the troughs and in defect over the crests of the waves. The actual velocity of the waves relatively to the water (c) is here less than that of free waves of the given wave-length, i.e.,

$$\sqrt{\{g/k + T'k\}}.$$

But when the actual velocity c is greater than that of free waves of the given wave-length, $(g + T'k^2 - kc^2)$ is negative, and then the excess of pressure is to be found over the crests, and the defect of pressure over the troughs of the waves. In the case of transition, when c coincides with the velocity of the free waves, the term in h must be retained, and it shows that the place of maximum pressure is now at that *shoulder* of the wave where the water in its forward motion is falling.

In general, when the pressure along the surface is arbitrary, we must have recourse to Fourier's theorem. Thus

$$\frac{\varpi}{\rho} = \phi(x) = \frac{1}{\pi} \int_0^\infty dk \int_{-\infty}^{+\infty} \phi(v) \cos k(v - x)\, dv, \quad \dots\dots\dots(11)$$

which is of the form (9).

We now suppose that the abnormal pressure is confined to a very narrow strip at $x = 0$, so that $\phi(v) = 0$, except when v is very small. In this case (11) may be written

$$\frac{\varpi}{\rho} = \frac{1}{\pi} \int \phi(v)\, dv \int_0^\infty dk \cos kx = \frac{1}{\pi} \Phi \int_0^\infty dk \cos kx, \quad \dots\dots(12)$$

if we put Φ for $\int \phi(v)\, dv$.

The corresponding value of z, from (10), is

$$z = \frac{1}{\pi} \Phi \int_0^\infty dk \frac{(g + T'k^2 - kc^2)\cos kx + hc \sin kx}{(g + T'k^2 - kc^2)^2 + h^2c^2}; \quad \dots\dots(13)$$

and this gives the form of surface assumed by the running water when subjected to a small excess of pressure acting over a narrow strip at the origin.

Before entering upon the general integration and interpretation of (13), it may be well to point out its application in the case where the water is originally at rest $(c = 0)$. The formula (13) then reduces to

$$\frac{1}{\pi} \Phi \int_0^\infty dk \frac{\cos kx}{g + T'k^2} = \frac{\Phi\, e^{\mp \sqrt{(g/T')}\, x}}{2 \sqrt{(gT')}},$$

the upper sign being taken when x is positive, and the lower when x is negative.

This solution of the statical problem may of course be obtained independently from the differential equation

$$- T' \frac{d^2 z}{dx^2} + gz = \frac{\varpi}{\rho}.$$

In the subsequent treatment of (13) it will conduce to brevity if we put unity for c and T', symbols which can always be restored when desirable from considerations of dimensions. We have, then, to consider

$$\int_0^\infty dk \, \frac{(g - k + k^2) \cos kx + h \sin kx}{(g - k + k^2)^2 + h^2}, \quad \dots\dots\dots\dots\dots(14)$$

and it will assume different forms according as the roots of

$$g - k + k^2 = 0$$

are real or imaginary. For the present, we will take the former alternative, which is equivalent to supposing that the velocity of the water exceeds the minimum velocity of propagation of free waves. We assume, accordingly, that

$$g - k + k^2 = (k_1 - k)(k_2 - k),$$

where

$$k_1 k_2 = g, \quad k_1 + k_2 = 1.$$

The quantities k_1, k_2 are positive, and we will suppose them to be in ascending order of magnitude. We thus replace (14) by

$$\int_0^\infty dk \, \frac{(k_1 - k)(k_2 - k) \cos kx + h \sin kx}{(k_1 - k)^2 (k_2 - k)^2 + h^2}, \quad \dots\dots\dots\dots(15)$$

and of this integral we shall require only the limiting form when $h = 0$, as we do not propose to consider in general the effect of finite dissipative forces. On this understanding, the first part of the integral may be replaced at once by

$$\int_0^\infty dk \, \frac{\cos kx}{(k_1 - k)(k_2 - k)} = \frac{1}{k_2 - k_1} \left\{ \int_0^\infty \frac{\cos kx \, dk}{k_1 - k} - \int_0^\infty \frac{\cos kx \, dk}{k_2 - k} \right\} \dots\dots(16)$$

The integrals which make up (16) are even functions of x, i.e., they take the same arithmetical values whether x be positive or negative. For distinctness, we will suppose that x is positive. Now

$$\int_0^\infty \frac{\cos kx \, dk}{k_1 - k} = \int_{-\infty}^{k_1 x} \frac{\cos (k_1 x - u) \, du}{u} = \cos k_1 x \int_{-\infty}^{k_1 x} \frac{\cos u \, du}{u} + \sin k_1 x \int_{-\infty}^{k_1 x} \frac{\sin u \, du}{u},$$

in which

$$\int_{-\infty}^{k_1 x} \frac{\cos u \, du}{u} = -\int_{k_1 x}^\infty \frac{\cos u \, du}{u} = \operatorname{ci}(k_1 x),$$

$$\int_{-\infty}^{k_1 x} \frac{\sin u \, du}{u} = \tfrac{1}{2}\pi + \int_0^{k_1 x} \frac{\sin u \, du}{u} = \tfrac{1}{2}\pi + \operatorname{si}(k_1 x).$$

The functions ci and si may be regarded as known functions, and have been fully tabulated by Glaisher*. Thus

$$\int_0^\infty \frac{\cos kx\, dk}{k_1 - k} = \cos k_1 x \; \text{ci}\,(k_1 x) + \sin k_1 x \left\{ \tfrac{1}{2}\pi + \text{si}\,(k_1 x) \right\}, \;\ldots\ldots (17)$$

and

$$\int_0^\infty \frac{\cos kx\, dk}{(k_1 - k)(k_2 - k)} = \frac{1}{k_2 - k_1}\left[\begin{array}{l} \cos k_1 x \; \text{ci}\, k_1 x + \sin k_1 x \left(\tfrac{1}{2}\pi + \text{si}\, k_1 x \right) \\ - \cos k_2 x \; \text{ci}\, k_2 x - \sin k_2 x \left(\tfrac{1}{2}\pi + \text{si}\, k_2 x \right) \end{array} \right]. \;\ldots (18)$$

When $\quad\quad k_1 x = \infty, \quad\quad\quad \text{ci}\, k_1 x = 0, \quad\quad\quad\quad \text{si}\, k_1 x = \tfrac{1}{2}\pi\,;$

when $\quad\quad\; k_1 x = 0, \quad\quad\quad\; \text{ci}\, k_1 x = -\infty, \quad\quad\quad\; \text{si}\, x = 0.$

In the latter case the limiting form for ci $(k_1 x)$ is

$$\text{ci}\,(k_1 x) = \gamma + \tfrac{1}{4}\log_e (k_1 x)^4 - \tfrac{1}{2}\frac{k_1^2 x^2}{1\,.\,2} + \ldots\,; \quad\ldots\ldots\ldots (19)$$

so that, when $x = 0$,

$$(18) = \frac{1}{k_2 - k_1}\log\frac{k_1}{k_2},$$

which is finite in all cases.

When $k_2 = k_1$, (18) changes its form and is replaced by

$$-\frac{d}{dk}\left\{ \cos kx \; \text{ci}\, kx + \sin kx \left(\tfrac{1}{2}\pi + \text{si}\, kx \right) \right\},$$

that is,

$$x \left\{ \sin kx \; \text{ci}\, kx - \cos kx \left(\tfrac{1}{2}\pi + \text{si}\, kx \right) - 1/kx \right\}. \quad\ldots\ldots\ldots (20)$$

We have now to consider the second part of (15), that is, the limit when $h = 0$ of

$$\int_0^\infty \frac{h \sin kx\, dk}{(k_1 - k)^2 (k_2 - k)^2 + h^2}. \quad\ldots\ldots\ldots\ldots\ldots\ldots (21)$$

With respect to this, it is evident that the only elements of the integral which contribute to the limiting value are those for which the denominator vanishes with h, i.e., those lying in the immediate neighbourhood of the roots k_1 and k_2. Thus, as k passes through k_1 we may put $(k_2 - k)$ equal to $(k_2 - k_1)$, and as k passes through k_2 we may put $(k_1 - k)$ equal to $(k_1 - k_2)$. Hence the limit of (21) is the same as the limit of

$$\int_0^\infty \frac{h \sin kx\, dk}{(k_1 - k)^2 (k_2 - k_1)^2 + h^2} + \int_0^\infty \frac{h \sin kx\, dk}{(k_1 - k_2)^2 (k_2 - k)^2 + h^2},$$

or the same as the limit of

$$\frac{1}{k_2 - k_1}\int_0^\infty \frac{h' \sin kx\, dk}{(k_1 - k)^2 + h'^2} + \frac{1}{k_2 - k_1}\int_0^\infty \frac{h'' \sin kx\, dk}{(k_2 - k)^2 + h''^2},$$

* "Tables of the Numerical Values of the Sine-integral, Cosine-integral, and Exponential-integral," *Phil. Trans.* 1870.

in which, k_2 being greater than k_1, h' and h'' are positive, and are supposed ultimately to vanish.   Now

$$\int_0^\infty \frac{h' \sin kx\, dk}{(k_1 - k)^2 + h'^2} = \sin k_1 x \int_{-\infty}^{k_1} \frac{h' \cos ux\, du}{u^2 + h'^2} - \cos k_1 x \int_{-\infty}^{k_1} \frac{h' \sin ux\, du}{u^2 + h'^2} \; ;$$

and

$$\lim. \int_{-\infty}^{k_1} \frac{h' \sin ux\, du}{u^2 + h'^2} = \lim. \int_{-\infty}^{+\infty} \frac{h' \sin ux\, du}{u^2 + h'^2} = 0,$$

$$\lim. \int_{-\infty}^{k_1} \frac{h' \cos ux\, du}{u^2 + h'^2} = \lim. \int_{-\infty}^{+\infty} \frac{h' \cos ux\, du}{n^2 + h'^2}$$

$$= \lim. \int_{-\infty}^{+\infty} \frac{\cos (h' xv)\, dv}{1 + v^2} = \lim. \pi e^{-h' x} = \pi.$$

Accordingly, the limit of (21) is

$$\frac{\pi \sin k_1 x + \pi \sin k_2 x}{k_2 - k_1}, \quad \ldots\ldots\ldots\ldots\ldots\ldots(22)$$

and retains the same form whether x be positive or negative.

It is evident that in the case of equal roots (22), unlike (18), becomes infinite, so that the retention of h is necessary for a practical result.   It is not difficult to show that, when h is very small,

$$\int_0^\infty \frac{h \sin kx\, dk}{(k_1 - k)^4 + h^2} = \frac{\pi \sin k_1 x}{\sqrt{(2h)}}, \quad \ldots\ldots\ldots\ldots\ldots(23)$$

which therefore represents for this case the leading term of the complete expression (15).

Combining (18) and (22), we see that, when x is large and positive, the value of (15) is

$$\frac{2\pi \sin k_1 x}{k_2 - k_1}, \quad \ldots\ldots\ldots\ldots\ldots\ldots(24)$$

and that, when x is large and negative, the value of (15) is

$$\frac{2\pi \sin k_2 x}{k_2 - k_1}. \quad \ldots\ldots\ldots\ldots\ldots\ldots(25)$$

On both sides of the place of disturbance, the surface is covered with waves whose free velocity is that of the water.   On the down-stream side (x positive) the wave-length is the greater of the two which satisfy the condition ($k_1 < k_2$); on the up-stream side it is the smaller.   In the immediate neighbourhood of the place of disturbance the form is a little more complicated, and is best understood from a drawing.

When the roots of $g - k + k^2 = 0$ are imaginary, which happens when the velocity of the water is less than that of any free wave, the analytical expressions change their form.   The *second* part of (14), written separately in (21), vanishes when $h = 0$, the denominator being always finite. For the first part we have, in place of (16),

$$\int_0^\infty \frac{\cos kx \, dk}{g - k + k^2} = \int_0^\infty \frac{\cos kx \, dk}{(k - \tfrac{1}{2})^2 + g - \tfrac{1}{4}}$$

$$= \cos \tfrac{1}{2} x \int_{-\frac{1}{2}}^\infty \frac{\cos ux \, du}{u^2 + g - \tfrac{1}{4}} - \sin \tfrac{1}{2} x \int_{-\frac{1}{2}}^\infty \frac{\sin ux \, du}{u^2 + g - \tfrac{1}{4}}, \ldots \ldots (26)$$

in which $g - \tfrac{1}{4}$ is positive.

So far as I have been able to learn, the integrals in (26), or others equivalent to them, have not been tabulated. On this side, therefore, the solution of our problem is incomplete, but fortunately this is not the case to which the most interest attaches. It is probable that the disturbance is limited to the immediate neighbourhood of the origin.

For the numerical calculation, it will be convenient to write (17) in the form

$$\cos k_1 x \ \text{ci} \ k_1 x + \sin k_1 x \ (\text{si} \ k_1 x - \tfrac{1}{2}\pi), \ldots \ldots \ldots \ldots (27)$$

$$+ \pi \sin k_1 x,$$

of which the part (27) vanishes when x is great enough. The value of (27) as a function of kx is shown by curve A (fig. 1). It is negative throughout, and infinite when $kx = 0$.

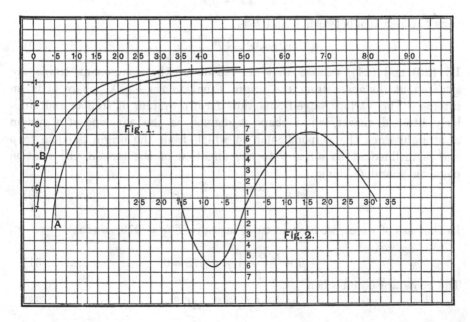

Fig. 1.

Fig. 2.

The form of the standing wave produced by the local application of pressure to the surface depends upon the velocity of the water. To take a case, we will suppose that this is such that the wave-lengths before and behind are in the ratio of $1 : 2$, so that $k_2 = 2k_1$. The value of

$$\cos k_1 x \text{ ci } k_1 x + \sin k_1 x \, (\text{si } k_1 x - \tfrac{1}{2}\pi)$$
$$- \cos 2k_1 x \text{ ci } 2k_1 x - \sin 2k_1 x \, (\text{si } 2k_1 x - \tfrac{1}{2}\pi) \ldots \ldots (28)$$

is shown by curve B (fig. 1), and the ordinates are to have the same value when x is negative as when x is positive. The part near the origin is filled in from the approximate analytical value

$$- \log_e 2 + \tfrac{1}{2}\pi k_1 x.$$

The wave-form is now easily deduced, and is shown in fig. 2. On the positive side we are to add to (28) $2\pi \sin k_1 x$, and on the negative side we are to add $2\pi \sin 2k_1 x*$.

We now pass to the consideration of the effect of a pressure localized near a point, instead of distributed along a line. The wave-form is to be found by the superposition of an infinite series of systems similar to (24), (25), at various degrees of obliquity (θ), and of such wave-lengths that

$$v_0 \cos \theta = v,$$

v being the velocity perpendicular to the wave-front in each case, and v_0 the velocity of the water (previously denoted by c). Now

$$v^2 = kT' + g/k ;$$

thus the relation between k and θ is

$$v_0{}^2 \cos^2 \theta = kT' + g/k. \quad \ldots \ldots \ldots \ldots \ldots (29)$$

By (23) and (24), we see that the crests of the component trains are situated at distances from the origin equal to $(m + \tfrac{1}{4})\lambda$, where m is an integer. The various wave-fronts thus form a system of similar and similarly situated curves, whose shape is defined as the envelope of a system of straight lines, the perpendicular on which from the origin is equal to p and is inclined at an angle θ to the direction of the stream, the relation between p and θ being

$$p^2 + \frac{v_0{}^2 \cos^2 \theta}{g} p + \frac{T'}{g} = 0. \quad \ldots \ldots \ldots \ldots \ldots (30)$$

In the case of water, we have in C.G.S. measure $T'/g = {\cdot}073$, so that

$$p^2 + \frac{v_0{}^2 \cos^2 \theta}{g} p + {\cdot}073 = 0. \quad \ldots \ldots \ldots \ldots \ldots (31)$$

The roots are equal when

$$v_0{}^2 \cos^2 \theta \, / \, g = 2 \sqrt{({\cdot}073)}. \quad \ldots \ldots \ldots \ldots \ldots (32)$$

---

* [1899. Two sentences, vitiated by an error pointed out to me by Lord Kelvin, are here omitted.]

The case proposed for consideration is that in which the asymptotes include an angle of 90°, so that the maximum value of θ is 45°. Substituting this in (32), we find, $v_0^2/g = 1\cdot081$, and thus

$$p = \cdot5405 \cos^2 \theta \pm \sqrt{\{\cdot5405^2 \cos^4 \theta - \cdot0730\}}. \quad \ldots\ldots\ldots\ldots(33)$$

From this equation we may calculate any number of corresponding values of p and θ, and thus draw the tangents of which the required curves are the envelopes. The annexed table contains a few such pairs of values, sufficient for an indication of the forms of the curves :—

θ	p_1	p_2	$\cos \theta$	$p_1/\cos \theta$	$p_2/\cos \theta$
0°	1·0086	·0724	1·0000	1·0086	·0724
9°	·9800	·0744	·9877	·9922	·0753
18°	·8964	·0814	·9511	·9425	·0856
27°	·7625	·0957	·8910	·8557	·1074
36°	·5823	·1253	·8090	·7196	·1549
45°	·2702	·2702	·7071	·3821	·3821

The two last columns give the intercepts on the axis, by means of which and the value of $\cos \theta$ the lines are more readily drawn than from the perpendiculars themselves. The result is shown in fig. 3.

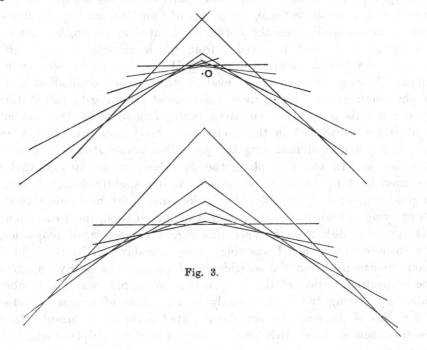

Fig. 3.

110.

ACOUSTICAL OBSERVATIONS.—V.

[*Philosophical Magazine*, XVII. pp. 188—194, 1884.]

Smoke-jets by Intermittent Vision.

In the second series of these observations (*Phil. Mag.* 1879 [vol. I. p. 406]) I proved that when stationary sonorous waves occupy the region surrounding a sensitive flame, the action of sound in causing the flame to flare manifests itself when the burner is situated at a loop, but not when the burner is situated at a node; from which we infer that the effects are due to a lateral disturbance causing the issuing jet to bend from its course. During the same year I made a stroboscopic examination of a jet of phosphorus-smoke issuing from a drawn-out glass nozzle, and disturbed by the neighbourhood of a vibrating tuning-fork of pitch 256. So much light is necessarily lost in this method of observation that some precaution is required in illuminating the jet. Two points should be especially attended to. In the first place, the eye must be so situated that the scattered light by which the jet is seen is but slightly deflected from its original course; and, secondly, the background must be thoroughly dark. By carrying out adequately this system of illumination, and by so choosing the revolving disk that the apertures bore a not too small proportion to the entire circumference, I was able to see tolerably well by the light of a good gas-flame. When the coincidence of periods was nearly approached, the serpentine motion of the jet previous to rupture was clearly observable. By placing the nozzle exactly in the plane of symmetry between the prongs of the fork, the law above stated could be confirmed. In this position there was but little effect; but the slightest displacement caused an early rupture.

Smoke-jets and Resonators.

In order to exalt the sensitiveness of jets to notes of moderate pitch, I found the use of resonators advantageous. These may be of Helmholtz's pattern; but suitably selected wide-mouth bottles answer the purpose. What is essential is that the jet should issue from the nozzle in the region of rapid reciprocating motion at the mouth of the resonator, and in a transverse direction. I usually placed the resonator's mouth uppermost, so that the jets were horizontal.

Good results were obtained at a pitch of 256. When two forks of about this pitch, and slightly out of tune with one another, were allowed to sound simultaneously, the evolutions of the smoke-jet in correspondence with the audible beats were very remarkable. By gradually raising the pressure at which the smoke is supplied, in the manner usual in these experiments, a high degree of sensitiveness may be attained, either with a drawn-out glass nozzle or with the steatite pin-hole burner used by Tyndall. In some cases (even at pitch 256) the combination of jet and resonator proved almost as sensitive to sound as the ear itself.

The behaviour of the sensitive jet does not depend upon the smoke-particles, whose office is merely to render the effects more easily visible. I have repeated these observations without smoke by simply causing air-jets from the same nozzles to impinge upon the flame of a candle placed at a suitable distance. In such cases, as has been pointed out by Tyndall, the flame acts merely as an indicator of the condition of the otherwise invisible jet. Even without a resonator the sensitiveness of such jets to hissing sounds may be taken advantage of to form a pretty experiment.

The combination of jet, resonator, and flame shows sometimes a tendency to *speak* on its own account; but I did not succeed in getting a well-sustained sound. Such as it is, the effect probably corresponds to one observed by Savart and Plateau with water-jets breaking up under the operation of the capillary tension, and when resolved into drops impinging upon a solid obstacle, such as the bottom of a sink, in mechanical connexion with the nozzle from which the jet originally issues. In virtue of the connexion, any regular cycle in the mode of disintegration is able, as it were, to propagate itself.

The increased and more discriminating sensitiveness obtainable by use of resonators is turned to account in the arrangement of flame described in the *Proceedings* of the Cambridge Philosophical Society for November 8, 1880. [Vol. I. p. 500.]

In this case the resonator takes the form of a tube, one of whose ends opens in the gas-chamber close to the nozzle. The other end is closed by a cork, whose position can be adjusted so as to vary the pitch. I see from my note-book that, on the evening of Dec. 4, 1879, I found the flame nearly as sensitive as the ear to vibrations of frequency 512; but I have not always been equally successful in subsequent attempts to recover this degree of delicacy.

With the very acute sounds, to which alone the high-pressure gas-flame (lighted at the burner) is sensitive, little can be expected from the use of resonators.

Jets of Coloured Liquid.

In the hope of being able to make better observations upon the transformations of unstable jets, I next had recourse to coloured water issuing under water. In this form the experiment is more manageable than in the case of smoke-jets, which are difficult to light, and liable to be disturbed by the slightest draught. Permanganate of potash was preferred as a colouring agent, and the colour may be discharged by mixing with the general mass of liquid a little acid ferrous sulphate. The jets were usually projected downwards into a large beaker or tank of glass, and were lighted from behind through a piece of ground glass.

The notes of maximum sensitiveness of these liquid jets were found to be far graver than for smoke-jets or for flames. Forks vibrating from 20 to 50 times per second appeared to produce the maximum effect, to observe which it is only necessary to bring the stalk of the fork into contact with the table supporting the apparatus. The general behaviour of the jet could be observed without stroboscopic appliances by causing the liquid in the beaker to vibrate from side to side under the action of gravity. The line of colour proceeding from the nozzle is seen to become gradually more and more sinuous, and a little further down presents the appearance of a rope bent backwards and forwards upon itself. I have followed the process of disintegration with gradually increasing frequencies of vibrational disturbance from 1 or 2 per second up to about 24 per second, using electro-magnetic interruptors to send intermittent currents through an electro-magnet which acted upon a soft-iron armature attached to the nozzle. At each stage the pressure at which the jet is supplied should be adjusted so as to give the right degree of sensitiveness. If the pressure be too great, the jet flares independently of the imposed vibration, and the transformations become irregular: in the contrary case the phenomena, though usually observable, are not so well marked as when a suitable adjustment is made. After a little practice it is possible to interpret pretty well what is seen directly; but in order to have before the eye an image of what is really going on, we must have recourse to intermittent vision.

The best results are obtained with two forks slightly out of tune, one of which is used to effect the disintegration of the jet, and the other (by means of perforated plates attached to its prongs) to give an intermittent view. The difference of frequencies should be about one per second. When the means of obtaining uniform rotation are at hand, a stroboscopic disk may be substituted for the second fork. It was, in fact, with the use of such a disk, driven by a water-engine, that the drawing (fig. 1) was made by Mrs Sidgwick in August 1880. It is hardly necessary to say that these appearances are difficult to reproduce in drawings, and that the result must be regarded merely as giving a general idea of what is actually

Fig. 1.

observed. The upper part of the jet is seen sufficiently steadily to be pretty accurately copied; but further down true periodicity is lost, and no steady impression is produced upon the eye.

The carrying out of these observations, especially when it is desired to make a drawing, is difficult unless we can control the plane of the bendings. In order to see the phases properly it is necessary that the plane of bendings should be perpendicular to the line of vision; but with a symmetrical nozzle this would occur only by accident. The difficulty may be got over by slightly nicking the end of the drawn-out glass nozzle at two opposite points. In this way the plane of bending is usually rendered determinate, being that which includes the nicks, so that by turning the nozzle round its axis the sinuosities of the jet may be properly presented to the eye.

Occasionally the jet appears to divide itself into two parts imperfectly connected by a sort of sheet. This appears to correspond to the duplication of flames and smoke-jets under powerful sonorous action, and to be due to what we may regard as the broken waves taking alternately different courses.

Fish-tail Burners.

"Experiments upon jets from fish-tail burners*.—As with gas, so with smoke and coloured water, these are sensitive, and when much excited throw out tall streamers in the perpendicular plane. I have not yet fully succeeded in tracing the genesis of these, but believe them due to the rupture or collision of the sinuosities which are formed in the quickly-moving part of the sheet. When the sheet, seen broadways on, is excited by slow vibrations, a line of deepened colour is seen to descend, and presently becomes very deep. This means that the sheet is so far bent over as to be seen tangentially."

Even with the best arrangements as to sensitiveness and intermittent vision, the appearances presented by these jets are somewhat difficult to interpret and to reproduce in a drawing. The jets shown in figs. 2—5 issued from flattened glass nozzles, and are of the same character as those given by fish-tail burners. In fig. 2 the flat side is presented to the

Fig. 2.   Fig. 3.   Fig. 4.   Fig. 5.

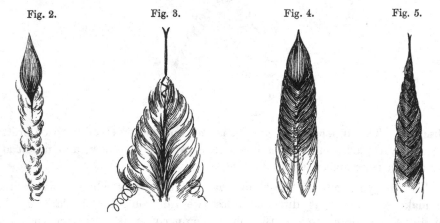

observer; in fig. 3 the sheet (if undisturbed) would be seen edgeways. The complication arises, partly at any rate, from the different degrees of sensitiveness of different parts of the sheet, from which it results that one part reaches disruption and loses its periodicity, while another is yet in the earlier stages of the transformation. In figs. 2 and 3 the jet is under the influence of a vibration sufficiently powerful to cause it to flare in a regular manner; in figs. 4 and 5 the vibration is less powerful, and the transformations stop short of the final stage.

* *Laboratory Note-book*, Dec. 12, 1879.

Influence of Viscosity.

It has already been noticed that the notes appropriate to water-jets are far graver than for air-jets from the same nozzles. Moreover, the velocities suitable in the former case are much less than in the latter. This difference relates not, as might perhaps be at first supposed, to the greater density, but to the smaller viscosity of the water, measured of course kinematically. It is not difficult to see that the density, presumed to be the same for the jet and surrounding fluid, is immaterial, except of course in so far as a denser fluid requires a greater pressure to give it an assigned velocity. The influence of fluid viscosity upon these phenomena is explained in a former paper on the Stability or Instability of certain Fluid Motions*; and the laws of dynamical similarity with regard to fluid friction, laid down by Prof. Stokes†, allow us to compare the behaviour of one fluid with another. The dimensions of the kinematic coefficient of viscosity are those of an area divided by a time. If we use the same nozzle in both cases, we must keep the same standard of length; and thus the times must be taken inversely, and the velocities directly, as the coefficients of viscosity. In passing from air to water the pitch and velocity are to be reduced some ten times. But, in spite of the smaller velocity, the water-jet will require the greater pressure behind it, inasmuch as the densities differ in a ratio exceeding 100 : 1.

Guided by these considerations, I made experiments to try whether the jets would behave differently in warm and cold water. At temperatures respectively about 130° F. and 52° F., the difference was found to be extremely well marked. "With a drawn-out glass nozzle, a pressure of $1\frac{1}{2}$ inch was enough with hot water to cause flaring, whereas perhaps $3\frac{1}{2}$ inches were necessary with the cold water. At one inch the jet in cold water was dead, but in hot water was still quite active ‡."

These experiments were resumed at Cambridge in April and May 1880 by Mrs Sidgwick, with use not only of hot and cold water but also of mixtures of alcohol and water, whose viscosity is known to be much greater than that of water alone. In order to retard cooling, and thus to diminish convection-currents, the experimental beaker was placed within a larger one, and supported at the rim only, so as to be surrounded by a jacket of warm air. The liquid intended to form the jet was placed in a narrow

* *Math. Soc. Proc.* Feb. 12, 1880. [Vol. I. p. 474.]

† *Camb. Phil. Trans.* 1850, "On the Effect of Internal Friction of Fluids on the Motion of Pendulums," § 5. See also Helmholtz, *Wied. Ann.* Bd. VII. p. 337 (1879), or Reprint, vol. I. p. 891.

‡ *Laboratory Note-book*, Jan. 20, 1880. Prof. Osborne Reynolds has availed himself of differences of temperature in order to vary the viscosity, in some recent important observations upon the cognate subject of the flow of water in tubes, *Proc. Roy. Soc.* March 15, 1883.

glass jar about 10 inches high, and the head was adjusted by raising or lowering the jar and by varying the amount of liquid. The communication between the two vessels was by a glass syphon, whose lower end was drawn out so as to form a suitable nozzle of about $\frac{1}{40}$ inch diameter (fig. 6). The transparent tube was advantageous on account of the more ready detection of air-bubbles, the presence of which, especially near the nozzle, is a source of disturbance. The apparatus stood in front of a window, supported on a stone table carried by the walls of the building, and the sensitiveness of the jet was usually tested by dropping upon the table a large nail through a height of about 2 inches. Observations were made of the greatest pressure that the jet would bear, in the absence of

Fig. 6.

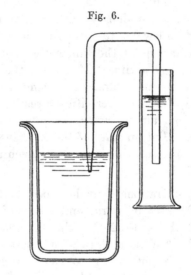

external disturbance, without flaring before reaching the bottom of the beaker, and also of the least pressure at which the jet was sensitive. In the case of the mixture of alcohol and water in equal parts, a modified arrangement was necessary in order to obtain sufficient head.

With plain water the colour was given by permanganate of potash, and was discharged, as soon as the jet was broken up, by ferrous sulphate previously added to the liquid in the beaker. In some of the more delicate experiments it was found necessary to bring the densities to a more exact equality by the addition of indifferent saline material to the jet, but in most cases this precaution is superfluous. For the jets of methylated alcohol and water, permanganate was found unsuitable, and was replaced by soluble aniline blue.

The following table will give an idea of the results of a large number of observations carefully repeated on different occasions:

	Maximum pressure borne without flaring before reaching bottom	Minimum pressure required for sensitiveness
Meth. spirit and water, 1 : 1	Between 21 inches and 30 inches	About 21 inches
Meth. spirit and water, 1 : 3	About 12 inches	From 7 to $8\frac{3}{4}$ inches
Meth. spirit and water, 1 : 7	From 8 to 9 inches	About $2\frac{1}{2}$ inches
Cold water 12°·5 Cent.	$4\frac{1}{2}$ inches	$2\frac{1}{4}$ inches
Hot water, 61° Cent.	About $\frac{1}{2}$ inch	About $\frac{1}{2}$ inch
Hot water, 84° Cent.	About $\frac{1}{4}$ inch	About $\frac{1}{4}$ inch

It will be seen that the effect of varying the viscosity is very distinct. In the extreme cases a jet which will not bear a pressure of more than $\frac{1}{4}$ inch without flaring when the liquid is water nearly at the boiling-point, requires from 20 to 30 inches to make it flare when we pass to the mixture of alcohol and water in equal proportions. It is probable that with the hot water the results are somewhat disturbed by the impossibility of getting the liquid perfectly quiet owing to the formation of convection-currents; but the differences are so great that the general conclusion cannot be disturbed by this circumstance. The comparisons between the cold water and the alcoholic mixtures, which were all used at the temperature of the room, are entirely independent of this source of error.

111.

ON THE MEASUREMENT OF THE ELECTRICAL RESISTANCE BETWEEN TWO NEIGHBOURING POINTS ON A CONDUCTOR.

[Camb. Phil. Soc. Proc., v. pp. 133, 134, 1884.]

As an alternative to the method of Matthiessen and Hockin, the following process may be used with good effect for the above purpose. It is founded upon the combination of resistances described in a paper upon the determination of the ohm by Lorenz's method (*Phil. Trans.* 1883 [Art. 94]), by which a small *effective* resistance is obtained from elements of moderate and accurately measurable value.

The main current γ passes principally through a shunt of resistance a, but partly through a small resistance b and a large resistance c arranged in series. The terminals of a galvanometer of resistance g are connected to the extremities of b. If g were infinite, the difference of potentials at its electrodes would be

$$\frac{ab}{a+b+c}\gamma,$$

so that $ab/(a+b+c)$ is the effective resistance of the combination. For example, if $a=1$, $b=1$, $c=98$, the effective resistance is $\frac{1}{100}$, and notwithstanding its smallness is susceptible of accurate determination. Suppose now that the main current traverses also a German silver strip (*Proc.,* Nov. 26, 1883 [Art. 107]) provided with tongues between which we require to know the resistance. It is evident that by adjustment of c the combination may be made to give the same effect upon the galvanometer as the German silver strip, so that the required result would be readily obtained from the above formula. If c is taken from a resistance-box, we may find the effects, one greater and one less than that of the strip, corresponding to resistances c_0

and $c_0 + 1$, whence the value that would give exactly the same effect is deduced by interpolation. In order to guard against disturbance from thermo-electricity the readings should be taken by *reversal* of the battery, and to eliminate the effects of varying current the combination and the strip should be interchanged as rapidly as possible.

In practice the resistance of the galvanometer could not usually be treated as infinite, and the interpretation of the results is a little more complicated. In the case of the combination it may be shewn that the current through the galvanometer is

$$\frac{ab\gamma}{g(a+b+c)+b(a+c)}.$$

By putting a infinite, or otherwise, we see that the corresponding current for the strip is $x\gamma(g+x)$, if x be the required resistance between the tongues. Equating these, we find

$$x = \frac{abg}{g(a+b+c)+bc}.$$

This method has recently been tested in the Cavendish Laboratory by Messrs Shackle and Ward, and the results appear to shew that even with so moderate a main current as ·2 ampère, the sensitiveness is sufficient, the mean of a few readings being probably correct to $\frac{1}{1000}$.

112.

ON THE ELECTRO-CHEMICAL EQUIVALENT OF SILVER, AND
ON THE ABSOLUTE ELECTROMOTIVE FORCE OF CLARK
CELLS. By LORD RAYLEIGH, D.C.L., F.R.S., AND MRS H. SIDGWICK.

[*Philosophical Transactions*, 175, pp. 411—460, 1884.]

§ 1. IN former communications* to the Royal Society we have investigated the absolute unit of electrical resistance, and have expressed
it in terms of the B.A. unit and of a column of mercury at 0° of known
dimensions. The complete solution of the problem of absolute electrical
measurement involves, however, a second determination, similar in kind,
but quite independent of the first. In addition to resistance, we require
to know some other electrical quantity, such as current or electromotive
force. So far as we are aware, all the methods employed for this purpose
define, in the first instance, an electrical current; but as a current cannot,
like a resistance, be embodied in any material standard for future use,
the result of the measurement must be recorded in terms of some effect.
Thus, several observers have determined the quantity of silver deposited,
or the quantity of water decomposed, by the passage of a known current
for a known time. In this case the definition relates not so much to
electric current as to electric quantity. A more direct definition of the
unit current, and one which may perhaps be of practical service for the
measurement of strong currents of 50 ampères or more, would be in terms
of the rotation of the plane of polarisation of sodium light, which traverses
a long column of bisulphide of carbon enveloped by the current a given
number of times†.

Other observers have expressed their results as a measurement of the
electromotive force of a standard galvanic cell. In this case it is neces-

* *Proceedings*, April 12, 1881 [vol. II. p. 1]; *Phil. Trans.* 1882, Part II. [vol. II. p. 38]
and 1883, Part I. [vol. II. p. 155].

† See *Camb. Phil. Proc.* Nov. 26, 1883 [vol. II. p. 237].

sary to assume a knowledge of resistances. The known current in passing a known resistance gives rise to a known electromotive force, which is compared with that of the cell.

In the present communication are detailed the experiments that we have made to determine the electro-chemical equivalent of silver, and the electromotive force of standard Clark cells. As regards the choice of *silver* there is not much room for a difference of opinion. The difficulties to be overcome in the use of a water voltameter are much greater. *Copper* is, indeed, employed in ordinary laboratory practice and for commercial purposes; but it is decidedly inferior to silver, both on account of its tendency to oxidise when heated in the air, and also because it changes weight in contact with copper sulphate solution without the passage of an electric current. Dr Gore* has made observations upon this subject, and our own experience has shown that no constancy of weight is to be found under these circumstances. Silver, on the other hand, seems to be entirely unaffected by contact with neutral solution of the nitrate.

§ 2. The readiest method of measuring currents is, perhaps, that followed by Kohlrausch, both in his earlier† and in his recent‡ work upon this subject, viz., to refer the current to the earth's horizontal magnetic intensity (H) with an absolute galvanometer. The constant of the galvanometer is readily found from the data of construction with the necessary accuracy, and there is no doubt that in a well-equipped magnetic observatory the method is satisfactory. But the determination of H is no such easy matter, and its continual fluctuations must be registered by an auxiliary instrument. Many of the results obtained in past years do not appear to be very trustworthy, though Kohlrausch and Wild, who has discussed the sources of error in an elaborate manner, are of opinion that a high degree of accuracy is attainable. When, however, a current determination is the only object, the exclusion of this element seems to be desirable, except for rough purposes, when a sufficiently accurate value of H can be assigned without special experiment.

§ 3. Of the arrangements which may be adopted for measuring the mechanical action between a fixed and a mobile conductor conveying the same current, the one that is best known is Weber's electro-dynamometer§. Two fixed coils may be arranged on Helmholtz's principle, so as to give at the centre a very uniform field of force, in which the movable coil is suspended bifilarly. In the equilibrium position the planes of the coils are perpendicular, but under the influence of the current they tend

* *Nature*, Feb. 1, 1883 ; Feb. 15, 1883.
† *Pogg. Ann.* Bd. cxlix. S. 170, 1873.
‡ *Ber. der Phys.—Med. Ges. zu Würzberg*, 1884.
§ Maxwell's *Electricity*, § 725.

to become parallel, and the deflection produced may be taken as a measure of the square of the current. The constant of the instrument, so
far as dependent upon the dimensions of the large coils, can be readily
determined; the difficulty is to measure with sufficient accuracy the dimensions of the small coil, and to determine the force of restitution
corresponding to a given rotation. The latter element is usually obtained
indirectly from the moment of inertia of the suspended parts and from
the time of vibration. If the small coil contain a large number of turns
in several layers, its constant is very difficult to determine by direct
measurement. If, indeed, we could trust to the inextensibility of the
wire, as some experimenters have thought themselves able to do, the
mean radius could be accurately deduced from the total length of wire,
and from the number of turns; but actual trial has convinced us that
fine wire stretches very appreciably under the tension necessary for
winding a coil satisfactorily. It is possible that the difficulty might
be satisfactorily met by an electrical determination of the area of the
windings after the method given by Maxwell*, or that employed in the
present investigation.

§ 4. In the researches of Joule and Cazin the electromagnetic action
is a simple attraction or repulsion, and can be evaluated directly by
balancing it against known weights. This method has been followed by
Mascart in his recent important work upon this subject†. A long solenoid
is suspended vertically in the balance, and is acted upon by a flat coaxal
coil of much larger radius, whose mean plane coincides with that of the
lower extremity of the solenoid. If the solenoid is uniformly wound, it is
equivalent to a simple magnet, whose poles are condensed at the terminal
faces. The electromagnetic action then depends upon $(M - M_0)$, where M
is the coefficient of mutual induction between the fixed coil and the
lowest winding of the solenoid, and M_0 the corresponding, much smaller,
quantity for the uppermost winding.

This arrangement, though simple in conception, does not appear to us
to be the one best adapted to secure precise results. It is evident that a
large part of the solenoid is really ineffective; those turns which lie nearly
in the plane of the flat coil being but little attracted, as well as those
which lie towards the further extremity. The result calculated from the
total length of wire (even if this could be trusted), the length of the
solenoid, and the number of turns, has an appearance of accuracy which
is illusory, unless it can be assumed that the distribution of the wire
over the length is strictly uniform. In order to save weight, it would
appear that all the turns of the suspended coil should operate as much as

* *Electricity*, § 754. McKichan, *Phil. Trans.* 1873, p. 425. See also Kohlrausch, *Wied. Ann.*
Bd. xviii. 1883.

† *Journal de Physique*, March, 1882.

possible, that is, that the suspended coil should be compact and should be placed in the position of maximum effect*.

§ 5.   Neglecting for the time the small corrections of the second order rendered necessary by the sensible dimensions of the sections, let us consider the attraction between two coaxal coils of mean radii A and a, situated at distance x.   If M be the coefficient of mutual induction for the central turns, n, n', the number of windings in the two coils, i the current which passes through both, the attraction is

$$nn'i^2 \frac{dM}{dx} \dots\dots\dots\dots\dots\dots\dots\dots\dots(1)$$

In this expression i^2 is already of the dimensions of a force, and M is linear.   Accordingly dM/dx, though a function of A, a, and x, is itself a pure number, and independent of the absolute dimensions of the system.   Its value is a question only of the *ratios* a/A, x/A.   If we write $dM/dx = \pi f (A, a, x)$, and consider the variation of f as a function of the three linear quantities, the coefficients in the equation

$$\frac{df}{f} = \lambda \frac{dA}{A} + \mu \frac{da}{a} + \nu \frac{dx}{x} \dots\dots\dots\dots\dots\dots(2)$$

are subject to the relation

$$\lambda + \mu + \nu = 0. \dots\dots\dots\dots\dots\dots(3)$$

If the coils are placed at such a distance apart that the attraction is a maximum, $\nu = 0$, and the calculation is independent of small errors in the value of x.   Under these circumstances $\lambda + \mu = 0$, so that proportional errors in A and a affect the result in the same degree and in opposite directions.   In other words, the attraction becomes practically a function of the ratio a/A only.

To this feature we attach great importance.   The ratio of galvanometer constants can be accurately determined by the purely electrical process of Bosscha without linear measurement of either, and from this ratio we can pass to that of the mean radii by the introduction of certain small corrections of the second order.

In this way all that is necessary for the absolute determination of currents can be obtained without measurements of length, or of moments of inertia, or even of absolute angles of deflection.   The forces are, however, evaluated in gravitation measure, so that the final result requires a knowledge of gravity at the place of observation; but except through this quantity there is no reference to the units of space or time.

§ 6.   The final calculation of the attraction is best made with the use of elliptic functions; but useful information, sufficient for a general idea of the conditions and for the design of the apparatus, may be derived from the series developed in Maxwell's *Electricity*, § 699.   If B, b be the distances

---

* *Brit. Assoc. Report*, 1882, p. 445 [vol. II. p. 126].

of two coaxal coils of radii A and a from a point on the axis taken as origin, and $C^2 = A^2 + B^2$, we have

$$M = \frac{\pi^2 A^2 a^2}{C^3} \left\{ 1.2 + 2.3 \frac{Bb}{C^2} + 3.4 \frac{(B^2 - \frac{1}{4}A^2)(b^2 - \frac{1}{4}a^2)}{C^4} \right.$$

$$+ 4.5 \frac{Bb(B^2 - \frac{3}{4}A^2)(b^2 - \frac{3}{4}a^2)}{C^6} + 5.6 \frac{(B^4 - \frac{3}{2}B^2 A^2 + \frac{1}{8}A^4)(b^4 - \frac{3}{2}b^2 a^2 + \frac{1}{8}a^4)}{C^8}$$

$$\left. + 6.7 \frac{Bb(B^4 - \frac{5}{2}B^2 A^2 + \frac{5}{8}A^4)(b^4 - \frac{5}{2}b^2 a^2 + \frac{5}{8}a^4)}{C^{10}} + \dots \right\},$$

$$\frac{dM}{db} = \frac{\pi^2 A^2 a^2}{C^4} \left\{ 1.2.3 \frac{B}{C} + 2.3.4 \frac{(B^2 - \frac{1}{4}A^2)b}{C^3} \right.$$

$$+ 3.4.5 \frac{B(B^2 - \frac{3}{4}A^2)(b^2 - \frac{1}{4}a^2)}{C^5} + 4.5.6 \frac{(B^4 - \frac{3}{2}B^2 A^2 + \frac{1}{8}A^4)(b^2 - \frac{3}{4}a^2)b}{C^7}$$

$$\left. + 5.6.7 \frac{B(B^4 - \frac{5}{2}B^2 A^2 + \frac{5}{8}A^4)(b^4 - \frac{3}{2}b^2 a^2 + \frac{1}{8}a^4)}{C^9} + \dots \right\} \quad \dots\dots\dots(4)^*$$

in which a, b are supposed to be small relatively to A, B. If we limit ourselves to the first term, which we may do when a/A is small, we see that so far as it depends upon the small coil the effect is proportional to the area. The position of maximum effect for given coils is found [see below] by making B/C^5 a maximum, which leads to $B = \frac{1}{2}A$; so that to obtain the greatest attraction the distance of the coils must be equal to half the radius of the larger.

In the present measurements there were *two* equal fixed coils, one on either side of the small coil. If we take the origin midway between, the terms of odd order in b ultimately disappear in virtue of the symmetry, and we may write

$$\frac{dM}{db} = \pi^2 \frac{A^2 a^2}{C^4} \left\{ 1.2.3 \frac{B}{C} + 3.4.5 \frac{B(B^2 - \frac{3}{4}A^2)}{C^5}(b^2 - \frac{1}{4}a^2) \right.$$

$$\left. + 5.6.7 \frac{B(B^4 - \frac{5}{2}b^2 A^2 + \frac{5}{8}A^4)}{C^9}(b^4 - \frac{3}{2}b^2 a^2 + \frac{1}{8}a^4) + \dots \right\}. \quad \dots(5)$$

There would be some advantage in a disposition of the coils such that $B^2 - \frac{3}{4}A^2 = 0$, for then the attraction would be in a high degree independent of the position of the suspended coil†. In this case

$$\frac{dM}{db} = 6\pi^2 \frac{a^2}{A^2} \times \cdot 2138. \quad \dots\dots\dots\dots\dots\dots(6)$$

* [1899. The equation for M, as well as additional terms in that for dM/db, is now inserted.]

† [1899. This was the arrangement adopted for the Board of Trade standard gauge. The coefficient of b^2 in (5) is proportional to

$$B^2 - \frac{3}{4}A^2 - \frac{21a^2 (B^4 - \frac{5}{8}B^2 A^2 + \frac{5}{8}A^4)}{4C^4}.$$

If this vanishes, the first approximation for the ratio of B to A gives, as above, $B^2 - \frac{3}{4}A^2 = 0$. A second approximation is

$$B^2 = \frac{3}{4}A^2 \left(1 - \frac{77a^2}{49A^2} \right).$$

It is not unimportant to remark that independence of b^2 carries with it a corresponding

If, on the other hand, we take $B^2 = \frac{1}{4} A^2$, we find from the first term

$$\frac{dM}{db} = 6\pi^2 \frac{a^2}{A^2} \times \cdot 2862, \quad \dots\dots\dots\dots\dots\dots\dots(7)$$

showing a not unimportant increase of effect. To the second order of approximation [see below] the distance between the fixed coils ($2B$), corresponding to the maximum effect upon a small coil suspended at their centre, is given by

$$B = \frac{1}{2} A \left(1 - \frac{9}{10} a^2/A^2\right) \quad \dots\dots\dots\dots\dots\dots(8)$$

so that when a^2/A^2 is sensible the fixed coils should be somewhat closer than when a^2/A^2 is negligible. For the actual apparatus used a^2/A^2 is very sensible, and the ideal state of things was only imperfectly approached. The coils of the dynamometer used for the "fixed coils" conform to the relation $B^2 = \frac{1}{4} A^2$, and are not adjustable. It will be seen later that but little is practically lost by the slight imperfection of the arrangements in this respect.

Formula (7) is sufficient for the preliminary estimate of the attraction to be expected, and from (5) we can form an idea of the exactitude necessary in the adjustment of the suspended coil. Thus if b be not zero, the correcting factor is, when $B = \frac{1}{2} A$,

$$1 - 3\cdot2\, b^2/A^2. \quad \dots\dots\dots\dots\dots\dots\dots(9)$$

With the actual apparatus an error in b of one millimetre alters the attraction by only $\frac{1}{20,000}$.

[1899. It may be well to exhibit the approximate values of λ, μ, ν in (2). If we make $b = 0$, retaining the two first terms of (5), we see that f may be considered to be proportional to

$$\frac{A^2 B\, a^2}{C^5} \left\{1 - \frac{5\,(B^2 - \frac{3}{4} A^2)\, a^2}{2 C^4}\right\},$$

in which $C^2 = A^2 + B^2$. Hence

$$\frac{df}{f} = \frac{dA}{A} \left\{\frac{2B^2 - 3A^2}{C^2} + \frac{5A^2 a^2}{2C^4} \left(\frac{3}{2} + \frac{4B^2 - 3A^2}{C^2}\right)\right\}$$

$$+ \frac{dB}{B} \left\{\frac{A^2 - 4B^2}{C^2} - \frac{5B^2 a^2}{2C^4} \left(2 - \frac{4B^2 - 3A^2}{C^2}\right)\right\}$$

$$+ \frac{da}{a} \left\{2 - \frac{5\,(4B^2 - 3A^2)\, a^2}{4C^4}\right\}.$$

independence of *lateral* displacements. For if we consider the value of the attraction (parallel to the axis) for a coil moved without rotation whose centre is at the point x, y, z, we recognise that it satisfies Laplace's equation in these coordinates. If x, y, z be measured from the central position, and the attraction be expanded in powers of these quantities, the terms of the first order vanish by symmetry and those of the second order will be proportional to $(2x^2 - y^2 - z^2)$, if x be the coordinate parallel to the axis. Independence of x^2, viz. b^2, involves accordingly independence of y^2 and z^2. The variable part of the attraction thus becomes a quantity of the *fourth* order in the displacements.]

According to the method adopted, a is not measured, but instead a/A. If this be called α, we have $da/a = d\alpha/\alpha + dA/A$; and

$$\frac{df}{f} = \frac{d\alpha}{\alpha}\left\{2 - \frac{5\,(4B^2 - 3A^2)\,a^2}{4C^4}\right\} + \left\{\frac{dA}{A} - \frac{dB}{B}\right\}\left\{\frac{4B^2 - A^2}{C^2} + \frac{5B^2a^2(5A^2 - 2B^2)}{2C^6}\right\}.$$

To secure independence of dA and dB, we have as a first approximation, $B^2 = \tfrac{1}{4}A^2$. The second approximation is

$$\frac{4B^2 - A^2}{C^2} + \frac{36\,a^2}{25A^2} = 0,$$

whence, as above,

$$B = \tfrac{1}{2}A\left(1 - \frac{9}{10}\frac{a^2}{A^2}\right).$$

If the relation be actually $B = \tfrac{1}{2}A$, we have

$$\frac{df}{f} = \left\{2 + \frac{8}{5}\frac{a^2}{A^2}\right\}\frac{d\alpha}{\alpha} + \frac{36}{25}\frac{a^2}{A^2}\left\{\frac{dA}{A} - \frac{dB}{B}\right\};$$

and this agrees with values found below for the actual experiment in which $A = 2\cdot42\,a$.]

§ 7. It may be convenient to carry through the rough theory so as to show the dependence of the current upon the quantities actually measured. Thus

$$\text{Force of attraction} = hnn'i^2a^2/A^2,$$

where h is written for $6\pi^2 \times \cdot2862$. If the ratio of the galvanometer constants of the coils be β, we have

$$a^2/A^2 = \beta^2 n'^2/n^2,$$

whence

$$\text{Force} = h\beta^2 i^2 n'^3/n,$$

and

$$i = \beta^{-1}h^{-\frac{1}{2}}n^{\frac{1}{2}}n'^{-\frac{3}{2}}\,(\text{Force})^{\frac{1}{2}}. \quad\ldots\ldots\ldots\ldots\ldots(10)$$

We may observe that an error in the number of windings, or, which comes to the same thing, a defect of insulation, produces a more serious effect in the case of the suspended than in the case of the fixed coils. The error in the ratio of the galvanometer constants enters proportionately, but the error in the weighings is halved.

Full details of the coils are given later. It will be sufficient here to say that the radius of the large coils is about 25 centims., and that of the suspended coil about 10 centims. The total number of windings on the fixed coils is 450, and on the suspended coil 242. The current usually employed was about $\tfrac{1}{3}$ ampère, and the double attraction was about the weight of one gram*.

* The actual apparatus was not adapted to the measurement of currents much exceeding $\tfrac{1}{2}$ ampère. The flexible copper connexions of the suspended coil would take an ampère, but the

§ 8.   The double attraction is spoken of, inasmuch as the readings were always taken by *reversal* of the current in the fixed coils, for which purpose (fig. 1, *E*) a suitable key was provided.   The difference of the weights required to balance the suspended parts in the two cases gives twice the force of attraction between the suspended coil and the fixed coils, independently of the action upon the former of any other part of the circuit, and of terrestrial or other permanent magnetism.   The cur-

Fig. 1.

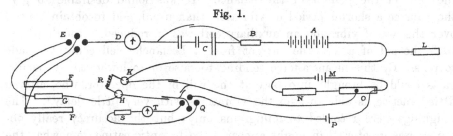

rent was supplied from about 10 either Grove or secondary cells *A*, and traversed in succession a rough tangent galvanometer *D* (convenient for a preliminary test of the strength and direction of the current), two or more silver voltameters in series *C*, the suspended coil *G*, and then (of course, in opposite directions) the two fixed coils *F*.   The weights necessary for balance (in the same position of the key) alter somewhat, both on account of variation in the electric current and also from the formation of air currents, due to a slight progressive warming of the suspended coil.   By recording the times of each weighing we can plot two curves (§ 24), from which we can find what would have been at any moment the weighing in either position of the key.   The difference of ordinates gives us what we should have observed, were it possible to make both measurements simultaneously.   The whole duration of an experiment was from three-quarters of an hour to two hours, measured by a chronometer, and as a weighing could be taken about every five minutes there was ample material for the construction of the curves.   What we require for comparison with the deposited silver is the mean current, whereas what we should obtain directly from the curves represents the *square* of the current.   The whole interval is divided into periods (usually of fifteen minutes), and the difference of ordinates corresponding to the middle

---

suspended coil itself is unduly heated by the passage of an ampère for more than a few minutes. Had it been desirable to use stronger currents, it would, of course, have been possible to do so by increasing the gauge of the wire.   The grooves in which the wire is wound being given, it is evident that a proportional increase of the current and of the section of the wire leave both the heating and the electromagnetic effects unaltered.   In this way the apparatus might easily be modified, so as to take currents of 3 or 4 ampères, the only other change that would be required being a multiplication of the flexible leading wires, several of which might be arranged in parallel.   But for the determination of the electro-chemical equivalent of silver, the currents actually used were quite strong enough.

of the periods is taken from the curves. The mean square root of the numbers thus obtained gives us a result to which the rate of silver deposit should be proportional.

§ 9. The use of a balance for the measurement of electromagnetic attraction involves some special arrangements. The suspended coil must in every case be brought to rest in its proper position, corresponding to the zero of the pointer of the balance. It was found desirable to give the balance a shorter period of vibration than usual, and to obtain control over the arc of vibration an auxiliary coil was introduced, through which, with the aid of a key, the current from a Leclanché cell could be made to pass. By this means a force tending to raise or to lower the suspended parts could be brought into play at the will of the operator, who, after a little practice, is able to stop the vibrations with very little delay*. The weighings were recorded to milligrams only; but the accuracy really obtained was greater than might appear, since by anticipating somewhat the change in progress it was possible to note the *time* at which the balance demanded an integral number of milligrams.

The current was led into the suspended coil by means of fine flexible copper wires. To diminish the force conveyed by these to the suspended parts, they were bent so as to place themselves naturally in the required positions before the final solderings were made. It is important, however, to observe that no assumption is made as to the equality of these forces before and during the passage of the current. Under its influence the fine wires are no doubt sensibly warmed, but this effect and any consequent alterations in the mechanical properties are the same in both sets of readings, the *only* change relating to the direction of the current in the fixed coils.

This point is the more important since the balance is not used in these experiments in quite the normal manner. In ordinary weighings there is no force in operation upon the pans but gravity, and this vertical force is transferred to the beam. In the present application the "pan" is not quite free and is subjected to forces which may have a small horizontal component. In virtue of the freedom of rotation about the knife-edge suspending the pan, these forces are transferred without change to the beam. The horizontal component would, however, produce little effect in any case, since in the horizontal position of the beam its direction would pass very nearly through the knife-edge supporting the beam. The weights in the other scale-pan give rise to a strictly vertical force. We shall thus be doubly secured against error if we provide that the force to be measured (due to the reversal of the current in the fixed coils) is strictly

---

* See "Suggestions for Facilitating the Use of a Delicate Balance." *Brit. Assoc. Report*, 1883 [vol. II. p. 226].

vertical, and that the horizontal force, if sensible, remains unaltered in passing from one direction of the current to the other. These objects are attained when the coils are carefully levelled, and when the readings are always taken for a definite position of the suspended coil conveying a constant current.

§ 10. The suspended coil is wound upon an ebonite ring (§ 13), and is supported by three screws upon a light brass triangle hanging in the balance by a stout copper wire. The fixed coils are those of the dynamo-meter, described in Maxwell's *Electricity*, § 725, and in Latimer Clark's paper (*Phil. Trans.*, 1874, Part I). In setting up the apparatus the ebonite coil is first suspended, and the dynamometer coils are levelled, and adjusted laterally until concentric with it. This is tested by carrying round a metal piece making five contacts with the upper ring of the dynamometer, and provided with a pointer just reaching inwards to the circumference of the ebonite coil. The piece in question may be described as a sort of three-legged stool, standing upon the upper horizontal face of the dynamometer ring and carrying below two studs which are pressed outwards into contact with the inner cylindrical face of the ring. As the piece is carried round the pointer describes a circle coaxal with the dynamometer rings. To level the ebonite ring, the distance is calculated by which its upper surface should be below the upper surface of the (upper) dynamometer ring, and a pointer attached to a straight rule is so adjusted that when the rule is laid upon its edge along the upper face of the dynamometer ring the pointer should just scrape the upper face of the ebonite ring. By applying this test at three points the ebonite ring is brought to occupy the desired position. These adjust-ments were made in the first instance by our assistant, Mr G. Gordon, and subsequently examined by ourselves. With a little care the neces-sary accuracy is attained without difficulty, for, it is scarcely necessary to say, all the errors due to maladjustment are of the second order. When in use the suspended parts are protected from currents of air by a suitable paper casing.

Examination showed that the insulation of the various parts was satis-factory. Twenty-five cells of a De la Rue's battery failed to show any appreciable leakage between the wire and the rings of the dynamometer coils, though the capacity of the *condenser* thus formed was very noticeable.

§ 11. The test for leakage from winding to winding of a coil is a more difficult matter. The ebonite ring was first wound on August 9, 1882, and its galvanometer constant was compared with that of one coil of the dynamometer by Mr J. M. Dodds. The result agreed very ill with the measurements taken during the winding, and led to the suspicion that several turns were short-circuited by a false contact. The matter was put to a further test in two ways. A second coil of the same dimensions was wound with the same number of turns; and the two coils were placed co-axally close together, and so connected in series that a current would circulate opposite ways. The circuit was completed by a galvanometer of long period. Under these circumstances when one pole of a very long steel magnet is thrust suddenly through the opening, there should be no effect observable if the insulation is good; but if any of the turns of one of the coils are short circuited the other coil will of course have the advantage, and the galvanometer will indicate a current in the corresponding direction. It was found in fact that the second coil preponderated, and that 13 extra turns had to be put upon the first coil to obtain the balance. With proper precautions this method of testing seems satisfactory, being approximately independent of the equality of mean radii of the coils compared.

A second test was suggested and executed by Mr Glazebrook. The two coils retaining a fixed position, the ratios of the self-inductions of each to the mutual induction of the pair were determined by Maxwell's method*. These ratios, which should have been nearly equal, were found to differ considerably in the direction which showed a deficiency in the self-induction of the ebonite coil.

After this it was no longer doubtful that the coil was defective. In unwinding it more than one bad place was detected, although the original winding had been carefully done under our own eyes. The ring was rewound with fresh wire on Nov. 30, 1882; and we were so much impressed with the necessity of a thorough check upon the insulation that we devised a delicate test similar, as we afterwards found, to one that had already been successfully used by Graham Bell†. Four similar coils of fine wire, wound upon wood, and of the same mean diameter as the ebonite coil, were arranged so as to form a Hughes induction balance. The lower coils form a primary circuit, and are connected with a microphone clock or other source of variable current. The upper coils and associated telephone form a secondary current. The distance between the upper and lower coils is such as to allow the insertion of the ebonite coil between them, suitable support being provided for it to guard against displacement of the principal coils. If the distances of the four coils are adjusted by screw-motions to an exact balance, so that no sound is audible in the telephone (held at some distance away), the introduction of a tertiary circuit between one primary and secondary causes a revival of sound whose intensity depends upon the conductivity, &c., of the

---

* *Electricity and Magnetism*, § 756.

† "Upon the Electrical Experiments to determine the Location of the Bullet in the Body of the late President Garfield," &c. A paper read before the American Association for the Advancement of Science, August, 1882.

tertiary circuit. If the tertiary circuit consists of a single turn of wire, such as that on the ebonite ring, the sound heard is quite loud, and remains audible when a resistance of about 1 ohm is included. A single circlet of copper wire ·004 inch diameter gives a very distinct sound. When the ebonite coil, with ends unconnected, is introduced, the sound is audible, but much less than that from the fine copper circlet. Part of this effect may be attributed to its finite capacity as a condenser, in virtue of which sound might be heard in any case; but it is probable that the insulation is in reality somewhat imperfect. The closing of the circuit through a megohm gives a distinct augmentation of sound; and thus it is evident that the insulation, if not perfect, is at any rate abundantly sufficient for the purposes of the present investigation.

The current weighing apparatus was set up in February, 1883, and worked satisfactorily from the first. Apart from errors in the constant of the instrument, the determination of the mean value of a current of (say) half an hour's duration should easily be correct to $\frac{1}{10,000}$.

The fixed coils.

§ 12. These are the coils of the dynamometer constructed by the Electrical Committee of the British Association (see § 10). The mean radii of the two coils and the dimensions of the sections are very nearly identical, and for our purpose it is unnecessary to note anything but the mean. The following are derived from the dimensions recorded in Professor Maxwell's handwriting in the laboratory note-book :—

$$A = \text{mean radius} = 24\cdot81016$$
$$2B = \text{distance of mean planes} = 25\cdot00$$
$$2h = \text{radial dimension of section} = 1\cdot29$$
$$2k = \text{axial} \qquad \text{,,} \qquad \text{,,} \qquad = 1\cdot50$$

the unit in each case being the centimetre.

The number of turns of wire on each coil is 225.

The above values are those employed in the calculations of the present investigation, and they can be only partially verified without unwinding the wire. Owing, however, to the final result being comparatively independent of A and B, even a rough verification is not without value. The distance parallel to the axis from outside to outside of the grooves in which the wire is wound can be found pretty accurately with callipers, and was determined to be 10·433 inches. From inside to inside of the grooves the corresponding distance was 9·252 inches. The mean of these is the distance of mean planes, which is thus 9·8425 inches, or 25·000 centims. exactly. This element is, therefore, verified with abundant accuracy. The half difference of the two numbers above given represents the axial dimension of the section, and comes out 1·5024 centims., practically identical with 1·50 centims. The mean radius and the radial dimension of the section are not now accessible to measurement, but the outside circumference agrees sufficiently well with that calculated from the recorded dimensions to serve as a verification.

The number of turns has to be taken entirely upon trust; but the use of the method given in Maxwell's *Electricity*, § 708, makes a mistake in this respect very unlikely. Moreover, the electrical comparisons to be detailed later (§ 14) verify the *equality* of the number of windings on the two coils.

The resistance of each coil is about 14½ B.A. units, and both coils are well insulated from the frame on which they are wound.

The suspended coil.

§ 13. This consisted of 242 turns of copper wire insulated with silk saturated with paraffine wax, and was wound upon an ebonite ring supplied by Messrs Elliotts. The weight of the ring was 135 grms., and its section is shown full size in the adjoining figure (fig. 2). The weight of the wire

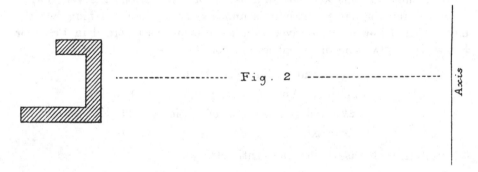

Fig. 2

Axis

was 440 grms., so that the total weight to be carried in the balance was about 575 grms. The mean diameter of the coil of wire, as determined from the inside and outside circumferences, was 8·090 inches; but it cannot be so determined with sufficient accuracy, and the result is not used in the calculation. It agrees perhaps about as well as could be expected with that deduced electrically by comparison with the large coil.

The radial dimension of the section $(2h') = $ ·9690 centim.

The axial   „   „   „ $(2k') = 1·3843$ centims.

The difficulties experienced in respect of the insulation, and the tests applied, have already been related (§ 11).

The electrical comparison of radii (§ 14) gave for the ratio of the dynamometer radius A to that of the suspended coil a

$$2\cdot42113,$$

whence

$$a = 10\cdot2473 \text{ centims.}$$

The mean radius thus determined is not necessarily that corresponding to the geometrical centre of the section, as it allows for any inequality in the distribution of the windings.

The resistance of the coil is about $10\frac{1}{2}$ ohms.

Determination of mean radius of suspended coil.

§ 14. This quantity cannot be determined advantageously by direct measurement, but its ratio to that of the large coils can be deduced from the ratio of the galvanometer-constants of the coils, and this ratio can be accurately determined by the electrical method introduced by Bosscha *.

It may be shown † that for all purposes we may take the mean radius and mean plane of a coil to correspond with the circle passing through the *centre of density* of the windings. If the windings are distributed with absolute uniformity, this point coincides with the geometrical centre of the section; otherwise there may be an appreciable distinction. The corrections of the second order, which in consequence of the finiteness of the section must be introduced in calculating the effects of the coil, have the same values as if the density of the windings were absolutely, instead of merely approximately, uniform.

For example, the galvanometer-constant G_1 is related to the mean radius A (as above defined) and to the radial and axial dimensions of the section, $2h$, $2k$, according to ‡

$$G_1 = \frac{2\pi}{A}\left(1 + \tfrac{1}{3}\frac{h^2}{A^2} - \tfrac{1}{2}\frac{k^2}{A^2}\right).$$

If, therefore, we can determine for two coils the ratio of galvanometer constants, it is a simple matter to infer therefrom the ratio of mean radii.

In Bosscha's method the two coils to be compared are arranged approximately in the plane of the magnetic meridian, so that their axes and mean planes coincide, and a very small magnet with attached mirror is delicately suspended at the common centre. If the current from a battery be divided between the coils, connected in such a manner that the magnetic effects

---

* *Pogg. Ann.* xciii. p. 402, 1854.
† *Camb. Phil. Proc.* Feb. 12, 1883 [vol. ii. p. 184].
‡ See Maxwell's *Electricity*, § 700.

are opposed, it is possible by adding resistances to one or other of the branches in multiple arc to annul the magnetic force at the centre, so that the same reading is obtained whichever way the battery current may circulate. The ratio of the galvanometer constants is then simply the ratio of the resistances in multiple arc.

To obtain this ratio in an accurate manner, the two branches already spoken of are combined with two standard resistances so as to form a Wheatstone's balance. Of these resistances both must be accurately known, and one at least must be adjustable. The electromagnetic balance is first secured by variation of the resistance associated with one of the given coils, which resistance does not require to be known. During this operation the galvanometer of the Wheatstone's bridge is short-circuited. Afterwards the galvanometer is brought into action, and the resistance balance is adjusted. The ratio of the galvanometer constants is thus equal to the ratio of the known resistances. The two adjustments may be so rapidly alternated as to eliminate any error due to changes of temperature in the copper wires.

The above comparison was carried out for each of the two coils of the dynamometer, and the coil wound on the ebonite ring, called for shortness the ebonite coil. On account of the smallness of the latter some care is necessary in the adjustments, which, however, do not require to be described in detail. It will be sufficient to refer to the description of the adjustments when the ebonite coil was suspended, and to mention that the errors arising from maladjustment (all of course of the second order) could hardly affect the final ratio by more than $\frac{1}{10,000}$. The length of the magnet was $\frac{1}{10}$ inch, and the error due to neglecting it could not exceed $\frac{1}{10,000}$. To the magnet was attached a light silvered glass mirror, such as are employed in Thomson's galvanometers, and it was protected from air currents by a glass cell. The readings were taken by observing the motion of a spot of light thrown upon a scale in the usual way.

The electrical connexions are shown in the adjoining figure (fig. 3). The current from a large Daniell cell A, after passing the reversing key B, divides itself at C between the brass coil of the dynamometer D and the ebonite coil E. The remaining terminals of these coils are led into mercury cups F and H, into which also dip the terminals of the bridge galvanometer g. With the ebonite coil is associated a resistance box N. The other branches of the balance were (in one arrangement) composed of a coil of 10 units in multiple arc with which was placed a high resistance box K, and three coils combined in series whose values were about 24, 1, 1 units, making together 26. All these coils were of the standard pattern, and their values had been already carefully determined. From the cup L the current passed back to the key B.

The high resistance box K gives a fine adjustment by which the ratio
of resistances can be brought to the required value. The smallest resistance actually used here was 4000 units. While the electromagnetic

Fig. 3.

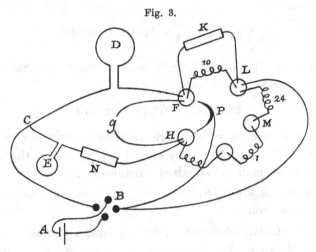

balance was under observation a horse-shoe piece of stout copper rod P,
connected with the key as shown in the figure, was inserted in the
cups F, H. By this means these cups are brought accurately to the
same potential, and nearly all the current is diverted from the standard
resistance coils.

The determination of the electromagnetic balance is rendered more
troublesome by the fact that the first motion of the magnet on the
reversal of the current is influenced by induction, and cannot be used as
a test. No attempt was made actually to complete the adjustment, but
by preliminary trials resistances from N differing by about $\frac{1}{20}$ unit were
found, such that the effects observed were reversed in passing from one
to the other. From the magnitude of these effects the required result is
obtained by interpolation. At the beginning and end of a series the *two*
ratios of resistances were determined by use of K, the horse-shoe P being
of course withdrawn; and the mean of the initial and final values (which
usually differed extremely little) was employed in the reduction.

As an example, we may take some observations on Sept. 5, 1883, with
the coil of the dynamometer marked B. The difference of readings on reversal of the battery in a given manner was taken alternately with certain
resistances from N, which we may call a and b. The results were

with a + ·7, + ·3, + 1·3, + 1·0, mean + ·8;
with b − 8·4, − 8·4, − 8·5, − 9·5, mean − 8·7.

Now with a the resistance from K, associated with the [10], and necessary
for the resistance balance, had to be such that (at a standard temperature)

the resultant resistance of this branch was 9·97772; while with b the re-sultant resistance had to be 9·99182. The resistance that would have been required here, if N had been accurately adjusted for the electromagnetic balance, is thus

$$9·97772 + \frac{·8}{9·5} \times ·01410 = 9·97890.$$

The resistance in the other branch was 25·95648, so that the ratio of galvanometer constants is determined to be

$$25·95648/9·97890 = 2·60113.$$

It will be seen that even with a single cell the sensitiveness was such that the errors of reading could scarcely exceed $\frac{1}{10,000}$; indeed, the weakest part of the arrangement is in the standard resistances.

With use of the above resistance coils the values obtained for coil B on three occasions were

2·60087,   2·60098,   2·60113,   mean 2·60099.

As a further check, the experiment was repeated with a different com-bination of resistance coils. The 26 was replaced by 13, made up of three singles and of the same [10], while the [10] was replaced by a [5]. Two experiments gave

2·60046,   2·60026,   mean 2·60036.

The mean result of the two arrangements is thus 2·60067. The difference is about $\frac{1}{4000}$, and would be explained by an error of $\frac{1}{8000}$ in the value of the [10]*.

For coil A of the dynamometer the ratio of galvanometer constants was found in like manner to be 2·60072, the close agreement of which with 2·60067 is a verification of the winding and insulation of the coils. For the further calculations we require only the mean, and we therefore take as the ratio of galvanometer constants for the ebonite coil and a coil of the dynamometer

2·60070.

The accuracy obtained in the above determinations is doubtless quite sufficient for the purposes of the present investigation, but if it were desired to push the power of the method to its limit it would be neces-sary to design the coils so that the ratio should be (approximately) expressible by very simple numbers. If in the present case, for example, we were content to sacrifice one-fifth of the number of turns on the ebonite coil, the ratio could be made to approach that of 2 : 1. The

---

* For the methods used to find the values of the [24], &c., reference must be made to former papers.

standard resistances might then be composed of three equal resistance coils, which could be more accurately combined and tested than the more complicated combinations that we were obliged to use. In such a case the limit of accuracy could probably depend upon the difficulty of adjusting the coils under comparison and the suspended magnet to their proper places. It is scarcely necessary to say that care must be exercised in the disposition of the leading wires, and that the direct action of the current in the principal coils upon the needle of the bridge galvanometer must be tested, and, if necessary, allowed for.

We have now to deduce the ratio of mean radii. For the ebonite coil the correcting factor is

$$1 + \tfrac{1}{3}h'^2/a^2 - \tfrac{1}{2}k'^2/a^2 = 1 + \cdot000741 - \cdot002269.$$

For the dynamometer coil

$$1 + \tfrac{1}{3}h^2/A^2 - \tfrac{1}{2}k^2/A^2 = 1 + \cdot000225 - \cdot000457.$$

Thus

$$A/a = \tfrac{225}{242} \times 2\cdot60070 \times 1\cdot001296 = 2\cdot42113;$$

and from this when A is known the value of a can be deduced (§ 13).

Calculation of attraction.

§ 15. The attraction between two coaxal circular currents of strength unity, of which the radii are A, a, and distance of planes is B, is (Maxwell, § 701)

$$\frac{\pi B \sin \gamma}{\sqrt{(Aa)}} \{2F_\gamma - (1 + \sec^2 \gamma) E_\gamma\}, \quad \dots\dots\dots\dots\dots(1)$$

where F_γ and E_γ denote the complete elliptic integrals of the first and second kind whose modulus is $\sin \gamma$. The value of $\sin \gamma$ itself is

$$\sin \gamma = \frac{2\sqrt{(Aa)}}{\sqrt{\{(A + a)^2 + B^2\}}}. \quad \dots\dots\dots\dots\dots(2)$$

The functions F_γ and E_γ were tabulated by Legendre. In an Appendix [p. 327] will be found a table of

$$\sin \gamma \{2F_\gamma - (1 + \sec^2 \gamma) E_\gamma\}, \quad \dots\dots\dots\dots\dots(3)$$

calculated with seven-figure logarithms from those of Legendre for the purpose of the present and similar investigations. It has been carefully checked, and it is hoped is free from error, except of course in the last place.

The value of (1), with omission of the factor π, is denoted by $f(A, a, B)$, and, as has already been explained, it is a function of no dimensions. To calculate it for the central windings of the fixed and suspended coils, we have first to find γ from (2). With the data already given $\gamma = 58°\ 57\frac{448}{180}'$, whence with use of the table

$$f(A, a, B) = 1\cdot044576.$$

This multiplied by π, by the product of the numbers of terms in the two coils, and by the square of the strength of the current, gives very nearly the force of attraction, but a correction is required for the finite dimensions of the sections. The quadruple integration over the two areas may be effected by suitably combining various values of f corresponding to the central turn of one section and to the middle of one of the linear boundaries of the other. (See Maxwell's *Electricity*, 2nd edition, § 706, Appendix II.) We find

$$\left. \begin{array}{l} f(A+h,\ a,\ B) = \cdot992719 \\ f(A-h,\ a,\ B) = 1\cdot098740 \end{array} \right\} \text{ sum } 2\cdot091459$$

$$\left. \begin{array}{l} f(A,\ a+h',\ B) = 1\cdot158576 \\ f(A,\ a-h',\ B) = \cdot937866 \end{array} \right\} \text{ sum } 2\cdot096442$$

$$\left. \begin{array}{l} f(A,\ a,\ B+k) = 1\cdot024612 \\ f(A,\ a,\ B-k) = 1\cdot059526 \end{array} \right\} \text{ sum } 2\cdot084138$$

$$\left. \begin{array}{l} f(A,\ a,\ B+k') = 1\cdot026306 \\ f(A,\ a,\ B-k') = 1\cdot058569 \end{array} \right\} \text{ sum } 2\cdot084875$$

The sum of the eight values is $8\cdot356914$. From this we subtract $2 \times f(A, a, B)$, viz., $2\cdot089152$, and divide by 6; whence for the mean value of f applicable to the sections as a whole

$$f = 1\cdot044627,$$

differing, as it turns out, extremely little from $f(A, a, B)$.

From the values given we see that f increases very sensibly as B diminishes, so that, as was expected, the distance between the fixed and the suspended coil, or between the two fixed coils, is too great to realise fully the advantageous condition of things described as the ideal, in which f would be approximately independent of variations in B.

To express the actual variations of f as a function of A, a, B, we write

$$\frac{df}{f} = \lambda \frac{dA}{A} + \mu \frac{da}{a} + \nu \frac{dB}{B};$$

and we obtain sufficiently accurate values of λ, μ, ν from those of f already given. Thus

$$\lambda = \frac{f(A+h,\ a,\ B) - f(A-h,\ a,\ B)}{f(A,\ a,\ B)} \div \frac{2h}{A} = -1\cdot95.$$

In like manner $\mu = +2\cdot23$, $\nu = -\cdot28$; so that

$$\frac{df}{f} = -1\cdot95\frac{dA}{A} + 2\cdot23\frac{da}{a} - \cdot28\frac{dB}{B}.$$

In the present investigation, however, a is not measured directly, but by comparison with A. If we write $a/A = \alpha$, so that

$$\frac{d\alpha}{\alpha} = \frac{da}{a} - \frac{dA}{A},$$

and eliminate da/a, we have

$$\frac{df}{f} = 2\cdot23\,\frac{d\alpha}{\alpha} + \cdot28\,\frac{dA}{A} - \cdot28\,\frac{dB}{B},$$

which is the equation by which the suitability of the proportions is to be judged. It will be seen that the stress is thrown upon the measurement of α, and that the errors of A and B enter to the extent of only about one quarter. If the proportions had been those described as ideal, the coefficients of dA/A and dB/B would have been zero.

It must not be forgotten that the error of f itself is halved in the final result, which thus involves the errors of A and B only after division by 8.

If the current be i, and the number of turns in the fixed and suspended coils, n, n', the attraction or repulsion is measured by

$$\pi n n' i^2 f.$$

This is expressed in absolute units. To find the value in gravitation units we must divide by g. If m be the observed difference of weights in air necessary to counterpoise the suspended coil when the current is reversed in the fixed coils,

$$\pi n n' i^2 f = \tfrac{1}{2} m g \times \cdot99986,$$

the last factor representing the "correction to vacuum" rendered necessary by the finite density of the brass weights.

The value of g at Cambridge is taken to be $981\cdot2282$. Introducing this and the numerical values of n, n', f, already given, we find

$$i = \mu \sqrt{m}, \quad \text{where } \mu = \cdot0370484.$$

The silver voltameters.

§ 16. The arrangement adopted for the voltameters is similar to that recommended originally by Poggendorff. The deposits are formed upon metallic basins (usually of platinum) charged with a neutral 15 per cent. solution of pure silver nitrate. They are prepared by careful cleaning with nitric acid and distilled water with subsequent ignition. After complete cooling in a desiccator, they are weighed to $\frac{1}{10}$ milligrm. in a delicate balance with trustworthy weights. The anode, by which the current enters the voltameter, is formed of fine silver sheet, suspended by platinum wire in a horizontal position near the top of the solution. In order to protect the cathode from disintegrated silver, which in our experience is invariably formed upon the anode, the latter is wrapped round with pure filter paper, secured at the back with a little sealing-wax. This arrangement appears to us for several reasons preferable to the vertical suspension of the electrodes in the form of flat plates. In the latter arrangement the deposited metal usually aggregates itself upon the edges and corners of the kathode with a tendency to looseness. Again the solution rapidly loses its uniformity. At the kathode the solution becomes impoverished and at the anode it becomes concentrated. With vertical plates the strong solution soon collects itself at the bottom, and the weak solution at the top, so as to give rise to considerable variation of density. It is true that the horizontal position of the electrodes necessitates the use of a porous wrapping, which would increase the difficulty of determining the loss of weight at the anode. M. Mascart appears to have succeeded in determining this loss, but the disintegration which we have always met with rendered the attempt on our parts hopeless. It is possible that something may depend upon the mechanical condition of the metal, but as to this we cannot speak with confidence. The blackish powder left upon the anode has at first the appearance of being due to chemical impurity, but it occurs with anodes of the highest quality of silver, and is completely soluble in nitric acid.

In our earlier trials, dating from October, 1882, we were much impressed with the importance of obtaining sufficient coherence in the deposit to guard against risk of loss in the washing and subsequent manipulations. The addition of a very small proportion of *acetate* of silver was found to be in this respect a great improvement, affording a deposit less crystalline in appearance and of much closer texture; and in consequence nearly all our experiments during the first year were conducted with solutions containing sensible quantities of acetate. In order to detect whether anything depended upon the "density" of the current, two platinum basins of different sizes were employed, the area of deposit being in about the proportion of 2 : 1, but no distinct systematic difference was observed. When the deposits were completed the basins were rinsed several times with distilled water, and then allowed to soak over night. The next day after more rinsings they were dried in a hot air closet at about 160° C., and after standing over another night in a desiccator were carefully weighed. Repetition of these weighings after intervals of standing in the desiccators showed that they were correct to $\frac{1}{10}$ milligrm., so that as the total weights of deposit amounted to 2 or 3 grms., a high degree of accuracy in the final evaluation of the ratio of deposit to current was expected. Discrepancies, however, presented themselves of an amount much greater than we had been prepared for, and they were of such a character as to show that the disturbing causes were to be sought in the behaviour of the voltameters and not in the current weighing apparatus. Thus it was found that the numbers obtained on the same occasion from the two voltameters in series, through which exactly the same quantity of electricity had passed, were liable to as great a disagreement as the numbers derived from experiments on different days.

§ 17.   At this stage the question presented itself as to whether the deposits were really pure silver. Two or three gravimetric analyses by conversion into chloride, conducted both by ourselves and by Mr Scott, to whose advice and assistance we have been constantly indebted throughout these investigations, having favoured the idea that the deposits were not quite pure, we arranged for a systematic volumetric analysis of all the deposits. The bulk of the metal after solution in pure nitric acid having been thrown down with a known quantity of chloride of sodium in strong solution, the titration was completed with weak ($\frac{1}{1000}$) salt solution from a burette in the usual manner. The bottle containing the solution was enclosed in a dark box and lighted in the manner recommended by Stas, with a convergent beam of yellow light which had passed through a flask containing chromate of potash. Towards the close of the operation the effect of the addition of two drops of solution (containing $\frac{1}{10}$ milligrm. of salt) becomes difficult of observation unless the liquid be very thoroughly cleared. At this stage we found it convenient to filter off about half the liquid into another bottle, through a funnel plugged with (purified) cotton wool. As soon as the pores are penetrated by the chloride of silver the filtration is effective, and yet so rapid that but little time is lost by the adoption of this procedure. The two drops of chloride solution are added to the liquid thus filtered, and *shaken up* so as to effect a complete mixture, and the bottle is then placed so that the cone of light traverses the *body* of the liquid. After an interval varying from a few seconds to several minutes the cloudiness develops itself, and the delay gives an indication of how nearly the point is approached. Before each test the filtrate is of course returned to the stock bottle and thoroughly shaken up. The operation is complete when the last addition of two drops gives no effect after a quarter of an hour. There is no difficulty in determining in this way the necessary quantity of salt to $\frac{1}{10}$ milligrm., and the point may be recovered any number of times after addition of small known quantities of silver.

In the interpretation of the results we placed no trust in the purity of the NaCl, nor depended upon any assumption as to the ratio of NaCl to Ag, but made comparison with the numbers obtained from precisely similar determinations with substitution for the electro-deposits of equal weights of silver of the highest quality, supplied by Messrs Johnson and Matthey. A large number of such comparisons showed that there was no difference that could be depended upon between the two kinds of silver; there was, indeed, a slight indication of inferiority in the deposits, to the extent of perhaps $\frac{1}{6000}$, but not more than might plausibly be attributed to the greater risk of loss in dissolving the deposits from off the platinum basins. The standard silver was dissolved without transference in the bottle used for the subsequent analysis, and thus under more advantageous conditions than were possible in the manipulation of the deposits.

§ 18.   Table I. [p. 308] gives the results of a laborious series of determinations made with solutions containing more or less acetate. It will be seen that up to August 16 the numbers in the final column are fairly concordant, and they rather narrowly escaped being accepted as satisfactory. In the month of November, however, the experiments were continued with a fresh stock of depositing solution (probably containing less acetate), when a systematic change became apparent in the direction of smaller deposits. From the first we had taken, as we thought, full precautions to secure adequate washing out of the silver salt, and special experiments had proved that the weights were not appreciably changed by further washing with pure water, or by resoaking in the depositing solution with a second washing and drying conducted like the first. Nevertheless the appearance of the deposits under the microscope was such as to suggest a doubt whether a complete elimination of the salt from its pores was possible with any amount of washing, and the evidence of the analyses was felt not to be decisive, inasmuch as the deficiency to be found in this way would correspond to only about one-third of the weight of salt actually present. According to this view the diminution in the weight of the deposits after August 16 was due to a more complete washing out of the salt, rendered possible by the more open texture of the deposits, and we proceeded to test the behaviour of pure nitrate solutions. The result was a further small, but distinct, diminution in the weights, as shown in Table II., and we were now convinced that the use of acetate had been a great mistake, costing us six months' almost fruitless labour. When the deposits are taken upon the concave surface of a bowl, they are coherent enough for convenient manipulation without the aid of acetate. The danger of the retention of salt or other impurity is far greater than of loss of metal, and this danger is aggravated by the acetate. Indeed it would be scarcely too much to say that the danger is converted into a certainty, for from the fine pores of these deposits it seems almost impossible to remove the salt effectually.

It is evident that, in spite of the retention of a small quantity of salt, a satisfactory conclusion might be reached were there any means of estimating its amount. Theoretically the analysis for silver, as many times effected, is adequate to this purpose, since the difference of the total weight of the (impure) deposit, and of the metal found on analysis, would represent the NO_3 of the salt. But the circumstances are so disadvantageous that no satisfactory result could be looked for without an extraordinary, and perhaps impossible, perfection of manipulation. A direct test for nitric acid is not applicable; but at a sufficiently high temperature the silver nitrate would be decomposed, so that the loss of weight incurred on heating to redness (after previous thorough drying at, say, 160° C.) would represent the NO_3. Unfortunately this method is difficult to carry out thoroughly without injury to the platinum basins, inasmuch as silver and platinum begin to alloy at a red heat. But an exposure for five minutes to a heat just short of redness does not seriously damage the basins, and appears to be nearly, if not quite, sufficient to drive off the last traces of NO_3. With a pure nitrate depositing solution, and with the treatment for elimination of the salt presently to be described, there was sometimes no loss on heating (Table II.), but perhaps more often the balance indicated a loss of one or two-tenths of a milligram. With respect to the interpretation of this, it is difficult to say whether or not it ought to be regarded as due to traces of salt retained in spite of all the washings. If so, the true weight of deposit is smaller still by nearly twice the apparent loss; but it is very possible that there may be traces of grease liable to be burnt off at a red heat, so that the loss in question cannot with confidence be attributed to nitrate. On this account the real amount of the deposit remains somewhat uncertain to nearly half a milligram.

With respect to the procedure best adapted to eliminate the salt from the pores of the deposit, it is evident that the difficulty is to cause any displacement of the liquid in the interior. It was thought that this object might to some extent be attained by rapid alternations of temperature, and for this purpose the basins were (after thorough rinsing) passed backwards and forwards between cold and boiling distilled water. Recourse was had also to soaking in alcohol, somewhat diluted. Still wet with the alcohol, the basins were plunged into boiling water with the idea of promoting disturbance inside the cavities of the deposit. After a course of treatment of this kind the basins were filled and allowed to stand over night so as to give free play to diffusion. They were then rinsed a few times, and placed in the air closet to be dried at 160° C.

§ 19. In order to meet the difficulty of the alloying of silver and platinum at a temperature high enough to decompose with certainty the last traces of silver nitrate, we made, at the suggestion of Professor Dewar, several attempts to replace platinum by silver bowls. One evident objection to the silver is the impossibility of removing the deposit with nitric acid, so as to restore the original condition of the bowl. But a more serious difficulty arises from the want of constancy in the weight of a silver bowl (without deposit) when strongly heated. On more than one occasion a *gain* of a milligram or two was observed after heating in a porcelain basin over an alcohol flame. We have reason to believe that this effect depends upon the presence of traces of copper. In order to test the question we carefully cleaned and dried at 160° a piece of the highest quality of silver, such as was used latterly for the anodes. The weight was now 28·1628, and after heating to redness for a quarter of an hour over a naked alcohol flame *fell* to 28·1619. On another occasion a loss of 2 milligrms. was observed under similar circumstances. On the other hand, a parallel experiment with a less pure sample of silver, known to contain a small quantity of copper, gave after the first heating to redness a *gain* of 3 milligrms., followed by a further gain of 2 milligrms. after a second heating.

These changes are, however, insignificant compared to that observed by Mr Scott, who heated one of our large silver basins in a porcelain bowl for a long time over a Bunsen gas-flame. After two nights' treatment the weight had risen from 57·3008 to 57·4521. Mr Scott traced the increase in his case to the formation of silver sulphate, but it does not appear possible that this can be the explanation of the changes observed by us. The matter appears worthy of the further attention of chemists; but for our purposes the conclusion is that, for the present at any rate, platinum is preferable to silver. With suitable precautions, the platinum basins may be heated to redness without changing more than $\frac{1}{10}$ milligrm.

§ 20. In some of our later experiments (*e.g.*, those on January 30, April 2) we included a voltameter, charged with a higher proportion of acetate, in order to exaggerate the errors that we had met with, in the hope of better detecting their origin. When the nitrate solution is nearly saturated with acetate, the deposit is of a beautiful snow-white appearance, and almost always 5 or 7 milligrms. too heavy. On the second weighing, after heating to the verge of redness, a loss revealed itself, whose amount usually agreed fairly well with the view that the original excess of weight was due to nitrate, reduced to metal by the second heating.

§ 21. In the hope of obtaining better evidence as to the cause of the anomalous weights, and also with the view of confirming our results by the substitution for nitrate of some other salt of silver, we have made several observations on deposits from *chlorate* of silver. The salt was prepared for us by Mr Scott from chlorate of barium, and was found to give as good deposits as the nitrate. The chlorate was used in a nearly saturated 10 per cent. solution*, and also in a 5 per cent. solution. Voltameters charged with nitrate were included in the same circuit, so that the comparison was made under the most favourable conditions. The results (Table II.) show an exceedingly good agreement, and constitute perhaps the most accurate verification which Faraday's law of electrolysis has as yet received.

But the second object which we had in view in using the chlorate has not been attained. The idea was to get a too heavy deposit by addition of acetate, and then after washing and weighing as usual, to dissolve up the metal with nitric acid and test for chlorine. If chlorate were present, and were the cause of the excessive weight, it should on strong heating be resolved into chloride, whose presence might be detected. Preliminary experiments showed that as little as $\frac{1}{10}$ milligrm. of silver chloride could be rendered evident. The deposits were dissolved in nitric acid, and strongly supersaturated with pure ammonia. After standing for some time with frequent stirring, the solution was diluted, and again rendered acid with nitric acid. The deposits from chlorate, which we had reason to regard as pure, stood the test almost perfectly, the amount of chloride of silver present being less than $\frac{1}{20}$ milligrm. If one drop of the dilute NaCl ($\frac{1}{20}$ milligrm.) were added to the solution in its alkaline condition, the cloud formed on acidification was perfectly evident after a minute or two when examined in Stas' box. When a piece of fused silver chloride weighing 3 milligrms. was added to the alkaline solution, it dissolved after about half an hour, and gave a dense milkiness on addition of nitric acid.

The application of this method to deposits from chlorate and acetate, which the balance showed to be several milligrams too heavy, has given the unlooked-for result that no corresponding quantity of chloride was present. Something more than a mere trace was indeed detected, but of amount probably not exceeding $\frac{1}{2}$ milligrm. The deposit from chlorate and acetate of April 2, and another which does not appear in the table as the current weighings were not taken successfully, in which the excess was about 7 milligrms. were both treated in this way with similar results. The loss of weight on strong heating appears to exclude the supposition that though chlorate was present it escaped decomposition, and thus we seem almost driven to the conclusion that the redundant matter is principally acetate, although the proportion of acetate to chlorate in the solution is a small one.

§ 22. We have had occasion to examine another point relating to the chemistry of electrolysis, of which the result may here be recorded. In

---

* The tendency to crystallise upon the anode is an objection to the use of the strong solutions, and probably makes itself the more felt in consequence of the paper wrapping, which impedes the free circulation of the liquid.

our earlier experiments we used anodes containing an appreciable quantity of copper. The copper evidently tended to accumulate in the solution, becoming after a time apparent by its colour even when neutral; on addition of ammonia a distinct blue was struck. We were desirous of ascertaining whether under these circumstances there is danger of the deposits becoming contaminated. A distinctly blue solution was prepared, in which the proportion of copper to silver was considerable, and a deposit made. The texture was very much modified by the action of the copper, and the appearance was such that it was difficult to believe that the weight could be more than a small fraction of that of the simultaneous deposit from a pure silver solution. Some of the metal, which adhered very loosely, was lost in the washing, but the weights agreed to within a few milligrams. On dissolution in nitric acid and supersaturation with ammonia the solution showed no trace of colour, although about $\frac{1}{10,000}$ of copper can thus be detected.

§ 23. In the absolute measurements the determination of the interval (never less than three-quarters of an hour) between the first passage of the current through the voltameters and its final cessation could readily be effected with sufficient accuracy (probably to $\frac{1}{10,000}$), but a slight correction is called for in order to take account of the loss of time incurred at each operation of the reversing key which controlled the direction of the current in the fixed coils (§ 8). To obtain the necessary data for this correction the main current was led through a few turns of wire surrounding a reflecting galvanometer. The resulting deflection is independent of the position of the key, but at the moment of reversal the current is interrupted, and the spot of light falls back towards zero. From a comparison of the amount of this falling back with that of the steady deflection, in conjunction with observations of the period of vibration, it is easy to deduce the time of interruption. It proved to be less than $\frac{1}{10}$ second, and was so nearly constant that after sufficient experience had been gained further observations were judged to be unnecessary. The connexions for this purpose are accordingly not shown in the diagram (fig. 1).

§ 24. In order more fully to explain the procedure in taking a deposit it will be advisable to give the details of one experiment. Thus on March 10, 1884, the current, roughly regulated to the desired value with the aid of the tangent galvanometer, was allowed to pass through the coils of the current-weighing apparatus for about half an hour. The electromotive force of the storage cells (when in good order) remains almost perfectly constant during an experiment, but the gradual warming of the copper conductors causes a slight falling off of current. On the present occasion the preparatory current was a little stronger than that ultimately used, so as to produce a slight overheating. During this time the three platinum voltameters, previously cleaned and weighed, were charged with solution of silver nitrate; and the pure silver anodes, wrapped in filter paper, were adjusted to their places at the top of the liquid. As will be seen from Table II., two of the bowls were charged with solution of normal strength (15 per cent.), and the other with solution of double this strength. When all was ready, the current, previously running along a shunt, was caused to pass through the voltameters at $4^h 17^m$ by the chronometer. The weights required to bring the pointer of the current-weighing balance to zero, with the corresponding times, are given in Table III. In the second column the first number means

TABLE III.

Time	Weight	Time	Weight
h. m. s.		h. m. s.	
4 19 30	7·694	4 25 0	6·795
4 32 15	7·698	4 40 20	6·791
4 42 50	7·699	4 50 30	6·790
4 53 10	7·699	4 56 30	6·789
		5 1 15	6·789

that at the moment in question the weight required to balance the suspended coil, as acted upon electromagnetically, was 7·694 grms., or rather 577·694 grms., but the 570 grms. being never moved need not be recorded. In this position of the reversing key the electromagnetic force increased the apparent weight of the suspended coil. The other set of readings, in which the magnetic force tended to lift the coil, are given in the fourth column. At $5^h 2^m$ the circuit was interrupted.

From the numbers above given two curves are constructed (fig. 4), representing what would have been observed in either position of the

Fig. 4.

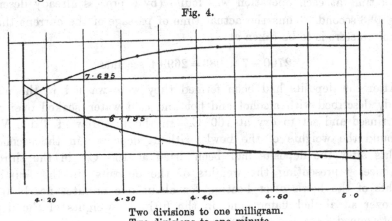

7·695

6·795

4·10   4·20   4·30   4·40   4·50   5·0   5·10

Two divisions to one milligram.
Two divisions to one minute.

key during the whole course of the experiment. To effect the integration of the current, the whole time, 45^m, is divided into nine periods of 5^m each, and the magnitude of the current at the middle of each period is taken to represent its value throughout the period. A more elaborate evaluation could easily have been applied, but was superfluous. The difference of ordinates at the middles of the periods gives the difference of weights in the second column of Table IV., and the mean of the square roots of these differences, viz. ·95171, is the square root of the difference of weights corresponding to the *mean current*.

TABLE IV.

Time	Difference of weight	Square root of Difference
h. m. s.		
4 19 30	·897	·9471
4 24 30	·900	·9487
4 29 30	·904	·9508
4 34 30	·906	·9518
4 39 30	·908	·9529
4 44 30	·908	·9529
4 49 30	·909	·9534
4 54 30	·910	·9539
4 59 30	·910	·9539
Mean	..	·95171

The whole time of deposit was 2700 seconds, but from this a deduction has to be made for the time lost in operating the reversing key. The loss of time at each operation was found (by a process already described) to be ·083 second. Thus the actual time of passage of the current through the voltameters is to be taken at

$$2700 - 7 \times ·083 = 2699·4 \text{ seconds.}$$

After the deposits had been formed they were washed in the manner already described with alcohol and hot and cold water, soaked over night, then rinsed and set to dry at 160° C. In the first row of Table V. will be found the weights of the bowls without deposits; in the second the weights after the deposits had been dried at 160° C.; in the third the differences representing the weights of the deposits; in the fourth the weights of the bowls after heating for about five minutes nearly to redness over an alcohol flame; and in the fifth the weights of the deposits as determined from the previous row.

TABLE V.—Deposits of March 10, 1884.

	Large bowl I. Pure nitrate Normal strength	Small bowl II. Pure nitrate Double strength	Small bowl III. Pure nitrate Normal strength
Before deposit	80·4490	17·2985	21·8789
After deposit, first weighing	81·5138	18·3628	22·9434
Gain	1·0648	1·0643	1·0645
After strong heating . . .	81·5135	18·3627	22·9433
Gain	1·0645	1·0642	1·0644

Mean 1·0644 grms.

To obtain numbers which, though of no absolute significance, allow of the comparison of experiments made on different occasions, we may divide ·95171 (the square root of the difference of the current weighings) by the amount of silver deposited per second. Thus for March 10 we have

$$\cdot 95171 \div \frac{1\cdot 0644}{2699\cdot 4} = 2413\cdot 7.$$

The magnitude of the current was about ·4 ampère, and the areas of deposit about 37 sq. centims. for the small bowls, and about 75 sq. centims. for the large bowl.

The whole resistance of the current-weighing apparatus and of the voltameters is about 42 ohms, so that sufficient current can be obtained from 10 small Grove cells, or from a rather less number of cells of a secondary battery.

§ 25. The tables in which are embodied the results of these protracted experiments will not now require much explanation. Those of Table I. are certainly erroneous on account of the presence of acetate (§ 18), and no weight is given to them in calculating a final result. For the same reason those deposits in Table II. which were prepared from solutions to which acetate had been added for the purpose of investigating the nature of the disturbance thereby produced, are of course excluded. The weights adopted for the silver deposits are those found after strong heating (nearly to redness) for about five minutes, no distinction being made between the deposits from chlorate and from nitrate of silver. The final mean 2414·45 expresses the square root of the difference of current weighings in grams divided by the rate of silver deposit in grams per second.

If we consider separately the deposits from chlorate of silver (without addition of acetate), we get as the mean number corresponding to the above 2414·3, in almost perfect agreement.

The deposits made on March 25 were *twice* strongly heated with intermediate weighing. Similar tests have been applied in other cases not recorded in the tables.

TABLE I.

Date, 1883	Solution	Bowl	Weight of deposits in grams	Weight adopted as that of the silver deposited	Duration of observation	Duration in seconds, corrected	Mean square root of double attraction in grams	$\sqrt{m}$ / deposit per sec.
					h. m.			
May 1 . .		Large platinum . Small platinum .	2·6338 } 2·6340 }	2·6339	2 0	7198·5	·88142	2408·9
„ 4 . .	15 parts of nitrate of silver dissolved in 100 parts of water and filtered through acetate of silver	Large platinum . Small platinum .	2·9162 } 2·9154 }	2·9158	2 0	7198·4	·97580	2409·0
„ 8 . .		Large platinum . Small platinum .	2·6887 } 2·6878 }	2·6882	1 45	6298·6	1·02889	2410·7
„ 12 . .		Large platinum .	2·7218	2·7218	2 0	7198·9	·91147	2410·7
August 16 .		Large platinum . Small platinum .	2·1441 } 2·1420 }	2·1430	1 15	4499·02	1·14765	2409·4
November 5 .	A new one similar to the last	Large platinum . Small platinum .	2·3675 } 2·3659 }	2·3667	1 25	5099·08	1·11947	2411·9
„ 13 .		Large platinum . Small platinum .	4·0251 } 4·0233 }	4·0242	2 18	8278·67	1·17268	2412·5
„ 19 .		Large platinum . Small platinum .	3·0237 } 3·0233 }	3·0235	1 45	6298·83	1·15773	2411·9

TABLE 11.

Date, 1883 and 1884	Bowl	Solution	Weight of deposits in grams — After drying at about 160°	Weight of deposits in grams — After heating to verge of redness	Weight adopted as that of the silver deposited	Duration of observation (h. m. s.)	Duration in seconds, corrected	Mean square root of double attraction in grams.	√m deposit per sec.	From chlorate alone
November 29	Large platinum	Nitrate 15 to water 100	3·0166 / 3·0165	··	3·01655	1 45 0	6298·67	1·15608	2413·9	
	Small	,, ,, ,,	2·9907	2·9901						
December 4	Large	,, ,, ,,	2·9902	2·9895	2·9698	1 40 0	5998·67	1·20321	2414·1	
January 7	Large	,, 15 to water 100, but filtered several times through acetate	2·3731	··	2·3731	1 25 0	5098·75	1·12371	2414·4	
January 30	Small silver	Nitrate 15 to water 100	2·3316	2·3287	2·3229	1 15 0	4499·08	1·24716	2415·5	
	,, silver	,, ,, ,,	2·3230 / 2·3229	2·3229						
February 18	Large platinum	Nitrate 7½ to water 100	2·3484 / 2·3483	2·3483 / 2·3481 / 2·3481	2·3482	1 25 0	5099·08	1·11213	2415·0	
	Small	Nitrate 15 to water 100								
February 22	Large	,, ,, ,,	2·3977	2·3975	3·2972	2 1 0	7258·75	1·09667	2414·4	
	Small	Nitrate 30 to water 100	3·2966 / 3·2979	3·2975 / 3·2975						
February 29	Large	Nitrate 15 to water 100	2·2698 / 2·2693	2·2696 / 2·2691	2·2695	1 30 0	5399·17	1·01497	2414·6	
	Small	Nitrate 30 to water 100	2·2701	2·2699						
March 5	Large	,, ,, ,,	1·2247 / 1·2247	1·2245 / 1·2246	1·2245	0 45 0	2699·33	1·09544	2414·8	
	Small	Nitrate 15 to water 100	1·2248	1·2245 / 1·2245						
March 10	Large	Nitrate 7½ to water 100	1·0648 / 1·0645	1·0642 / 1·0644	1·0644	0 45 0	2699·42	·95172	2413·7	
	Small	Nitrate 15 to water 100								
		Nitrate 30 to water 100								
		Nitrate 15 to water 100								
March 14	Large	,, ,, ,,	2·2897 / 2·2892	2·2896 / 2·2892	1·2894	0 45 0	2699·50	1·15311	2414·2	
	Small	Nitrate 15 to water 100 with nitric acid added	1·2893	1·2893						
March 25	Large	Chlorate 10 to water 100	1·5306	{ 1·5305 / 1·5303 }	1·52085	1 0 0	3599·08	1·02629	2414·4	2414·1
	Small	Nitrate 15 to water 100	1·5298	1·5295 / 1·5295						
March 29	Large	Chlorate 10 to water 100	1·5302	{ 1·5297 / 1·5297 }	1·4531	1 0 0	3599·00	·97489	2414·6	2414·6
	Large	Chlorate 5 to water 100	1·4530 / 1·4532	1·4529 / 1·4530						
	Small	Nitrate 15 to water 100	1·4533	1·4533						
	Small	Chlorate 5 to water 100								
April 2	Large	Chlorate 5 to water 100, filtered several times through acetate	1·7300	1·7287	1·72325	1 12 59½	4378·58	·95018	2414·3	2414·2
	Small	Nitrate 15 to water 100	1·7232	1·7232						
	,,	Chlorate 5 to water 100	1·7234	1·7233						
								Mean · ·	2414·45	2414·3

It should be stated that every determination since November, 1883, in which the manipulations were successfully conducted, is included in the table, and that nothing is excluded except in consequence of a decision made before the result was known.   In one or two cases the current was too irregular to give good weighings of the suspended coil, and then the observations were not reduced with the view of obtaining absolute results, although the comparison of the silver deposits in different bowls might still be of interest.   This happened on an occasion already alluded to when acetate and chlorate of silver were used in combination.

The results of Table II. agree together about as well as could be expected, the extreme difference from the mean being $\frac{1}{2500}$.   It must be remembered that apart from the difficulties of manipulating the silver deposits errors may arise in the determination of the current, whose mean value has to be deduced from observations relating to only a part of the whole time involved.   A small fluctuation in the strength of the current, lasting for a short time only, may thus escape detection.   There is also an error involved in the determination of the time of electrolysis, which may altogether amount to nearly half a second on a total in some cases as low as 2700 seconds.   When so many experiments are made we must expect the cases to arise in which the small errors, due to various causes, are accumulated in the result.

§ 26.   We may now calculate the results of our experiments in absolute measure.   In the notation of § 15 we have, as the relation between the current i and the difference of weighings observed in air m,

$$i = \mu \sqrt{m}, \qquad \text{where } \mu = \cdot037048.$$

If w be the electro-chemical equivalent of silver in C.G.S. measure, viz., the quantity of silver in grams deposited per second by the unit C.G.S. current, then the rate of deposit by current i is $w \cdot i$, or $w \cdot \mu \cdot \sqrt{m}$.   Now, by the table this rate of deposit is $\sqrt{m}./2414\cdot45$; so that

$$w = \frac{1}{2414\cdot45 \times \cdot037048} = \cdot\mathbf{0111794}.$$

In terms of practical units we have as the quantity of silver in grams deposited per ampère per hour

$$1\cdot11794 \times 10^{-3} \times 3600 = \mathbf{4\cdot0246}.$$

The number found by Kohlrausch in his recent experiments is

$$w = \cdot011183,$$

while that found by Mascart* is

$$w = \cdot01124.$$

* Journal de Physique, March, 1882.

The agreement between Kohlrausch and ourselves is perhaps as good as could be expected, and would be diminished almost to nothing were we to take in our experiments the weights as found after drying at 160° C., viz., before the strong heating. The account hitherto published by Kohlrausch is only an abstract, and does not explain how the deposits were treated*.

§ 27.   Considering that the silver voltameter may now be used satisfactorily for the standardising of current-measuring instruments, we have made some experiments in order to ascertain the limits within which the method is applicable. With regard to the strength of the nitrate solution there is considerable latitude when the currents are weak, e.g., not exceeding ¼ ampère. In such cases a 4 per cent. solution may be used satisfactorily in our voltameters. However, for practical purposes at the present time the object will usually be to measure stronger currents, and then it is advisable to keep the solution up to 15 or 30 per cent. If the solution is too weak in relation to the density of current, the deposit has a tendency to looseness, and is liable to grow up in an irregular manner, so as to meet the anode. In a 3-inch platinum bowl such a solution will allow of a current of about 1 ampère for a period of an hour. The strongest current which we have been able to use with a single voltameter is about 2 ampères, and for this purpose we employed a solution containing one part of salt to two parts of water. It is probable that the deposit would have deteriorated if the current had been allowed to flow for much longer than a quarter of an hour, but in that time an ample amount (about 2 grms.) is obtained. The practical conclusion is that currents not exceeding 1½ ampère may be conveniently measured in a 3-inch voltameter by using a strong solution, and by stopping the operation after about a quarter of an hour. A shorter time than this would hardly allow of sufficiently precise measurement when a high degree of accuracy is aimed at. For purposes where an error of ⅛ per cent. is admissible, a duration of five minutes (300 seconds) would be sufficient, and under these circumstances a stronger current would be unobjectionable.

It will be seen that the application of this method to the measurement of such currents as are usually passed through incandescent lamps presents no difficulty, and we hope that it may be generally adopted as a control upon the indications of instruments depending for their trustworthiness upon the constancy of springs or of steel magnets. The anodes should be composed of fine silver sheet (about ⅛ inch thick), such as is sold for five shillings per ounce, and should not approach the sides of the bowl too closely. As there need be no waste of metal, the expense of silver as compared with copper should not be allowed to stand in the

* See Notes.

way of its use. For practical purposes it will be unnecessary to take some of the precautions which we thought incumbent upon us. After rinsing a few times with distilled water the deposit may be left to soak for an hour or so, and then after another rinsing dried over a spirit lamp. After the lapse of another hour it may be weighed, with a risk of error not exceeding a few tenths of a milligram.

When still stronger currents have to be dealt with, the silver volta-meter is less convenient. Platinum bowls of large size are not usually met with, but two or three may be combined in parallel without much trouble. In one of our experiments the same current was passed suc-cessively through a single voltameter, and through two arranged in parallel. The deposit in the single bowl, thrown down in 13 minutes, was 2·2327 grms. Those in the other bowls were 1·0114 and 1·2215, alto-gether 2·2329, agreeing almost precisely. In this way with three bowls, such as we have used, in parallel, there would be no difficulty in measuring currents up to 5 ampères.

§ 28.   The second branch of our subject is the evaluation of the electro-motive force of standard galvanic cells. Enough has been said as to the means employed for measuring electric currents in absolute measure. If a current, after passing the current-weighing apparatus, is made to traverse a known resistance, it will generate at the extremities of that resistance a known electromotive force. By suitably accommodating to one another the magnitude of the resistance and the strength of the current, the electromotive force may be made to balance that of a standard cell, whose force is thus determined. The difficulty of the matter relates principally to the preparation and definition of the standard cells, and in order to test the constancy of the cells it is desirable to extend both the absolute determinations and the comparisons of various cells over a considerable range of time.

Before describing further the arrangements adopted for the absolute measurements, it will be convenient to consider the comparisons of E.M.F., which were always made by the method of compensation, in order to diminish as far as possible the currents actually passed through the cells under examination. The main circuit consisted of two Leclanché cells M, and two resistance boxes N, O (joined by a short stout wire) of 10,000 ohms each (fig. 1). Of this resistance a variable and adjustable proportion was included between the points of derivation, and (by use of the second box) the total was in all cases made up to 10,000. Thus, in compensating a single Clark cell the resistance from the first box might be 4900, and from the second 5100. By this means the constancy of the main current is secured. The derived branch includes the cell or cells to be tested (P), a mercury reversing key (Q), and a galvanometer (T), with which is associated a resistance (S) of 10,000 ohms. The galvanometer itself was of the Thomson pattern, and had a resistance of about 200 ohms. By the substitution of an instrument with a longer wire and of resistance up to 10,000, a greater degree of sensitiveness might have been obtained, but with careful reading of the galvanometer scale the arrangements were sufficient for the purpose, and would indicate the E.M.F. to about $\frac{1}{10,000}$. In the preliminary trials a simple contact key with platinum studs was used in the galvanometer branch with the idea that shorter contacts would thus suffice. But, probably from thermoelectric disturbance, the readings thus obtained were not so consistent as with the mercury reversing key, and the smallness of the currents actually allowed to pass rendered the longer contacts unobjectionable. From the data already given it will be seen that a current of 10^{-8} ampères was sensible, and no disturbance could be expected from currents 100 times, or more, greater than this. In order to test whether the connexions were rightly made, the first observation was usually taken with a still higher resistance in the galvanometer branch, which could easily be effected by causing the current to pass through the body of one of the observers from hand to hand. If by accident too large a current was allowed to pass through a cell, no further use was made of that cell until the next day*. It must be mentioned that great care was taken, and was necessary, in respect of the insulation of the various parts. For instance, no correct results were obtainable when the Leclanché's stood upon the (tiled) floor, if at the same time other parts of the combination were touched with the hand. A sheet of paraffined paper interposed proved a remedy. In this matter we have had several disagreeable lessons, and we cannot too strongly emphasise our advice to take too many rather than too few precautions.

When two cells under comparison differ by a considerable fraction, they may be compared separately with the Leclanché's, or rather expressed in terms of the current afforded by the Leclanché's through 10,000 ohms. Thus, on Dec. 3, 1883, in order to balance Clark No. 1 (see below) 4926 were required between the points of derivation. When a standard Daniell of Raoult's pattern was substituted for the Clark, the number required was 3798. In terms of No. 1 Clark the E.M.F. of the Daniell is thus 3798/4926, or ·7710. At the end of a series of comparisons it is proper to repeat the observation of the first standard cell, in order to check the constancy of the current supplied by the Leclanché's. In our experience there was usually no appreciable variation.

When the cells to be compared are nearly alike, it is better in the second observation to express the *difference* of forces by setting the second

---

* Experiments detailed later (§ 31) show that the precautions observed in this respect were more stringent than were really necessary.

cell to act against the first. Thus, the force of Clark No. 1 being expressed as before by 4926, the corresponding resistance for the excess of the force of Clark 1 over Clark 3 was 2 ohms. Hence, in terms of Clark 1 the force of Clark 3 is ·9996, and the result is less liable to error than if the comparisons of each with the Leclanché's were effected separately.

§ 29. Of the first batch of Clark's which were compared together from November, 1883, onwards, No. 1 was set up near the beginning, and Nos. 2, 3, 4, 5, towards the end of October. They were prepared generally according to the directions given by Dr Alder Wright*, to whom we have been indebted for advice and for samples of some of the materials. The saturated solution of zinc sulphate was nearly neutral. The metallic zinc was bought as pure from Messrs Hopkin and Williams. The mercurous sulphate was from the same source, and the metallic mercury was redistilled in the laboratory. We did not consider it desirable to take precautions against the presence of air, thinking that it was sure to find an entrance sooner or later.

Four new cells, Nos. 6, 7, 8, 9, were set up from the same materials on January 10, 1884. It will be seen from the table that when a fortnight old they differed but little from the first batch.

In preparing these cells the most troublesome part of the process was found to be the casting of the zincs. The metal, melted in a porcelain crucible, was sucked up into a previously heated tube of hard glass, but the operation required some address, and there was considerable waste of zinc from oxidation and otherwise. It occurred to us to try whether equally, or perhaps still more, satisfactory results might not be obtained by substitution for the solid metal of an amalgam of zinc. For this purpose a form of cell, called for brevity the H-cell, was contrived, and is shown full size (fig. 5). One of the legs is charged with the amalgam of zinc (B), the other with pure mercury (C), covered with a layer of mercurous sulphate (D). The whole is then filled up above the level of the cross tube with saturated zinc sulphate (E), and a few crystals are added. Evaporation is prevented by corks (F), closing the upper ends of the tubes. Electrical contact with the amalgam and with the pure mercury is made by platinum wires (A), sealed into the glass.

A preliminary experiment in which both legs of a cell were charged with amalgam (the mercurous sulphate being dispensed with) having shown that the E.M.F. was independent of the excess of undissolved zinc, two cells, H_1, H_2, were set up on February 12, 1884, and submitted to various tests, such as stirring up the amalgam with a glass rod. The amalgam was prepared from pure mercury and the same zinc as before. Subsequently, on March 6, six more cells were charged with a somewhat different treatment.

* *Phil. Mag.* July, 1883.

The sulphate of zinc was from another sample and contained appreciable quantities of iron. Moreover, the amalgam was differently prepared. The mercury and zinc were shaken up together in a bottle with a little acid, after which the acid was washed out by shaking with several changes of water, until litmus paper was no longer reddened. Into each cell, in addition to the fluid amalgam, there was dropped a piece of solid zinc from the bottle. The same mercurous sulphate as before was employed, but the washing with distilled water was dispensed with. The three remaining cells of this pattern H_9, H_{10}, H_{11}, were charged on March 12, 1884, with a third sample of zinc sulphate.

Fig. 5.

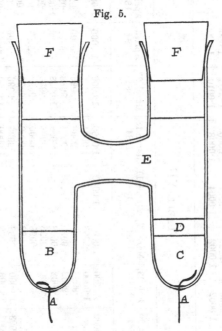

The agreement among themselves and the constancy of the H-cells has been all that could be wished; but some modification in preparation will be desirable, for it has been found that the amalgam tends to harden into compact lumps, the expansion of which is liable to burst the cells. From this cause H_3, H_4, H_7, succumbed at a comparatively early stage. It is probable that the addition of solid zinc to the fluid amalgam had better be omitted, but on this and other points we hope to make further investigation. The H pattern lends itself conveniently to experiment, as it is possible by withdrawing the corks to make any desired addition to the contents. On more than one occasion the contents of each leg have been vigorously stirred, without the slightest change in the E.M.F.

Since the first draft of this memoir was written two new batches of cells of the ordinary pattern have been prepared with different materials.

TABLE VI.

	Nov. 6, 1883	Nov. 9, 1883	Nov. 12, 1883	Nov. 14, 1883	Nov. 20, 1883	Nov. 22, 1883	Nov. 30, 1883	Dec. 3, 1883	Dec. 5, 1883	Dec. 11, 1883	Dec. 12, 1883
Clark 1	1·0000	1·0000	1·0000	1·0000	1·0000	1·0000	1·0000	1·0000	1·0000	1·0000	1·0000
,, 2	1·0015	1·0016	1·0006	1·0002	1·0006	1·0008	1·0014	1·0010	1·0010	1·0006	1·0008
,, 3	1·0001	1·0000	·9988	·9990	·9990	·9996	1·0000	·9996	·9996	·9994	·9996
,, 4	1·0008	1·0008	·9996	·9996	·9994	1·0000	1·0004	1·0000	1·0000	·9994	·9998
,, 5	1·0012	1·0016	1·0002	1·0002	1·0004	1·0008	1·0010	1·0000	1·0004	1·0000	1·0002

TABLE VII.

	Jan. 25, 1884	Jan. 28, 1884	Feb. 16, 1884	Feb. 20, 1884	Feb. 23, 1884	March 7, 1884	March 11, 1884	March 18, 1884	March 27, 1884	April 3, 1884	April 25, 1884	May 8, 1884	May 27, 29, 30, 1884	June 11, 1884
Clark 1	1·0000	1·0000	1·0000	1·0000	1·0000	1·0000	1·0000	1·0000	1·0000	1·0000	1·0000	1·0000	1·0000	1·0000
,, 2	1·0005	1·0000	·9996		1·0000	·9998		1·0004	·9996	1·0000	·9998	1·0000	·9998	1·0000
,, 4	·9999	·9997	·9994		·9993	1·0000		·9998	·9988	·9998	·9980	·9994	·9996	·9990
,, 5	·9998		·9998		·9992	·9990		·9994	1·0008		1·0006	1·0006	In ice.	1·0009
,, 6	1·0005		1·0008		1·0002	1·0008		1·0008	·9996		·9994	·9996		
,, 7	·9997		·9998		·9996	·9996		·9998	·9998		·9998	·9999		
,, 8	1·0001		1·0000			·9198		1·0002	1·0000		1·0000	1·0000	1·0000	1·0000
,, 9	1·0003		·9998	·9993	1·0000	1·0002		·9996	·9996	·9996	·9994	·9994	1·0001	1·0001
H_1			·9988	1·0003	1·0003	·9998		1·0000	1·0006	1·0001	·9998			
H_2			1·0006			1·0002		1·0008	1·0008					
H_3						1·0010	1·0010	1·0007	1·0006		1·0006	1·0006	1·0006	1·0004
H_4						1·0008	1·0008	1·0007	1·0006		1·0006	1·0006	1·0006	1·0006
H_5						1·0008	1·0008	1·0007	1·0008					
H_6						1·0008	1·0008	1·0007	1·0008		1·0006	1·0006	1·0008	
H_7						1·0008		1·0006	1·0005		1·0002	1·0006		
H_8						1·0008		1·0003	1·0006		1·0003	1·0005	·9998	1·0003
H_9								1·0005	1·0005		1·0002	1·0005	1·0006	1·0006
H_{10}								1·0005					1·0006	1·0006
H_{11}														

TABLE VIII.

1884

	May 8	May 12	May 14	May 15	May 16	May 17	May 19	May 20	May 21	May 22
Clark 1 . .	1·0000	1·0000	1·0000	1·0000	1·0000	1·0000	1·0000	1·0000	1·0000	1·0000
,, 10 . .	1·0200	1·0150	1·0022*	1·0022*	1·0022	1·0016*	1·0010*	1·0006	1·0004	1·0002
,, 11 . .	1·0132	1·0116	1·0104	1·0096	1·0010*	1·0010*	1·0010	1·0008	1·0008	1·0008
,, 12 . .	1·0072	1·0124	..	1·0118	1·0110	1·0104	1·0092	1·0080	1·0072	1·0050
,, 13 . .	1·0030	uncertain		1·0150						

1884

	May 23	May 24	May 26	May 27	May 28	May 29	May 30	June 2	June 4	June 11
Clark 1 . .	1·0000	1·0000	1·0000	1·0000	1·0000	1·0000	1·0000	1·0000	1·0000	1·0000
,, 10 . .	1·0003	1·0002	1·0002	1·0003	1·0002	1·0002*	1·0002	..	1·0003	1·0003
,, 11 . .	1·0006	1·0004	1·0004	1·0006	1·0004	1·0002*	1·0004	..	1·0003	1·0003
,, 12 . .	1·0026	1·0010	1·0006	1·0007	1·0004	1·0002*	1·0004	..	1·0001	1·0004
,, 13 . .	1·0090	..	..	..	1·0080	1·0008*	1·0008	..	1·0005	1·0003
,, 14 . .				1·0240	1·0214*	1·0158*		1·0126	1·0132	1·0100
,, 15 . .				1·0240	1·0230	1·0220		1·0195	1·0148	1·0092
,, 16 . .				1·0090	1·0144	1·0134		1·0098	·9997	1·0109
,, 17 . .				broken						
,, 18 . .				1·0210	1·0194	1·0168	1·0014		1·0006	1·0000
,, 19 . .				1·0230	1·0200	1·0168	1·0032		1·0006	1·0006

* For continuations of these tables, see notes.

In this case the zincs were used as supplied, without re-casting*, and the mercurous sulphate, though distinctly acid, was not washed. The first batch (10, 11, 12, 13) were set up on May 7, and the second batch (14, 15, 16, 17, 18, 19) on May 26.

§ 30.   Tables VI., VII., VIII. show the results of most of the comparisons, the value of every cell on each day being expressed in terms of Clark No. 1. It will be seen that there are durable differences between cells of the same batch, but that these do not much exceed $\frac{1}{1000}$. There are also changes of small amount in the force of a given cell, part of which is perhaps attributable to a difference of temperature coefficient. Moreover the actual temperatures may possibly have differed a few tenths of a degree in the case of various cells, many of which stood some feet apart. Clark No. 3 does not appear in Table VII., since on January 25 it was found to be short circuited. During the later comparisons, Nos. 6 and 7 were unavailable, having been diverted to another use.

The two last batches took a longer time than usual (about three weeks) to reach their normal values. It will be seen from Table VIII. that when first set up these cells were too strong by as much as 1 or 2 per cent. It was thought that the process of settling down might be quickened by closing the circuit occasionally for some minutes, through a resistance of 1000 ohms, and the asterisk in the table indicates that on the day previous to the comparison the cell in question had been so treated for about ten minutes. When once the settling down is completed, further short circuiting appear to be without effect.

§ 31.   Some observers having laid great stress upon the importance of guarding Clark cells from the passage of sensible currents, we give a specimen of the results of some tests to which we have subjected a few of the cells, in order to find out how much care was really necessary in their use to avoid polarisation. The accompanying Table IX. shows the variations of E.M.F. of Clark No. 6 on April 28, when very rudely treated. The other connexions remaining as usual, the poles of the cell were joined through a resistance-box, by means of which the cell could be short circuited with any external resistance from 0 to infinity. The numbers entered (such as 4994) are proportional to the difference of potential between the poles, being in fact the resistance between the points of derivation on the Leclanché circuit. It will be seen that in the course of a quarter of an hour the cell recovers, to within a few ten-thousandths of its value, from the effects of being short circuited for several minutes through such resistances as 1000 ohms. From the electromotive forces *during* the short circuiting it appears that the internal resistance is high, nearly as much as 300 ohms.

---

* The surface of the metal was brightened with file and sand-paper.

The manner in which the Clark cells have borne the tests applied to them justifies the hope that they may be found generally available as standards of E.M.F. But further experience is necessary as to the effect of various modes of preparation, and it is to be hoped that this may soon be forthcoming. As used by us, the process is so simple that no one need be deterred from setting up cells for himself.

TABLE IX.

Time	Resistance between Poles	E.M.F.
h. m.		
3 35	∞	4994
3 47	∞	4994
3 53	Changed from ∞ to 10,000	
3 56	10,000	4851
3 41	10,000	4853
4 59	Changed from 10,000 to ∞	
5 2	∞	4990
5 15	∞	4991
5 47	∞	4992
6 3	Changed from ∞ to 1000	
6 5	1,000	3990
6 11	1,000	3860
6 13	Changed from 1000 to ∞	
6 19	∞	4990
6 25	∞	4991
6 29	Changed from ∞ to 500	
6 34	Changed from 500 to ∞	
6 36	∞	4985
6 37	∞	4988
6 52	∞	4991

§ 32. Experiments on Daniell cells gave only a moderately good result. Raoult's form was employed, in which the zinc and copper solutions are placed in separate beakers, the connexion being only through a Y-tube charged with zinc sulphate and tied over the ends with bladder. One electrode was of pure zinc amalgamated with pure mercury, and the other of copper freshly coated electrolytically. The zinc and copper solutions were both of sp. gr. 1·1.

TABLE X.

	November 30, 1883	December 3, 1883	December 5, 1883	December 11, 1883	December 12, 1883
Clark No. 1	1·0000	1·0000	1·0000	1·0000	1·0000
Daniell . .	·7702	·7710	·7705	·7698	·7702

The Daniell cell has of course to be charged freshly on each occasion, and is thus far less convenient in use than the Clark's, which stand for months always ready for use. The temperature of the cells at the time of the comparisons tabulated was about 16° C.

Through the kindness of the inventor, we have had the opportunity of comparing some De La Rue cells with the Clark's. The cells are of a somewhat modified construction, the atmospheric oxygen being excluded by a layer of paraffine oil. They were set up some days before the comparisons, and short-circuited for five minutes in order to start the chemical action.

We found

> No. 1 De La Rue = ·7510 Clark.
> No. 2   „   = ·7512   „
> No. 3   „   = ·7382   „
> No. 4   „   = ·7458   „
> Mean   „   = ·7465   „

Mr De La Rue (*Phil. Trans.*, vol. CLXIX., Part I.) found a result decidedly smaller, the explanation of which is to be sought in the fact that in his experiments the cells were making a current of about $\frac{1}{1000}$ ampère, whereas in ours the electromotive force is measured when no current passes.

It may be useful to record also a comparison between our Clark's and a new form of Daniell, introduced by Sir W. Thomson. This cell is charged with zinc sulphate of sp. gr. 1·02, and with saturated solution of copper sulphate. The zinc is not amalgamated. According to Sir W. Thomson's directions, the circuit of the cell is closed through 250 ohms, and the E.M.F. measured is that between the poles under these conditions. After the current had been running for about an hour and a half, the E.M.F., which had been increasing, became fairly constant, and its value was then ·743 in terms of Clark No. 1. The comparison was made on April 8, 1884*.

§ 33. We now pass to the description of the method adopted for the absolute determinations. The current, after leaving the current-weighing apparatus, is caused to traverse a wire of known resistance R, whose stout copper terminals rest on the copper bottoms of suitable mercury cups H, K (fig. 1). To these cups are brought also the terminals of the derived branch, in which are included the galvanometer and the standard cell.

On account of the strength of the currents (about $\frac{1}{3}$ ampère) the resistance required to be of special construction in order to avoid too great heating.

Two ebonite rods were held in a parallel position by a frame of wood, and round these uncovered german silver wire was wrapped so as to be

* See notes.

exposed to the air as much as possible. The rods are about a foot apart, and are grooved, the better to keep the wire in its place. The resistance is about 4 B.A., and was determined with the aid of a *five* and a *single**. At $17°·6$ its value is 4·00699 B.A.

Even this resistance-wire heats sensibly when the current of $\frac{1}{3}$ ampère is passed through it for more than a few seconds. The increment of resistance was determined by observations taken immediately after the passage for some minutes of a stronger current (about 1 ampère). In this way it was found that for the currents usually employed a correcting factor 1·00041 must be introduced to take account of the heating, independently of course of the correction necessary for the difference between $17°·6$ and the temperature of the atmosphere at the time of an absolute determination.

§ 34. In order to obtain the balance of electromotive forces two distinct methods have been followed. In the earlier determinations there was no electromotive force in the derived branch except that of the standard cell, and the adjustment was effected by variation of a comparatively high auxiliary resistance from a box, placed in multiple arc with the [4]. The readings were taken by reversal of the galvanometer connexions at a mercury commutator, and the small outstanding galvanometer displacement was allowed for with the aid of observations of the effect of a known change in the auxiliary resistance. In this way could be determined the auxiliary resistance, and from it (by addition of conductivities) the effective resistance between the points of derivation necessary for a balance with the actual current. The value of the current at the moment in question is deduced from the curves representing the two sets of current weighings (§ 24). In the course of half-an-hour several almost independent determinations of the electromotive force could be completed.

This method is the simplest, and could usually be made to work satisfactorily. It is, however, open to the objection that if the current changes rapidly we must either allow for a considerable galvanometer displacement or else alter the auxiliary resistance. But the latter change reacts upon the principal current, and renders the current weighing curves discontinuous, thereby increasing the difficulty of specifying the value of the current at the moment of observation.

§ 35. In the second method the resistance between the points of derivation is the [4] simply, and compensation is made in the galvanometer branch by the introduction of a graduated E.M.F. (fig. 1). The arrangement is in fact almost the same as in the comparison of two cells by the method of difference (§ 28), one of the cells being replaced by the

---

* For the methods used to ascertain the value of the *five* the reader is referred to former papers.

resistance [4] traversed by the main current. As the apparatus for these comparisons was always ready for use, this method was, under the circumstances of the case, really more convenient than the other, and was employed in the later determinations. The procedure will be best understood from an example.

On March 29, 1884, determinations of silver and of electromotive force were made simultaneously, so that the same set of current weighings might serve for both purposes. Accordingly the main current traversed the three voltameters, the current weighing apparatus and the resistance [4]. In the derived branch (fig. 1) were the standard cell No. 4 Clark, the galvanometer with its commutator, and coils from a resistance box, through which passed the current from the two Leclanché cells (§ 28). If the compensation between the Clark and the difference of potentials at the terminals of the [4] were incomplete the balance could be restored by the introduction of a graduated part of the E.M.F. of the Leclanché's, the value of which, in terms of the Clark, is found by a subsequent experiment, in which the [4] is excluded. It will be understood that the Leclanché's worked in a perfectly constant manner, the whole resistance in circuit being always made up to 10,000 ohms (in addition to that of the cells themselves). If E be the E.M.F. of the Clark, ρ the resistance (traversed by the current of the Leclanché's) which must be used to get a balance when the [4] is excluded, r the resistance actually required during a set of measurements when [4] is connected, then the electromotive force actually compensating the action of [4] is $E(1 - r/\rho)$.

At the beginning of the proceedings on March 29 the main current was stronger than that required for the simple compensation of E, so that to get a balance at the galvanometer the Leclanché's would have had to be reversed. At 18^m from the commencement the current had fallen to the point of compensation with $r = 0$. At 28^m balance required $r = 20$ B.A., at 34^m $r = 37$, and at 48^m $r = 90$. To take these observations, the easiest way is to overshoot the point somewhat, and then continually reversing the galvanometer to note the time of passing through the balance. From the curves representing the current weighings, the double force of attraction at the above times were found to be ·9645, ·956, ·9495, ·931, expressed in grams. This is what has been denoted by m (§ 26), and the corresponding current is

$$i = ·037048 \sqrt{m}.$$

§ 36. The resistance R between the points of derivation must be expressed in absolute measure, if we wish E to be so expressed. But for comparison with the results of other observers it will be convenient to keep this question apart and, in the first instance, to express our electromotive forces as if the B.A. unit were correct. Any factor (such as ·9867) which may be adopted to express the B.A. unit in terms of the ohm will enter also into the expression of E in true volts.

At the atmospheric temperature $13°\cdot1$ the value of the [4] is $3\cdot9998$ B.A., whence

$$R = 4\cdot00143 \text{ B.A.},$$

correction being made for the heating effect of the current.

The formula for E is

$$E = \cdot037048\, R \,.\, \sqrt{m} \,.\, \frac{\rho}{\rho - r}.$$

The value of ρ (on the occasion in question) was 4999 B.A., and this completes the data for the evaluation of E. The four values corresponding to the above observations are

$$1\cdot4559, \quad 1\cdot4553, \quad 1\cdot4553, \quad 1\cdot4566,$$

giving as mean

$$E = 1\cdot4558 \text{ B.A. volts.}$$

This result is for No. 4 at a temperature of $13°\cdot1$. The value of No. 4 in terms of No. 1 at the time in question was about $\cdot9998$, so that we should have found for No. 1

$$E = 1\cdot4561 \text{ B.A. volts.}$$

We have still to reduce to the standard temperature of $15°$. The coefficient originally given by Latimer Clark is $1\cdot0006$ per degree centigrade. Wright and Thomson* found a smaller number, viz., $1\cdot00041$, and with this our results were first reduced. Later, however, we found reason to suspect that the actual change was greater than this, and accordingly made some special observations to clear up the doubt. One cell (No. 6) was mounted in a large test tube, the gutta-percha-covered leading wires being brought through a tightly-fitting indiarubber cork, and was kept constantly at $0°$ centigrade by being surrounded with ice. With this No. 1 at the temperature of the room was compared from day to day, with the result that its temperature coefficient is about the double ($1\cdot00082$) of that given by Wright and Thomson. A similar result was found by Helmholtz†, who remarks that the effect of temperature may vary according to the preparation of the cell.

Using this number to reduce the result of March 29, we have to subtract $\cdot0022$, thus obtaining

$$E = 1\cdot4539 \text{ B.A. volts}$$

as the electromotive of No. 1 Clark at $15°$.

---

* *Phil. Mag.* July, 1883, p. 36.

† *Sitzungsber. d. Kön. Akad. d. Wiss. zu Berlin*, February, 1882.

§ 37. This determination and twelve others, made at intervals from Oct., 1883, to April, 1884, are exhibited in Table XI.* They are all

TABLE XI.

I. Date, 1883 and 1884	II. Cell used	III. Temperature	IV. E.M.F. in B.A. volts	V. E.M.F. relative to No. 1	VI. E.M.F. of No. 1	VII. Correction to 15°	VIII. E.M.F. in B.A. volts corrected to 15°
October 23 .	Clark No. 1	15·9°	1·4542	1·0000	1·4542	+ ·0010	1·4552
November 20	,, No. 2	15·3	1·4549	1·0006	1·4540	+ ·0004	1·4544
,, 21	,, No. 1	14·9	1·4543	1·0000	1·4543	− ·0002	1·4541
,, 22	,, No. 1	14·9	1·4533	1·0000	1·4533	− ·0002	1·4531
December 4	,, No. 1	15·8	1·4524	1·0000	1·4524	+ ·0010	1·4534
,, 11	,, No. 1	17·2	1·4524	1·0000	1·4524	+ ·0026	1·4550
,, 12	,, No. 2	15·8	1·4549	1·0008	1·4537	+ ·0010	1·4547
January 28 .	,, No. 2	15·0	1·4541	1·0000	1·4541	+ ·0000	1·4541
March 20 . .	,, No. 4	15·8	1·4533	·9998	1·4536	+ ·0010	1·4546
,, 25 . .	,, No. 1	13·5	1·4560	1·0000	1·4560	− ·0018	1·4542
,, 29 . .	,, No. 4	13·1	1·4558	·9998	1·4561	− ·0022	1·4539
April 2 . .	,, No. 1	16·1	1·4524	1·0000	1·4524	+ ·0014	1·4538
,, 7· . .	,, No. 1	15·5	1·4535	1·0000	1·4535	+ ·0006	1·4541
Mean . . .	..	15·3	..	..	..	..	1·4542

deduced from observations with the current-weighing apparatus. It will be seen that there is little or no evidence of any progressive change. The casual fluctuations are of course partly due to errors of observation, but it would seem are principally to be attributed to real variations of electromotive force of the same kind as appear in the Tables VI., VII., VIII., showing the relative values of the various cells. The mean temperature at the times of the determinations differs so little from 15°, that the final number for that temperature is almost independent of the temperature coefficient.

We may take as applicable with but little error to all the cells of this type that have been experimented upon

$$E = 1·454 \text{ B.A. volts at } 15°.$$

The value for the H-cells would be a little higher. (See Tables.)

The corresponding number found by Mr Latimer Clark was

$$E = 1·457 \text{ B.A. volts,}$$

so that the difference between us is small, and perhaps even dependent upon variations in the materials or construction of the cells.

To express our results in true volts we have only to introduce the factor expressive of the B.A. unit in terms of the ohm. If in accordance with our

* For continuation of Table XI. see notes.

own determinations we take

$$1 \text{ B.A. unit} = \cdot 9867 \text{ ohm},$$

we shall have as the value of a Clark cell at 15°

$$E = 1 \cdot 435 \text{ volt.}$$

§ 38. It has been mentioned that on March 29 silver deposits were made at the same time as the observations of E.M.F. One object of this was to exemplify the procedure which will probably be in future the most convenient for the determination of E.M.F. when the very highest accuracy is not required. It is evident that if we assume a knowledge of the electro-chemical equivalent of silver, the weights obtained in a given time on March 29 will lead to a determination of E.M.F., *independently of the current weighings*. We propose to exhibit the method of calculation, ignoring altogether the use of the current-weighing apparatus, whose only effect will be that of a resistance of about 40 ohms. If W be the weight of silver deposited in the time t, w the electro-chemical equivalent, we have as the relation between W and E,

$$W = \frac{wE}{R} \int \left(1 - \frac{r}{\rho}\right) dt = \frac{wEt}{\rho R} \left(\rho - \int \frac{r \, dt}{t}\right).$$

On this occasion $W = 1 \cdot 4531$ grms., $t = 3599$ seconds, $R = 4 \cdot 0014$ B.A., $\rho = 4999$ B.A., as before. If w be assumed, the only other element required for the evaluation of E is

$$\int \frac{r \, dt}{t},$$

viz., the mean value of r necessary for a balance of E.M.F. during the time that the current ran through the voltameters. To get this the actual observations of r are plotted, the times being taken as abscissæ, and a curve constructed representing the value of r throughout the course of the experiment*. From this curve the ordinates are measured, which correspond to the middle of every five minutes' period. The values of r thus obtained are

TABLE XII.

Time	r	Time	r
m.		m.	
$2\frac{1}{2}$	$- 22$	$32\frac{1}{2}$	$+ 32$
$7\frac{1}{2}$	$- 16$	$37\frac{1}{2}$	$+ 48$
$12\frac{1}{2}$	$- 10$	$42\frac{1}{2}$	$+ 66$
$17\frac{1}{2}$	$- 2$	$47\frac{1}{2}$	$+ 86$
$22\frac{1}{2}$	$+ 8$	$52\frac{1}{2}$	$+112$
$27\frac{1}{2}$	$+18$	$57\frac{1}{2}$	$+140$

Mean $= + 38 \cdot 3$.

* In the formation of the curve use was made of observations in which the galvanometer balance was incomplete, the value of the scale divisions being approximately known.

The rapid falling-off of the current towards the end of the hour is believed to be due to the formation of crystals upon the anodes of the cells charged with silver chlorate.   The value of

$$\rho - \int \frac{r \, dt}{t}$$

being thus found to be 4960·7, the calculation of E may be completed. Taking $w = 1·1180 \times 10^{-2}$, we get

$$E = 1·4562 \text{ B.A. volts,}$$

as the electromotive force of No. 4 Clark at 13°·1.

On April 2 an equally satisfactory result was found from the silver deposits without use of the current weighings.   It will be seen that in this way anyone may determine the E.M.F. of his standard battery with a very moderate expenditure of trouble and without the need of any special apparatus.   So large a resistance in the main circuit as in the above example, due to the idle coils of the current-measuring apparatus, is not necessary, but some resistance in addition to R and that of the battery and voltameters would probably be advisable.   Otherwise the magnitude of the current would be too sensitive to the resistance of the volta-meters, which cannot be included in the circuit until the experiment actually begins.   In the preliminary adjustments the resistance of the voltameters should be represented by an estimated equivalent of wire resistance, and this should not be too large a fraction of the whole.   In our case the resistance of the three voltameters charged with nitrate solution of 15 per cent. was a little under two ohms, and the condi-tions under which we worked would be sufficiently imitated by a circuit containing, besides the [4] and the voltameters, an extra resistance of 10 ohms.   A battery of three or four Grove cells would then be sufficient for the generation of the current.

APPENDIX (see § 15).

TABLE of the [logarithm] of $\sin \gamma \{2F_\gamma - (1 + \sec^2 \gamma) E_\gamma\}$ from $\gamma = 55°$ to $\gamma = 70°$.

° ′		° ′		° ′	
55 0	$\bar{1}$·9198899	60 0	·1786408	65 0	·4433405
55 6	$\bar{1}$·9250674	60 6	·1838431	65 6	·4487720
55 12	$\bar{1}$·9302440	60 12	·1890478	65 12	·4542107
55 18	$\bar{1}$·9354198	60 18	·1942546	65 18	·4596565
55 24	$\bar{1}$·9405945	60 24	·1994636	65 24	·4651097
55 30	$\bar{1}$·9457677	60 30	·2046748	65 30	·4705707
55 36	$\bar{1}$·9509400	60 36	·2098887	65 36	·4760395
55 42	$\bar{1}$·9561123	60 42	·2151058	65 42	·4815165
55 48	$\bar{1}$·9612837	60 48	·2203260	65 48	·4870015
55 54	$\bar{1}$·9664536	60 54	·2255491	65 54	·4924944
56 0	$\bar{1}$·9716227	61 0	·2307753	66 0	·4979956
56 6	$\bar{1}$·9767918	61 6	·2360045	66 6	·5035052
56 12	$\bar{1}$·9819605	61 12	·2412367	66 12	·5090234
56 18	$\bar{1}$·9871288	61 18	·2464720	66 18	·5145504
56 24	$\bar{1}$·9922966	61 24	·2517106	66 24	·5200861
56 30	$\bar{1}$·9974637	61 30	·2569525	66 30	·5256304
56 36	·0026304	61 36	·2621981	66 36	·5311838
56 42	·0077970	61 42	·2674478	66 42	·5367469
56 48	·0129635	61 48	·2727014	66 48	·5423195
56 54	·0181298	61 54	·2779585	66 54	·5479017
57 0	·0232962	62 0	·2832194	67 0	·5534935
57 6	·0284628	62 6	·2884843	67 6	·5590948
57 12	·0336297	62 12	·2937533	67 12	·5647060
57 18	·0387966	62 18	·2990263	67 18	·5703278
57 24	·0439638	62 24	·3043035	67 24	·5759599
57 30	·0491317	62 30	·3095854	67 30	·5816022
57 36	·0542999	62 36	·3148717	67 36	·5872550
57 42	·0594684	62 42	·3201621	67 42	·5929188
57 48	·0646364	62 48	·3254571	67 48	·5985936
57 54	·0698062	62 54	·3307575	67 54	·6042795
58 0	·0749769	63 0	·3360628	68 0	·6099767
58 6	·0801480	63 6	·3413729	68 6	·6156851
58 12	·0853198	63 12	·3466879	68 12	·6214051
58 18	·0904926	63 18	·3520081	68 18	·6271370
58 24	·0956665	63 24	·3573335	68 24	·6328810
58 30	·1008414	63 30	·3626642	68 30	·6386371
58 36	·1060175	63 36	·3680004	68 36	·6444054
58 42	·1111950	63 42	·3733422	68 42	·6501859
58 48	·1163737	63 48	·3786896	68 48	·6559791
58 54	·1215535	63 54	·3840425	68 54	·6617852
59 0	·1267346	64 0	·3894014	69 0	·6676045
59 6	·1319170	64 6	·3947666	69 6	·6734371
59 12	·1371009	64 12	·4001380	69 12	·6792833
59 18	·1422865	64 18	·4055155	69 18	·6851433
59 24	·1474739	64 24	·4108993	69 24	·6910170
59 30	·1526636	64 30	·4162893	69 30	·6969043
59 36	·1578552	64 36	·4216858	69 36	·7028058
59 42	·1630486	64 42	·4270894	69 42	·7087220
59 48	·1682439	64 48	·4324998	69 48	·7146529
59 54	·1734412	64 54	·4379166	69 54	·7205985

EXPLANATION OF FIGURES.

Fig. 1. *A.* Principal battery of Grove's or storage cells.
       B. Resistance for adjustment of current.
       C. Voltameters.
       D. Rough tangent galvanometer.
       E. Reversing key of current-weighing apparatus.
       F. Fixed coils.
       G. Suspended coil.
       H, K. Mercury cups, into which dip the terminals of resistance *R.*
       L. Earth connexion.
       M. Leclanché's of E.M.F. compensator.
       N, O. Resistance-boxes of same.
       P. Standard galvanic cell.
       Q. Galvanometer commutator.
       S. Associated resistance of 10,000 ohms.
       T. Galvanometer.

Fig. 2. Section of ebonite ring (full size).

Fig. 3, § 14. Connexions for comparison of galvanometer constants.
           A. Daniell cell.
           B. Mercury reversing key.
           C. Point where current divides.
           D. Coil of electro-dynamometer.
           E. Ebonite coil.
           F, H, L, M. Mercury cups.
           G. Bridge galvanometer.
           K. Resistance-box in multiple arc with [10].
           P. Short circuiting piece to connect *F* and *H.*
           N. Resistance added to *E.*

Fig. 4, § 24. Curves of current weighings. In the original drawing two divisions along the line of abscissæ represent one minute, and two divisions along the line of ordinates represent one milligram. Of these divisions every tenth only is shown in the Figure.

Fig. 5, § 29. *H*-pattern of Clark cell.
           A. Platinum wires sealed through glass.
           B. Amalgam of zinc.
           C. Pure mercury.
           D. Mercurous sulphate.
           E. Saturated solution of zinc-sulphate.
           F. Corks.

NOTES.

(Added December, 1884.)

Note to § 25.

In order to investigate the effect (if any) of temperature upon the amount of silver deposits, we have made experiments in which volta-meters maintained at different temperatures were exposed to the same current. The results, exhibited in the accompanying table, show a small but apparently real *increase* in the weight of the deposit as the temperature rises. Had the effect been in the other direction, we should have been disposed to attribute it to imperfections of manipulation, for the deposits from the warm solutions were always coarser and looser in texture than the corresponding deposits (upon the same area) from the cold solutions.

1884	After usual washing and drying at 160°			After heating to verge of redness			Excess of hot over cold
	Hot bowl (about 50°)	Bowl at temperature of room (15°)	Cold bowl (4°)	Hot bowl	Bowl at temperature of room	Cold bowl	
May 27	..	..	..	2·3915	..	2·3905	·0010
June 4	2·0230	..	2·0220	2·0229	..	2·0221	·0008
July 22	1·9050	..	1·9043	1·9049	..	1·9043	·0006
July 31	1·9438	1·9432	1·9430	1·9440	1·9432	1·9431	·0009

The solution was a 15 per cent. solution of pure nitrate of silver, and the anodes were of pure metal. The current was about $\frac{1}{2}$ ampère, and passed for rather more than an hour.

The results here disclosed diminish, of course, the chemical significance of the number given as representing the electro-chemical equivalent of silver, but the variation is so small at ordinary laboratory temperatures that the use of the silver voltameter as a means of defining electric quantity is not practically interfered with.

Note to § 26.

M. Mascart (*Journal de Physique*, t. iii.; Juillet, 1884) has recently revised the calculation of the constant of his apparatus, by which revision the final number is altered from ·01124 to ·011156.

Note to § 27.

Although there can be no doubt that silver is greatly preferable to copper for the electrolytic measurement of currents, we have thought that it might be useful to make a few comparisons of the two metals, so as to allow copper to be referred to on an emergency with as much success as the nature of the case admits. The copper deposits were taken in the same way as the silver upon platinum bowls, the anodes being wrapped in filter-paper and suspended at the top of the liquid. On account of the tendency to oxidation it is not advisable to allow the copper deposits to soak for a long time. They were washed in boiling water for about half-an-hour, and then dried off in the hot closet at 150°. The solutions were made from sulphate, bought as pure, no acid being added. Of the four bowls I., II. are large and somewhat deep, III., IV. are shallow saucers about 3 inches in diameter. In the large bowls the area of deposit was about 32 sq. centims., in the smaller about 25 sq. centims. The strength of current on the first two occasions was about $\frac{1}{3}$ ampère, on the last about $\frac{3}{4}$ ampère, thus representing the circumstances for the measurement of the current through an incandescent lamp.

Date, 1884	Bowl	Solution	Weight of deposits	Mean	Ratio of copper to silver	Equivalent of copper (silver = 108)
Nov. 20 ..	I.	Silver nitrate 15 per cent.	1·3874	1·3872		
,, ..	III.	..	1·3870			
					·2937	31·72
,, ..	II.	Copper sulph. sp. gr. 1·174	·4065	·4074		
,, ..	IV.	..	·4082			
Nov. 27 ..	II.	Silver nitrate 15 per cent.	1·0523	1·0522		
,, ..	IV.	..	1·0522			
					·2934	31·69
,, ..	III.	Copper sulph. sp. gr. 1·115	·3094	·3087		
,, ..	I.	..	·3081			
Dec. 11 ..	II.	Silver nitrate 15 per cent.	3·0489	3·0488		
,, ..	IV.	..	3·0487			
					·2938	31·74
,, ..	III.	Copper sulph. sp. gr. 1·115	·8956	·8959		
,, ..	I.	..	·8962			
Mean ..	..	..	..	..	·2936	31·72

Multiplying ·2936 by 4·0246 we get 1·182 grms. as the amount of copper deposited per ampère per hour.

Note to § 30.

Observations made at intervals since this paper was read may here be given in continuation of Tables VII. and VIII.

	June 26	July 14	July 21, 22	Aug. 6	Oct. 8	Oct. 28	Nov. 14	Dec. 5
Clark 1 . .	1·0000	1·0000	1·0000	1·0000	1·0000	1·0000	1·0000	1·0000
,, 4 . .	·9998	1·0000	1·0000	1·0004	·9996	·9997	0·0000	0·0000
,, 5 . .	·9997	1·0007	·9998	1·0006	·9990	·9994	·9993	·9997
,, 8 . .	·9997	·9998	·9998	·9996	·9997	·9998	1·0000	·9996
,, 9 . .	1·0000	1·0000	1·0002	·9998	1·0000	1·0002	1·0002	·9999
,, 10 . .	1·0003	1·0003	1·0003	1·0003	1·0003	1·0003	1·0003	1·0003
,, 11 . .	1·0003	1·0004	1·0003	1·0007	1·0003	1·0003	1·0003	1·0004
,, 12 . .	1·0004	1·0004	1·0004	1·0007	1·0003	1·0003	1·0003	1·0003
,, 13 . .	1·0002	1·0002	1·0002	1·0000	1·0002	1·0002	1·0003	1·0001
,, 14 . .	1·0026	1·0003	1·0003	1·0002	1·0003	1·0003	1·0003	1·0003
,, 16 . .	1·0004	1·0003	1·0004	1·0002	1·0002	1·0003	1·0003	1·0003
,, 18 . .	·9974	·9920	·9900	·9860	·9800	·9810	·9760	unsteady
,, 19 . .	1·0004	1·0000	1·0000	·9997	1·0000	1·0002	1·0001	·9999
H_5	1·0005	1·0004	1·0005	1·0005	1·0003	1·0003	1·0004	1·0006
H_6	1·0007	1·0004	1·0005	1·0005	1·0004	1·0002	1·0003	. .
H_{10}	1·0004	1·0005	1·0005	1·0003	1·0003	1·0003	1·0004	1·0004
H_{11}	1·0004	1·0005	1·0005	1·0003	1·0002	1·0003	1·0003	1·0003
H_{12}	. .	. .	1·0030	1·0004	1·0003	1·0003	1·0003	. .
H_{13}	. .	. .	1·0009	1·0003	1·0002	1·0003	1·0003	1·0001

Some H-cells have been set up by Mr Threlfall, with amalgams of known composition, varying from $\frac{1}{32}$ zinc to $\frac{1}{5}$ zinc by weight. The duration of the test has as yet been scarcely adequate, but it appears that the smaller quantity of zinc is sufficient.

Note to § 32.

Comparisons of standard Daniell cells of the Post Office pattern sent me by Mr Preece have been made on several days, but did not give satisfactory results. The E.M.F. rises about 1 per cent. during the half-hour following the placing of the zincs and porous cells in the working compartment, and the two specimens differed about $2\frac{1}{2}$ per cent. The mean values were about 1·081 and 1·056 true volts.

Note 1 to § 37.

An examination of the recent comparisons of cells of different ages will probably lead to the conclusion that no important absolute change of E.M.F.

can have occurred during the thirteen months; but since the cells have been employed as standards for the determination of electric currents in various experiments, *e.g.*, for the determination of the constant of magnetic rotation (*Proc.*, June, 1884), it seemed desirable to supplement Table XI. with observations of later date. Two further absolute determinations have accordingly been made on November 21 and November 27, 1884, by the method of § 38, with the following results:—

TABLE XI. (*continued*).

Date	Cell used	Temperature	E.M.F. in B.A. volts	Correction to 15°	E.M.F. in B.A. volts corrected to 15°
November 21	Clark No. 1 .	13·7°	1·4548	− ·0016	1·4532
,, 27	,, No. 1 .	13·4	1·4555	− ·0019	1·4536
Mean . .	. .	. .	. .	. .	1·4534

The difference between 1·4534 and the mean of Table XI., viz., 1·4542, would indicate a fall of about $\frac{1}{2000}$, but the determinations are hardly precise enough to warrant us in regarding this fall as an established fact.

Note 2 to § 37.

Two further determinations of the E.M.F. of Clark cells have been published since this paper was communicated to the Royal Society. They both depend upon the evaluation of currents by means of silver, as in § 38.

A. v. Ettingshausen (*Zeitschrift für Elektrotechnik*, 1884, xvi. Heft) finds at 15°·5 the value 1·433 volt, using Kohlrausch's (second) value of the electro-chemical equivalent.

Again (*Amer. Journ. Sci.*, Nov., 1884) Mr Carhart obtains 1·434 volt. This appears to correspond to a temperature of 18°.

These results are satisfactory as tending to show that Clark cells may be set up in different places and by different hands so as to give nearly identical E.M.F.

113.

PRESIDENTIAL ADDRESS.

[British Association Report, pp. 1—23. Montreal, 1884.]

LADIES AND GENTLEMEN—

IT is no ordinary meeting of the British Association which I have now the honour of addressing. For more than fifty years the Association has held its autumn gathering in various towns of the United Kingdom, and within those limits there is, I suppose, no place of importance which we have not visited. And now, not satisfied with past successes, we are seeking new worlds to conquer. When it was first proposed to visit Canada, there were some who viewed the project with hesitation. For my own part, I never quite understood the grounds of their apprehension. Perhaps they feared the thin edge of the wedge. When once the principle was admitted, there was no knowing to what it might lead. So rapid is the development of the British Empire, that the time might come when a visit to such out-of-the-way places as London or Manchester could no longer be claimed as a right, but only asked for as a concession to the susceptibilities of the English. But seriously, whatever objections may have at first been felt soon were outweighed by the consideration of the magnificent opportunities which your hospitality affords of extending the sphere of our influence and of becoming acquainted with a part of the Queen's dominion which, associated with splendid memories of the past, is advancing daily by leaps and bounds to a position of importance such as not long ago was scarcely dreamed of. For myself, I am not a stranger to your shores. I remember well the impression made upon me, seventeen years ago, by the wild rapids of the St Lawrence, and the gloomy grandeur of the Saguenay. If anything impressed me more, it was the kindness with which I was received by yourselves, and which I doubt not will be again extended not merely to myself, but to all the English members of the Association. I am confident that those who have made up their minds to cross the ocean will not repent their decision, and that, apart altogether from scientific interests, great advantage may be expected from this visit. We Englishmen ought to know more than we do of matters relating to the Colonies, and anything which tends to bring the various parts of the Empire into closer contact can hardly be overvalued. It is pleasant to think that this Association is the means of furthering an object which should be dear to the hearts of all of us; and I venture to say that a large proportion of the visitors to this country will be astonished by what they see, and will carry home an impression which time will not readily efface.

To be connected with this meeting is, to me, a great honour, but also a great responsibility. In one respect, especially, I feel that the Association might have done well to choose another President. My own tastes have led me to study mathematics and physics rather than geology and biology, to which naturally more attention turns in a new country, presenting as it does a fresh field for investigation. A chronicle of achievements in these departments by workers from among yourselves would have been suitable to the occasion, but could not come from me. If you would have preferred a different subject for this address, I hope, at least, that you will not hold me entirely responsible.

At annual gatherings like ours the pleasure with which friends meet friends again is sadly marred by the absence of those who can never more take their part in our proceedings. Last year my predecessor in this office had to lament the untimely loss of Spottiswoode and Henry Smith, dear friends of many of us, and prominent members of our Association. And now, again, a well-known form is missing. For many years Sir W. Siemens has been a regular attendant at our meetings, and to few indeed have they been more indebted for success. Whatever the occasion, in his Presidential Address of two years ago, or in communications to the Physical and Mechanical Sections, he had always new and interesting ideas, put forward in language which a child could understand, so great a master was he of the art of lucid statement in his adopted tongue. Practice with Science was his motto. Deeply engaged in industry, and conversant, all his life, with engineering operations, his opinion was never that of a mere theorist. On the other hand, he abhorred rule of thumb, striving always to master the scientific principles which underlie rational design and invention.

It is not necessary that I should review in detail the work of Siemens. The part which he took, during recent years, in the development of the dynamo machine must be known to many of you. We owe to him the practical adoption of the method, first suggested by Wheatstone, of throwing into a shunt the coils of the field magnets, by which a greatly improved steadiness of action is obtained. The same characteristics are observable throughout—a definite object in view and a well-directed perseverance in overcoming the difficulties by which the path is usually obstructed.

These are, indeed, the conditions of successful invention. The world knows little of such things, and regards the new machine or the new method as the immediate outcome of a happy idea. Probably, if the truth were known, we should see that, in nine cases out of ten, success depends as much upon good judgment and perseverance as upon fertility of imagination. The labours of our great inventors are not unappreciated, but I doubt whether we adequately realise the enormous obligations under which we lie. It is no exaggeration to say that the life of such a man as Siemens is spent in the public service; the advantages which he reaps for himself being as nothing in comparison with those which he confers upon the community at large.

As an example of this it will be sufficient to mention one of the most valuable achievements of his active life—his introduction, in conjunction with his brother, of the Regenerative Gas Furnace, by which an immense economy of fuel (estimated at millions of tons annually) has been effected in the manufacture of steel and glass. The nature of this economy is easily explained. Whatever may be the work to be done by the burning of fuel, a certain *temperature* is necessary. For example, no amount of heat in the form of boiling water, would be of any avail for the fusion of steel. When the products of combustion are cooled down to the point in question, the heat which they still contain is useless as regards the purpose in view. The importance of this consideration depends entirely upon the working temperature. If the object be the evaporation of water or the warming of a house, almost all the heat may be extracted from the fuel without special arrangements. But it is otherwise when the temperature required is not much below that of combustion itself, for then the escaping gases carry away with them the larger part of the whole heat developed. It was to meet this difficulty that the regenerative furnace was devised. The products of combustion, before dismissal into the chimney, are caused to pass through piles of loosely stacked fire-brick, to which they give up their heat. After a time the fire-brick, upon which the gases first impinge, becomes nearly as hot as the furnace itself. By suitable valves the burnt gases are then diverted through another stack of brickwork, which they heat up in like manner, while the heat stored up in the first stack is utilised to warm the unburnt gas and air on their way to the furnace. In this way almost all the heat developed at a high temperature during the combustion is made available for the work in hand.

As it is now several years since your presidential chair has been occupied by a professed physicist, it may naturally be expected that I should attempt some record of recent progress in that branch of science, if indeed such a term be applicable. For it is one of the difficulties of the task that subjects as distinct as Mechanics, Electricity, Heat, Optics and Acoustics, to say nothing of Astronomy and Meteorology, are included under Physics. Any one of these may well occupy the life-long attention of a man of science, and to be thoroughly conversant with all of them is more than can be expected of any one individual, and is probably incompatible with the devotion of much time and energy to the actual advancement of knowledge. Not that I would complain of the association sanctioned by common parlance. A sound knowledge of at least the principles of general physics is necessary to the cultivation of any department. The predominance of the sense of sight as the medium of communication with the outer world, brings with it dependence upon the science of optics; and there is hardly a branch of science in which the effects of *temperature* have not (often without much success) to be reckoned with. Besides, the neglected borderland between two branches of knowledge is often that which best repays cultivation, or, to use a metaphor of Maxwell's, the greatest benefits may be derived from a cross fertilisation of the sciences. The wealth of material is an evil only from the point of view of one of whom too much may be expected. Another difficulty incident to the task, which must be faced but cannot be overcome, is that of estimating rightly the value, and even the correctness, of recent work. It is not always that which seems at first the most important that proves in the end to be so. The history of science teems with examples of discoveries which attracted little notice at the time, but afterwards have taken root downwards and borne much fruit upwards.

One of the most striking advances of recent years is in the production and application of electricity upon a large scale—a subject to which I have already had occasion to allude in connection with the work of Sir W. Siemens. The dynamo machine is indeed founded upon discoveries of Faraday now more than half a century old; but it has required the protracted labours of many inventors to bring it to its present high degree of efficiency. Looking back at the matter, it seems strange that progress should have been so slow. I do not refer to details of design, the elaboration of which must always, I suppose, require the experience of actual work to indicate what parts are structurally weaker than they should be, or are exposed to undue wear and tear. But with regard to the main features of the problem, it would almost seem as if the difficulty lay in want of faith. Long ago it was recognised that electricity derived from chemical action is (on a large scale) too expensive a source of mechanical power, notwithstanding the fact that (as proved by Joule in 1846) the conversion of electrical into mechanical work can be effected with great economy. From this it is an evident consequence that electricity may advantageously be obtained from mechanical power; and one cannot help thinking that if the fact had been borne steadily in mind, the development of the dynamo might have been much more rapid. But discoveries and inventions are apt to appear obvious when regarded from the standpoint of accomplished fact; and I draw attention to the matter only to point the moral that we do well to push the attack persistently when we can be sure beforehand that the obstacles to be overcome are only difficulties of contrivance, and that we are not vainly fighting unawares against a law of Nature.

The present development of electricity on a large scale depends, however, almost as much upon the incandescent lamp as upon the dynamo. The success of these lamps demands a very perfect vacuum—not more than about one-millionth of the normal quantity of air should remain,—and it is interesting to recall that, twenty years ago, such vacua were rare even in the laboratory of the physicist. It is pretty safe to say that these wonderful results would never have been accomplished had practical applications alone been in view. The way was prepared by an army of scientific men whose main object was the advancement of knowledge, and who could scarcely have imagined that the processes which they elaborated would soon be in use on a commercial scale and entrusted to the hands of ordinary workmen.

When I speak in hopeful language of practical electricity, I do not forget the disappointment within the last year or two of many over-sanguine expectations. The enthusiasm of the inventor and promoter are necessary to progress, and it seems to be almost a law of nature that it should overpass the bounds marked out by reason and experience. What is most to be regretted is the advantage taken by speculators of the often uninstructed interest felt by the public in novel schemes by which its imagination is fired. But looking forward to the future of electric lighting, we have good ground for encouragement. Already the lighting of large passenger ships is an assured success, and one which will be highly appreciated by those travellers who have experienced the tedium of long winter evenings unrelieved by adequate illumination. Here, no doubt, the conditions are in many respects especially favourable. As regards space, life on board ship is highly concentrated; while unity of management and the presence on the spot of skilled engineers obviate some of the difficulties that are met with under other circumstances. At present we have no experience of a house-to-house system of illumination on a great scale and in competition with cheap gas; but preparations are already far advanced for trial on an adequate scale in London. In large institutions, such as theatres and factories, we all know that electricity is in successful and daily extending operation.

When the necessary power can be obtained from the fall of water, instead of from the combustion of coal, the conditions of the problem are far more favourable. Possibly the severity of your winters may prove an obstacle, but it is impossible to regard your splendid river without the thought arising that the day may come when the vast powers now running to waste shall be bent into your service. Such a project demands of course the most careful consideration, but it is one worthy of an intelligent and enterprising community.

The requirements of practice react in the most healthy manner upon scientific electricity. Just as in former days the science received a stimulus from the application to telegraphy, under which everything relating to measurement on a small scale acquired an importance and development for which we might otherwise have had long to wait, so now the requirements of electric lighting are giving rise to a new development of the art of measurement upon a large scale, which cannot fail to prove of scientific as well as practical importance. Mere change of scale may not at first appear a very important matter, but it is surprising how much modification it entails in the instruments, and in the processes of measurement. For instance, the resistance coils on which the electrician relies in dealing with currents whose maximum is a fraction of an ampère, fail altogether when it becomes a question of hundreds, not to say thousands, of ampères.

The powerful currents, which are now at command, constitute almost a new weapon in the hands of the physicist. Effects, which in old days were rare and difficult of observation, may now be produced at will on the most conspicuous scale. Consider for a moment Faraday's great discovery of the 'Magnetisation of Light,' which Tyndall likens to the Weisshorn among mountains, as high, beautiful, and alone. This judgment (in which I fully concur) relates to the scientific aspect of the discovery, for to the eye of sense nothing could have been more insignificant. It is even possible that it might have eluded altogether the penetration of Faraday, had he not been provided with a special quality of very heavy glass. At the present day these effects may be produced upon a scale that would have delighted their discoverer, a rotation of the plane of polarization through 180° being perfectly feasible. With the aid of modern appliances, Kundt and Röntgen in Germany, and H. Becquerel in France, have detected the rotation in gases and vapours, where, on account of its extreme smallness, it had previously escaped notice.

Again, the question of the magnetic saturation of iron has now an importance entirely beyond what it possessed at the time of Joule's early observations. Then it required special arrangements purposely contrived to bring it into prominence. Now in every dynamo machine, the iron of the field-magnets approaches a state of saturation, and the very elements of an explanation of the action require us to take the fact into account. It is indeed probable that a better knowledge of this subject might lead to improvements in the design of these machines.

Notwithstanding the important work of Rowland and Stoletow, the whole theory of the behaviour of soft iron under varying magnetic conditions is still somewhat obscure. Much may be hoped from the induction balance of Hughes, by which the marvellous powers of the telephone are applied to the discrimination of the properties of metals, as regards magnetism and electric conductivity.

The introduction of powerful alternate-current in machines by Siemens, Gordon, Ferranti, and others, is likely also to have a salutary effect in educating those so-called practical electricians whose ideas do not easily rise above ohms and volts. It has long been known that when the changes are sufficiently rapid, the phenomena are governed much more by induction, or electric inertia, than by mere resistance. On this principle much may be explained that would otherwise seem paradoxical. To take a comparatively simple case, conceive an electro-magnet wound with two contiguous wires, upon which acts a given rapidly periodic electro-motive force. If one wire only be used, a certain amount of heat is developed in the circuit. Suppose now that the second wire is brought into operation in parallel—a proceeding equivalent to doubling the section of the original wire. An electrician accustomed only to constant currents would be sure to think that the heating effect would be doubled by the change, as much heat being developed in each wire separately as was at first in the single wire. But such a conclusion would be entirely erroneous. The total current, being governed practically by the self-induction of the circuit, would not be augmented by the accession of the second wire, and the total heating effect, so far from being doubled, would, in virtue of the superior conductivity, be halved.

During the last few years much interest has been felt in the reduction to an absolute standard of measurements of electro-motive force, current, resistance, etc., and to this end many laborious investigations have been undertaken. The subject is one that has engaged a good deal of my own attention, and I should naturally have felt inclined to dilate upon it, but that I feel it to be too abstruse and special to be dealt with in detail upon an occasion like the present. As regards resistance, I will merely remind you that the recent determinations have shown a so greatly improved agreement, that the Conference of Electricians assembled at Paris, in May, have felt themselves justified in defining the ohm for practical use as the resistance of a column of mercury of 0° C., one square milli-metre in section, and 106 centimetres in length—a definition differing by a little more than one per cent. from that arrived at twenty years ago by a committee of this Association.

A standard of resistance once determined upon can be embodied in a 'resistance coil,' and copied without much trouble, and with great accuracy. But in order to complete the electrical system, a second standard of some kind is necessary, and this is not so easily embodied in a permanent form. It might conveniently consist of a standard galvanic cell, capable of being prepared in a definite manner, whose electro-motive force is once for all determined. Unfortunately, most of the batteries in ordinary use are for one reason or another unsuitable for this purpose, but the cell introduced by Mr Latimer Clark, in which the metals are zinc in contact with saturated zinc sulphate and pure mercury in contact with mercurous sulphate, appears to give satisfactory results. According to my measurements, the electro-motive force of this cell is 1·435 theoretical volts.

We may also conveniently express the second absolute electrical measurement necessary to the completion of the system by taking advantage of Faraday's law, that the quantity of metal decomposed in an electrolytic cell is proportional to the whole quantity of electricity that passes. The best metal for the purpose is silver, deposited from a solution of the nitrate or of the chlorate. The results recently obtained by Professor Kohlrausch and by myself are in very good agreement, and the conclusion that one ampere flowing for one hour decomposes 4·025 grains of silver, can hardly be in error by more than a thousandth part. This number being known, the silver voltameter gives a ready and very accurate method of measuring currents of intensity, varying from $\frac{1}{10}$ ampère to four or five ampères.

The beautiful and mysterious phenomena attending the discharge of electricity in nearly vacuous spaces have been investigated and in some degree explained by De La Rue, Crookes, Schuster, Moulton, and the lamented Spottiswoode, as well as by various able foreign experimenters. In a recent research Crookes has sought the origin of a bright citron-coloured band in the phosphorescent spectrum of certain earths, and after encountering difficulties and anomalies of a most bewildering kind, has succeeded in proving that it is due to yttrium, an element much more widely distributed than had been supposed. A conclusion like this is stated in a few words, but those only who have undergone similar experience are likely to appreciate the skill and perseverance of which it is the final reward.

A remarkable observation by Hall of Baltimore, from which it appeared that the flow of electricity in a conducting sheet was disturbed by mag-

netic force, has been the subject of much discussion. Mr Shelford Bidwell
has brought forward experiments tending to prove that the effect is of
a secondary character, due in the first instance to the mechanical force
operating upon the conductor of an electric current when situated in a
powerful magnetic field. Mr Bidwell's view agrees in the main with
Mr Hall's division of the metals into two groups according to the direction of the effect.

Without doubt the most important achievement of the older generation of scientific men has been the establishment and application of the
great laws of Thermo-dynamics, or, as it is often called, the Mechanical
Theory of Heat. The first law, which asserts that heat and mechanical
work can be transformed one into the other at a certain fixed rate, is now
well understood by every student of physics, and the number expressing
the mechanical equivalent of heat resulting from the experiments of Joule,
has been confirmed by the researches of others, and especially of Rowland.
But the second law, which practically is even more important than the
first, is only now beginning to receive the full appreciation due to it.
One reason of this may be found in a not unnatural confusion of ideas.
Words do not always lend themselves readily to the demands that are
made upon them by a growing science, and I think that the almost
unavoidable use of the word equivalent in the statement of the first law
is partly responsible for the little attention that is given to the second.
For the second law so far contradicts the usual statement of the first,
as to assert that equivalents of heat and work are not of equal value.
While work can always be converted into heat, heat can only be converted
into work under certain limitations. For every practical purpose the work
is worth the most, and when we speak of equivalents, we use the word
in the same sort of special sense as that in which chemists speak of
equivalents of gold and iron. The second law teaches us that the real
value of heat, as a source of mechanical power, depends upon the temperature of the body in which it resides; the hotter the body in relation to
its surroundings, the more available the heat.

In order to see the relations which obtain between the first and the
second law of Thermo-dynamics, it is only necessary for us to glance at
the theory of the steam-engine. Not many years ago calculations were
plentiful, demonstrating the inefficiency of the steam-engine on the basis
of a comparison of the work actually got out of the engine with the
mechanical equivalent of the heat supplied to the boiler. Such calculations took into account only the first law of Thermo-dynamics, which deals
with the equivalents of heat and work, and have very little bearing upon
the practical question of efficiency, which requires us to have regard also
to the second law. According to that law the fraction of the total energy

which can be converted into work depends upon the relative temperatures
of the boiler and condenser; and it is, therefore, manifest that, as the
temperature of the boiler cannot be raised indefinitely, it is impossible to
utilise all the energy which, according to the first law of Thermo-dynamics,
is resident in the coal.

On a sounder view of the matter, the efficiency of the steam-engine is
found to be so high, that there is no great margin remaining for improvement.   The higher initial temperature possible in the gas-engine opens out
much wider possibilities, and many good judges look forward to a time
when the steam-engine will have to give way to its younger rival.

To return to the theoretical question, we may say with Sir W. Thomson,
that though energy cannot be destroyed, it ever tends to be dissipated,
or to pass from more available to less available forms.   No one who has
grasped this principle can fail to recognise its immense importance in the
system of the Universe.   Every change—chemical, thermal, or mechanical—
which takes place, or can take place, in Nature does so, at the cost of
a certain amount of available energy.   If, therefore, we wish to inquire
whether or not a proposed transformation can take place, the question
to be considered is whether its occurrence would involve dissipation of
energy.   If not, the transformation is (under the circumstances of the
case) absolutely excluded.   Some years ago, in a lecture at the Royal
Institution *, I endeavoured to draw the attention of chemists to the importance of the principle of dissipation in relation to their science, pointing
out the error of the usual assumption that a general criterion is to be
found in respect of the development of heat.   For example, the solution
of a salt in water is, if I may be allowed the phrase, a downhill transformation.   It involves dissipation of energy, and can therefore go forward;
but in many cases it is associated with the absorption rather than with
the development of heat.   I am glad to take advantage of the present
opportunity in order to repeat my recommendation, with an emphasis
justified by actual achievement.   The foundations laid by Thomson now
bear an edifice of no mean proportions, thanks to the labours of several
physicists, among whom must be especially mentioned Willard Gibbs and
Helmholtz.   The former has elaborated a theory of the equilibrium of
heterogeneous substances, wide in its principles, and we cannot doubt far-
reaching in its consequences.   In a series of masterly papers Helmholtz has
developed the conception of *free energy* with very important applications
to the theory of the galvanic cell.   He points out that the mere tendency
to solution bears in some cases no small proportion to the affinities more
usually reckoned chemical, and contributes largely to the total electromotive force.   Also in our own country Dr Alder Wright has published
some valuable experiments relating to the subject.

* Vol. I. p. 238.

From the further study of electrolysis we may expect to gain improved views as to the nature of the chemical reactions, and of the forces concerned in bringing them about. I am not qualified—I wish I were—to speak to you on recent progress in general chemistry. Perhaps my feelings towards a first love may blind me, but I cannot help thinking that the next great advance, of which we have already some foreshadowing, will come on this side. And if I might without presumption venture a word of recommendation, it would be in favour of a more minute study of the simpler chemical phenomena.

Under the head of scientific mechanics it is principally in relation to fluid motion that advances may be looked for. In speaking upon this subject I must limit myself almost entirely to experimental work. Theoretical hydro-dynamics, however important and interesting to the mathematician, are eminently unsuited to oral exposition. All I can do to attenuate an injustice, to which theorists are pretty well accustomed, is to refer you to the admirable reports of Mr Hicks, published under the auspices of this Association.

The important and highly practical work of the late Mr Froude in relation to the propulsion of ships is doubtless known to most of you. Recognising the fallacy of views then widely held as to the nature of the resistance to be overcome, he showed to demonstration that, in the case of fair-shaped bodies, we have to deal almost entirely with resistance dependent upon skin friction, and at high speeds upon the generation of surface waves by which energy is carried off. At speeds which are moderate in relation to the size of the ship, the resistance is practically dependent upon skin friction only. Although Professor Stokes and other mathematicians had previously published calculations pointing to the same conclusion, there can be no doubt that the view generally entertained was very different. At the first meeting of the Association which I ever attended, as an intelligent listener, at Bath in 1864, I well remember the surprise which greeted a statement by Rankine that he regarded skin friction as the only legitimate resistance to the progress of a well-designed ship. Mr Froude's experiments have set the question at rest in a manner satisfactory to those who had little confidence in theoretical prevision.

In speaking of an explanation as satisfactory in which skin friction is accepted as the cause of resistance, I must guard myself against being supposed to mean that the nature of skin friction is itself well understood. Although its magnitude varies with the smoothness of the surface, we have no reason to think that it would disappear at any degree of smoothness consistent with an ultimate molecular structure. That it is connected with fluid viscosity is evident enough, but the *modus operandi* is still obscure.

Some important work bearing upon the subject has recently been published by Professor O. Reynolds, who has investigated the flow of water in tubes as dependent upon the velocity of motion and upon the size of the bore. The laws of motion in capillary tubes, discovered experimentally by Poiseuille, are in complete harmony with theory. The resistance varies as the velocity, and depends in a direct manner upon the constant of viscosity. But when we come to the larger pipes and higher velocities with which engineers usually have to deal, the theory which presupposes a regularly stratified motion evidently ceases to be applicable, and the problem becomes essentially identical with that of skin friction in relation to ship propulsion. Professor Reynolds has traced with much success the passage from the one state of things to the other, and has proved the applicability under these complicated conditions of the general laws of dynamical similarity as adapted to viscous fluids by Professor Stokes. In spite of the difficulties which beset both the theoretical and experimental treatment, we may hope to attain before long to a better understanding of a subject which is certainly second to none in scientific as well as practical interest.

As also closely connected with the mechanics of viscous fluids, I must not forget to mention an important series of experiments upon the friction of oiled surfaces, recently executed by Mr Tower for the Institution of Mechanical Engineers. The results go far towards upsetting some ideas hitherto widely admitted. When the lubrication is adequate, the friction is found to be nearly independent of the load, and much smaller than is usually supposed, giving a coefficient as low as $\frac{1}{1000}$. When the layer of oil is well formed, the pressure between the solid surfaces is really borne by the fluid, and the work lost is spent in shearing, that is, in causing one stratum of the oil to glide over another.

In order to maintain its position, the fluid must possess a certain degree of viscosity, proportionate to the pressure; and even when this condition is satisfied, it would appear to be necessary that the layer should be thicker on the ingoing than on the outgoing side. We may, I believe, expect from Professor Stokes a further elucidation of the processes involved. In the meantime, it is obvious that the results already obtained are of the utmost value, and fully justify the action of the Institution in devoting a part of its resources to experimental work. We may hope indeed that the example thus wisely set may be followed by other public bodies associated with various departments of industry.

I can do little more than refer to the interesting observations of Prof. Darwin, Mr Hunt, and M. Forel on Ripplemark. The processes concerned would seem to be of a rather intricate character, and largely dependent upon fluid viscosity. It may be noted indeed that most of the still obscure phenomena of hydro-dynamics require for their elucidation a better comprehension of the laws of viscous motion. The subject is one which offers peculiar difficulties. In some problems in which I have lately been interested, a circulating motion presents itself of the kind which the mathematician excludes from the first when he is treating of fluids destitute altogether of viscosity. The intensity of this motion proves, however, to be independent of the coefficient of viscosity, so that it cannot be correctly dismissed from consideration as a consequence of a supposition that the viscosity is infinitely small. The apparent breach of continuity can be explained, but it shows how much care is needful in dealing with the subject, and how easy it is to fall into error.

The nature of gaseous viscosity, as due to the diffusion of momentum, has been made clear by the theoretical and experimental researches of Maxwell. A flat disc moving in its own plane between two parallel solid surfaces is impeded by the necessity of shearing the intervening layers of gas, and the magnitude of the hindrance is proportional to the velocity of the motion and to the viscosity of the gas, so that under similar circumstances this effect may be taken as a measure, or rather definition, of the viscosity. From the dynamical theory of gases, to the development of which he contributed so much, Maxwell drew the startling conclusion that the viscosity of a gas should be independent of its density,—that within wide limits the resistance to the moving disc should be scarcely diminished by pumping out the gas, so as to form a partial vacuum. Experiment fully confirmed this theoretical anticipation,—one of the most remarkable to be found in the whole history of science, and proved that the swinging disc was retarded by the gas, as much when the barometer stood at half an inch as when it stood at thirty inches. It was obvious, of course, that the law must have a limit, that at a certain point of exhaustion the gas must begin to lose its power; and I remember discussing with Maxwell, soon after the publication of his experiments, the whereabouts of the point at which the gas would cease to produce its ordinary effect. His apparatus, however, was quite unsuited for high degrees of exhaustion, and the failure of the law was first observed by Kundt and Warburg, as pressures below 1 mm. of mercury. Subsequently the matter has been thoroughly examined by Crookes, who extended his observations to the highest degrees of exhaustion as measured by MacLeod's gauge. Perhaps the most remarkable results relate to hydrogen. From the atmospheric pressure of 760 mm. down to about $\frac{1}{2}$ mm. of mercury the viscosity is sensibly constant. From this point to the highest vacua, in which less than one-millionth of the original gas remains, the coefficient of viscosity drops down gradually to a small fraction of its original value. In these vacua Mr Crookes regards the gas as having assumed a different, ultra-gaseous, condition; but we must remember that the phenomena have relation to the other circumstances of the case, especially the dimensions of the vessel, as well as to the condition of the gas.

Such an achievement as the prediction of Maxwell's law of viscosity has, of course, drawn increased attention to the dynamical theory of gases. The success which has attended the theory in the hands of Clausius, Maxwell, Boltzmann and other mathematicians, not only in relation to viscosity, but over a large part of the entire field of our knowledge of gases, proves that some of its fundamental postulates are in harmony with the reality of Nature. At the same time, it presents serious difficulties; and we cannot but feel that while the electrical and optical properties of gases remain out of relation to the theory, no final judgment is possible. The growth of experimental knowledge may be trusted to clear up many doubtful points, and a younger generation of theorists will bring to bear improved mathematical weapons. In the meantime we may fairly congratulate ourselves on the possession of a guide which has already conducted us to a position which could hardly otherwise have been attained.

In Optics attention has naturally centred upon the spectrum. The mystery attaching to the invisible rays lying beyond the red has been fathomed to an extent that, a few years ago, would have seemed almost impossible. By the use of special photographic methods Abney has mapped out the peculiarities of this region with such success that our knowledge of it begins to be comparable with that of the parts visible to the eye. Equally important work has been done by Langley, using a refined invention of his own based upon the principle of Siemens' pyrometer. This instrument measures the actual energy of the radiation, and thus expresses the effects of various parts of the spectrum upon a common scale, independent of the properties of the eye and of sensitive photographic preparations. Interesting results have also been obtained by Becquerel, whose method is founded upon a curious action of the ultra-red rays in enfeebling the light emitted by phosphorescent substances. One of the most startling of Langley's conclusions relates to the influence of the atmosphere in modifying the quality of solar light. By the comparison of observations made through varying thicknesses of air, he shows that the atmospheric absorption tells most upon the light of high refrangibility; so that, to an eye situated outside the atmosphere, the sun would present a decidedly bluish tint. It would be interesting to compare the experimental numbers with the law of scattering of light by small particles given some years ago as the result of theory*. The demonstration by Langley of the

---

* Conf. vol. I. p. 95.

inadequacy of Cauchy's law of dispersion to represent the relation between refrangibility and wave-length in the lower part of the spectrum must have an important bearing upon optical theory.

The investigation of the relation of the visible and ultra-violet spectrum to various forms of matter has occupied the attention of a host of able workers, among whom none have been more successful than my colleagues at Cambridge, Professors Liveing and Dewar. The subject is too large both for the occasion and for the individual, and I must pass it by. But, as more closely related to Optics proper, I cannot resist recalling to your notice a beautiful application of the idea of Doppler to the discrimination of the origin of certain lines observed in the solar spectrum. If a vibrating body have a general motion of approach or recession, the waves emitted from it reach the observer with a frequency which in the first case exceeds, and in the second case falls short of, the real frequency of the vibrations themselves. The consequence is that, if a glowing gas be in motion in the line of sight, the spectral lines are thereby displaced from the position that they would occupy were the gas at rest—a principle which, in the hands of Huggins and others, has led to a determination of the motion of certain fixed stars relatively to the solar system. But the sun is itself in rotation, and thus the position of a solar spectral line is slightly different according as the light comes from the advancing or from the retreating limb. This displacement was, I believe, first observed by Thollon; but what I desire now to draw attention to is the application of it by Cornu to determine whether a line is of solar or atmospheric origin. For this purpose a small image of the sun is thrown upon the slit of the spectroscope, and caused to vibrate two or three times a second, in such a manner that the light entering the instrument comes alternately from the advancing and retreating limbs. Under these circumstances a line due to absorption within the sun appears to tremble, as the result of slight alternately opposite displacements. But if the seat of the absorption be in the atmosphere, it is a matter of indifference from what part of the sun the light originally proceeds, and the line maintains its position in spite of the oscillation of the image upon the slit of the spectroscope. In this way Cornu was able to make a discrimination which can only otherwise be effected by a difficult comparison of appearances under various solar altitudes.

The instrumental weapon of investigation, the spectroscope itself, has made important advances. On the theoretical side, we have for our guidance the law that the optical power in gratings is proportional to the total number of lines accurately ruled, without regard to the degree of closeness, and in prisms that it is proportional to the thickness of glass traversed. The magnificent gratings of Rowland are a new power in the hands of the spectroscopist, and as triumphs of mechanical art seem to be little short of perfection. In our own report for 1882, Mr Mallock has described a machine, constructed by him, for ruling large diffraction gratings, similar in some respects to that of Rowland.

The great optical constant, the velocity of light, has been the subject of three distinct investigations by Cornu, Michelson, and Forbes. As may be supposed, the matter is of no ordinary difficulty, and it is therefore not surprising that the agreement should be less decided than could be wished. From their observations, which were made by a modification of Fizeau's method of the toothed wheel, Young and Forbes drew the conclusion that the velocity of light *in vacuo* varies from colour to colour, to such an extent that the velocity of blue light is nearly two per cent. greater than that of red light. Such a variation is quite opposed to existing theoretical notions, and could only be accepted on the strongest evidence. Mr Michelson, whose method (that of Foucault) is well suited to bring into prominence a variation of velocity with wave-length, informs me that he has recently repeated his experiments with special reference to the point in question, and has arrived at the conclusion that no variation exists comparable with that asserted by Young and Forbes. The actual velocity differs little from that found from his first series of experiments, and may be taken to be 299,800 kilometres per second.

It is remarkable how many of the playthings of our childhood give rise to questions of the deepest scientific interest. The top is, or may be understood, but a complete comprehension of the kite and of the soap-bubble would carry us far beyond our present stage of knowledge. In spite of the admirable investigations of Plateau, it still remains a mystery why soapy water stands almost alone among fluids as a material for bubbles. The beautiful development of colour was long ago ascribed to the interference of light, called into play by the gradual thinning of the film. In accordance with this view the tint is determined solely by the thickness of the film, and the refractive index of the fluid. Some of the phenomena are however so curious, as to have led excellent observers like Brewster to reject the theory of thin plates, and to assume the secretion of various kinds of colouring matter. If the rim of a wine-glass be dipped in soapy water, and then held in a vertical position, horizontal bands soon begin to show at the top of the film, and extend themselves gradually, downwards. According to Brewster these bands are not formed by the 'subsidence and gradual thinning of the film,' because they maintain their horizontal position when the glass is turned round its axis. The experiment is both easy and interesting; but the conclusion drawn from it cannot be accepted. The fact is that the various parts of the film cannot quickly alter their thickness, and hence when the glass is rotated they re-arrange themselves in order of superficial density, the thinner parts floating up over, or through, the thicker parts. Only thus can the tendency be satisfied for the centre of gravity to assume the lowest possible position.

When the thickness of a film falls below a small fraction of the length of a wave of light, the colour disappears and is replaced by an intense blackness. Professors Reinold and Rücker have recently made the remarkable observation that the whole of the black region, soon after its formation, is of uniform thickness, the passage from the black to the coloured portions being exceedingly abrupt. By two independent methods they have determined the thickness of the black film to lie between seven and fourteen millionths of a millimetre; so that the thinnest films correspond to about one-seventieth of a wave-length of light. The importance of these results in regard to molecular theory is too obvious to be insisted upon.

The beautiful inventions of the telephone and the phonograph, although in the main dependent upon principles long since established, have imparted a new interest to the study of Acoustics. The former, apart from its uses in every-day life, has become in the hands of its inventor, Graham Bell, and of Hughes, an instrument of first-class scientific importance. The theory of its action is still in some respects obscure, as is shown by the comparative failure of the many attempts to improve it. In connection with some explanations that have been offered, we do well to remember that molecular changes in solid masses are inaudible in themselves, and can only be manifested to our ears by the generation of a to and fro motion of the external surface extending over a sensible area. If the surface of a solid remains undisturbed, our ears can tell us nothing of what goes on in the interior.

In theoretical acoustics progress has been steadily maintained, and many phenomena, which were obscure twenty or thirty years ago, have since received adequate explanation. If some important practical questions remain unsolved, one reason is that they have not yet been definitely stated. Almost everything in connection with the ordinary use of our senses presents peculiar difficulties to scientific investigation. Some kinds of information with regard to their surroundings are of such paramount importance to successive generations of living beings, that they have learned to interpret indications which, from a physical point of view, are of the slenderest character. Every day we are in the habit of recognising, without much difficulty, the quarter from which a sound proceeds, but by what steps we attain that end has not yet been satisfactorily explained. It has been proved* that when proper precautions are taken we are unable to distinguish whether a pure tone (as from a vibrating

* Conf. vol. I. pp. 277, 314.

tuning-fork held over a suitable resonator) comes to us from in front or from behind. This is what might have been expected from an *à priori* point of view; but what would not have been expected is that with almost any other sort of sound, from a clap of the hands to the clearest vowel sound, the discrimination is not only possible but easy and instinctive. In these cases it does not appear how the possession of two ears helps us, though there is some evidence that it does; and even when sounds come to us from the right or left, the explanation of the ready discrimination which is then possible with pure tones, is not so easy as might at first appear. We should be inclined to think that the sound was heard much more loudly with the ear that is turned towards than with the ear that is turned from it, and that in this way the direction was recognised. But if we try the experiment, we find that, at any rate with notes near the middle of the musical scale, the difference of loudness is by no means so very great. The wave-lengths of such notes are long enough in relation to the dimensions of the head to forbid the formation of anything like a sound shadow in which the averted ear might be sheltered.

In concluding this imperfect survey of recent progress in physics, I must warn you emphatically that much of great importance has been passed over altogether. I should have liked to speak to you of those far-reaching speculations, especially associated with the name of Maxwell, in which light is regarded as a disturbance in an electro-magnetic medium. Indeed, at one time, I had thought of taking the scientific work of Maxwell as the principal theme of this address. But, like most men of genius, Maxwell delighted in questions too obscure and difficult for hasty treatment, and thus much of his work could hardly be considered upon such an occasion as the present. His biography has recently been published, and should be read by all who are interested in science and in scientific men. His many-sided character, the quaintness of his humour, the penetration of his intellect, his simple but deep religious feeling, the affection between son and father, the devotion of husband to wife, all combine to form a rare and fascinating picture. To estimate rightly his influence upon the present state of science, we must regard not only the work that he executed himself, important as that was, but also the ideas and the spirit which he communicated to others. Speaking for myself as one who in a special sense entered into his labours, I should find it difficult to express adequately my feeling of obligation. The impress of his thoughts may be recognised in much of the best work of the present time. As a teacher and examiner he was well acquainted with the almost universal tendency of uninstructed minds to elevate phrases above things: to refer, for example, to the principle of the conservation of energy for an explanation of the persistent rotation of a fly-wheel, almost in the style of the doctor in *Le Malade Imaginaire,* who explains the fact that opium sends you to sleep by its soporific virtue. Maxwell's endeavour was always to keep the facts in the foreground, and to his influence, in conjunction with that of Thomson and Helmholtz, is largely due that elimination of unnecessary hypothesis which is one of the distinguishing characteristics of the science of the present day.

In speaking unfavourably of superfluous hypothesis, let me not be misunderstood. Science is nothing without generalisations. Detached and ill-assorted facts are only raw material, and in the absence of a theoretical solvent, have but little nutritive value. At the present time and in some departments, the accumulation of material is so rapid that there is danger of indigestion. By a fiction as remarkable as any to be found in law, what has once been published, even though it be in the Russian language, is usually spoken of as 'known,' and it is often forgotten that the rediscovery in the library may be a more difficult and uncertain process than the first discovery in the laboratory. In this matter we are greatly dependent upon annual reports and abstracts, issued principally in Germany, without which the search for the discoveries of a little-known author would be well-nigh hopeless. Much useful work has been done in this direction in connection with our Association. Such critical reports as those upon Hydro-dynamics, upon Tides, and upon Spectroscopy, guide the investigator to the points most requiring attention, and in discussing past achievements contribute in no small degree to future progress. But though good work has been done, much yet remains to do.

If, as is sometimes supposed, science consisted in nothing but the laborious accumulation of facts, it would soon come to a stand-still, crushed, as it were, under its own weight. The suggestion of a new idea, or the detection of a law, supersedes much that had previously been a burden upon the memory, and by introducing order and coherence facilitates the retention of the remainder in an available form. Those who are acquainted with the writings of the older electricians will understand my meaning when I instance the discovery of Ohm's law as a step by which the science was rendered easier to understand and to remember. Two processes are thus at work side by side, the reception of new material and the digestion and assimilation of the old; and as both are essential, we may spare ourselves the discussion of their relative importance. One remark, however, should be made. The work which deserves, but I am afraid does not always receive, the most credit is that in which discovery and explanation go hand in hand, in which not only are new facts presented, but their relation to old ones is pointed out.

In making oneself acquainted with what has been done in any subject, it is good policy to consult first the writers of highest general reputation. Although in scientific matters we should aim at independent judgment, and not rely too much upon authority, it remains true that a good deal must often be taken upon trust. Occasionally an observation is so simple and easily repeated, that it scarcely matters from whom it proceeds; but as a rule it can hardly carry full weight when put forward by a novice whose care and judgment there has been no opportunity of testing, and whose irresponsibility may tempt him to 'take shots,' as it is called. Those who have had experience in accurate work know how easy it would be to save time and trouble by omitting precautions and passing over discrepancies, and yet, even without dishonest intention, to convey the impression of conscientious attention to details. Although the most careful and experienced cannot hope to escape occasional mistakes, the effective value of this kind of work depends much upon the reputation of the individual responsible for it.

In estimating the present position and prospects of experimental science, there is good ground for encouragement. The multiplication of laboratories gives to the younger generation opportunities such as have never existed before, and which excite the envy of those who have had to learn in middle life much that now forms part of an undergraduate course. As to the management of such institutions there is room for a healthy difference of opinion. For many kinds of original work, especially in connection with accurate measurement, there is need of expensive apparatus; and it is often difficult to persuade a student to do his best with imperfect appliances when he knows that by other means a better result could be attained with greater facility. Nevertheless it seems to me important to discourage too great reliance upon the instrument maker. Much of the best original work has been done with the homeliest appliances; and the endeavour to turn to the best account the means that may be at hand develops ingenuity and resource more than the most elaborate determinations with ready-made instruments. There is danger otherwise that the experimental education of a plodding student should be too mechanical and artificial, so that he is puzzled by small changes of apparatus much as many school-boys are puzzled by a transposition of the letters in a diagram of Euclid.

From the general spread of a more scientific education, we are warranted in expecting important results. Just as there are some brilliant literary men with an inability, or at least a distaste practically amounting to inability, for scientific ideas, so there are a few with scientific tastes whose imaginations are never touched by merely literary studies. To save these from intellectual stagnation during several important years of their lives is something gained; but the thorough-going advocates of scientific education aim at much more. To them it appears strange, and almost monstrous, that the dead languages should hold the place they do in general education; and it can hardly be denied that their supremacy is the result of routine rather than of argument. I do not, myself, take up the extreme position. I doubt whether an exclusively scientific training would be satisfactory; and where there is plenty of time and a literary aptitude I can believe that Latin and Greek may make a good foundation. But it is useless to discuss the question upon the supposition that the majority of boys attain either to a knowledge of the languages or to an appreciation of the writings of the ancient authors. The contrary is notoriously the truth; and the defenders of the existing system usually take their stand upon the excellence of its discipline. From this point of view there is something to be said. The laziest boy must exert himself a little in puzzling out a sentence with grammar and dictionary, while instruction and supervision are easy to organise and not too costly. But when the case is stated plainly, few will agree that we can afford so entirely to disregard results. In after life the intellectual energies are usually engrossed with business, and no further opportunity is found for attacking the difficulties which block the gateways of knowledge. Mathematics, especially, if not learned young, are likely to remain unlearned. I will not further insist upon the educational importance of mathematics and science, because with respect to them I shall probably be supposed to be prejudiced. But of modern languages I am ignorant enough to give value to my advocacy. I believe that French and German, if properly taught, which I admit they rarely are at present, would go far to replace Latin and Greek from a disciplinary point of view, while the actual value of the acquisition would, in the majority of cases, be incomparably greater. In half the time usually devoted, without success, to the classical languages, most boys could acquire a really serviceable knowledge of French and German. History and the serious study of English literature, now shamefully neglected, would also find a place in such a scheme.

There is one objection often felt to a modernised education, as to which a word may not be without use. Many excellent people are afraid of science as tending towards materialism. That such apprehension should exist is not surprising, for unfortunately there are writers, speaking in the name of science, who have set themselves to foster it. It is true that among scientific men, as in other classes, crude views are to be met with as to the deeper things of Nature; but that the life-long beliefs of Newton, of Faraday, and of Maxwell, are inconsistent with the scientific habit of mind, is surely a proposition which I need not pause to refute. It would be easy, however, to lay too much stress upon the opinions of even such distinguished workers as these. Men, who devote their lives to investigation, cultivate a love of truth for its own sake, and endeavour instinctively to clear up, and not, as is too often the object in business and politics, to obscure a difficult question. So far the opinion of a scientific worker may have a special value; but I do not think that he has a claim, superior to that of other educated men, to assume the attitude of a prophet. In his heart he knows that underneath the theories that he constructs there lie contradictions which he cannot reconcile. The higher mysteries of being, if penetrable at all by human intellect, require other weapons than those of calculation and experiment.

Without encroaching upon grounds appertaining to the theologian and the philosopher, the domain of natural science is surely broad enough to satisfy the wildest ambition of its devotees. In other departments of human life and interest, true progress is rather an article of faith than a rational belief; but in science a retrograde movement is, from the nature of the case, almost impossible. Increasing knowledge brings with it increasing power, and great as are the triumphs of the present century, we may well believe that they are but a foretaste of what discovery and invention have yet in store for mankind. Encouraged by the thought that our labours cannot be thrown away, let us redouble our efforts in the noble struggle. In the Old World and in the New, recruits must be enlisted to fill the place of those whose work is done. Happy should I be if, through this visit of the Association, or by any words of mine, a larger measure of the youthful activity of the West could be drawn into this service. The work may be hard, and the discipline severe; but the interest never fails, and great is the privilege of achievement.

114.

A LECTURE EXPERIMENT ON INDUCTION.

[*British Association Report*, p. 632, 1884.]

IT is well known that an electro-magnet, interposed in the circuit of an alternate current machine, diminishes the effect far more than in a degree corresponding to the resistance of the additional wire. This behaviour of an electro-magnet may be exhibited to an audience in an instructive manner, by use of a helix wound with two contiguous wires (such as are commonly used for large instruments), one of which is included in the circuit of a De Meritens machine and a few incandescent lamps. If the circuit of the second wire be open, the introduction of a few stout iron wires into the helix causes a very marked falling off in the incandescence. On closing the second circuit, currents develope themselves in it of such a kind as to compensate the self-induction, and the lights recover their brilliancy. Even without iron, the effect of closing the second circuit is perceptible, provided the degree of incandescence be suitable.

An arrangement suitable for illustrating the same phenomenon with currents of small intensity was described in *Nature* for May 23, 1872. [Art. 20, vol. I. p. 167.]

115.

ON TELEPHONING THROUGH A CABLE.

[*British Association Report*, pp. 632, 633, 1884.]

The principles of this subject were laid down thirty years since by Thomson, but the author had not met with an application to the circumstances of the telephone.

A periodic variation of potential, imposed at one end, is propagated along the line in accordance with the law

$$v = e^{-\sqrt{\frac{n}{2k}} \cdot x} \cos\left(nt - \sqrt{\frac{n}{2k}} \cdot x\right),$$

in which $n/2\pi$ is the frequency of the electrical vibration. For Atlantic cables the constant k, depending upon the resistance and the capacity, has in c.g.s. measure such a value as 2×10^{16}. The distance, in traversing which the amplitude is reduced in the ratio $e : 1$, is given by

$$x = \sqrt{\frac{2k}{n}} = \frac{2 \times 10^8}{\sqrt{n}} \text{ centimetres.}$$

If we take a pitch rather more than an octave above that of middle c, we have $n = 3{,}600$, $\sqrt{n} = 60$, so that

$$x = 3 \times 10^6 \text{ centimetres} = 20 \text{ miles approximately.}$$

A distance of twenty miles would thus reduce the intensity of sound to almost a tenth, an operation which could not be often repeated without rendering it inaudible. With such a cable the practical limit would not be likely to exceed fifty miles, more especially as the easy intelligibility of speech requires the presence of notes still higher than is supposed in the above numerical example.

116.

ON A GALVANOMETER WITH TWENTY WIRES.

[*British Association Report*, p. 633, 1884.]

GALVANOMETERS suitable for currents of an ampère or two are most accurately standardised by means of the silver voltameter, but this method ceases to be convenient when the current to be dealt with rises above five ampères. The present instrument is a kind of differential galvanometer, provided with two electrically distinct coils, whose constants are in ratio of ten to one. A current of one ampère through one coil thus balances a current of ten ampères through the other. If the first be measured in terms of silver, the second serves to standardise an instrument suitable for the larger current.

The novelty consists in the manner in which the ten to one ratio is secured. Twenty pieces of No. 17 cotton-covered wire, being cut to equal lengths of about eight feet, were twisted closely together, two and two, so as to form ten pairs, which ten pairs were again in their turn twisted slightly together so as to form a rope. In each of the two circuits there are ten wires. In one, that intended for the larger current, these wires are in parallel; in the other circuit the ten wires are in series. Of each of the original twists one wire belongs to the parallel and one to the series group. Now the two wires forming an original twist are equally effective upon a needle suspended in any reasonable situation with respect to them, and thus if the ten wires in parallel have the same resistance, the circuit formed by the ten wires in series will be precisely ten times as effective as the circuit formed by the ten wires in parallel. This is independent of the disposition of the ten original pairs, but by winding them loosely into a rope we gain an additional security in case the ten parallel wires, though of the same length and cut from the same hank, should have slightly different resistances. If all the twenty wires could be assumed to have equal efficiency in deflecting the needle, the equality of resistances of the wires in parallel would be of no moment.

The rope is bent into a single circle of about a foot diameter with leads two feet long. At this distance the necessary junctions can be effected without fear of disturbance. The electrodes for the heavy currents are formed of parallel copper strips, separated by an insulating layer, and the current is brought up through twisted leads as in Sir W. Thomson's graded galvanometers. In the case of the smaller current, which embraces the needle ten times, so much precaution is not required.

After the wires in parallel had been soldered up, but while those destined to be joined in series were still disconnected, insulation tests were made between each wire of the series group and the other wires of that group as well as the group in parallel. The resistance between each series wire and the parallel group was about $2\frac{1}{2}$ megohms, and (as might be expected) about twice as much between any pair of wires of the series group.

It will be seen that when, in the use of the instruments, two currents are balancing one another, every one of the twenty wires carries the same current. In the actual instrument this current might amount, without undue heating, to four ampères, so that the heavy current would be 40 ampères. If it be not thought necessary to deal with currents heavier than 10 ampères, the gauge of wire might be reduced, a change which would facilitate the winding of the rope.

The magnet and mirror should be of the kind used in reflecting galvanometers, and may be hung at the centre of the circle.

117.

ON CLARK'S STANDARD CELLS.

[*British Association Report*, pp. 651, 652, 1884.]

In the hope of finding a clue as to the origin of some of the minor anomalies of Clark's cells, I have made experiments upon the E.M.F. of combinations, in which two different strengths of zinc amalgam take the place of the zinc and pure mercury of the Clark cell. No mercurous sulphate is employed, the liquid being simply a saturated solution of zinc sulphate.

If the same kind of amalgam be used for both poles, the symmetry is complete, and there should be no E.M.F. But if we take for one pole a strong, but fluid, amalgam, and for the other the same amalgam diluted with an equal volume of pure mercury, we find a very sensible E.M.F., the strong amalgam corresponding to the zinc of the ordinary Clark. In my experiment the E.M.F. was ·004 Clark, and remained pretty constant from day to day. In another cell the same strong amalgam was used for one pole, and for the other pole was diluted with three times its volume of pure mercury. In this case the E.M.F. was ·009 Clark.

If we replace the diluted amalgam with pure mercury, we obtain (without mercurous sulphate) nearly the full E.M.F. of the Clark cell, but, as might be expected, the force is very unsteady. From this it would seem that the function of the mercurous sulphate in the usual form of cell is to retain the purity of the mercury, and that the E.M.F. is largely due to the affinity of mercury for zinc.

118.

ON THE CONSTANT OF MAGNETIC ROTATION OF LIGHT IN BISULPHIDE OF CARBON.

[*Philosophical Transactions*, 176, pp. 343—366, 1885.]

1. THE phenomenon, to which the present investigation relates, is Faraday's discovery of the "Magnetisation of Light," or in more usual language the rotation of the plane of polarisation of light in traversing certain media exposed to powerful magnetic force. One of the characteristics of this rotation is that it takes place in the same absolute direction whichever way the light may be travelling, differing in this respect from the rotation which occurs without the operation of magnetic force in quartz and many organic liquids. Advantage of this property has been taken by Faraday and others in order to magnify the effect. By reflecting the light backwards and forwards it is possible to make it traverse several times a field of force whose length is limited.

A consequence remarkable from the theoretical point of view is the possibility of an arrangement in which the otherwise general optical law of reciprocity shall be violated. Consider, for example, a column of diamagnetic medium exposed to such a force that the rotation is 45°, and situated between two Nicols whose principal planes are inclined to one another at 45°. Under these circumstances light passing one way is completely stopped by the second Nicol, but light passing the other way is completely transmitted. A source of light at one point A would thus be visible at a second point B, when a source at B would be invisible at A; a state of things *at first sight* inconsistent with the second law of thermodynamics.

2. It is known that the rotation may be considered to be due to the propagation at slightly different velocities of the two circularly polarised

components into which plane polarised light may be resolved; and it is interesting to consider what difference of velocity our instrumental appliances enable us to detect. A retardation, amounting to one wavelength (λ), of one circularly polarised component relatively to the other would correspond to a rotation of the plane of polarisation through $180°$. If we can observe a rotation of one *minute*, we are in a position to detect a retardation of $\lambda/10800$. If l be the thickness traversed, v and $v + \delta v$ the two velocities of propagation, the relative retardation is $l\delta v/v$. To take an example, suppose that $l = 20$ inches, $\lambda = \frac{1}{40,000}$th inch; so that if $\delta v/v$ exceed 10^{-8}, the fact might be detected*. It appears therefore that we are able to observe extraordinarily minute relative differences in the velocities of propagation of the two circularly polarised rays.

3. The laws of the phenomenon were investigated in detail by Verdet, who proved experimentally that in a given medium the rotation between any two points on a ray of light of given kind is proportional to the difference of magnetic potential at those points. When the path of the ray is singly or doubly curved, the rotation is to be estimated upon principles similar to those applicable to *twist*† in curved rods‡.

4. Absolute determinations of magnetic rotation in bisulphide of carbon have been made by Gordon§, and by H. Becquerel ‖, whose results differ, however, by about 9 per cent. The former obtained his magnetic force by means of an electric current circulating a great many times round the column of CS_2. This column being a good deal longer than the coil, the electro-magnetic effect is approximately determined by the strength of the current and the number of turns. Of these data the first was found by a comparison with H (the horizontal component of terrestrial magnetism). The number of windings in the coil was determined, not by a simple counting, but *à posteriori* by an electrical process.

In M. Becquerel's experiments the magnetic force was that of the earth acting on a column of CS_2 more than 3 metres in length. The very small effect (obtained by reversal of the apparatus in azimuth) was augmented by causing the light to pass the tube 3 or 5 times, but even with 5 passages the double rotation amounted to only about 30 minutes. M. Becquerel regards his determination for sodium light as accurate to

* Camb. Nat. Sci. Trip. Ex., 1883.

† Thomson and Tait's *Natural Philosophy*, §§ 119—123.

‡ When polarized light passes from one medium to another, e.g., from air to glass, the plane of polarisation is in general twisted without the operation of any magnetic force. This effect, however, depends upon a part of the light being diverted by reflection, and would disappear if the transition from one medium to the other were gradual, i.e., occupied a stratum a few wave-lengths thick. (See *Proc. Math. Soc.* vol. xi. No. 159.) [Art. 63, vol. i. p. 460.]

§ *Phil. Trans.* 1877, p. 1.

‖ *Ann. d. Chimie*, 1882.

within 1 per cent., which would be indeed a wonderful result considering the smallness of the rotation.

5. It is important to observe that great care is required in order to define with sufficient accuracy the kind of light employed. Since the rotation is approximately proportional to λ^{-2}, a change from one sodium line to the other would make a difference of two parts per thousand. Both of the above-mentioned experimenters started with white light. Gordon threw a spectrum upon a screen perforated with a slit, the position of which was adjusted to correspond with the thallium line; while Becquerel corrected his results indirectly by a subsequent comparison between the effects of the more mixed light used by him and that emitted by sodium.

Considering that the employment of white light involved very elaborate arrangements for analysis (according to wave-length), in order to avoid errors exceeding in magnitude those likely to be encountered in the polarimetric or electric determinations, I decided to use light actually emitted from sodium vapour. The sodium chloride was held by a spoon of platinum gauze in the flame of a small ordinary Bunsen burner (fig. 1, A). As in

Fig. 1.

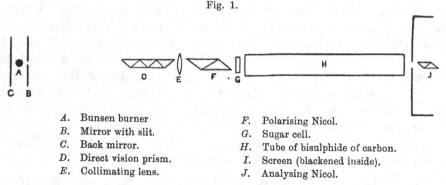

A. Bunsen burner	*F.* Polarising Nicol.
B. Mirror with slit.	*G.* Sugar cell.
C. Back mirror.	*H.* Tube of bisulphide of carbon.
D. Direct vision prism.	*I.* Screen (blackened inside).
E. Collimating lens.	*J.* Analysing Nicol.

Mr Glazebrook's optical investigations, the evaporation of the salt and the temperature of the flame were stimulated by a jet of oxygen gas brought in laterally and caused to play round the gauze*.

* [1899. Fox Talbot's early optical work is so little known that I am tempted to quote in full his short note, in which probably this valuable method "of obtaining homogeneous light of great intensity" is first described.

"As it is a desideratum in optical science to procure *perfectly* homogeneous light of sufficient brightness for many important experiments, I am glad to be able to communicate a method which in a satisfactory manner supplies that deficiency.

"It is only requisite to place a lump of common salt upon the wick of a spirit-lamp and *to direct a stream of oxygen gas* from a blow-pipe upon the salt. The light emitted is quite homogeneous, and of dazzling brightness. If instead of common salt we use the various salts of strontian, barytes, &c., we obtain the well-known coloured flames, which are characteristic of those substances, with far more brilliancy than by any other method with which I am acquainted." (*Phil. Mag.* III. p. 35, 1833.)

At the close of the experiments I examined the light thus obtained with a powerful spectroscope, and found that under the influence of the oxygen the originally narrow bright lines dilate almost to the point of contact, thus forming a bright field upon which the dark D-lines are seen with beautiful definition. Although the distribution of light appeared to be tolerably symmetrical, it is a question to what degree of accuracy the mean quality of this light can be identified with that coming from midway between the D-lines. Probably we shall be safe in estimating that the error from this cause is well below $\frac{1}{1000}$.

The bright part of the flame being much larger than is required, a screen (B), perforated with a slit, may conveniently be interposed. In this course there are two advantages. It allows us to purify the light from rays of other refrangibilities (of which there is always a sensible accompaniment, both red and blue) by use of a direct-vision prism (D). Again, by making this screen of looking-glass, from which a narrow strip of silvering is removed, and by backing the flame with a parallel mirror (C), we gain by repeated reflections to and fro, an important increase of illumination. The success of the polarimetry is very dependent upon the intensity of the light, but there must be also a reasonable steadiness. Several arrangements of flame which at first promised well failed in the latter requirement.

6. The rays from the slit, after purification by the direct vision prism, are rendered parallel by a collimating lens (E) and pass into the polarising Nicol (F). The polarimeter employed is on the principle of Laurent, but according to a suggestion of Poynting* the half-wave plate of quartz is replaced by a cell (G) containing syrop, so arranged that the two halves of the field of view are subjected to small rotations differing by about 2°. The difference of thicknesses necessary is best obtained by introducing into the cell a piece of thick glass, the upper edge of which divides the field into two parts. The upper half of the field is thus rotated by a thickness of syrop equal to the entire width of the cell (say $\frac{1}{2}$ inch), but in the lower half of the field part of the thickness of syrop is replaced by glass, and the rotation is correspondingly less. With a pretty strong syrop a difference of 2° may be obtained with a glass $\frac{3}{16}$ inch [inch = 2·54 cm.] thick. For the best results the operating boundary should be a true plane nearly perpendicular to the face. The pieces used by me, however, were not worked, being simply cut with a diamond from thick plate glass; and there was usually no difficulty in finding a part of the edge sufficiently flat for the purpose, *i.e.*, capable of exhibiting a field of view sharply divided into two parts. I had expected to be troubled with depolarisation, especially in the thick glass, but a small piece thus cut

* *Phil. Mag.*, July, 1880.

out of a large plate is relieved from most of the strain to which it was originally subject. Probably more care would be required in experiments where a strong white light could be used; but by previously testing the rather thin plates used for the sugar cell and for closing the CS_2 tube, I was able to secure a field of view either half of which under the actual circumstances could be made quite dark by suitable orientation of the analysing Nicol.

By this use of sugar, half-shade polarimeters may be made of large dimensions at short notice and at very little cost. The syrop should be filtered (hot) through paper, and the cell must be closed to prevent evaporation.

7. On leaving the sugar cell the light entered the column of bisulphide of carbon (H). To contain the liquid two tubes of brass were employed at various times, the ends being closed with plates of worked glass cemented to the metal with a mixture of glue and treacle. Near one end these tubes were provided with a lateral (vertical) branch, closed with a cork, through which passed the stem of the thermometer used for observing the temperature of the CS_2. The length of the larger tube (used in Series I. and II.) was 31·591 inches, and the diameter about $1\frac{1}{2}$ inch. The length of the smaller tube (used in Series III.) was 29·765 inches, and the diameter 1 inch.

When, as in Series I., it was wished to cause the light to traverse the tube more than once, mirrors were necessary at the ends of the tube. They consisted of plates of thin looking-glass, from which part of the silvering was removed, and by means of a little glycerine they were brought into optical contact with the plates by which the tube was closed. This arrangement was simple, and had the further advantage of practically annulling some troublesome reflections; but the want of means of adjustment rendered it necessary that the closing plates should themselves be pretty accurately parallel.

8. The internal diameter of the ebonite tube, upon which the helix was wound (§ 13), was about $1\frac{7}{8}$ inch, and it was intended to utilise the annular space between the ebonite and the brass as a jacket, through which water at the temperature of the room might be made to circulate. This arrangement, however, failed utterly. Within about 10 minutes of the closing of the circuit of the helix, the definition was lost, and nothing further could be done until after a long interval of repose. The water-jacket was then abolished, and the available space filled with paper wrapped pretty tightly round the tube. This effected a great improvement, enhanced still further in the later experiments of Series III., in which, by reduction of the diameter of the tube, a wider space became available for heat insulation. The disturbance by conduction of heat from the wire to the CS_2 remained, however, the worst feature of the experi-

ments, and could not be obviated without a fundamental alteration in the apparatus. Probably the best arrangement would be a water-jacket next the wire, and a good thickness of paper or other insulator between the water and the CS_2.

9. The bisulphide of carbon was purified by treatment with corrosive sublimate and grease with subsequent distillation (according to the procedure advocated by Becquerel), until most of the unpleasant odour had disappeared. The transparency is much greater than is readily (if at all) obtainable with water, provided proper precautions are taken to avoid exposure to light. After being acted upon by light, the CS_2 attacks brass and becomes rapidly opaque. In this respect it would be an advantage to replace the metal tube by one of glass.

10. The analyser consisted, in some experiments, of a Nicol (J), and in others of a double image prism, and was mounted in a circle made by the Cambridge Scientific Instrument Company. In order that a rotation of the plane of polarisation may be correctly indicated by the difference of the two circle readings, it is necessary that the axis of rotation should coincide with the direction of the light. This requirement is, however, not very easily satisfied. At the commencement of a series of experiments the adjustment was made with the aid of a telescope and cross wires temporarily substituted for the Nicol, but during the course of a set of readings the passage of heat into the liquid tended to make the upper strata warmer than the lower, and thus to bend the rays into a different direction. It is known* that the error arising from maladjustment in this respect is in great part eliminated by reading the Nicol always in both the positions (differing by about 180°) which give extinction, or (in the half-shade arrangement) equality of illumination. This plan was constantly followed, but it is not clear that the whole error can be thus got rid of. It occurred to me that another term in the harmonic expansion of the error would be destroyed by use of a double image prism read in *four* positions distant about 90°. Experiment showed that in spite of the glare of the unextinguished image, good readings could be obtained after a little practice, and the comparison of the results arrived at in this way tends to show that the error is not wholly eliminated in the mean of two readings taken in positions differing by 180°. But the matter could be much better investigated with a simplified apparatus and the use of a strong white light.

In Series II. and III., when the light traversed the tube but once, no magnification was necessary, and the eye was applied immediately behind

* " Zur Theorie des Polaristrobometer und des drehenden Nicols." V. D. Sande Bakhuyzen, *Pogg. Ann.* t. cxlv. 259. 1872.
" Notes on Nicol's Prism," Glazebrook, *Phil. Mag.* October, 1880.

the analyser. In Series I., the apparent magnitude of the field was much less, and an opera-glass, magnifying about twice, was employed between the analyser and the eye.

11. The setting of the Nicol (or double-image prism) by adjustment of the match between the two parts of the field presented by the half-shade apparatus was facilitated by a device that may be found useful. "In addition to the principal helix, the tube was embraced by an auxiliary coil of insulated wire, through which could be led the current from a Leclanché cell. This current was controlled by a reversing key under the hand of the observer, who was thus able to *rock* the plane of polarisation backwards and forwards through a small angle about its normal position. The amount of the rocking being suitably chosen, the comparison of the three appearances (two with auxiliary current, and one without) serves to exclude some imperfect matches that might otherwise have been allowed to pass*."

12. Apart from the effect of heat upon the CS_2, the working of the optical parts was fairly satisfactory. The following zero readings taken without the current on June 4, 1884, will give an idea of the sort of accuracy attained. The analyser was a double image prism, and was read in all four positions, the circuit being made three times.

TABLE I.

	° ′	° ′	° ′	° ′
	103 2	193 4	283 0	13 2
	102 55	193 5	283 2	12 59
	102 58	193 3	283 2	13 4
Mean	102 58	193 4	283 1	13 2
Subtract . . .	90	180	270	
	12 58	13 4	13 1	13 2

It appears that an error of 3 or 4 minutes may occur in a single setting.

13. I now pass to the description of the electrical arrangements. The magnetic force depends upon the helix and upon the strength of the current, and we will take these elements in order.

* "Preliminary Note on the Constant of Electro-magnetic Rotation of Light in Bisulphide of Carbon," *Proc. Roy. Soc.* vol. xxxvii. p. 146 (June 19, 1884).

The helix.

The wire is wound upon an ebonite tube, the outside surface of which was turned true in the lathe, and is kept in its place laterally by ebonite flanges screwed upon the tube. The distance between the flanges, equal to the length of the helix, is 9·990 inches; but the tube itself projects some inches beyond the flanges, and when it was desired to use an internal water-jacket, could be further prolonged by additional lengths of brass tube.

In order to give better opportunity for testing the insulation, on which the correctness of the final results is entirely dependent, it was decided to wind on two wires simultaneously, which should be in contact with one another throughout their entire lengths. The operation was performed on December 14–15, 1883, with triply-covered wire of diameter about $\frac{1}{20}$ inch, and no particular difficulty was experienced. The revolutions of the ebonite tube, mounted in the lathe, were taken with all care by an engine counter, and amounted to 1842, so that the total number of windings is 3684. The internal diameter of the helix is 2·188 inches, and the external diameter is 4·13 inches. [inch = 2·54 cm.]

By endeavouring to force a current from one wire to the other we obtain a very severe, though of course not absolutely complete, test of the insulation. The resistance between the two wires varied with the hygrometric condition of the silk, which was not impregnated with paraffin. At first it was not much over 2 megohms, but latterly reached 6 or 8 megohms, and was thus abundantly sufficient.

14. As a further test observations were made of the external effect of the helix upon a suspended magnet, when a powerful current was passed in one direction through the first wire, and in the opposite direction through the second. If the positions of the two wires could be treated as identical, the external effect ought everywhere to vanish. In consequence, however, of the fact that one wire lies throughout on the same side of the other, the compensation could not be expected to be complete, except when the suspended magnet is equidistant from the two ends. Experiment with the magnet of a reflecting galvanometer showed that the effect, in fact, varied as the magnet was displaced, but even in the symmetrical position there was a perceptible outstanding differential effect. In order to eliminate the influence of other parts of the circuit, the readings referred only to the deflection of the needle as the current was reversed in the *helix*; and the scale of sensitiveness was obtained by repeating the observations after altering the connexions of the two wires, so that the current circulated the same way round both, and after insertion of a high resistance by which the intensity of the current was

reduced in a known proportion. From this it appeared that the differential effect of the two wires (with a given current) was $\frac{1}{2500}$ of the combined effect.

This fraction is tolerably small, but I had expected to find it smaller still. It seems probable that the incompleteness of compensation is due to a small difference ($\frac{1}{5000}$) in the mean diameter of the windings in the two cases. To throw light upon this I took careful measures of the resistances of the two wires. Although they had originally formed one length, their resistances differed by as much as $\frac{1}{70}$th part, that of the wire which had shown itself *least* effective being 7·075 B.A., and of the other 6·965. If, as it seems plausible to do, we attribute the difference of resistance to difference of diameter, this actual difference must amount to $\frac{1}{2800}$ inch. The mean diameter of the windings is about three inches; and if the two wires were wound upon a smooth cylinder of this diameter, the difference in the diameter of the windings would be $\frac{1}{8400}$ of the whole. As this estimate would be increased were we to take into account the fact that each winding really sits upon two windings of the layer underneath, and that these cannot be practically in actual contact, we may perhaps consider the small anomalous differential effect upon the external magnet to be sufficiently explained by the observed difference of resistances.

Correction for finite length.

15. If the tube were infinitely long, the difference of potentials at its ends due to the unit current in one winding would be 4π. But on account of the finiteness of the length a correction is required, whose approximate amount is given in Gordon's paper.

Fig. 2.

Considering, in the first place, one layer of windings of radius Aa, we know that the external effect is the same as would be produced by a uniform distribution of imaginary magnetic matter over the ends, positive (say) over Aa and negative over Bb, the superficial density being equal to the number (m) of windings per unit length. The potential at L of the matter on Aa is $2\pi m\,(La - LA)$, or approximately

$$\pi m \left(\frac{Aa^2}{LA} - \tfrac{1}{4}\frac{Aa^4}{LA^3} \right).$$

Similarly the potential at L for the matter on Bb is

$$-\pi m \left(\frac{Aa^2}{LB} - \tfrac{1}{4}\frac{Aa^4}{LB^3} \right);$$

so that altogether the potential at L for this layer of windings is

$$\pi m AB \left(\frac{Aa^2}{LA \cdot LB} - \frac{Aa^4}{4} \frac{LB^2 + LA \cdot LB + LA^2}{LA^3 \cdot LB^3} \right),$$

in which mAB denotes the whole number of windings in the layer. This result has now to be integrated so as to represent the effect of the helix, whose inner and outer radii we may call Aa_1 and Aa_2. The mean value of Aa^2 is

$$\frac{Aa_2^3 - Aa_1^3}{3(Aa_2 - Aa_1)},$$

and that of Aa^4 is

$$\frac{Aa_2^5 - Aa_1^5}{5(Aa_2 - Aa_1)}.$$

Thus, if n be the whole number of windings on the helix, the difference of potential from L to M corresponding to the unit current is

$$4n\pi - 4n\pi \left[\frac{Aa_2^3 - Aa_1^3}{12\, a_2 a_1} \left(\frac{1}{LA \cdot LB} + \frac{1}{MA \cdot MB} \right) \right.$$
$$\left. - \frac{Aa_2^5 - Aa_1^5}{80\, a_2 a_1} \left(\frac{LB^2 + LA \cdot LB + LA^2}{LA^3 \cdot LB^3} + \frac{MA^2 + MA \cdot MB + MB^2}{MB^3 \cdot MA^3} \right) \right].$$

In the present case

$$Aa_2 = 2 \cdot 065 \text{ (inches)}, \qquad Aa_1 = 1 \cdot 094, \qquad a_2 a_1 = \cdot 971,$$

from which we get

$$\frac{Aa_2^3 - Aa_1^3}{12\, a_2 a_1} = \cdot 6433, \qquad \frac{Aa_2^5 - Aa_1^5}{80\, a_2 a_1} = \cdot 4632.$$

In the remainder of the calculation we have to distinguish the two tubes. For the first

$$LA = MB = 10 \cdot 800 \text{ inches}, \qquad LB = MA = 20 \cdot 790 \text{ inches};$$

and for the second

$$LA = MB = 9 \cdot 887 \text{ inches}, \qquad LB = MA = 19 \cdot 877 \text{ inches}.$$

Hence for the first tube we have

$$4n\pi (1 - \cdot 00573 + \cdot 00006) = 4n\pi \times \cdot 99433*;$$

and for the second

$$4n\pi (1 - \cdot 00655 + \cdot 00008) = 4n\pi \times \cdot 99353,$$

the correction for finite length thus somewhat exceeding one-half per cent.

---

* In the Preliminary Note the reducing factor for this tube was given as $\cdot 99449$. The alteration is due to the use of more precise data in place of some quite rough measurements in round numbers on which, by an oversight, the first calculation was founded.

16. We have now obtained the difference of potential at the ends of the column of CS_2 due to the passage through the helix of unit current. It yet remains to describe the means adopted for the measurement of the actual current in absolute measure.

In a former paper, "On the Electro-chemical Equivalent of Silver, and on the Absolute Electromotive Force of Clark Cells*," it was shown how the E.M.F. of a Clark cell was obtained by comparison with the difference of potentials at the extremities of a wire of known resistance, due to the passage of a current known either directly from its effect upon a current measuring apparatus, or indirectly through the deposition of silver. For the purposes of the present investigation this process was reversed, the Clark cell itself being treated as a standard of E.M.F., by which to determine the value of the current, which traversed the known resistance, and also the helix by which the magnetic rotation was produced. The arrangements differed so little from those elaborately described in the paper referred to, that it seems unnecessary to enter into the matter at length. If the reader will refer to Fig. (1), [p. 285], he will understand the electrical connexions, and he may suppose the current-measuring apparatus, EGF, replaced by the magnetising helix. In point of fact this helix was situated in another room at a distance from the E.M.F. compensator and its galvanometer T. The direction of the current in the helix was reversed by a mercury key of the rocker pattern, and care had to be taken that at this moment the galvanometer contact Q was open. The general nature of the arrangement will be sufficiently understood when it is said that the want of balance between the E.M.F. of the Clark and that at the terminals of the resistance R was made up by E.M.F., taken from an auxiliary circuit, the value of which was afterwards expressed in terms of the Clark. Denoting the force thus added or subtracted by r, upon a scale according to which the force of the Clark was ρ, the actual difference of potential at the terminals of R may be written

$$\left(1 \pm \frac{r}{\rho}\right) \times \text{Clark.}$$

17. As it was intended to use currents of about one ampère, the resistance R was made about [1½] ohms. The construction was somewhat similar to that of the [4] described in § 33 of the former paper, but on account of the increase in the current to be carried, three wires of German silver were used in parallel. The amount of heating was unimportant for the purposes of the present investigation.

The value of the [1½] was determined by comparison with a combination of three standard units, one (taking the whole current), and two in parallel (giving the ½). At 13° the resistance is 1·4945 B.A. At 15°,

* *Phil. Trans.* 1884, Part II. §§ 35, 36, 38. [Art. 112, vol. II. p. 278.]

TABLE II.—July 25, 1884.

Time (h. m.)	r	Circle reading +	Time (h. m.)	r	Circle reading −	Time (h. m.)	r	Circle reading +	Time (h. m.)	r	Circle reading −
6 3½	1525	261° 44′	6 5	1515	269° 19′	6 7½	1500	351° 56′	6 8½	1490	359° 24′
6 14½	1450	261 45	6 15½	1445	269 18	6 18½	1425	351 56	6 20½	1415	359 22
6 29½	1365	261 47	6 31½	1355	269 16	6 33	1350	351 57	6 35	1340	359 23
Mean	..	261 45·3	..	..	269 17·7	..	..	351 56·3	..	..	359 23·0
6 10	1480	81 48	6 11	1475	89 23	6 12½	1465	171 49	6 13½	1460	179 22
6 23	1400	81 48	6 24½	1395	89 19	6 26	1385	171 53	6 27	1380	179 20
6 37½	1325	81 53	6 39	1320	89 16	6 40	1315	171 55	6 41	1310	179 18
Mean	..	81 49·7	..	..	89 19·3	..	..	171 52·3	..	..	179 20·0

One passage. Double-image prism.

$$\rho = 7017.$$

Temperature of Clark and $R = 17°\cdot6$. Mean temperature of $CS_2 = 18°\cdot3$.

which was adopted as the standard temperature for R and for the Clark, we have

$$R = 1.4958 \text{ B.A.}$$

18.   In consequence of the heating of the copper wires, the current (usually obtained from secondary cells) fell off somewhat rapidly during a set of observations, and it was found convenient to take readings of the E.M.F. compensator simultaneously with the adjustment of the polarimeter.   The former readings were taken by myself and the latter by Mrs Sidgwick, while the flame (at which the optical observer should not look) was regulated by an assistant, who also recorded the circle readings.

The procedure will be most easily explained by an example, for which purpose I take at random the observations of July 25, recorded in Table II.

It will be seen that the cycle consisted of eight readings, four with positive and four with negative rotation of the plane of polarisation, and that this cycle is repeated three times.

The three readings under any one head vary in consequence of the diminution of the current as well as from errors of observation.   The value of ρ was

at the beginning .... $\rho = 7018$

at the end.......................... $\rho = 7016$

Mean .... $\rho = 7017$

Thus in the first observation at $6^{\text{h}} 3\frac{1}{2}^{\text{m}}$, when the circle reading was $261° 44'$, the difference of potentials at the extremities of the [$1\frac{1}{2}$] was $\left(1 + \dfrac{1525}{7018}\right) \times$ Clark I., the temperature of Clark I. and of the [$1\frac{1}{2}$] being $17°\cdot6$.

For the mean double rotations in the four positions of the double-image prism we have

$$269\ 17.7 - 261\ 45.3 = 7\ 32.4$$
$$359\ 23.0 - 351\ 56.3 = 7\ 26.7$$
$$89\ 19.3 -\ \ 81\ 49.7 = 7\ 29.6$$
$$179\ 20.0 - 171\ 52.3 = 7\ 27.7$$
$$\text{Mean} \ldots\ldots\ldots\ 7\ 29.1$$

Since all the effects are proportional to the current, it is sufficient to compare the mean rotation with the mean value of r, viz., 1413; so that the double rotation $7° 29'\cdot1$, or $449''\cdot1$, corresponds to a difference of potentials equal to

$$\left(1 + \frac{1413}{7017}\right) \times \text{Clark I.} = \frac{8430}{7017} \times \text{Clark I.}$$

The double rotation that would have been found if the current had been just strong enough to balance Clark I. (at the actual temperature) is

$$\frac{7017}{8430} \times 449'\cdot1 = 373'\cdot8.$$

19. This result is a function of the temperatures of the cell and of R as well as of the CS_2; and it is rather unfortunate that all three temperature corrections tell in the same direction. A rise of the thermometer involves a rise in R and a fall in the force of the standard cell, so that on both accounts the current giving the balance is diminished. At the same time the smaller current acts less advantageously in producing rotation in consequence of the properties of the CS_2. It will be convenient to postpone the last correction, and take first the corrections for temperature in R and the E.M.F. of Clark, which relate rather to the machinery for measuring the current, and which can be made from data obtained in previous investigations. For this purpose 15° C. is adopted as the standard temperature; and the proportional corrections per degree are ·00082 for the E.M.F. of Clark and ·00044 for the R, making altogether ·00126 per degree. For the observations of July 25, the correction is therefore

$$+ 2\cdot6 \times \cdot00126 \times 373'\cdot8 = + 2\cdot6 \times \cdot471 = + 1'\cdot2.$$

If we take as a standard current that which in traversing R at 15° would balance Clark I. at 15°, the double rotation of July 25 reduced so as to correspond with the standard current will be

$$373'\cdot8 + 1'\cdot2 = 375'\cdot0.$$

This rotation corresponds to the temperature 18°·3 of the CS_2. To obtain comparable results we must reduce to a standard temperature, for which purpose we will select 18°. According to Bichat the rotation at $t°$ may be expressed by

$$1 - \cdot00104t - \cdot000014t^2,$$

the rotation at 0° being taken as unity. To obtain a more convenient formula, applicable in the neighbourhood of 18°, we may write $t = 18 + t'$. Thus

$$1 - \cdot00104t - \cdot000014t^2 = \cdot9767 - \cdot00154t' = \cdot9767\,(1 - \cdot00158t');$$

so that the coefficient for the correction is ·00158. Hence, if the CS_2 on July 25 had been at 18°, we should have had

$$375'\cdot0 + 375'\cdot0 \times \cdot00158 \times \cdot3 = 375'\cdot0 + \cdot592 \times \cdot3 = 375\cdot0 + \cdot2 = 375'\cdot2.$$

Thus reduced the results for the observations of different days should agree together.

TABLE III.

Series I.

Date, 1884	Actual mean rotation (2a)	Mean value of r	ρ	$2a\frac{\rho}{\rho\pm r}$	Temperature of Clark and R	Correction to 15°	Rotation corrected to standard current	Temperature of CS$_2$	Correction to 18°	Rotation corrected to 18°	Deviation from mean
May 5	1123·0	44	5032	1132·9	15·0	+0·0	1132·9	16·8	−2·2	1130·7	+2·2
,, 6	1086·7	205	5040	1132·8	15·6	+0·8	1133·6	17·2	−1·4	1132·2	+3·7
,, 7	...	...	...	1132·1	14·8	−0·3	1131·8	17·2	−1·4	1130·4	+1·9
,, 9	1035·0	402	5046	1124·7	16·5	+2·1	1126·8	18·0	−0·0	1126·8	−1·7
,, 10	1008·0	515	5033	1122·9	17·0	+2·8	1125·7	19·8	+3·2	1128·9	+0·4
,, 13	1009·0	502	5019	1121·1	17·8	+3·9	1125·0	20·0	+3·6	1128·6	+0·1
,, 15	1040·5	359	5028	1120·5	18·2	+4·5	1125·0	19·3	+2·3	1127·3	−1·2
,, 16	543·6	2582	5026	1117·9	18·8	+5·3	1123·2	20·3	+4·1	1127·3	−1·2
,, 19	589·0	2416	5050	1129·3	16·2	+1·7	1131·0	17·5	−0·9	1130·1	+1·6
,, 21	681·2	1998	5065	1125·0	16·4	+2·0	1127·0	17·2	−1·4	1125·6	−2·9
,, 23	664·8	2070	5062	1124·7	17·0	+2·8	1127·5	18·1	+0·2	1127·7	−0·8
,, 26	755·0	1667	5066	1125·3	17·1	+3·0	1128·3	18·0	−0·0	1128·3	−0·2
,, 29	...	...	...	1128·6	14·9	−0·1	1128·5	15·5	−4·5	1124·0	−4·5
,, 31	685·4	2850	7194	1135·1	14·4	−0·8	1134·3	15·0	−5·4	1128·9	+0·4
June 2	640·7	3116	7164	1133·9	15·7	+1·0	1134·9	15·9	−3·8	1131·1	+2·6
Mean	...	...	...	...	...	...	1129·0	17·7	...	1128·5	...

Series II.

Date, 1884	Actual mean rotation (2a)	Mean value of r	ρ	$2a\frac{\rho}{\rho\pm r}$	Temperature of Clark and R	Correction to 15°	Rotation corrected to standard current	Temperature of CS$_2$	Correction to 18°	Rotation corrected to 18°	Deviation from mean
June 3	518·6	2719	7121	375·3	17·5	+1·2	376·5	16·5	−0·9	375·6	−0·1
,, 5	543·0	3168	7122	375·8	15·8	+0·4	376·2	16·3	−1·0	375·2	−0·5
,, 6	393·5	323	7118	376·4	15·1	+0·0	376·4	15·6	−1·4	375·0	−0·7
,, 9	552·9	3275	7110	378·5	14·3	−0·3	378·2	15·9	−1·3	376·9	+1·2
Mean	...	...	...	...	...	...	376·8	16·1	...	375·7	...

TABLE IV.

Series III.

Date, 1884	Actual mean rotation (2a)	Mean value of r	ρ	$\frac{2a\rho}{\rho \pm r}$	Temperature of Clark and R	Correction to 15°	Rotation corrected to standard current	Temperature of CS₂	Correction to 18°	Rotation corrected to 18°	Deviation from mean
July 23 .	447·2	1358	7017	374·7	18·4°	+1·6	376·3	19·2	+0·7°	377·0	+1·2
,, 24 .	449·6	1410	7019	374·4	18·0	+1·4	375·8	18·9	+0·5	376·3	+0·5
,, 25 .	449·1	1413	7017	373·8	17·6	+1·2	375·0	18·3	+0·2	375·2	-0·4
,, 28 .	448·2	1390	7017	374·1	17·1	+1·0	375·1	17·9	-0·1	375·0	-0·8
Aug. 2 .	389·7	319	6998	372·7	19·1	+1·9	374·6	20·7	+1·6	376·2	+0·4
,, 4 .	343·9	516	7001	371·3	19·0	+1·9	373·2	20·1	+1·3	374·5	-1·3
,, 5 .	530·2	2963	7005	372·6	19·0	+1·9	374·5	20·8	+1·7	376·2	+0·4
Mean .	...	...	...	...	...	...	374·9	19·4	...	375·8	...

20. The results of all the observations (other than preliminary) which were thought worthy of reduction are exhibited in the accompanying tables, grouped in three series. In Series I., II. the first tube was employed; the principal difference between them being that in Series I. the light traversed the tube *three times*, while in Series II. the light passed but *once*. It will be seen that in Series I. the actual double rotation varied from about 9° to 19°, and the currents from about ½ ampère to 1 ampère. In Series II. stronger currents were usually passed, amounting to about 1½ ampère, but the rotation was only about 9°. The extreme deviation from the mean is only about ·4 per cent., if we exclude the observations of May 29, which owing to interruptions and other causes were marked as unsatisfactory before reduction.

The Nicol was used as analyser in Series I., and on June 3 of Series II. The remaining observations of Series II. and the whole of Series III. were taken with a double-image prism, read in all four positions as already explained by the example of July 25.

For the observations of Series III. the second tube was employed, with some improvements in the provision against the communication of heat. The diminished diameter of the tube was the inducement to pass the light but once, though it would have been possible to work with three passages. But when the rays skirt the walls of the tube, there is more disturbance from heat; and, indeed, generally the advantage of augmented rotation is in great measure paid for by greater sensitiveness to deviation from optical uniformity.

Not only does the communication of heat disturb the definition, but it tends also to render the actual temperature uncertain. During some of the more protracted sets of readings with the stronger currents there was a rise of nearly 2° in the temperature of the CS_2; and, although this rise was carefully watched, it is difficult to feel confident that the effective mean temperature can be determined with a less error than say ⅛ of a degree. Such an error would correspond to about $\frac{1}{2000}$ in the final number. To avoid increasing the uncertainty under this head the readings were often concluded, although the definition still remained satisfactory.

If the apparatus were to be designed afresh I should endeavour to guard more adequately against these disturbances, and it might then be possible to use five passages with advantage, more especially if by increasing the weight of the coil it were practicable to bring the double rotation up to about 90°. The determination of such a rotation with the double-image prism would be free in high degree from the polarimetric errors considered in § 10. But it is doubtful whether in the present state of science the additional accuracy would repay the labour involved.

21. It only remains now to work out the results in absolute measure. And first as to the value of the standard current, defined as that which, flowing through the [1½] at 15°, balances Clark I. at the same temperature. This value in ampères is expressed by dividing the E.M.F. of Clark I. in B.A. volts (see Table XI. of former paper [p. 324]) by the resistance of the [1½] in B.A. units.   Hence the standard current is

$$\frac{1\cdot4542}{1\cdot4958} = \cdot9722 \text{ ampère} = \cdot09722 \text{ C.G.S.}$$

If the tube were infinitely long, the difference of magnetic potentials at its ends would be $4n\pi$; but in the case of the actual tubes we have to introduce the correcting factors ·99433 and ·99353 (§ 15). Thus for the first tube, if x be the (single) rotation in minutes corresponding to difference of potential 1 C.G.S., the whole actual double rotation for a single passage of the light will be

$$2 \times \cdot09722 \times 4\pi n \times \cdot99433 \times x.$$

From Series I. at 18° this quantity is found to be $\frac{1}{3} \times 1128\cdot5$, or 376·2, so that

$$x = \frac{376\cdot2}{2 \times \cdot09722 \times 4\pi n \times \cdot99433} = \cdot04203.$$

In like manner from Series II. we get

$$x = \frac{375\cdot7}{2 \times \cdot09722 \times 4\pi n \times \cdot99433} = \cdot04198.$$

For the second tube used in Series III. we have to employ a slightly different correction for finite length. We have

$$x = \frac{375\cdot8}{2 \times \cdot09722 \times 4\pi n \times \cdot99353} = \cdot04202.$$

The results of Series I. and III. are thus in precise agreement, while that of Series II. is about $\frac{1}{1000}$ lower. Ascribing a somewhat less importance to Series II. in consequence of the smaller number of sets of observations, we may take as the final result of the investigation

$$x = \cdot04202,$$

which gives the rotation in minutes in bisulphide of carbon at 18°, corresponding to a difference of potential equal to 1 C.G.S. It should be noticed that the mean temperature of the observations was so nearly 18° that the result as given depends scarcely at all upon Bichat's formula for the dependence of the rotation upon temperature.

22. M. Becquerel gives as his result for 0° C. ·0463 minute. To find the rotation at 18°, this must be multiplied by ·9767 according to Bichat's

formula: and as Becquerel's observations were in fact made at about 18°, this reduction does not introduce, but rather removes, an extraneous element. Thus according to Becquerel—

$$x = \cdot 0452 \text{ minute,}$$

differing by about 7 per cent. from the value found by me.

The comparison with Gordon is more uncertain, inasmuch as his observations were made on light of the refrangibility of the thallium line. The corrected* result for this light is in circular measure $1 \cdot 5238 \times 10^{-5}$, or $\cdot 05238$ minute. To pass to sodium we may use a formula given by Becquerel† and Verdet, according to which the rotation for different wave lengths (λ) is proportional to $\mu^2 (\mu^2 - 1) \lambda^{-2}$, μ being the refractive index. At this rate the $\cdot 05238$ minute for thallium would be $\cdot 04163$ minute for sodium. The temperature was not directly observed by Gordon, but was estimated to be about 13° C. Assuming this to be correct, the value for 18° would be $\cdot 0413$ minute, or about 2 per cent. *less* than according to my determinations.

APPENDIX.

Notes on Polarimetry in general.

The problem of the polarimeter is how best to render evident the rotation through a small angle θ of the plane of polarisation of light of brightness h. The effect of the rotation is to introduce light of amplitude $h^{\frac{1}{2}} \sin \theta$, or $h^{\frac{1}{2}} \theta$, polarised in the perpendicular plane, and it is this which must be made to produce a recognisable change. By the use of a Nicol, or double-image prism, adjusted to the original plane, the light of brightness $h\theta^2$ may be isolated, but, as will be proved presently, this is not the best method of rendering its existence evident.

From the preceding mode of statement it is clear that the accuracy obtainable in determining the plane of polarisation increases indefinitely with the brightness of the light, and is in fact proportional to the *square root* of that brightness‡. Again we see that little is to be expected from such devices as that of Fizeau, in which the rotation is magnified by causing the light to pass obliquely through a pile of glass plates. The brightness of the light polarised in the perpendicular plane ($h\theta^2$) can only be diminished by such treatment, and the increase of rotation, being due merely to weakening of the first component, is of no value.

---

* Mr Gordon's result was originally given at double its proper value.

† *Ann. d. Chim.* t. XII. 1877, p. 78.

‡ This point is insisted upon in an excellent paper by Lippich (*Wien. Ber.* 85, 9 Feb. 1882), which has lately come to my notice.

The arrangements to be adopted depend for their justification upon the physiological law of the perception of differences of brightness. If dE denote the difference of sensations, corresponding to two degrees of brightness, H and $H + dH$, we have*

$$dE = A \frac{dH}{H + H_0},$$

in which H_0 is a certain constant brightness, supposed to depend chiefly upon the proper or internal light of the eye, but to which may be added the effect of light diffused by imperfect translucency of the optical apparatus. If dE denote the smallest perceptible difference, the value of dE/A is in favourable circumstances as low as $\frac{1}{50}$ or $\frac{1}{100}$, which means that with a sufficient total brightness differences of this amount may be apparent to observation.

Let us now consider the values of dE corresponding to different methods of procedure. If the analysing Nicol be adjusted for extinction of the original light, the comparison is between the brightness which cannot be got rid of (H_0) and ($H_0 + h\theta^2$)†. Near the limit of discrimination, to which case we may confine our attention, $h\theta^2$ is small relatively to H_0, and thus we may take

$$dE = A \frac{h\theta^2}{H_0}.$$

The procedure just considered is that which would naturally be adopted to render evident a small quantity of light of given amount, viz., to isolate it and compare it with the best attainable darkness. But in the present problem the circumstances are peculiar in that we are able to deal with phases. Now if we regard the amplitude (α) of the feeble light as given, putting $\alpha^2 = h\theta^2$, we may produce more effect from it by combining it with other light in the same phase of amplitude (β) than by isolating it. The comparison is then between brightnesses $(\alpha + \beta)^2$ and β^2, or as α is very small, between $\beta^2 + 2\alpha\beta$ and β^2. Thus

$$dE = A \frac{2\alpha\beta}{\beta^2 + \beta_0^2},$$

in which β_0^2 is written H_0.

The light of amplitude β is obtained in the simplest possible manner by merely rotating the analysing Nicol through a small angle, and the only question is how to exhibit the comparison light, which shall not be affected when β is changed to $(\beta + \alpha)$. For this purpose we may divide the field of view into two halves with an oblique mirror in which is seen

* Helmholtz: *Physiologische Optik*, § 27.
† We may imagine the presentation of the two brightnesses to be consecutive, or more favourably that both are seen at once, half the field of view being occupied by a black body seen after reflection in an oblique mirror, whose edge forms the dividing line.

380 ON THE CONSTANT OF MAGNETIC [118

by reflection a feeble light, of the same colour and coming ultimately from the same source.

It is possible that an instrument upon this principle might be made to work satisfactorily*, but the half-shade polarimeters of Jellet and Laurent seem to be in most respects preferable. In them the comparison is between $(\beta + \alpha)^2$ and $(\beta - \alpha)^2$, so that

$$dE = A\,\frac{4\alpha\beta}{\beta^2 + \beta_0{}^2},$$

representing twice as great a sensibility. The only thing to be said upon the other side is that the division line in these instruments can hardly be made as invisible as the sharp edge of a mirror may be.

In these formulæ β may be chosen at pleasure by suitable adjustments of the polarising arrangements. In order to get the best result, dE must be made a maximum by variation of β, α and β_0 being treated as constants. The maximum occurs when $\beta = \beta_0$, and its value in the last case is

$$dE = A\,\frac{2\alpha}{\beta_0}.$$

Taking $dE/A = \frac{1}{25}$, which is probably about as small as can be expected in practice, we have for the least perceptible value of α

$$\alpha = \frac{1}{50}\,\beta_0;$$

whereas without the half-shade arrangement, and with a Nicol simply set to extinction of the original light,

$$\frac{h\theta^2}{H_0} = \frac{\alpha^2}{\beta_0{}^2} = \frac{1}{25},$$

so that

$$\alpha = \frac{1}{5}\,\beta_0.$$

According to these numbers the half-shade arrangement would have a tenfold superiority, a result not fully borne out in practice. In explanation of this it is important to notice that the procedure in the absence of a half-shade arrangement would in reality be very different from what we have tacitly supposed. The experienced operator, in setting a Nicol to the position of maximum extinction, does not judge merely by the degree of darkness attained in the final position, but displacing the analyser alternately in opposite directions, he estimates the position which lies midway between those which give similar revivals of light on the two sides; or, endeavouring to retain in his memory a certain degree of

---

* Readings would of course be taken in both the positions (one on either side of extinction) which give a match with the comparison light.

brightness, he may take actual readings on both sides, of which the mean will correspond to the desired position. In this way the fundamental advantage of the half-shade method is in a sense attained, the only difference being that the brightnesses to be compared are seen consecutively after a short interval of time, instead of almost simultaneously; and even this difference becomes less important when the line dividing the field of view of the half-shade apparatus is so coarse that it cannot be rendered invisible.

The carrying out of this method is facilitated by a device which is worthy of trial. The Nicol may be mounted loosely, so as to be capable of turning through a small angle (2 or 3 degrees) between two stops. These stops are rigidly attached to a rotating piece carrying the vernier, and it is to the position of this piece (and not that of the Nicol) to which the readings relate. In taking an observation the piece is turned until the degree of brightness is unaltered, when the Nicol is put over from the one stop to the other. It is probable that under these advantageous conditions more favourable results than hitherto would be obtained with an undivided field of view.

In the application of the polarimeter, with which the present paper is mainly concerned, the free play of the Nicol is advantageously replaced by an equivalent rocking of the plane of polarisation itself through a small angle on either side of its normal position, produced by the action of an auxiliary electric current, embracing the experimental tube a moderate number of times, and reversed at pleasure by a suitable key under the hand of the observer.

In these discussions it has been convenient to take as a basis the fractional difference of brightnesses which can be recognised on simple presentation to the eye*, but it must be remembered that if suitable precautions are taken to avoid asymmetry, there is no theoretical limit of final accuracy. Thus in ordinary photometry with a divided field (*e.g.*, Bunsen's grease-spot photometer), the match must not be approached from one side only. By combining a large number of observations in which the match is approached as much from one side as from the other, a degree of accuracy may be practically attained far beyond that corresponding to the difference of brightness which can be directly recognised by the eye. It is not necessary actually to take readings on the two sides, though it is sometimes desirable to do so; the essential point is to secure symmetry. Time may be saved by the plan of providing means for instantaneous displacements of given amount on either side, as was

---

* August, 1885. I find that the sensitiveness of the eye to small differences of brightness is subject to very rapid fatigue. Even a few seconds' gazing is often enough to obliterate a distinction quite apparent at first, and appreciable again after a little repose. This defect is a great obstacle to the further improvement of photometric methods.

done in the experiments of the present paper by the auxiliary reversible current.

In practical applications of the polarimeter we have almost always to determine, not so much a particular plane of polarisation as the *rotation* of this plane, due to electromagnetic action, to the substitution of syrop for water, etc., and it appears that the measurement of this angle must be affected with a possible error, double of the error possible in the determination of a single plane. M. Becquerel, indeed, in his interesting memoir upon the rotation in bisulphide of carbon under the terrestrial magnetic force*, describes a procedure by which, as he considers, the error may be reduced. By the introduction of a half-wave plate, adjusted so that its principal section coincides nearly with the plane of first polarisation, the angle of rotation is, as it were, reflected by the former plane, and the difference of readings taken with and without the plate is the double of the real angle of rotation. If ϵ be the greatest angular error possible in determining a single plane, M. Becquerel shows that the error in setting the plate cannot exceed ϵ, from which he argues that the whole error possible in determining the double angle of rotation is only 3ϵ, or $\frac{3}{2}\epsilon$ upon the single angle. It appears, however, that the error of adjustment of the half-wave plate enters *doubly* into the result, so that the whole error possible in determining the double angle of rotation rises to 4ϵ, and the use of the half-wave plate gives no advantage.

One other point may be considered in conclusion. In determinations of rotation by magnetic force, the effect to be measured may be multiplied (as Faraday showed), by causing the light to be reflected backwards and forwards at the ends of the tube. Against this augmentation of the angle of rotation we must set the loss in the section of the beam, and the waste of light in reflection and by absorption. Putting out of sight for the moment the alteration in the section of the beam, we may easily determine the most advantageous number of passages as dependent upon magnitude of rotation and intensity of light. If r be the factor by which the original intensity must be multiplied, in order to express the intensity after a single passage and reflection, r^n will express the intensity after n such passages and reflections. The accuracy of the determination will thus be proportional to $nr^{\frac{1}{2}n}$, which is a maximum when $r = e^{-2/n}$. The values of r corresponding to n equal to 1, 3, 5, 7, ..., are ·135, ·514, ·670, ·752, ..., so that 3 or 5 passages will usually give the best result.

The argument in favour of a moderate use only of the principle of reflection is strengthened when we take into account the diminution in the section of the beam. The already contracted aperture is seen at a

---

* *Ann. d. Chim.* t. cci. p. 323 ; 1882.

greater distance (proportional to n), so that the apparent magnitude of the field of view is rapidly narrowed. Under these circumstances the comparisons cannot be made with the usual accuracy. If we have recourse to a telescope we can indeed restore the apparent magnitude, but (usually) only at expense of the illumination, since the aperture of the telescope is limited. If the available aperture do not exceed $\frac{1}{4}$ inch, *any* degree of magnification involves a loss of brightness. The importance of these considerations depends upon the length and diameter of the tube; but the tendency of the discussion is to show that more than five passages can rarely be desirable, and that in many cases three passages ought to be preferred to five. If there is any exception, it will be when powerful white light (as from the sun) is available, or when it is possible by use of a larger number of passages to bring the whole rotation up to $90°$ or $180°$, in which cases, as has already been noticed, the angle may be determined with peculiar advantage.

POSTSCRIPT.

(October, 1885.)

An important paper* has recently been communicated to the French Academy† by M. Becquerel, in which he abandons his former result (§ 4), obtained with the aid of terrestrial magnetic force, in favour of a number agreeing more nearly with that given by Gordon and myself. In the new experiments a long column of CS_2 was employed, encompassed by a spiral conveying a current, the effect of which is shown to depend upon the magnitude of the current and upon the number of turns, in approximate independence of other circumstances. M. Becquerel speaks of this method as new, but it is in reality that employed by Gordon in 1877‡. Most of the complication in Gordon's memoir relates to the determination of the current, and especially to the circumstance that the number of turns in the spiral was not ascertained (as it should have been) during construction, *but subsequently by electrical processes.* When the number of turns and the current are known, there is no difference between the procedure of Gordon and Becquerel and that of the present memoir.

There is a pretty close resemblance between M. Becquerel's recent work and mine. In both a soda flame is used as the source of light, and in both the number of windings on the helices is ascertained during construction. In the current determinations, M. Becquerel used a galvanometer as an intermediate standard, while I employed for the same purpose a Clark's cell, the ultimate standard being a silver voltameter

---

* *Ann. d. Chim.* Oct. 1885.   † *C. R.*, June 2, 1885.
‡ See his equation (24), p. 15.

(and in my case a current-weighing apparatus). Inasmuch as M. Becquerel uses the same number as that which I obtained for the electro-chemical equivalent of silver, there should be no difference between us in the estimation of currents.

In M. Becquerel's experiments the temperature of the CS_2 was usually about 0° C., and he reduces his results to that standard temperature. He regards Bichat's formula as confirmed by his observations. According to this my result for 18° would become ·04302′; whereas M. Becquerel obtains ·04341′, nearly 1 per cent. higher. I am at a loss to understand the cause of this discrepancy. M. Becquerel estimates that his result should be correct to $\frac{1}{800}$, about the same degree of accuracy which I also had hoped to have attained. So far as I can judge, I should consider that in respect of current measurement the advantage lay with me, but that on the optical side M. Becquerel's arrangements were probably superior.

M. Becquerel repeats his proposal* to found upon his value of the constant a method for current measurement. I had considered this question at (I believe) an earlier date; and the less sanguine view expressed in the following paragraph seems to be justified by the discrepancies between the results of various observers at various times as to the value of the constant in bisulphide of carbon :—

"Another method, available with the strong currents which are now common, depends upon Faraday's discovery of the rotation of the plane of polarisation by magnetic force. Gordon found 15°† as the rotation due to the reversal of a current of 4 ampères circulating about 1000 times round a column of bisulphide of carbon. With heavy glass, which is more convenient in ordinary use, the rotation is somewhat greater. With a coil of 100 windings we should obtain 15° with a current of 40 ampères; and this rotation may easily be tripled by causing the light to traverse the column three times, or what is desirable with so strong a current, the thickness of the wire may be increased and the number of windings reduced. With the best optical arrangements the rotation can be determined to one or two minutes, but in an instrument intended for practical use such a degree of delicacy is not available. One difficulty arises from the depolarising properties of most specimens of heavy glass. Arrangements are in progress for a redetermination of the rotation in bisulphide of carbon ‡."

* C. R. t. xcviii. p. 1253; 1884.

† Jan. 1884. In a note recently communicated to the Royal Society (Proceedings, Nov. 15, 1883), Mr Gordon points out that, owing to an error in reduction, the number given by him for the value of Verdet's constant is twice as great as it should be. The rotations above mentioned must therefore be halved, a correction which diminishes materially the prospect of constructing a useful instrument upon this principle.

‡ From the Proceedings of the Cambridge Philosophical Society for Nov. 26, 1883. See also Nature, Dec. 13, 1883.

119.

OPTICS.

[Encyclopædia Britannica, XVII. 1884.]

Optics, Geometrical. The subject of optics is so extensive that some subdivision of it is convenient if not necessary. Under the head of Light will be found a general sketch accompanied by certain developments. The wave theory and those branches of the subject which are best expounded in connexion with it are reserved for treatment in a later volume. The object of the present paper is to give some account of what is generally called geometrical optics,—a theoretical structure based upon the laws of reflexion and refraction. We shall, however, find it advisable not to exclude altogether the conceptions of the wave theory, for on certain most important and practical questions no conclusions can be drawn without the use of facts which are scarcely otherwise interpretable. Indeed it is not to be denied that the too rigid separation of optics into geometrical and physical has done a good deal of harm, much that is essential to a proper comprehension of the subject having fallen between the two stools.

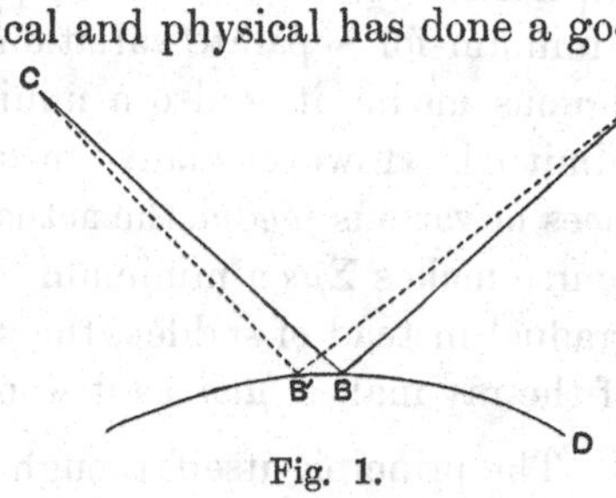

Systems of Rays in General.—In the investigation of this subject a few preliminary propositions will be useful.

Fig. 1.

If a ray *AB* (fig. 1) travelling in a homogeneous medium suffer reflexion at a plane or curved surface *BD*, the total path between any two points *A*, *C* on the ray is a minimum, *i.e.* *AB + BC* is less along the actual path than it would be if the point *B* were slightly varied.

For a variation of *B* in a direction perpendicular to the plane of reflexion (that of the diagram) the truth of this statement is at once evident. For a small variation *BB'* in the plane of reflexion we see that the difference

R. II.

25

$AB' - AB$ is equal to the projection of BB' upon AB, and that the difference $CB - CB'$ is equal to the projection of BB' upon BC. These projections are equal, since by the law of reflexion AB and BC are equally inclined to BB', and thus the variation of the total path, $AB' + B'C - (AB + BC)$, vanishes.

A corresponding proposition holds good in the case of refraction. If we multiply the distances travelled in the first and second media respectively by the refractive indices appropriate to the media, the quantity so obtained is a minimum for the actual path of the ray from any point to any other. It is sufficient to consider the case of a variation of the point of passage in the plane of refraction.

In the first medium (fig. 2) $\mu AB' - \mu AB = \mu BB' \cos ABD$, and in the second medium $\mu' CB - \mu' CB' = \mu' BB' \cos CBD$. The whole variation of the quantity in question is therefore

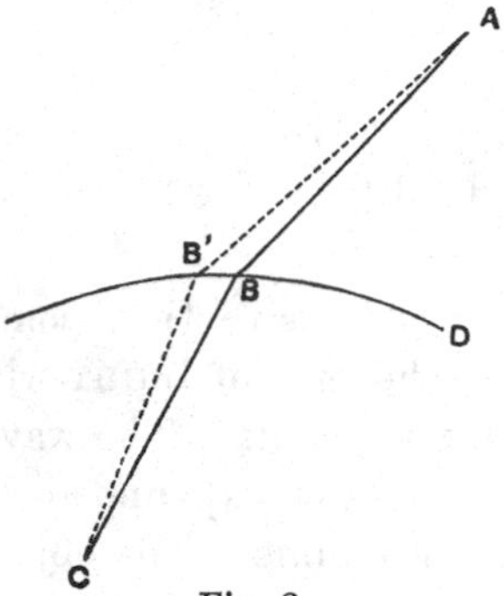

$$BB' (\mu \cos ABD - \mu' \cos CBD).$$

Now by the law of refraction the sines of the angles of incidence and refraction are in the ratio $\mu' : \mu$, and accordingly

$$\mu \cos ABD - \mu' \cos CBD = 0.$$

In whichever direction, therefore, the point of transition be varied, the variation of the quantity

Fig. 2.

under consideration is zero. It is evident that the second proposition includes the first, since in the case of reflexion the two media are the same.

The principle of the superposition of variations now allows us to make an important extension. If the quantity, which we may denote by $\Sigma\mu s$, be a minimum for separate variations of all the points of passage between contiguous media, it is also a minimum even when simultaneous variations are admitted. However many times a ray may be reflected or refracted at the surfaces of various media, the actual path of the ray between any two points of its course makes $\Sigma\mu s$ a minimum. Even if the variations of refractive index be gradual instead of sudden, the same principle holds good, and the actual path of the ray makes $\int\mu ds$, as it would now be written, a minimum.

The principle itself, though here deduced from the laws of reflexion and refraction, is an immediate consequence of the fundamental suppositions of the wave-theory of light, and if we are prepared to adopt this point of view we may conversely deduce the laws of reflexion and refraction from the principle. The refractive index μ is inversely proportional to the velocity of propagation, and the principle simply asserts that in passing from any point to any other the light follows the shortest course, that is, the course of earliest arrival.

If two points be such that rays issuing from one of them, and ranging through a finite angle, converge to the other after any number of reflexions and refractions, the value of $\Sigma\mu s$ from one focus to the other must be the same for all the rays.

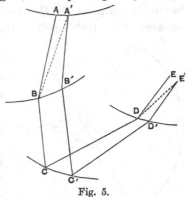

Fig. 3.

Thus, in order to condense rays issuing from one point S upon a second point H by a single reflexion (fig. 3), the reflecting surface must be such that $SP + HP =$ const., $i.e.$ must be an ellipsoid of revolution with S and H foci.

Again, if it be required to effect the same operation by a single refraction at the surface of a medium whose index is μ, we see that the surface (fig. 4) must be such that

$$SP + \mu HP = \text{const.}$$

Fig. 4.

If S be at an infinite distance, $i.e.$ if the incident rays be parallel, the surface is an ellipsoid of revolution with H for focus, and of eccentricity $\mu^{-1}\,(\mu > 1)$.

Another important proposition, obvious from the point of view of the wave-theory, but here requiring an independent proof, was enunciated by Malus. It asserts that a system of rays, emanating originally from a point, retains always the property of being normal to a surface, whatever reflexions or refractions it may undergo in traversing singly-refracting media.

Suppose that $ABCDE$, $A'B'C'D'E'$ ... (fig. 5) are rays originally normal to a surface AA', which undergo reflexions or refractions at BB', CC', &c. On every ray take points E, E', &c., such that $\Sigma\mu s$ is the same along the courses AE, $A'E'$, &c. We shall prove that the rays in the final medium are normal to the surface EE'. For by hypothesis $\Sigma\mu s$ along $ABCDE$ is the same as along $A'B'C'D'E'$, and, by the property proved above to attach to every ray, $\Sigma\mu s$ reckoned along the neighbouring hypothetical course $A'BCDE'$ is the same as along $A'B'C'D'E'$. Hence $\Sigma\mu s$ along $A'BCDE'$ is the same as along $ABCDE$, or (on subtraction of the common part) the same along $A'B$, DE' as along AB, DE. But since AB is perpendicular to AA', the value along $A'B$ is the same as along AB, and therefore the value along DE' is the same as along DE; or, since the index is the same, $DE = DE'$, that is, EE' is perpendicular to DE. The same may be proved for every point E' which lies infinitely near E, and thus the surface EE' is perpendicular to the ray DE, and by similar reasoning to every other ray of the system. It follows that reflexions and refractions cannot deprive

25—2

a system of rays of the property of being normal to a surface, and it is evident that a system issuing from a point enjoys the property initially.

Consecutive rays do not in general intersect one another; but if we select rays which cut the orthogonal surface along a *line of curvature*, we meet with ultimate intersection, the locus of points thus determined being a *caustic* curve to which the rays are tangents. Other lines of curvature of the same set give rise to similar caustic curves, and the locus of these curves is a caustic surface to which every ray of the system is a tangent. By considering the other set of lines of curvature we obtain a second caustic surface. Thus every ray of the system touches two caustic surfaces.

In the important case in which the system of rays is symmetrical about an axis, the orthogonal surface is one of revolution. The first set of lines of curvature coincide with meridians. The rays corresponding to any one meridian meet in a caustic curve, and the surface which would be traced out by causing this to revolve about the axis is the first caustic surface. The second set of lines of curvature are the circles of latitude perpendicular to the meridians. The rays which are normal along one of these circles form a cone of revolution, and meet in a point situated on the axis of symmetry. The second caustic surface of the general theorem is therefore here represented by a portion of the axis.

The character of a limited symmetrical pencil of rays is illustrated in fig. 6, in which *BAC* is the orthogonal surface, and *HFI* the caustic curve having a cusp at *F*, the so-called geometrical focus. The distance *FD* between *F* and the point where the extreme ray *BHDG* cuts the axis

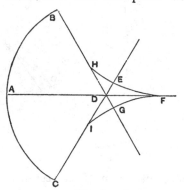

is called the longitudinal aberration. On account of the symmetry *FD* is an even function of *AB*. If the pencil be small, we may in general consider *FD* to be proportional to AB^2, although in particular cases the aberration may vanish to this order of approximation. Let us examine the nature of the sections at various points as they may be exhibited by holding a piece of paper in the solar rays converging from a common burning-glass of large aperture. In moving the paper towards the focus nothing special is observed

Fig. 6.

up to the position *HI*, where the caustic surface is first reached. A bright ring is there formed at the margin of the illuminated area, and this gradually contracts. At *D* the second caustic surface *DF* is reached, and a bright spot develops itself at the centre. A little farther back, at *EG*, the area of the illuminated patch is a minimum, and its boundary is called the least circle of aberration. Farther back still the outer boundary corresponding to the extreme rays begins to enlarge, although the circle of intersection with the caustic surface continues to contract. Beyond F the caustic surfaces are passed, and no part of the area is specially illuminated.

As a simple example of a symmetrical system let us take the case of parallel rays QR, OA (fig. 7), incident upon a spherical minor AR. By the law of reflexion the angle ORq = angle ORQ = angle qOR. Hence the triangle RqO is isosceles, and if we denote the radius of the surface OA by r, and the angle AOR by α, we have

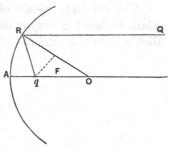

Fig 7.

$$Oq = \frac{r}{2 \cos \alpha}.$$

If F be the geometrical focus, $OF = AF = \frac{1}{2}r$. If α be a small angle, the longitudinal aberration $Fq = Oq - OF = \frac{1}{2}r(\sec \alpha - 1) = \frac{1}{4}\alpha^2 r$, in which $AR = r\alpha$.

Focal Lines.—In the general case of a small pencil of rays there is no one point which can be called the geometrical focus. Consider the corresponding small area of the orthogonal surface and its two sets of lines of curvature. Of all the rays which are contiguous to the central ray there are only two which intersect it, and these will in general intersect it at different points. These points may be regarded as foci, but it is in a less perfect sense than in the case of symmetrical pencils. Even if we limit ourselves to rays in one of the principal planes, the aberration is in general a quantity of the first order in the angle of the pencil, and not, as before, a quantity of the second order. If, however, we neglect this aberration and group the rays in succession according to the two sets of lines of curvature, we see that the pencil of rays passes through two focal lines perpendicular to one another and to the central ray, and situated at the centres of curvature of the orthogonal surface. At some intermediate place the section of the pencil is circular.

It happens not unfrequently that the pencil under consideration forms part of a symmetrical system, but is limited in such a manner that the central ray of the pencil does not coincide with the axis of the system. The plane of the meridian of the orthogonal surface is called the primary plane, and the corresponding focus, situated on the caustic surface, the primary focus. The secondary focus is on the axis of symmetry through which every ray passes. The distinction of primary and secondary is also employed when the system, though not of revolution, is symmetrical with respect to a plane passing through the central ray, this plane being considered primary.

The formation of focal lines is well shown experimentally by a plano-convex lens of plate-glass held at an obliquity of 20° or 30° in the path of the nearly parallel rays, which diverge from a small image of the sun formed by a lens of short focus. The convex face of the lens is to be turned towards the parallel rays, and a piece of red glass may be interposed to mitigate the effects of chromatic dispersion.

To find the position of the focal lines of a small pencil incident obliquely upon a plane refracting surface of index μ.

The complete system of rays issuing from Q (fig. 8) and refracted at the plane surface CA is symmetrical about the line QC drawn through Q perpendicularly to the surface. Hence, if QA be the central ray of the pencil, the

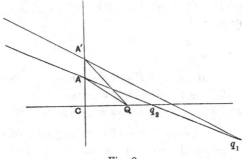

Fig. 8.

secondary focus q_2 lies at the intersection of the refracted ray with the axis. If ϕ be the angle of incidence, ϕ' of refraction, $AQ = u$, $Aq_2 = v_2$, then

$$\frac{v_2}{u} = \frac{\sin \phi}{\sin \phi'} = \mu. \quad \dots\dots\dots\dots\dots\dots\dots\dots(1)$$

To find the position of the primary focus q_1, let QA' be a neighbouring ray in the primary plane (that of the paper) with angles of incidence and refraction $\phi + \delta\phi$ and $\phi' + \delta\phi'$, $Aq_1 = v_1$. We have

$$AA' \cos \phi = u\,\delta\phi, \qquad AA' \cos \phi' = v_1\,\delta\phi' ;$$

moreover, by the law of refraction,

$$\cos \phi\, \delta\phi = \mu \cos \phi'\,\delta\phi' ;$$

and thus

$$\frac{v_1}{u} = \frac{\mu \cos^2 \phi'}{\cos^2 \phi}. \quad \dots\dots\dots\dots\dots\dots\dots(2)$$

If the refracting surface be curved, with curvature $1/r$, we get by similar reasoning

$$\frac{\mu \cos^2 \phi'}{v_1} - \frac{\cos^2 \phi}{u} = \frac{\mu \cos \phi' - \cos \phi}{r} ; \quad \dots\dots\dots\dots(3)$$

$$\frac{\mu}{v_2} - \frac{1}{u} = \frac{\mu \cos \phi' - \cos \phi}{r} ; \quad \dots\dots\dots\dots(4)$$

in which (1) and (2) are of course included as particular cases.

When the incidence is direct, $\cos\phi' = 1$, $\cos\phi = 1$, and $v_2 = v_1$. In this case (3) and (4) become

$$\frac{\mu}{v} - \frac{1}{u} = \frac{\mu - 1}{r}. \quad\dots\dots\dots\dots\dots\dots(5)$$

To find the positions of the focal lines of a pencil refracted obliquely through a plate of thickness t *and index* μ.

If ϕ be the angle of incidence (and emergence), ϕ' the angle of refraction of the ray $QAST$ (fig. 9), $Sq_1 = v_1$, $Sq_2 = v_2$, $AQ = u$, we get by successive applications of (1) and (2)

$$v_1 = u + \frac{t\cos^2\phi}{\mu\cos^2\phi'}; \quad\dots\dots\dots\dots\dots\dots(6)$$

$$v_2 = u + \frac{t}{\mu\cos\phi'}. \quad\dots\dots\dots\dots\dots\dots(7)$$

If the incidence be direct,

$$v_1 = v_2 = u + t/\mu. \quad\dots\dots\dots\dots\dots\dots(8)$$

Thus, if we interpose a plate between the eye and an object, the effect is to bring the object apparently nearer by the amount

$$\frac{(\mu - 1)t}{\mu}. \quad\dots\dots\dots\dots\dots\dots(9)$$

On this result is founded a method for determining the refractive index of materials in the form of plates. A set of cross wires is observed through a magnifying glass. On interposition of the plate the glass must be drawn back through a distance given by (9) in order to recover the focus. If we measure this distance and the thickness of the plate, we are in a position to determine the refractive index.

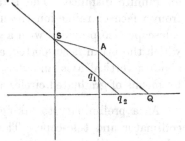

Fig. 9.

Prism.—By a prism is meant in optics a portion of transparent material limited by two plane faces which meet at a finite angle in a straight line called the edge of the prism. A section perpendicular to the edge is called a principal section.

Parallel rays, refracted successively at the two faces, emerge from the prism as a system of parallel rays. The angle through which the rays are bent is called the deviation.

The deviation depends upon the angles of incidence and emergence; but, since the course of a ray may always be reversed, the deviation is necessarily a symmetrical function of these angles. The deviation is consequently a maximum or a minimum when a ray within the prism is equally inclined to the two faces, in which case the angles of incidence and emergence are equal. It is in fact a minimum; and this position of the prism is described as the position of minimum deviation, and is usually adopted for the purposes of measurement.

The relation between the minimum deviation D, the angle of the prism i, and the refractive index μ is readily found. In fig. 10 the internal angles ϕ', ψ' are each equal to $\frac{1}{2}i$. The external angles ϕ, ψ are also equal, and are connected with ϕ' by the law of refraction $\sin\phi = \mu\sin\phi'$. The deviation is $2(\phi - \phi')$. Hence

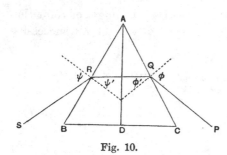

Fig. 10.

$$\mu = \frac{\sin\phi}{\sin\phi'} = \frac{\sin\frac{1}{2}(D+i)}{\sin\frac{1}{2}i};$$

and this is the formula by which the refractive index is usually determined, since both D and i can be measured with great precision.

The instrument now usually employed for this purpose is called a goniometer or spectrometer. Parallel rays are provided by a collimator, consisting of an object-glass and telescope-tube, by means of which the subject of examination, either a fine slit or a set of cross wires, is seen as if it were at an infinite distance. The parallel rays from the collimator, after reflexion from a face or refraction through the body of the prism, are received by a telescope also provided with a set of cross wires at its focus. The table upon which the prism is supported, as well as the telescope, are capable of rotation about a vertical axis, and the position of either can be read off at any time by means of graduated circles and verniers.

As a preliminary to taking an observation it is necessary to focus the collimator and telescope. The first step is to adjust the eye-lens of the telescope until the cross wires are seen distinctly and without effort. The proper position depends, of course, upon the eyesight of the observer, and is variable within certain limits in virtue of the power of accommodation. It is usually best to draw out the lens nearly to the maximum distance consistent with distinct vision. The telescope is now turned to a distant object and focused by a common motion of the cross wires and eye-lens, until both the object and the cross wires are seen distinctly at the same time. The final test of the adjustment is the absence of a relative motion when the eye is moved sideways across the eye-piece. The collimator is now brought opposite to the telescope and adjusted until the cross wires in its focus behave precisely like the distant object.

To measure the angle of a prism it may be placed with its edge vertical upon the table, in a symmetrical position with respect to the collimator (fig. 11). The telescope is then successively brought into such positions that the cross wires of the telescope coincide with the cross wires of the collimator when seen by reflexion in the two faces. The difference of the readings is twice the angle of the prism.

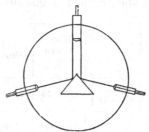

Fig. 11.

Another method is also often employed in which the telescope is held fixed and the prism is rotated. The angle between the two positions of the table found by use in succession of the two faces is the supplement of the angle of the prism.

Suppose next that we wish to determine D for the given prism and for sodium light. The slit of the collimator is backed by a sodium flame, the telescope is adjusted for direct vision of the slit, and the reading taken. The prism is now placed upon the table, and rotated until the deviation of the light from its original direction when seen through the prism is a minimum. The difference of the readings for the two positions of the telescope is the value of D. The angle to be observed may be doubled by using the deviation in both directions. In this case no direct reading in the absence of the prism is required.

The following table of indices of refraction is taken from Watt's *Dictionary of Chemistry*, article "Light."

Name of Substance	Index of Refraction	Name of Substance	Index of Refraction
Chromate of lead	2·50 to 2·97	Phosphoric acid	1·534
Diamond	2·47 to 2·75	Sulphate of copper	1·531 to 1·552
Phosphorus	2·224	Canada balsam	1·532
Glass of antimony	2·216	Citric acid	1·527
Sulphur (native)	2·115	Crown glass	1·525 to 1·534
Zircon	1·95	Nitre	1·514
Borate of lead	1·866	Plate glass	1·514 to 1·542
Carbonate of lead	1·81 to 2·08	Spermaceti	1·503
Ruby	1·779	Crown glass	1·500
Felspar	1·764	Sulphate of potassium	1·500
Tourmalin	1·668	Ferrous sulphate	1·494
Topaz (colourless)	1·610	Tallow ; wax	1·492
Beryl	1·598	Sulphate of magnesium	1·488
Tortoise-shell	1·591	Iceland spar	1·654
Emerald	1·585	Obsidian	1·488
Flint-glass	1·57 to 1·58	Gum	1·476
Rock-crystal	1·547	Borax	1·475
Rock-salt	1·545	Alum	1·457
Apophyllite	1·543	Fluorspar	1·436
Colophony	1·543	Ice	1·310
Sugar	1·535	Tabasheer	1·1115

A selection from some results given by Hopkinson[*], relating to Chance's glasses, may be useful to those engaged in the designing of optical instruments.

* *Proc. Roy. Soc.* June 1877.

D is the more refrangible of the pair of sodium lines; b is the most refrangible of the group of magnesium lines; (G) is the hydrogen line near G.

	Hard Crown a	Soft Crown	Extra Light Flint	Light Flint	Dense Flint	Extra Dense Flint	Double Extra Dense Flint
Specific Gravity	2·48575	2·55035	2·86636	3·20609	3·65865	3·88947	4·42162
B	1·513625	1·510916	1·536450	1·568558	1·615701	1·642874	1·701060
C	1·514568	1·511904	1·537673	1·570011	1·617484	1·644866	1·703478
D	1·517114	1·514591	1·541011	1·574015	1·622414	1·650388	1·710201
E	1·520331	1·518010	1·545306	1·579223	1·628895	1·657653	1·719114
b	1·520967	1·518686	1·546166	1·580271	1·630204	1·659122	1·720924
F	1·523139	1·520996	1·549121	1·583886	1·634748	1·664226	1·727237
(G)	1·527994	1·526207	1·555863	1·592190	1·645267	1·676111	1·742063
G	1·528353	1·526595	1·556372	1·592824	1·646068	1·677019	1·743204
h	1·530902	1·529359	1·560010	1·597332	1·651840	1·683577	1·751464
H_1	1·532792	1·531416	1·562760	1·600727	1·656219	1·688569	1·757785

To determine the index of refraction of a liquid it must of course be placed in a hollow prism, whose faces are formed of some transparent material, usually of glass. The following results of Dale and Gladstone show the influence of temperature upon the refracting power of some important liquids. They relate to the soda flame, or the line D in the solar spectrum.

Temperature	Bisulphide of Carbon	Water	Ether	Alcohol Absolute
0°	1·6442	1·3330	..	..
10°	1·6346	1·3327	1·3592	1·3658
20°	1·6261	1·3320	1·3545	1·3615
30°	1·6182	1·3309	1·3495	1·3578
40°	1·6103	1·3297	..	1·3536
50°	..	1·3280	..	1·3491
60°	..	1·3259	..	1·3437

Refractive Indices of Bisulphide of Carbon for the several Fixed Lines.

Temperature	A	B	D	E	F	G
11°	1·6142	1·6207	1·6333	1·6465	1·6584	1·6836
36°·5	1·5945	1·6004	1·6120	1·6248	1·6362	1·6600
Difference ...	0·0197	0·0203	0·0213	0·0217	0·0222	0·0236

The rapid alteration of refractive power with temperature is a serious obstacle to the use of bisulphide of carbon prisms for exact purposes. Not only does the dispersive power vary from day to day, but inequalities of temperature in the various parts of the liquid at any one moment disturb the optical uniformity, and are thus the cause of bad definition. A difference of 1° Cent. alters the index about as much as a change in the light from one of the two D lines to the other, so that a variation of one degree within the prism may be expected to prevent the satisfactory resolution of this double line.

Excellent results have recently been obtained by Liveing with prisms containing aqueous solution of iodide of potassium and mercury. This liquid can be brought up to a density as high as three times that of water, and gives a powerful dispersion. Some difficulty has, however, been experienced in finding a suitable cement for the faces. Bisulphide of carbon prisms are usually cemented with a mixture of glue and treacle.

For many purposes the deviation of the light in passing through an ordinary prism is objectionable. In such cases recourse may be had to *direct vision* prisms (fig. 12), in which two materials, usually flint and crown, are so combined that the refractions are equal and opposite for a selected ray, while the dispersions are as unequal as may be. The direct vision prism may be contrasted with the achromatic lens (see LIGHT). In the first the object is to obtain dispersion without refraction, and in the second to obtain refraction without dispersion.

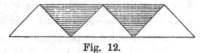

Fig. 12.

Compound prisms, composed of a flint between two crowns, are also made, in which the action of the crown is not carried so far as to destroy the deviation due to the flint. By this construction a larger angle is admissible for the more dispersive material, but it is not clear that any sufficient advantage is gained.

The principle of the compound prism is carried to its limit by employing media of *equal* refracting power for the part of the spectrum under examination. For this purpose bisulphide of carbon and flint glass may be chosen. With Chance's "dense flint" the refractions are the same, and the difference of dispersions is about as great as for "double-extra-dense flint" and crown. A dozen glass prisms of 90° may be cemented in a row on a strip of glass and immersed in a tube of bisulphide of carbon closed at the ends by glass plates. To vary the ray, which passes without deviation, ether may be mixed with the bisulphide*.

The formation of a pure spectrum, which may be either thrown upon a screen or photographic plate, or received at once by the eye armed with a

---

* See "Investigations in Optics," *Phil. Mag.* January 1880. [Vol. I. Art. 62.]

magnifier, has been explained under LIGHT. It sometimes happens that the object is not to see the spectrum itself, but to arrange a field of view uniformly illuminated with approximately homogeneous light. For this purpose the pure spectrum is received upon a screen perforated by a narrow slit parallel to the fixed lines. The light which passes this second slit (eye-slit) is approximately homogeneous. Suppose that it corresponds to the red of the spectrum. The eye, placed immediately behind the eye-slit, receives only red light, and, if focused upon the prism, sees a red field of view whose brightness is uniform if the light falling in different directions upon the original slit be uniform. To secure the fulfilment of the last condition we may use the light from an overcast sky, or that of the sun reflected from a large surface of white paper. If it be desired to work by artificial light, an Argand gas flame diffused by an opal globe will be found suitable. When the adjustments are correct the tint should be perfectly uniform. Any difference of colour on the two sides of the field of view is an indication that the screen is not in its proper place.

The most important application of this arrangement is to the investigation of compound colours, as carried out by Maxwell*. If light be admitted also through a second slit, displaced laterally from the position occupied by the first, a second spectrum overlapping the former will be thrown upon the screen, and a second kind of light will be admitted to the eye. In this way we may obtain a field of view lighted with a mixture of two or more spectrum colours, and we may control the relative proportions by varying the widths of the slits. For instance, by mixing almost any kind of red with any kind of green not inclining to blue we may match the brightest yellows, proving what so many find it difficult to believe, that yellow is a compound colour. In Maxwell's systematic examination of the spectrum, mixtures of three colours were used, and the proportions were adjusted so as to match the original white light incident upon the apparatus.

A similar arrangement (with one original slit) was employed by Helmholtz in his examination of a fundamental question raised by Brewster. The latter physicist maintained that there was abundant evidence to show that light of definite refrangibility was susceptible of further analysis by absorption, so that the colour of light (even of given brightness) could not be defined in terms of refrangibility or wave-length alone. The appearances which misled Brewster have since been explained as the effect of contrast or of insufficient purity. It is obvious that light, *e.g.*, from the red end of the spectrum, may be contaminated with light from some other part, say the yellow, in such proportion that though originally entirely preponderant it may fall into the second place under the action of a medium very much more transparent to yellow than to red. To obtain light of sufficient purity for

* "Theory of Compound Colours," *Phil. Trans.* 1860.

these experiments Helmholtz found it advisable to employ a double prismatic analysis. A spectrum is first thrown upon a screen perforated by a slit in the manner already described. The light which penetrates the second slit, already nearly pure, is caused to pass a second prism by the action of which any stray light is thrown aside. Using such doubly purified light, Helmholtz found the colour preserved, whatever absorbing agents were brought into play. Light of given refrangibility may produce a variety of *effects*, visual, thermal, or chemical, but (apart from polarization) it is not itself divisible into parts of different kinds. If yellow light produces the compound sensation of yellow, we are to seek the explanation in the constitution of the retina, and not in the divisibility of the light.

In all accurate work with the prism the use of a collimating lens to render the incident light parallel is a matter of necessity. If the incident rays diverge from a point at a finite distance, the pencil after emergence will be of a highly complicated character. There are, however, cases in which a collimator is dispensed with, and thus it is a problem of interest to find the foci of a thin pencil originally diverging from a point at a moderate distance. Even when a collimator is employed, the same problem presents itself whenever the focusing is imperfect. For the sake of simplicity the pencil is supposed to pass so near the edge of the prism that the length of path within the glass may be neglected in comparison with the distances of the foci.

We denote as usual the angles of incidence and emergence by ϕ, ψ, and the corresponding angles within the glass by ϕ', ψ'. The distance AQ from the edge of the prism to original source is denoted by u; the corresponding distances for the primary and secondary foci q_1, q_2 by v_1, v_2. By successive applications of the results already proved for a single refraction, we get

$$v_2 = u, \quad v_1 = \frac{\cos^2 \phi' \cos^2 \psi}{\cos^2 \phi \cos^2 \psi'} u; \quad \dots\dots\dots\dots(1)$$

so that

$$\frac{v_1}{v_2} = \frac{\cos^2 \phi' \cos^2 \psi}{\cos^2 \phi \cos^2 \psi'} = \frac{(\mu^2 - 1) \tan^2 \phi + \mu^2}{(\mu^2 - 1) \tan^2 \psi + \mu^2}. \quad \dots\dots\dots(2)$$

In order that the primary and secondary foci may coincide we must have $\psi = \phi$; that is to say, the ray must pass with minimum deviation. This is sometimes given as a reason why this arrangement should be adopted in spectroscopes; but in reality, since the slit is parallel to the edge of the prism, a slight elongation in this direction of the image of a point is without detriment to the definition. Hence a good image will be seen when the telescope is adjusted for the *primary* focus; and it is not clear that any improvement would arise from coincidence of the two foci, the question being in fact one of aberration. The position of minimum deviation is, however, usually adopted for the sake of definiteness, and sometimes it is convenient that the fixed lines and the extremities of the slit (or the markings produced by dust) should be in focus together.

The deviation is a symmetrical function of ϕ and ψ, and therefore is not altered by an interchange of these angles. The corresponding values of v are thus by (1) reciprocals, and their product is equal to u^2. This principle has been ingeniously applied by Schuster* to the adjustment for focus of the telescope and collimator of a spectroscope. The telescope is so placed that the deviation necessary to bring the object upon the cross wires is greater than the minimum, and the prism is adjusted in azimuth until the effect is produced, that position being chosen for which the angle of incidence is greater than the angle of emergence, so that v_1 is greater than u. After focusing the telescope the prism is turned into the other position which gives the same deviation, and the *collimator* is focused, the telescope remaining untouched. The prism is next brought back to the first position, and the telescope is again focused. A few repetitions of this operation, always focusing the telescope in the first position of the prism and the collimator in the second, will bring both into perfect adjustment for parallel rays.

Lenses.—The usual formula for the focal length of lenses (*Enc. Brit.* vol. XIV. p. 593),

$$\frac{1}{f} = (\mu - 1)\left(\frac{1}{r} - \frac{1}{s}\right), \quad \dots\dots\dots\dots\dots\dots(1)$$

ignores the fact that the various parts of a lens bounded by spherical surfaces have not the same focus, and is applicable in strictness only when the aperture is small. It is not necessary here to repeat the process by which (1) is usually obtained, but before passing on to give the formulæ for the aberration of lenses it may be well to exhibit the significance of (1) from the point of view of the wave-theory.

Taking the case of a convex lens of glass, let us suppose that parallel rays DA, EC, GB (fig. 13) fall upon the lens ACB, and are collected by it to a focus at F. The points D, E, G, equally distant from ACB, lie upon a front of the wave before it impinges upon the lens. The focus is a point at which the different parts of the wave

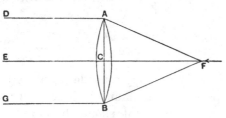

Fig. 13.

arrive at the same time, and that such a point can exist depends upon the fact that the propagation is slower in glass than in air. The ray ECF is retarded from having to pass through the thickness (t) of glass by the amount $(\mu - 1)t$. The ray DAF, which traverses only the extreme edge of the lens, is retarded merely on account of the crookedness of its path, and

* *Phil. Mag.* February 1879.

the amount of the retardation is measured by $AF - CF$. If F is a focus these retardations must be equal, or

$$AF - CF = (\mu - 1)t.$$

Now if y be the semi-aperture AC of the lens, and f be the focal length CF,

$$AF - CF = \sqrt{\{f^2 + y^2\}} - f = \tfrac{1}{2}y^2/f \quad \text{approximately,}$$

whence

$$f = \frac{y^2}{2(\mu - 1)t} \cdot \quad \dots\dots\dots\dots\dots\dots\dots\dots(2)$$

In the case of plate-glass $\mu - 1 = \tfrac{1}{2}$ nearly, and then the rule (2) may be thus stated: *the semi-aperture is a mean proportional between the focal length and the thickness.* The form (2) is in general the more significant, as well as the more practically useful, but we may of course express the thickness in terms of the curvatures and semi-aperture by means of

$$t = \tfrac{1}{2}y^2 \left(\frac{1}{r} - \frac{1}{s} \right).$$

In the preceding statement it has been supposed for simplicity that the lens comes to a sharp edge. If this be not the case we must take as the thickness of the lens the *difference* of the thicknesses at the centre and at the circumference. In this form the statement is applicable to concave lenses, and we see that the focal length is positive when the lens is thickest at the centre, but negative when the lens is thickest at the edge.

To determine practically the focal length of a convex lens we may proceed in several ways. A convenient plan is to set up a source of light Q (fig. 14) and a screen q at a distance exceeding four times the focal length, and to observe the two positions of the lens A, A' at which the source is in focus upon the screen.

Fig. 14.

These positions are symmetrically situated, and the distance between them is observed. Thus

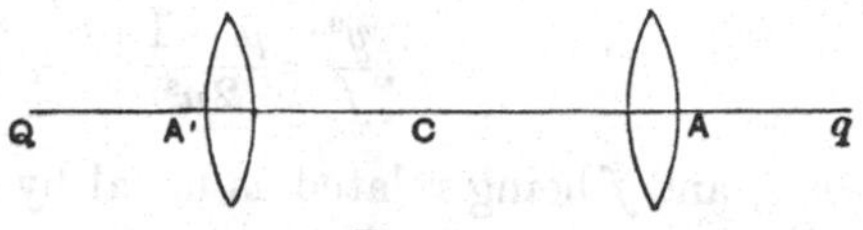

$$AQ = \tfrac{1}{2}Qq + \tfrac{1}{2}AA', \qquad\qquad Aq = \tfrac{1}{2}Qq - \tfrac{1}{2}AA'.$$

Now

$$\frac{1}{AQ} + \frac{1}{Aq} = \frac{1}{f},$$

so that

$$f = \frac{AQ \cdot Aq}{AQ + Aq} = \tfrac{1}{4}\frac{Qq^2 - AA'^2}{Qq}.$$

From the measured values of Qq and AA', f can be deduced.

If A and A' coincide, the conjugate foci Q and q are as close as possible to one another, and then $f = \tfrac{1}{4}Qq$.

The focal length on a concave lens may be found by combining it with a more powerful convex lens of known focus.

Aberration of Lenses.—The formula (1) determines the point at which a ray, originally parallel to the axis and at but a short distance from it, crosses the axis after passage through the lens. When, however, the ray considered is not quite close to the axis, the point thus determined varies with the distance y. In the case of a convex lens the ray DH (fig. 15), distant HC $(= y)$ from the axis, crosses it after refraction at a point F' which lies nearer

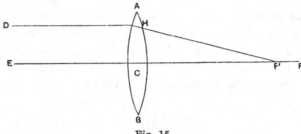

Fig. 15.

to the lens than the point F determined by (1), and corresponding to an infinitely small value of y. The distance $F'F$ is called the *longitudinal aberration* of the ray, and may be denoted by δf.

The calculation of the longitudinal aberration as dependent upon the refractive index (μ) and the anterior and posterior radii of the surfaces (r, s) is straight forward, but is scarcely of sufficient interest to be given at length in a work like the present. It is found that

$$- \delta f : \frac{y^2}{f} = \frac{\mu - 1}{2\mu^2} \left\{ \frac{1}{r^3} + \left(\frac{\mu + 1}{f} - \frac{1}{s} \right) \left(\frac{1}{f} - \frac{1}{s} \right)^2 \right\} f^3, \quad \ldots\ldots\ldots(3)$$

r, s, and f being related as usual by (1).

The first question which suggests itself is whether it is possible so to proportion r and s that the aberration may vanish. Writing for brevity R, S, F respectively for r^{-1}, s^{-1}, f^{-1}, and taking

$$G = \frac{\mu F}{\mu - 1}, \quad \text{so that} - S = (G/\mu) - R,$$

we get

$$\frac{1}{r^3} + \left(\frac{\mu + 1}{f} - \frac{1}{s} \right) \left(\frac{1}{f} - \frac{1}{s} \right)^2 = G \left\{ R^2 (\mu + 2) - RG (2\mu + 1) + \mu G^2 \right\}$$

$$= G \left\{ \sqrt{(\mu + 2)} \cdot R - \frac{(2\mu + 1) G}{2 \sqrt{(\mu + 2)}} \right\}^2 + \frac{\mu^3 F^3}{(\mu - 1)^3} \frac{4\mu - 1}{4 (\mu + 2)} \cdot \quad \ldots\ldots\ldots(4)$$

Since $\mu > 1$, both terms are of the same sign; and thus it appears that the *aberration can never vanish*, whatever may be the ratio of r to s. Under these circumstances all that we can do is to ascertain for what form of lens the aberration is a minimum, the focal length and aperture being given. For this purpose we must suppose that the first term of (4) vanishes, which gives

$$r = \frac{2(\mu+2)(\mu-1)}{\mu(2\mu+1)} f. \quad\ldots\ldots\ldots\ldots\ldots\ldots(5)$$

The corresponding value of $-s$ is

$$-s = \frac{2(\mu+2)(\mu-1)}{4+\mu-2\mu^2} f; \quad\ldots\ldots\ldots\ldots\ldots(6)$$

so that

$$-s : r = \frac{\mu(2\mu+1)}{4+\mu-2\mu^2}. \quad\ldots\ldots\ldots\ldots\ldots(7)$$

In the case of plate-glass $\mu = 1\cdot5$ nearly, and then from (5), (6), (7)

$$r = \frac{7}{12}f, \quad -s = \frac{7}{2}f, \quad -s : r = 6 : 1.$$

Both surfaces are therefore convex, but the curvature of the anterior surface (that directed towards the incident parallel rays) is six times the curvature of the posterior surface. By (3) the outstanding aberration is

$$\delta f = -\frac{15}{14}\frac{y^2}{f}. \quad\ldots\ldots\ldots\ldots\ldots\ldots(8)$$

The use of a plano-convex lens instead of that above determined does not entail much increase of aberration. Putting in (3) $s = \infty$, and therefore by (1) $r = \frac{1}{2}f$, we get

$$\delta f = -\frac{7}{6}\frac{y^2}{f}. \quad\ldots\ldots\ldots\ldots\ldots\ldots(9)$$

This is on the supposition that the curved side faces the parallel rays. If the lens be turned round so as to present the plane face to the incident light we have $r = \infty$, $-s = \frac{1}{2}f$, and then

$$\delta f = -\frac{9}{2}\frac{y^2}{f}, \quad\ldots\ldots\ldots\ldots\ldots\ldots(10)$$

nearly four times as great.

For a somewhat higher value of μ the plano-convex becomes the form of minimum aberration. If $s = \infty$ in (6), $4 + \mu - 2\mu^2 = 0$, whence $\mu = 1\cdot69$.

If μ be very great, we see from (5) and (6) that r and s tend to become identical with f.

For the general value of μ the minimum aberration corresponding to (7) is by (4)

$$-\delta f : \frac{y^2}{f} = \frac{\mu(4\mu-1)}{8(\mu-1)^2(\mu+2)}. \quad\ldots\ldots\ldots(11)$$

The right-hand member of (11) tends to diminish as μ increases, but it remains considerable for all natural substances. If $\mu = 2$,

$$- \delta f : \frac{y^2}{f} = \frac{7}{16}.$$

Oblique Pencils.—Hitherto we have supposed that the axis of the pencil coincides with the axis of the lens. If the axis of the pencil, though incident obliquely, pass through the centre of the lens, it suffers no deviation, the surfaces being parallel at the points of incidence and emergence. In this case the primary and secondary foci are formed at distances from the centre of the lens which can only differ from the distance corresponding to a direct pencil by quantities of the second order in the obliquity. Hence, if the obliquity be moderate, we may use the same formulæ for oblique as for direct pencils.

The consideration of excentrical pencils leads to calculations of great complexity, upon which we do not enter.

Chromatic Aberration.—The operation of simple lenses is much interfered with by the variation of the refractive index with the colour of the light. The focal length is decidedly less for blue than for red light, and thus in the ordinary case of white light it is impossible to obtain a perfect image, however completely the spherical aberration may be corrected. From the formula for the focal length we see that

$$\frac{\delta f}{f^2} = - \delta\mu \left(\frac{1}{r} - \frac{1}{s}\right) = - \frac{\delta\mu}{\mu - 1} \frac{1}{f},$$

so that

$$- \delta f = \frac{\delta\mu}{\mu - 1} f;$$

or the longitudinal chromatic aberration varies as the focal length and as the dispersive power of the material composing the lens. The best image will be formed at a position midway between the two foci, and the diameter d of the circle over which the rays are spread bears the same ratio to the semi-aperture of the lens (y) that δf bears to f. Hence

$$d = \frac{\delta\mu}{\mu - 1} y.$$

The diameter of the circle of chromatic aberration is thus proportional to the aperture and *independent of the focal length*; and, since the linear dimensions of the image are proportional to the focal length, the confusion due to chromatic aberration may be considered to be inversely as the focal length. Before the invention of the achromatic object-glass this source of imperfect definition was by far the most important, and, in order to mitigate its influence, telescopes were made of gigantic length. Even at the present day the images of large so-called achromatic glasses are sensibly impaired by secondary chromatic aberration, the effect of which is also directly as the aperture and inversely as the focal length.

Achromatic Object-glasses.—It has been shown in *Enc. Brit.* vol. XIV. p. 595, that the condition of achromatism for two thin lenses placed close together is

$$\frac{\delta\mu}{\mu-1}\frac{1}{f} + \frac{\delta\mu'}{\mu'-1}\frac{1}{f'} = 0, \quad\dots\dots\dots\dots\dots\dots(1)$$

in which f, f' are the focal lengths of the two lenses, and $\delta\mu/(\mu-1)$, $\delta\mu'/(\mu'-1)$ the dispersive powers of the two kinds of glass. In practice crown and flint glass are used, the dispersive power of the flint being greater than that of the crown. Thus f' is negative and numerically greater than f, so that the combination consists of a convex lens of crown and a concave lens of flint, the converging power of the crown overpowering the diverging power of the flint. When the focal length F of the combination is given, the focal lengths of the individual lenses are determined by (1) in conjunction with

$$\frac{1}{F} = \frac{1}{f} + \frac{1}{f'}. \quad\dots\dots\dots\dots\dots\dots\dots(2).$$

The matter, however, is not quite so simple as the above account of it might lead us to suppose. In consequence of what is called the irrationality of spectra, the ratio of dispersive powers of two media is dependent upon the parts of the spectrum which we take into consideration. Whatever two rays of the spectrum we like to select, we can secure that the compound lens shall have the same focal length for these rays, but we shall then find that for other rays the focal length is slightly different. In the case of a single lens the focal length continually diminishes as we pass up the spectrum from red to violet. By the use of two lenses the spectrum, formed as it were along the axis, is doubled upon itself. The focal length is least for a certain ray, which may be selected at pleasure. Thus in the ordinary achromatic lens, intended for use with the eye, the focal length is a minimum for the green, and increases as we pass away from the green, whether towards red or towards blue. Stokes has shown that the secondary colour gives a sharp test of the success of the achromatizing process.

"The secondary tints in an objective are readily shown by directing the telescope to a vertical line separating light from dark, such as the edge of a chimney seen in the shade against the sky, and covering half the object-glass with a screen having a vertical edge. So delicate is this test that, on testing different telescopes by well-known opticians, a difference in the mode of achromatism may be detected. The best results are said to be obtained when the secondary green is intermediate between green and yellow. This corresponds to making the focal length a minimum for the brightest part of the spectrum.

"To enable me to form a judgment as to the sharpness of the test furnished by the tint of the secondary green, as compared with the performance of an object-glass, I tried the following experiment. A set of parallel lines of increasing fineness was ruled with ink on a sheet of white paper, and a broader black object was laid upon it as well, parallel to the lines. The paper was placed, with the black lines vertical, at a considerable distance on a lawn, and was viewed through two opposed prisms, one of crown glass and the other of flint, of such angles as nearly to achromatize each other in the positions of minimum deviation, and then through a small telescope. The achromatism is now effected, and varied in character, by moving one of the prisms slightly in azimuth, and after each alteration the telescope was focused afresh to get the sharpest vision that could be had. I found that the azimuth of the prism was fixed within decidedly narrower limits by the condition that the secondary green should be of such or such a tint, even though no attempt was made to determine the tint otherwise than by memory, than by the condition that the vision of the fine lines should be as sharp as possible. Now a small element of a double object-glass may be regarded, so far as chromatic compensation is concerned, as a pair of opposed prisms; and therefore we may infer that the tint of the secondary green ought to be at the very least as sharp a test of the goodness of the chromatic compensation as the actual performance of the telescope*."

In the case of photographic lenses the conditions of the problem are materially different. It is usually considered to be important to secure "coincidence of the visual and chemical foci," so that the sensitive plate may occupy the exact position previously found by the eye for the ground glass screen. For this purpose the ray of minimum focus must be chosen further up in the spectrum. If, however, the object be to obtain the sharpest possible photographs, coincidence of visual and chemical foci must be sacrificed, the proper position for the sensitive plate being found by trial. The middle of the chemically-acting part of the spectrum, which will vary somewhat according to the photographic process employed, should then be chosen for minimum focus.

When the focal lengths of the component lenses have been chosen, it still remains to decide upon the curvatures of the individual faces. Between the four curvatures we have at present only two relations, and thus two more can be satisfied. One of these is given by the condition that the first term in the expression for the aberration—that proportional to the square of the aperture—shall vanish for parallel rays. As to the fourth condition, various proposals have been made. If equal and opposite curvatures are given to the second and third surfaces, the glasses may be cemented together, by which some saving of light is effected. Herschel proposed to make the

* *Proc. Roy. Soc.* June 1878.

aberration vanish for nearly parallel, as well as for absolutely parallel, rays. This leads to a construction nearly agreeing with that adopted by Fraunhofer.

The following results are given by Herschel * for the radii of the four surfaces, corresponding to various dispersive powers, and to mean refractive indices 1·524 (crown) and 1·585 (flint). The focal length of the combination is taken equal to 10, and, as well as the radii, is measured in arbitrary units; so that all the numbers in the table (with the exception of the first column) may be changed in any proportion.

Ratio of Dispersive Powers	Radius of First Surface +	Radius of Second Surface −	Radius of Third Surface −	Radius of Fourth Surface −	Focal Length of Crown Lens +	Focal Length of Flint Lens −
·50	6·7485	4·2827	4·1575	14·3697	5·0	10·0000
·55	6·7184	3·6332	3·6006	14·5353	4·5	8·1818
·60	6·7069	3·0488	3·0640	14·2937	4·0	6·6667
·65	6·7316	2·5208	2·5566	13·5709	3·5	5·3846
·70	6·8279	2·0422	2·0831	12·3154	3·0	4·2858
·75	7·0816	1·6073	1·6450	10·5186	2·5	3·3333

The general character of the combination is shown in fig. 16.

The radii of the first and fourth surfaces within practical limits are so nearly constant that Herschel lays down the following rule as in all probability sufficiently exact for use. A double object-glass will be free from aberration, provided the radius of the exterior surface of the crown lens be 6·720 and of the flint 14·20, the focal length of the combination being 10·000, and the radii of the interior surface being computed from these data, by the formulæ given in all elementary works on optics, so as to make the focal lengths of the two glasses in the direct ratio of their dispersive powers.

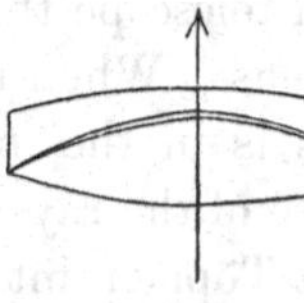

Fig. 16.

Numerous experiments have been made with the view of abolishing the secondary spectrum. Theoretically, if three different kinds of glasses are combined it will generally be possible to make the focal lengths of the combination equal for any *three* selected rays of the spectrum. Or the ingredients of one of the glasses may be mixed in such proportions as to suit the requirements of the problem when combined with crown. In this way Stokes has succeeded in constructing a small object-glass free from secondary

* *Phil. Trans.* 1821.

colour, but it is doubtful whether the practical difficulties could be overcome in the construction of a large object-glass, where alone the outstanding chromatic aberration is important.

The practical optician is not limited to spherical surfaces, and the final adjustment of the aberration of large object-glasses is controlled by the action of the polishing tool. It is understood that some of the best makers apply a local correction, according to the methods developed by Foucault for mirrors. The light from a natural or artificial star is allowed to fall upon the lens. At the focus is placed a small screen, which is gradually advanced so as to cut off the light. The eye is immediately behind the screen and is focused upon the lens. If there are no imperfections the illumination falls off very suddenly, the surface of the mirror passing from light to dark through a nearly uniform grey tint. If, however, from uniform aberration, or from local defects, any of the light goes a little astray, the corresponding parts of the surface will show irregularities of illumination during the passage of the screen, and in this manner a guide is afforded for the completion of the figuring.

Töpler* has developed the idea of Foucault into a general method for rendering visible very small optical differences. Instead of a mere point of light, it is advisable to use as source an aperture (backed by a bright flame) of sensible size, and bounded on one side by a straight edge. An image of this source is formed at a considerable distance by a lens of large aperture and free from imperfections, and in the plane of the image is arranged a screen whose edge is parallel to the straight edge of the image, and can be advanced gradually so as to coincide with it. Behind this screen comes a small telescope through which the observer examines the object placed near the lens. When the light is just cut off by the advancing screen, the apparatus is in the most sensitive state, and the slightest disturbance of the course of the rays is rendered evident. To show the delicacy of the arrangement Töppler introduced into the cone of light a small trough with parallel glass sides containing distilled water. A syphon dipped under the surface and discharged distilled water from another vessel, and it was found almost impossible so to control the temperatures that the issuing jet should remain invisible. Not only were sound-waves in air, generated by electric sparks, rendered visible, but their behaviour when reflected from neighbouring obstacles was beautifully exhibited.

An apparatus on this principle may often be employed with advantage in physical demonstrations,—for instance, for the exhibition of the changes of density in the neighbourhood of the electrodes of a metallic solution undergoing electrolysis. The smallest irregularity that could be rendered visible

* *Pogg. Ann.* cxxxi. 1867.

would be such as would retard transmitted light by a moderate fraction of the wave-length*.

In objectives for photographic use the requirements are in many respects different from those most important in the case of telescopes. A flat field, a wide angle of view—in some cases as much as 90°—freedom from distortion, and a great concentration of light are more important than a high degree of definition. As a rule, photographs are not subjected to the ordeal of a high magnifying power. Usually the picture includes objects at various distances from the camera, which cannot all be in focus at once. That the objects at one particular distance should be depicted with especial sharpness would often be rather a disadvantage than otherwise. A moderate amount of " diffusion of focus " is thus desirable, and implies residual aberration. In some lenses an adjustment is provided by means of which the diffusion of focus may be varied according to the circumstances of the case.

For landscapes and general purposes a so-called single lens is usually employed. This, however, for the sake of achromatism, is compounded of a flint and a crown cemented together; or sometimes three component lenses are used, the flint being encased in two crowns, one on each side. To get tolerable definition and flatness of field a stop must be added, whose proper place is some little distance in front of the lens.

For portraiture, especially before the introduction of the modern rapid dry plates, a brilliant image was a necessity. This implies a high ratio of aperture to focal length, which cannot be attained satisfactorily with any form of single lens. To meet the demand, Petzval designed the "portrait-lens," in which two achromatic lenses, placed at a certain distance apart, combine to form the image. This construction is so successful that the focal length is often no more than three times the available aperture. When stops are employed to increase the sharpness and depth of focus they are placed *between* the lenses.

Vision through a Single Lens.—A single lens may be used to improve the vision of a defective eye, or as a magnifying-glass. A normal eye is capable of focusing upon objects at any distance greater than about 8 inches. The eyes of a short-sighted person are optically too powerful, and cannot be focused upon an object at a moderate distance. The remedy is of course to be found in concave glasses. On the other hand, persons beyond middle life usually lose the power of seeing near objects distinctly, and require convex glasses.

* Even when the optical differences are not small it is well to remember that transparent bodies are only visible in virtue of a variable illumination. If the light falls equally in all directions, as it might approximately do for an observer on a high monument during a thick fog, the edge of (for example) a perfectly transparent prism would be absolutely invisible. If a spherical cloud, composed of absolutely transparent material, surround symmetrically a source of light, the illumination at a distance would not be diminished by its presence.

A not uncommon defect, distinct from mere short or weak sight, is that known as astigmatism. In such cases the focal length varies in different planes, and at no distance is the definition perfect. Many people, whose sight would not usually be considered inferior, are affected by astigmatism to a certain extent. If a set of parallel black lines ruled upon white paper be turned gradually round in its own plane, it will often be seen more distinctly and with greater contrast of the white and black parts in one azimuth than in another. When the focal line on the retina is parallel to the length of the bar, the definition (as in the case of the spectroscope) is not much prejudiced, but it is otherwise when the bars are turned through a right angle so as to be perpendicular to the focal line.

In extreme cases a remedy may be applied in the form of glasses of different curvatures in perpendicular planes, so adjusted both in form and position as to compensate the corresponding differences in the lens of the eye.

The use of a lens as a magnifier has been explained under MICROSCOPE. The simplest view of the matter is that the lens, consistently with good focusing, allows of a nearer approach, and therefore of a higher visual angle, than would otherwise be possible.

Telescope, &c.—In a large class of optical instruments an *image* of the original object is first formed, and this image is examined through a magnifier. If we use a single lens merely for the latter purpose, the field of view is very restricted. A great improvement in this respect may be effected by the introduction of a *field*-lens. The ideal position for the field-lens is at the focal plane of the object-glass. The image is then entirely uninfluenced, and the only effect is to bend round the rays from the margin of the field which would otherwise escape, and to make them reach the eye-lens, and ultimately the eye. If the field-lens and the eye-lens have nearly the same focal length an image of the object-glass will be formed upon the eye-lens and through this small image will pass every ray admitted by the object-glass and field-lens.

However, to obtain a sufficient augmentation of the field of view it is not necessary to give the field-lens the exact position above mentioned, and other considerations favour a certain displacement. For example, it is not desirable that dust and flaws on the field-lens should be seen in focus. In Huygens's eye-piece the field-lens is displaced from its ideal position *towards* the object-glass. In Ramsden's eye-piece, on the other hand, the focal plane of the object-glass is outside the system. This eye-piece has the important advantage that cross wires can be placed so as to coincide with the image as formed by the object-glass. The component lenses of a Ramsden's eye-piece are sometimes achromatic. For further particulars with diagrams, on the subject of eye-pieces, see MICROSCOPE.

In large telescopes the object-glass is often replaced by a mirror, which may be of speculum metal, or of glass coated chemically with a very thin layer of polished silver. The mirror presents the advantage (especially important for photographic applications) of absolute achromatism. On the other hand, more light is lost in the reflexion than in the passage through a good object-glass, and the surface of the mirror needs occasional re-polishing or re-coating. For fuller information see TELESCOPE.

The function of a telescope is to increase the "apparent magnitude" of distant objects; it does not increase the "apparent brightness." If we put out of account the loss of light by reflexion at glass surfaces (or by imperfect reflexion at metallic surfaces) and by absorption, and suppose that the magnifying power does not exceed the ratio of the aperture of the object-glass to that of the pupil, under which condition the pupil will be filled with light, we may say that the "apparent brightness" is absolutely unchanged by the use of a telescope. In this statement, however, two reservations must be admitted. If the object under examination, like a fixed star, have no sensible apparent magnitude, the conception of "apparent brightness" is altogether inapplicable, and we are concerned only with the total quantity of light reaching the eye. Again, it is found that the visibility of an object seen against a black background depends not only upon the "apparent brightness" but also upon the apparent magnitude. If two or three crosses of different sizes be cut out of the same piece of white paper, and be erected against a black background on the further side of a nearly dark room, the smaller ones become invisible in a light still sufficient to show the larger. Under these circumstances a suitable telescope may of course bring also the smaller objects into view. The explanation is probably to be sought in imperfect action of the lens of the eye when the pupil is dilated to the utmost. The author of this article has found that in a nearly dark room he becomes distinctly short-sighted, a defect of which there is no trace whatever in a moderate light*. If this view be correct, the brightness of the image on the retina is really less in the case of a small than in the case of a large object, although the so-called apparent brightnesses may be the same. However this may be, the utility of a night-glass is beyond dispute.

The general law that (apart from the accidental losses mentioned above) the "apparent brightness" depends only upon the area of the pupil filled with light, though often ill understood, has been established for a long time, as the following quotation from Smith's *Optics* (Cambridge, 1738), p. 113, will show.

"Since the magnitude of the pupil is subject to be varied by various degrees of light, let *NO* be its semi-diameter when the object *PL* is viewed by the naked eye from the distance *OP*; and upon a plane that touches the eye at O, let OK be the semi-diameter of the greatest area, visible through all the glasses to another eye at P, to be found as PL was; or, which is the same thing, let OK be the semi-diameter of the greatest area inlightened by a pencil of rays flowing from P through all the glasses; and when this area is not less than the area of the pupil, the point P will appear just as bright through all the glasses as it would do if they were removed; but if the inlightened area be less than the area of the pupil, the point P will appear less bright through the glasses than if they were removed in the same proportion as the inlightened area is less than the pupil. And these proportions of apparent brightness would be accurate if all the incident rays were transmitted through the glasses to the eye, or if only an insensible part of them were stopt."

Resolving Power of Optical Instruments.—According to the principles of common optics, there is no limit to the resolving power of an instrument. If the aberrations of a microscope were perfectly compensated it might reveal to us worlds within a space of a millionth of an inch. In like manner a telescope might resolve double stars of any degree of closeness. The magnifying power may be exalted at pleasure by increase of focal length and of the power of eye-pieces; and there are at any rate some objects, such as the sun, in dealing with which the accompanying loss of light would be an advantage rather than the contrary. How is it, then, that the power of the microscope is subject to an absolute limit, and that if we wish to observe minute detail on the over-lighted disk of the sun we must employ a telescope of large aperture? The answer requires us to go behind the approximate doctrine of rays, on which common optics is built, and to take into consideration the finite character of the wave-length of light.

A calculation based upon the principles of the wave-theory shows that, no matter how perfect an object-glass may be, the image of a star is represented, not by a mathematical point, but by a disk of finite size surrounded by a system of alternately dark and bright rings. Airy found that if the angular radius of the central disk (as seen from the centre of the object-glass) be θ, $2R$ the aperture, λ the wave-length, then

$$\theta = 1\cdot2197 \frac{\lambda}{2R},$$

showing that the definition, as thus limited by the finiteness of λ, increases with the aperture.

In estimating theoretically the resolving power of a telescope on a double star we have to consider the illumination of the field due to the superposition of the two independent images. If the angular interval between the components of the double star were equal to 2θ, the central disks would be just in contact. Under these conditions there can be no doubt that the star would appear to be fairly resolved, since the brightness of the external ring systems is too small to produce any material confusion, unless indeed the components are of very unequal magnitude. The diminution of star disks with increasing aperture was observed by W. Herschel; and in 1823 Fraunhofer formulated the law of inverse proportionality. In investigations extending over a long series of years, the advantage of a large aperture in separating the components of close double stars was fully examined by Dawes.

The resolving power of telescopes was investigated also by Foucault, who employed a scale of equal bright and dark alternate parts; it was found to be proportional to the aperture and independent of the focal length. In telescopes of the best construction the performance is not sensibly prejudiced by outstanding aberration, and the limit imposed by the finiteness of the waves of light is practically reached. Verdet has compared Foucault's results with theory, and has drawn the conclusion that the radius of the visible part of the image of a luminous point was nearly equal to half the radius of the first dark ring.

The theory of resolving power is rather simpler when the aperture is rectangular instead of circular, and when the subject of examination consists of two or more light or dark lines parallel to one of the sides of the aperture. Supposing this side to be vertical, we may say that the definition, or resolving power, is *independent of the vertical aperture,* and that a double line will be about on the point of resolution when its components subtend an angle equal to that subtended by the wave-length of light at a distance equal to the *horizontal aperture.*

The resolving power of a telescope with a circular or rectangular aperture is easily investigated experimentally. The best object is a grating of fine wires, about fifty to the inch, backed by a soda-flame. The object-glass is provided with diaphragms pierced with round holes or slits. One of these, of width equal, say, to one-tenth of an inch, is inserted in front of the object-glass, and the telescope, carefully focused all the while, is drawn gradually back from the grating until the lines are no longer seen. From a measurement of the maximum distance the least angle between consecutive lines consistent with resolution may be deduced, and a comparison made with the rule stated above.

Merely to show the dependence of resolving power on aperture it is not necessary to use a telescope at all. It is sufficient to look at wire-gauze backed by the sky, or by a flame, through a piece of blackened cardboard pierced by a needle and held close to the eye. By varying the distance the point is easily found at which resolution ceases; and the observation is as sharp as with a telescope. The function of the telescope is in fact to allow the use of a wider, and therefore more easily measurable, aperture. An interesting modification of the experiment may be made by using light of various wave-lengths.

In the case of the microscope the wave-theory shows that there must be an absolute limit to resolving power independent of the construction of the instrument. No optical contrivances can decide whether light comes from one point or from another if the distance between them do not exceed a small fraction of the wave-length. This idea, which appears to have been familiar to Fraunhofer, has recently been expanded by Abbe and Helmholtz into a systematic theory of the microscopic limit. See MICROSCOPE.

Similar principles may be applied to investigate the resolving power of spectroscopes, whether dispersing or diffracting. Consider for simplicity any combination of prisms, anyhow disposed, but consisting of one kind of glass. Let a be the width and μ the index of a parallel beam passing through, and let the thicknesses of glass traversed by the extreme rays on either side be t_2 and t_1. It is not difficult to see that, if the index be changed to $\mu + \delta\mu$, the rays will be turned through an angle θ given by

$$\theta = \frac{\delta\mu\,(t_2 - t_1)}{a}.$$

Now, if the two kinds of light correspond to a double line which the instrument can just resolve, we have $\theta = \lambda/a$, and thus

$$t_2 - t_1 = \lambda/\delta\mu,$$

a formula of capital importance in the theory of the dispersing spectroscope. In a well-constructed instrument, t_1, the smaller thickness traversed may be small or negligible, and then we may state the law in the following form :— the smallest thickness of prisms necessary for the resolution of a double line whose indices are μ and $\mu + \delta\mu$ is found by dividing the wave-length by $\delta\mu$.

As an example, let it be required to find the smallest thickness of a prism of Chance's " extra dense flint," necessary for resolution of the soda-lines.

By Cauchy's formula for the relation between μ and λ we have

$$\mu = A + B\lambda^{-2}, \qquad\qquad \delta\mu = -2B\lambda^{-3}\,\delta\lambda.$$

From the results given by Hopkinson for this kind of glass we find

$$B = \cdot 984 \times 10^{-10},$$

the unit of length being the centimetre. For the two soda-lines

$$\lambda = 5\cdot 889 \times 10^{-5}, \qquad\qquad \delta\lambda = \cdot 006 \times 10^{-5};$$

and thus the thickness t necessary to resolve the lines is

$$t = \frac{\lambda^4}{2B\delta\lambda} = \frac{10^{10}\lambda^4}{1\cdot 968\,\delta\lambda} = 1\cdot 02 \text{ centimetre,}$$

the meaning of which is that the soda-lines will be resolved if, and will not be resolved unless, the difference of thicknesses of glass traversed by the two sides of the beam amount to one centimetre. In the most favourable arrangement the centimetre is the length of the base of the prism. It is to be understood, of course, that the magnifying power applied is sufficient to narrow the beam ultimately to the diameter of the pupil of the eye; otherwise the full width would not be utilized.

The theory of the resolving power of a diffracting spectroscope, or grating, is even simpler. Whatever may be the position of the grating, a double line of wave-lengths λ and $\lambda + \delta\lambda$ will be just resolved provided

$$\frac{\delta\lambda}{\lambda} = \frac{1}{mn},$$

where n is the total number of lines in the grating, and m is the order of the spectrum under examination.

If a grating giving a spectrum of the first order and a prism of extra dense glass have equal power in the region of the soda-lines, the former must have about as many thousand lines as the latter has centimetres of available thickness.

The dispersion produced by a grating situated in a given manner is readily inferred from the resolving power. If a be the width of the beam after leaving the grating, the angle $\delta\theta$, corresponding to the limit of resolution, is λ/a, and thus

$$\frac{\delta\theta}{\delta\lambda} = \frac{mn}{a}.$$

Thus the dispersion depends only upon the order of the spectrum, the total number of lines, and the width of the emergent beam.

An obvious inference from the necessary imperfection of optical images is the uselessness of attempting anything like an absolute destruction of aberration. In an instrument free from aberration the waves arrive at the focal point in the same phase. It will suffice for practical purposes if the error of phase nowhere exceeds $\frac{1}{4}\lambda$. This corresponds to an error of $\frac{1}{8}\lambda$ in a reflecting and $\frac{1}{2}\lambda$ in a (glass) refracting surface, the incidence in both cases being perpendicular.

If we inquire what is the greatest admissible longitudinal aberration in an object-glass according to the above rule, we find

$$\delta f = \lambda \alpha^{-2},$$

α being the angular semi-aperture.

In the case of a single lens of glass with the most favourable curvatures, δf is about equal to $f\alpha^2$; so that α^4 must not exceed λ/f. For a lens of 3-feet focus this condition is satisfied if the aperture do not exceed 2 inches.

When parallel rays fall directly upon a spherical mirror the longitudinal aberration is only about one-eighth as great as for the most favourable-shaped single lens of equal focal length and aperture. Hence a spherical mirror of 3-feet focus might have an aperture of $2\frac{1}{2}$ inches, and the image would not suffer materially from aberration*.

On general optics the treatises most accessible to the English reader are Parkinson's *Optics* (3rd ed., 1870) and Glazebrook's *Physical Optics* (1883). Verdet's *Leçons d'optique physique* is an excellent work. Every student should read the earlier parts of Newton's *Optics*, in which are described the fundamental experiments upon the decomposition of white light.

[1900. To the above references may now be added Preston's *Theory of Light* and Mascart's *Traité d'Optique*.]

---

* For fuller information on the subject of the preceding paragraphs see Lord Rayleigh's papers entitled "Investigations in Optics," *Phil. Mag.* 1879, 1880. [Art. 62, vol. I. p. 415.]

120.

ÜBER DIE METHODE DER DÄMPFUNG BEI DER BESTIMMUNG DES OHMS.

[*Annalen der Physik und Chemie*, Band XXIV. pp. 214, 215, 1885.]

Mit grossem Interesse habe ich aus einer neueren Mittheilung in den Annalen ersehen, dass Hr. Wild im Anschluss an einen Vorschlag von Dorn seine Zahl für diese Werthe der Siemens'schen Einheit in Ohme 0,9462 auf 0,94315 corrigirt hat, wodurch die Differenz zwischen seiner Zahl und der von mir gefundenen 0,9415 auf etwa ein Drittel reducirt wird. Die Untersuchung von Wild scheint sehr sorgfältig ausgeführt worden zu sein, indess möchte ich doch die Aufmerksamkeit derer, welche an die Vorzüge der Dämpfungsmethode glauben, auf einige Punkte lenken.

Bei der theoretischen Untersuchung wird die Wirkung des Magnets als identisch mit der eines Solenoids angesehen, durch welches ein constanter Strom geleitet wird, während sie in der That mehr mit der eines mit einem Eisenkern versehenen Solenoides verglichen werden kann. Mir scheint die Einführung einer grossen Eisenmasse in den Multiplicator sehr sorgfältige Erwägungen zu verdienen. Selbst wenn man annimmt, dass der grössere Theil der Wirkung durch die Aenderungen gewisser Grössen, wie der Inductionscoëfficienten, compensirt werden kann, so kann doch ein kleines Residuum zurückbleiben infolge der Abweichung der Magnetisirung des Eisens von den einfachen Gesetzen. Ich will nicht behaupten, dass dies in der That der Fall ist, indess müssen diejenigen, welche die Dämpfungsmethode benutzen, das Gegentheil beweisen.

Ferner ist der Magnet ein Leiter der Electricität. Es ist nicht erwiesen, dass nicht galvanische Ströme von erheblicher Stärke in einem 36 mm. langen und 12 mm. dicken Stabe erzeugt werden können, welcher in einer vom Strom durchflossenen Spirale schwingt. Diese Ströme würden wahrscheinlich proportional der Schwingungsweite wirken; indess fühlte sich Wild veranlasst, andere Glieder dieser Art in seine Reductionsformeln einzuführen.

Eine unerfreuliche Seite in Wild's Untersuchungen ist die Nothwendigkeit einer Correction, welche $\frac{6}{1000}$ beträgt, und durch den Eisengehalt des Multiplicatordrahtes nöthig wird. Dieser Einfluss wird behandelt, wie wenn seine einzige Wirkung nur eine Kraft wäre, welche den Magnet stets zu der Symmetrieebene des Multiplicators zurückführt. Auch dieser Punkt scheint mir genauerer Erwägungen zu bedürfen.

Die Wichtigkeit, womöglich die Gründe der Abweichungen in den Resultaten der verschiedenen Forscher auf diesem Gebiete aufzufinden, möge die ausgesprochenen Zweifel entschuldigen. Ich hoffe, dass dadurch nicht die Meinung erweckt werde, als wenn ich meine eigene Untersuchung für unfehlbar hielte.

121.

ON THE THEORY OF ILLUMINATION IN A FOG.

[*Philosophical Magazine*, XIX. pp. 443—446, 1885.]

As a step towards a better understanding of the action of fog upon light, it seems desirable to investigate what the phenomena would be in the simplest case that can be proposed. For this purpose we may consider the atmosphere and the material composing the fog to be *absolutely transparent*, and also make abstraction from the influence of obstacles, among which must be included the ground itself.

Conceive a small source of radiation, *e.g.* an incandescent carbon filament, to be surrounded by a spherical cloud, of uniform density, or at any rate symmetrically disposed round the source, outside of which the atmosphere is clear. Since by hypothesis there is no absorption, whatever radiation is emitted by the source passes outward through the external surface of the cloud. The effect of the cloud is to cause diffusion, *i.e.* to spread the rays passing through any small area of the surface (which in the absence of the cloud would be limited to a small solid angle) more or less uniformly over the complete hemisphere.

Whether the total radiation passing outwards through the small area on the external surface of the cloud is affected by the existence of the cloud depends upon the circumstances of the case. If it be laid down that the total emission of energy from the source is given, then the presence of the cloud makes no difference in respect of the energy passing any element of the spherical area. But this supposition does not correspond to a constant temperature of the source, in consequence of the energy received back from the cloud by reflection. To keep the total emission of energy constant, we should have to suppose a rise of temperature increasing indefinitely with the size and density of the cloud.

Let us now suppose that the region under consideration is bounded upon all sides by a distant envelope of perfect reflecting-power. Then, whatever the density of the clouds which may wholly or partially occupy the enclosure, we know, by the second law of thermodynamics, that at every internal point there is radiation in every direction of the full amount corresponding to the temperature of the source. In one sense this conclusion holds good, even although the matter composing the cloud has the power of absorption. But in that case equilibrium would not be attained until the clouds themselves to the remotest parts had acquired the temperature of the source; whereas under the supposition of perfect transparency the temperature of the cloud is a matter of indifference; and equilibrium is attained in a time dependent upon that required by light to traverse the enclosure. So far we have made no supposition as to the distribution of the cloud; but we will now imagine a layer of such thickness as to allow only a very small fraction of the incident radiation to penetrate it, to line the interior of the reflecting envelope. This layer itself plays the part of a practically perfect reflector; and it is not difficult to see that the reflecting envelope hitherto conceived to lie beyond it may be removed without interfering with the state of things on the inner side of the layer of cloud. We thus arrive at the rather startling conclusion that at any distance from the source, and whatever the distribution of clouds, there is always in every direction the full radiation due to the temperature of the source, provided only that there lie outside a complete shell of cloud sufficiently thick to be impervious. And this state of things is maintained without (on the whole) emission of energy from the source.

Even if the material composing the cloud possesses absorbing-power for some kinds of radiation, e.g. for dark radiation, but is perfectly transparent to other kinds, e.g. luminous radiation, the general theorem holds good as respects the latter kinds; so that in the case supposed the light would still be everywhere the same as in a clear enclosure whose walls have throughout the same luminosity as the source. But in order to compensate the absorption of dark rays, the source must now be supplied with energy.

Some of the principles here enunciated have an acoustical as well as an optical application, and indeed first occurred to me some years ago in connection with Prof. Tyndall's investigations upon fog-signals. The effect of "acoustic clouds" analogous to fog (and unattended with absorption of energy), might be very different upon the report of a gun and upon the sustained sound of a syren, the latter being reinforced by reflection from the acoustic fog.

The theory presented in the present paper may be illustrated by the known solution of the comparatively simple problem of a pile of transparent plates*. If ρ denote the proportion of the incident light reflected at a single surface, then the proportion reflected $\phi(m)$, and transmitted $\psi(m)$, by a pile of m plates is given by

$$\frac{\phi(m)}{2m\rho} = \frac{\psi(m)}{1-\rho} = \frac{1}{1+(2m-1)\rho}.$$

From these expressions it is evident that, however small ρ may be, *i.e.* however feeble the reflection at a single surface, we have only to suppose m large enough in order that the reflection may be as complete, and the transmission as small, as we please. Such a pile may, under ordinary conditions, be regarded as impervious.

But now suppose that after passing the pile of m plates, the light is incident upon a second pile of n plates, and consider the intensity between the two piles, the original intensity being unity, as before. For the intensity of the light travelling in the original direction we have

$$\psi(m) + \psi(m) \cdot \phi(n) \cdot \phi(m) + \psi(m) \cdot \{\phi(n) \cdot \phi(m)\}^2$$
$$+ \psi(m) \cdot \{\phi(n) \cdot \phi(m)\}^3 + \ldots;$$

or on summation of the geometric series,

$$\frac{\psi(m)}{1 - \phi(n) \cdot \phi(m)}.$$

If we introduce the values of ϕ and ψ in terms of m, n, ρ, this becomes

$$\frac{2n\rho + 1 - \rho}{2(m+n)\rho + 1 - \rho}.$$

In like manner, for the light going the other way we have

$$\frac{\phi(m) \cdot \phi(n)}{1 - \phi(m) \cdot \phi(n)};$$

or in terms of m, n, ρ,

$$\frac{2n\rho}{2(m+n)\rho + 1 - \rho}.$$

When m and n are great, both expressions reduce to $n/(m+n)$; so that the light passing in the two directions is equally bright. Moreover, and this is the point to be especially noticed, however great m may be—that is, however impervious the first pile is, the light between the two piles may be made to approach the original light in brightness as nearly as we please, by sufficiently increasing the number of plates in the second pile; that is, the light between the piles may be made to be the same as if the first pile were removed. From this example we may understand more clearly how a very small quantity of light penetrating directly may be beaten backwards and forwards, as between two reflectors, until the original intensity is recovered.

* Stokes, *Proc. Roy. Soc.* vol. XI. p. 545 (1862).

122.

A MONOCHROMATIC TELESCOPE, WITH APPLICATION TO PHOTOMETRY.

[*Philosophical Magazine*, XIX. pp. 446, 447, 1885.]

THE purpose of this instrument is to exhibit external objects as they would be seen either with the naked eye, or through a telescope, if lighted with approximately monochromatic light; that is, to do more perfectly what is done roughly by a coloured glass.

The arrangement is not new, though I am not aware that it has ever been described. In 1870 I employed it for determinations of absorption, and, if my memory serves me right, I heard soon afterwards from Clerk-Maxwell that he also had used it. It is, indeed, a very slight modification of Maxwell's colour-box.

In the ordinary form of that instrument, white light admitted through a slit is rendered parallel by a collimating lens, dispersed by flint-glass prisms, and then brought to a focus at a screen, upon which accordingly a pure spectrum is formed. This screen is perforated by a second slit, immediately behind which the observer places his eye. It is evident that the light passing the aperture is approximately monochromatic, so that the observer, if he focuses his eye suitably, will see the prism illuminated with this kind of light. The only addition now required to convert the instrument into a monochromatic telescope is a lens placed just within the first slit, of such power as to throw an image of external objects upon the prism or diaphragm upon which the eye is focused. If desired, an eye-lens may be placed at the second slit; but this is not generally needed.

In the present instrument a direct-vision dispersing prism is used, so that the optical parts can be all disposed in a narrow box of nearly 3 feet in length. The lenses are all single lenses, and work sufficiently well. The slits are of such width that either coincides with the image of the other, and their relative position is so chosen that the mean refrangibility of the light is that corresponding to sodium. Objects seen through the instrument thus appear as if lighted by a sodium flame.

The principal object which I had in view in the construction of the instrument now exhibited was to see whether it could be made of service in the comparison of compound lights of somewhat different colours—a problem just now attracting attention in connection with electric lighting. It is scarcely necessary to say that a comparison of this kind is physically incomplete unless it extends to all the spectral components separately; but for commercial purposes such an extended comparison is too complicated, and indeed useless. Determinations at two points of the spectrum, as proposed by Capt. Abney, would certainly suffice for ordinary purposes; and in view of the convenience of expressing the result by a single number, it is not unlikely that people practically concerned in these matters will content themselves with a comparison at one point. It seems desirable that some convention should be arrived at without much further delay; so that two lights should be considered to be commercially equal, if they have the same intensity of, e.g., sodium or of thallium light. It will be understood that such a mode of estimation assumes that the intensity varies along the spectrum in a gradual manner; and this consideration may tell against the use of the sodium light as a standard, inasmuch as the component in question often predominates unduly in candle-flames from the actual presence of sodium *.

Whatever choice be made, an instrument like the present may be employed to make the desired selection; and it is applicable to any photometric arrangement. For my own experiments I have used the shadow-method, and find it possible to compare any bright sources, however different in colour. The only difficulty arises from the necessary enfeeblement of the light by selection, and this practically precludes observation with standard candles. With gas-flames and glow-lamps the light is sufficient.

* [1900. When this paragraph was written, I had an exaggerated idea of the difficulty of directly comparing the intensities of variously coloured lights.]

123.

SELF-INDUCTION IN RELATION TO CERTAIN EXPERIMENTS OF MR WILLOUGHBY SMITH AND TO THE DETERMINATION OF THE OHM.

[*Nature*, XXXII. p. 7, 1885.]

In a Lecture delivered by Mr Willoughby Smith before the Royal Institution in June last (see *Proceedings*) some experiments are detailed, which are considered to afford an explanation of discrepancies in the results of various investigators relating to the ohm, or absolute unit of electrical resistance. As having given more attention than probably any-one else in recent years to this subject, I should like to make a few remarks upon Mr Willoughby Smith's views, which naturally carry weight corresponding to the good service done by the author in this branch of science.

In the first series of experiments a primary circuit is arranged in connection with a battery and interrupter, and a secondary circuit in connection with a galvanometer and commutator of such a character that the make and break induced currents pass in the same direction through the instrument. Under these circumstances it is found that at high speeds the insertion of a copper plate between the primary and secondary spirals entails a notable diminution in the galvanometer deflection, and this result is regarded as an indication that the molecules of copper need to be polarised by the lines of force—an operation for which there is not time at the higher speeds. The orthodox explanation of the experiment would be that currents are developed by induction in the copper sheet, which thus screens the secondary spiral from the action of the primary, and the result is exactly what might have been anticipated from known electrical principles. I have the less hesitation in saying this, because as a matter of fact I did anticipate from theory the action of a combination very similar in character. The experiment is described in the *Philosophical Magazine* for May 1882 [vol. II. p. 99], and differs from Mr W. Smith's only in the substitution of a telephone for the galvanometer, and of a microphone for the interrupter, no reverser in the secondary circuit being required.   By the interposition of a thick copper sheet the sound is greatly enfeebled.

The second series of experiments were made with Faraday's "new magneto-electro machine," in which a copper disk rotates about its centre between the poles of a horse-shoe magnet.   The currents developed are examined with a galvanometer whose electrodes touch two points upon the disk—in Mr Smith's experiments one at the centre and the other at the circumference.   At low speeds the distribution is symmetrical with respect to that diameter of the disc which is passing at any moment between the poles; but as the speed is increased, a certain "drag" is observed, disturbing the symmetry.   This drag, or lagging, was noticed by Nobili in a very similar arrangement as long ago as 1833 (Wiedemann's *Electricity*, third edition, vol. IV. § 374), and is no doubt to be attributed to the induction of the currents upon themselves.

This question of self-induction is indeed a very important one in respect of certain methods for determining the ohm; but it certainly cannot be said to have been neglected, as Mr W. Smith seems to suggest.   Both in the original experiments of the British Association Committee with a coil revolving about a vertical axis, and in my own recent repetition of them, the self-induction of the coil is a most important feature, and may cause a displacement of the position of the maximum current from the plane of the magnetic meridian through as much as 20°.   In my paper (*Phil. Trans.* 1882, p. 661) [vol. II. Art. 80] I thought I had discussed the question at almost tedious length.

It is possible that Mr W. Smith had in his mind rather determinations by the method of Lorenz, in which Faraday's disk is used.   The arrangement here, however, differs in one very important respect from that of Mr W. Smith's experiments in that lines of force are symmetrically arranged in relation to the axis of rotation.   The consequence is that, however great the speed of rotation, there are no currents circulating in the disk, and therefore no question arises as to the self-induction of such currents.   What is observed is simply the difference of electrical potential between the centre and the circumference.   It is impossible to discuss the matter fully here, but the reader will find all that is necessary by way of explanation in the paper published in the *Phil. Trans.* ("Experiments by the method of Lorenz for the further Determination of the Absolute Value of the British Association Unit of Resistance," etc.).   My object in writing is to correct the inference, suggested by W. Smith's remarks, that the question of self-induction has been neglected by workers upon this subject.

124.

PROFESSOR TAIT'S "PROPERTIES OF MATTER *."

[*Nature*, XXXII. pp. 314, 315, 1885.]

THE subject of this excellent little book includes the Mechanical Properties of matter, and much that is usually treated under the head of Chemical Physics, such as Diffusion and Capillarity. It might be difficult to give a reason why electric and thermal conductivities of mercury, for example, should not be included among its properties as much as its density and its capillarity; but the distinction is convenient, and to some extent sanctioned by usage.

In the introductory chapters the author expounds some rather peculiar views with perhaps more insistence than is desirable in an elementary work. The word "force" is introduced apologetically, and with the explanation that, "as it does not denote either matter or energy, it is not a term for anything objective." No one will dispute the immense importance of the property of conservation, but the author appears to me to press his view too far. As Dr Lodge has already pointed out, if conservation is to be the test of existence, Prof. Tait himself does not exist. I forbear from speculating what Dr Lodge will say when he reads on p. 11 that "not to have its price is conclusive against objectivity."

Chapters IV. to VII. form an elementary treatise on Mechanics in which even the learned reader will find much that is interesting in the way of acute remark and illustration. Under the head of Gravitation are considered Kepler's laws, the experimental methods for determining the constant of gravitation ("the mean density of the earth"), and the attempts (such as Le Sage's) which have been made to explain the origin of gravitation.

The succeeding chapters on the deformation of solids and the compression of solids, liquids and gases, are perhaps the most valuable part

---

* "Properties of Matter." By Prof. Tait (Edinburgh, Black).

of the work, and will convey a much needed precision of ideas to many students of physics whose want of mathematical training deters them from consulting the rather formidable writings of the original workers in this field. The connection of Young's modulus of elasticity, applicable to a rod subject to purely longitudinal pull or push, with the more fundamental elastic constants expressing the behaviour of the body under hydrostatic pressure and pure shearing stress respectively is demonstrated in full. Prof. Tait remarks that "Young's treatment of the subject of elasticity is one of the few really imperfect portions of his great work (*Lectures on Natural Philosophy*). He gives the value of his modulus for water, mercury, air, &c.!" A deficiency of explanation must be admitted, but I am not sure that Young's ideas were really confused. The modulus for solids corresponds to a condition of no lateral force, that for liquid to no lateral extension. The distinction should certainly have been pointed out; but the moduli are really comparable in respect of very important effects, which Young probably had in his mind—viz. the propagation of sound along a bar of the solid in one case, and in the other through a fluid, whether unlimited or contained in an unyielding tube.

As a great admirer of Dr Young's work, I cannot resist adding that if in some respects his treatment of elasticity is defective, in others it is in advance of many modern writings. Witness the following passage*:— "There is however a limit beyond which the velocity of a body striking another cannot be increased without overcoming its resilience, and breaking it, however small the bulk of the first body may be, and this limit depends upon the inertia of the parts of the second body, which must not be disregarded when they are impelled with a considerable velocity. For it is demonstrable that there is a certain velocity, dependent on the nature of a substance, with which the effect of any impulse or pressure is transmitted through it; a certain portion of time, which is shorter accordingly as the body is more elastic, being required for the propagation of the force through any part of it; and if the actual velocity of any impulse be in a greater proportion to this velocity than the extension or compression, of which the substance is capable, is to its whole length, it is obvious that a separation must be produced, since no parts can be extended or compressed which are not yet affected by the impulse, and the length of the portion affected at any instant is not sufficient to allow the required extension or compression."

The theories of "bending" and of "torsion" are discussed in Chapter XI. When the section of the rod deviates from the circular form, the torsional problem becomes rather complicated; but a statement is given of some of

* [Lectures on Natural Philosophy, vol. i. p. 144.]

the interesting results of Saint Venant's investigations. In his treatment of the compression of solids and liquids, the author is able to make valuable contributions derived from his own experimental work.

In the chapter on "gases" a long extract is given from Boyle's *Defence of the Doctrine Touching the Spring and Weight of the Air*, in order to show how completely the writer had established his case in 1662. As to this there can hardly be two opinions, and Prof. Tait is fully justified in insisting upon his objections to "Mariotte's law." In Appendix IV. a curious passage from Newton is discussed, in which the illustrious author appears to speak of Mariotte sarcastically. It is proper that these matters should be put right; but Prof. Tait is hardly impartial enough himself to succeed in enlisting the complete sympathy of foreigners. Cases of glaring injustice should be rectified; but there will always be a tendency (from which Englishmen cannot claim to be exempt) to give a full measure of credit to one's own countrymen, if only because one is better informed concerning their labours.

There is one matter, suitable to an elementary work, which I should be glad to see included in a future edition, viz., the principle of dynamical similarity, or the influence of scale upon dynamical and physical phenomena. It often happens that simple reasoning founded upon this principle tells us nearly all that is to be learned from even a successful mathematical investigation, and in the very numerous cases in which such an investigation is beyond our powers, the principle gives us information of the utmost importance. An example will make this clear. The pitch of a tuning-fork of homogeneous steel is dependent upon the size and shape as well as upon the elastic quality of the material; but the matter is too difficult for rigorous mathematical treatment. If, however, it be asked, How does the pitch depend upon the *size* of the fork, the shape and material being given? we need no complicated mathematics at all. The principle of dynamical similarity tells us at once that the time of vibration is proportional to the linear dimension.

Another example might be taken from a reaction which Prof. Tait describes as specially complex—viz., collision. A glass ball drops upon a marble floor from a height of one foot. How does the size of the ball affect the strains during collision and the danger of rupture? The principle teaches that if the scale of time be altered in the same proportion as the scale of length, similarity is secured, so that the strains are equal at corresponding times and at corresponding places. Hence a larger ball is not more likely to break than a smaller one, unless in consequence of the greater *duration* of the strains. I feel sure that in Prof. Tait's hands this very important and fundamental principle might be made intelligible to the great mass of physical students.

It would lead us too far to refer in detail to the various subjects treated in the later chapters under capillarity, diffusion, osmose, transpiration, viscosity, &c., but there is one point that I should like to mention. The explanation on p. 249 of the behaviour under water of drops of ink and of solution of permanganate of potash assumes the existence of a capillary tension in the surface separating the two fluids. In my own experiments on jets with this very solution, I have never seen any tendency to break up into drops (as, according to Savart and Plateau, there would be in air) and have therefore supposed that the capillary force was *nil*, or at any rate very small. Moreover theory shows that the force depends entirely upon the suddenness of transition between two media, which suddenness must be broken down almost instantaneously when two miscible liquids come into contact. As the matter stands there seems to be here some discrepancy, which, perhaps, Prof. Tait could elucidate.

In his preface the author holds out hopes of further volumes on the same plan, dealing with dynamics, sound, and electricity. The readers of the present work will, I am sure, join in the wish that the appearance of these may be delayed no longer than is absolutely necessary.

125.

A THEOREM RELATING TO THE TIME-MODULI OF DISSIPATIVE SYSTEMS.

[*Report of the British Association*, pp. 911, 912. 1885.]

In the *Proceedings of the Mathematical Society* for 1873 [Art. 21], it is shown that the time of vibration of a conservative system fulfils a stationary condition, so that the time of vibration in any normal mode would remain unaltered, even though the system, by the application of suitable constraints, be made to vibrate in a mode slightly different. It is pretty evident that a similar theorem must obtain for the time-moduli of the normal modes of a dissipative system, but a formal statement may not be useless.

The class of systems referred to is that of which the mechanical properties depend upon two functions, one being the dissipation function F and the other either the kinetic energy T, or the potential energy V. As examples of the first case may be mentioned the subsidence of the small motion of a viscous fluid contained in a fixed envelope, and of free electric currents in a conductor. On the other hand, in the distribution of heat in a thermal conductor, or of electricity in a cable, the undissipated energy is usually regarded as potential. The argument is almost exactly the same whichever case be contemplated; to fix ideas we will take the former.

By suitable transformation the two quadratic functions T and F may be reduced to sums of squares of co-ordinates, and these co-ordinates are consequently called normal. Thus:—

$$T = \tfrac{1}{2} [1] \, \dot{\phi}_1{}^2 + \tfrac{1}{2} [2] \, \dot{\phi}_2{}^2 + \ldots$$

$$F = \tfrac{1}{2} (1) \, \dot{\phi}_1{}^2 + \tfrac{1}{2} (2) \, \dot{\phi}_2{}^2 + \ldots$$

in which all the coefficients $[1] \ldots (1) \ldots$ are positive.

The normal modes are those represented by the separate variation of the co-ordinates, and the corresponding differential equations are of the form :—

$$(s) \, \ddot{\phi}_s + [s] \, \dot{\phi}_s = 0,$$

whence

$$\phi_s = P e^{-pt},$$

where

$$p = (s)/[s].$$

If τ_s be the time-modulus, the time in which the motion is diminished in the ratio of $e : 1$, $\tau_s = p^{-1}$.

Suppose now that by suitable constraints an arbitrary type of motion is imposed upon the system, so that $\phi_1 = A_1 \theta$, $\phi_2 = A_2 \theta$, ... where A_1, A_2, &c. are given (real) coefficients. Then

$$T = \{\tfrac{1}{2} \, [1] \, A_1{}^2 + \tfrac{1}{2} \, [2] \, A_2{}^2 + \dots\} \, \dot{\theta}^2,$$
$$F = \{\tfrac{1}{2} \, [1] \, A_1{}^2 + \tfrac{1}{2} \, [2] \, A_2{}^2 + \dots\} \, \dot{\theta}^2;$$

and the equation of motion

$$\frac{d}{dt}\left(\frac{dT}{d\dot{\theta}}\right) + \frac{dF}{d\dot{\theta}} = 0$$

gives as the solution $\theta \propto e^{-pt}$, where

$$p = \frac{(1) \, A_1{}^2 + (2) \, A_2{}^2 + \dots}{[1] \, A_1{}^2 + [2] \, A_2{}^2 + \dots}.$$

It is evident that the value of p (and therefore of τ) is stationary when all but one of the coefficients A_1, A_2, &c., vanish, that is when the type coincides with one of those natural to the system.

From this theorem corollaries may be drawn as from the corresponding theorem for times of vibration. The greatest time-modulus can only be reduced by the application of constraint, and where the normal mode is difficult of calculation a good approximation to the greatest time-modulus may be had from a hypothetical type chosen so as not to deviate too widely from the real one. Any increase in T, or diminution in F, as a function of the co-ordinates entails in general an augmentation in all the time-moduli. In the case of free electric currents, already referred to as an example, this augmentation of time-moduli would result from the approximation of iron (treated as a non-conductor), or from an improvement (however local) in conductivity.

126.

ON THE ACCURACY OF FOCUS NECESSARY FOR SENSIBLY PERFECT DEFINITION.

[*Phil. Mag.* xx. pp. 354—358, 1885.]

IN my "Investigations in Optics*" I have examined the effect upon definition of small disturbances of the wave-surfaces from their proper forms. It follows, for instance, that the aberration of a plano-convex lens focusing parallel rays of homogeneous light is unimportant, so long as the fourth power of the angular semi-aperture does not exceed the ratio of the wave-length to the focal distance $\{\alpha^4 < (\lambda/f)\}$, a condition satisfied by a lens of 3 feet focus, provided that the aperture be less than 2 inches. I propose at present to apply similar principles to the question of focusing.

The most convenient point of view is that explained† for calculating the

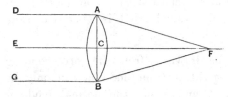

focal length of lenses. If the lens AB converges parallel rays to a focus at F, the retardation of the central ray EF, due to the substitution of a thickness t of glass for air, is $(\mu - 1)t$; and this must be equal to the retardation of the extreme rays passing the (sharp) edge of the lens, *i.e.* $AF - CF$. Thus, if $AC = y$, $FC = f$,

$$(\mu - 1)\,t = \sqrt{(f^2 + y^2)} - f = \tfrac{1}{2}y^2/f, \quad\ldots\ldots\ldots\ldots\ldots(1)$$

approximately, which gives the focal length in terms of the semi-aperture and the "thickness" of the lens.

If we suppose that μ varies,

$$\delta\mu\,.\,t = -\frac{1}{2}\frac{y^2}{f^2}\,\delta f, \quad\ldots\ldots\ldots\ldots\ldots\ldots(2)$$

giving the change of focus required to compensate the change of μ. Let us, however, inquire what is the state of things at the old focus. The secondary rays from the extreme boundary of the lens arrive with the same phase as before the change of index; but the central ray undergoes a relative retardation amounting to $\delta\mu\,.\,t$. This quantity tells us the discrepancy of phase;

---

* *Phil. Mag.* 1879 and 1880. [Vol. I. p. 435.]

† *Loc. cit.* p. [439].

and we know that if it is less than $\frac{1}{2}\lambda$, the agreement of phase is still good enough to give nearly perfect definition. Hence from (2) we see that a displacement δf from the true focus will not impair definition, provided

$$\delta f < \frac{f^2\lambda}{y^2} \quad \dots\dots\dots\dots\dots\dots\dots\dots\dots\dots(3)$$

It appears that the linear accuracy required is the same whatever the absolute aperture of the object-glass may be, provided that the *ratio* of aperture to focal length be preserved.

In some trials that I have made the diameter of the object-glass was $1\frac{1}{8}$ inch, and the focal length 12 inches [inch = 2·54 cm.]. Taking

$$\lambda = \tfrac{1}{40,000} \text{ inch,}$$

we get from (3)

$$\delta f < ·0115 \text{ inch,}$$

a result which corresponded very well with observation. The instruments employed were the collimator and telescope of a spectrometer, the object under examination being a slit backed with a soda-flame. A high-power eye-piece was used, and the telescope was adjusted until the edge of the slit and the wire in the eye-piece were seen well defined together. The instrument was unprovided with an easy focusing motion, so that it was not possible to try backwards and forwards conveniently. In this way the setting corresponded more closely to the suppositions of theory than if it were the result of comparisons between appearances at equal distances within and without the point chosen. It will be understood that there is no theoretical limit to the accuracy with which a focal point may be ultimately determined, if the lenses are good, and observations are multiplied with suitable precautions to avoid asymmetry.

In ten settings the extreme difference was only ·02 inch, showing that a displacement of ·01 inch from the true focal point was just recognizable.

By using various coloured flames, or by throwing a spectrum upon the slit of the apparatus, we may determine the focal length for different kinds of light. With proper achromatic lenses the differences should be pretty small, the minimum focal length corresponding to the yellow-green rays. It so happens that my instrument is far from properly compensated, and gives a fair primary spectrum, so that the difference of focus for yellow and green is very easily recognized. In the case of a single lens this method would give the dispersive power of the glass with fair accuracy. By comparison with the theory of the resolving power of prisms, we see that the dispersion is about as favourably determined with a lens as with a prism of equal thickness. In either case a change of index such that $\delta\mu \cdot t = \frac{1}{2}\lambda$ leaves the phase agreement nearly unaltered at the original points; but in other respects the circumstances are probably rather more favourable in the case of the prism.

It is generally considered that the most accurate way of focusing a small telescope is to move the eye across the eye-piece, altering the adjustment until there seems to be no relative motion of object and cross wires. I have tried this plan in an improved form in order to see whether a higher degree of accuracy of adjustment was really attainable, although theory seemed to show that no great advance was to be looked for. A heavy pendulum, executing complete vibrations in about two seconds, was fitted up in front of the telescope, and carried with it a screen perforated by a slit. The width of the slit was about a quarter of the entire aperture, and the oscillations were at first of such amplitude as just to bring the extreme edges of the lens into play. In the earlier experiments the slit of the collimator was backed by the clouds, a piece of green glass being interposed. This was before I had discovered the remarkably unachromatic character of the instruments, and I was puzzled to interpret the appearances presented. On one side of the focus the relative motion of the image was (as it should be) in the same direction as that of the pendulum, and on the other side in the opposite direction; but the transition was not well defined, and the image executed evolutions very visible to the observer, who at the same time was not able to describe them as swinging in one direction or the other. The effect upon the eye was remarkably unpleasant and fatiguing to watch; it disappeared when recourse was had to sodium light, and doubtless depended upon the variation of quality in the light. It may be noticed that spherical aberration would show itself by a swinging of the image in a period *half* that of the pendulum.

With the soda-flame the adjustment to focus by getting rid of the swinging motion was pretty accurate; but not much advantage was gained in comparison with a setting by simple inspection under full aperture. As before, the extreme difference in a set of ten was about ·02 inch.

The substitution of white for monochromatic light was instructive. In either extreme position of the oscillating slit the light was seen to be spread into a spectrum of moderate length, the blue and red being interchanged after each half period. Under these circumstances the cross wires can be made to maintain their position in that part of the spectrum only for which the telescope is focused. If, for example, it be the green of the spectrum, we may bring the cross wires to this position when the pendulum is at rest, and then, in spite of the oscillation, the position will be maintained. If, without altering the focus, we move the cross wires to another part of the spectrum, then, when the pendulum oscillates, the wires will be seen on a different part of the spectrum after each half period. In order to fix the new part of the spectrum upon the cross wires, a change of focus is demanded. This experiment would hardly succeed with properly compensated object-glasses, but it could be imitated with the aid of single lenses.

127.

ON AN IMPROVED APPARATUS FOR CHRISTIANSEN'S EXPERIMENT.

[*Philosophical Magazine*, xx. pp. 358—360, 1885.]

THE very beautiful experiment in question, described by C. Christiansen in Wiedemann's *Annalen* for November 1884, consists in immersing glass-powder in a mixture of benzole and bisulphide of carbon of such proportions that for one part of the spectrum the indices of the solid and of the fluid are the same. Being interested in this subject from having employed the same principle for a direct-vision spectroscope (*Phil. Mag.* January 1880, p. 53) [vol. I. p. 456], I have repeated Christiansen's experiment in a somewhat improved form, which it may be worth while briefly to describe, as the matter is one of great optical interest.

I must premise that the beauty of the effect depends upon the correspondence of index being limited to one part of the spectrum. Rays lying within a very narrow range of refrangibility traverse the mixture freely, but the neighbouring rays are scattered laterally, much as in passing ground glass. Two complementary colours are therefore exhibited, one by direct, and the other by oblique, light. In order to see these to advantage, there should not be much diffused illumination; otherwise the directly transmitted monochromatic light is liable to be greatly diluted. The prettiest colours are obtained when the undisturbed rays are from the green; but the greatest general transparency corresponds to a lower point in the spectrum.

The improvement referred to relates merely to the use of a flat-sided bottle to contain the preparation. In order to get a satisfactory result it is necessary that the sides of the containing vessel be pretty good optically. This condition may be satisfied with a built-up cell, but on account of the difficulty of finding a suitable cement, it is rarely that such cells remain in good order for any length of time. It occurred to me that a bottle might be made to answer the purpose, provided the precaution were taken of *using the same kind of glass for the bottle and for the powder.* The *outer* surfaces of the glass sides of the bottle can be worked flat, while the unavoidable irregularities of the inner surfaces are compensated by the liquid, which, being adjusted to have the same index as the powder, will have also the same index as the glass of the bottle.

The bottles that I have used* are about 3 inches high, $1\frac{1}{2}$ inch wide, and about $\frac{3}{4}$ inch thick, outside measurement. The outer surfaces are worked (like plate-glass), and not merely flattened upon a wheel, as is usual with ordinary perfume bottles. For my earlier trials I was provided with a piece of flint glass from the same pot as the bottles; but although the experiment succeeded well enough as regards the elimination of the internal irregularities of the walls, the glass-powder itself did not behave as well as I had seen plate-glass powder do. It appeared ultimately that the flint was not sufficiently homogeneous for the purpose, and another specimen of flint was also a partial failure, from the same cause; but a sample of optical flint, kindly supplied to me by Dr Hopkinson, gave excellent results.

It is more important that the powder should be homogeneous in itself than that it should correspond very accurately with the glass of the bottle. For ordinary purposes plate-glass powder (all, of course, from one piece) may be used in a bottle of soda-glass, or even of ordinary low flint. In preparing the powder great care is required to exclude dirt. With respect to the coarser grades there is no great difficulty, but the finer powder is apt to be contaminated with the substance of the mortar. I prefer to use one of iron, so that a magnet will remove the foreign matter. The elimination of fine dust is also facilitated by a blast of wind from bellows.

In order to get good definition it is necessary not only that the powder be homogeneous, but that the temperature be uniform; for, as Christiansen has shown, the transmitted ray rises rapidly in refrangibility with temperature. In order to secure homogeneity it is sometimes necessary to shake up the preparation, which (to prevent the formation of air-bubbles) is best done with a rather gentle motion while the bottle is held nearly horizontal. The proportion of liquids necessary varies with the temperature and with the kind of glass. Flint will require a higher proportion of bisulphide of carbon than plate-glass. If the transmitted light is too high in the spectrum, add more bisulphide of carbon; if too low, more benzole.

The experimenter, working in a room lighted in the ordinary manner, will be disposed at first to underrate the degree of homogeneity of the transmitted light. If the shutters be nearly closed, so that light finds entrance through a narrow slit only, a better judgment can be formed, which may be tested by prismatic analysis.

* Supplied by Messrs Harvey and Peak, Beak Street, Regent Street.

In an otherwise dark room lighted by a powerful soda-flame, it is interesting to remark how very slight a change in the critical colour manifests itself in the general appearance of surrounding objects seen through the preparation. When the ray of maximum transmission corresponds closely to that of soda, the powder is almost invisible, and objects are seen as through a clear medium. But so slight a change of temperature brings with it a hazy appearance, that it requires some care to obtain the best effect. It is desirable also to exclude by absorbing media the blue light which usually attends a soda-flame in very sensible degree.

[1900. In subsequent experiments (*Nature*, LX. p. 64, 1899) it has been found that the range of refrangibility of the light transmitted by a layer 15 to 20 mm. thick is but $2\frac{1}{2}$ times that embraced by the two *D*-lines.]

28—2

128.

OPTICAL COMPARISON OF METHODS FOR OBSERVING SMALL ROTATIONS.

[*Philosophical Magazine*, xx. pp. 360, 361, 1885.]

In order to measure very small rotations, *e.g.* of the suspended parts of a galvanometer or magnetometer, two methods are commonly employed. We may either observe with a magnifier the motion of a material pointer; or, following Gauss, cause the rotating parts to carry round a mirror in which a scale is seen by reflection. In a modification of Gauss's method, well known from Sir W. Thomson's galvanometers, the image of a dark or bright line is thrown objectively upon the scale. In deciding which arrangement to adopt in any particular case, various circumstances would have to be taken into account, but still a comparison of capabilities from a purely optical point of view is not without interest.

In the mirror method the optical limit depends upon the horizontal breadth of the mirror itself. The easiest road to the desired conclusion, as well as the most instructive, is by a direct application of the principles of the wave theory. To take the simplest case, we will suppose the mirror rectangular. Consider, then, a luminous point, and its image after reflection, whether in the focal plane of a telescope, or formed directly upon a scale. The optical work being perfect, the secondary rays from every part of the mirror agree in phase at the focal point. Now suppose that the mirror rotates through such an angle that one vertical edge advances a quarter of a wave-length ($\frac{1}{4}\lambda$), while the other retreats to the same amount, and consider the effect on the phase-relations at the point in question. It is evident that one extreme wave is accelerated and the other retarded by $\frac{1}{2}\lambda$, and that the phases are now distributed uniformly over a complete cycle. The result is therefore darkness; and the effect of the rotation has been to shift the image through half the width of the central bright band which, with accompanying fringes, is the representative in the image of a mathematical line*. Such a motion would be visible (with proper arrangements as to magnifying), but the limits of resolving power are being approached. It is to be noticed that the conclusion is independent of the focal length of the mirror and of the employment of a telescope. Provided of course that the full width of the mirror is really used, a motion of its vertical edges through $\frac{1}{4}\lambda$ may be made evident.

A comparison with the method by direct observation of a pointer is now easy; for, as has been proved by the researches of Abbe and Helmholtz, a motion of $\frac{1}{4}\lambda$ may be rendered evident in a very similar degree by direct application of a perfect microscope to the moving object. If, therefore, we suppose the length of the pointer to be equal to the half-width of the mirror, the two methods are optically upon a level. It is needless to say that it would be easy to give the pointer a great advantage in this respect; but the direct use of the microscope would often be interfered with by motions in the line of sight, making it impossible to preserve the focus. And besides this, it is sometimes necessary for the observer to remain at a distance.

* See "Investigations in Optics," *Phil. Mag.* 1879, 1880.   [Art. 62, vol. I.]

129.

ON THE THERMODYNAMIC EFFICIENCY OF THE THERMOPILE.

[*Philosophical Magazine*, xx. pp. 361—363, 1885.]

DURING the last few years the thoughts of many electricians have turned to the question of the possibility of replacing the dynamo by some development of the thermopile; and it is, I believe, pretty generally recognized that the difficulty in the way is the too free passage of heat by ordinary conduction from the hot to the cold junction. The matter may perhaps be placed in a clearer light by an actual calculation, accompanied by a rough numerical estimate applicable to the case of German silver and iron.

If t, t_0 denote the temperatures of the hot and cold junctions respectively, e the electromotive force per degree Centigrade, the whole electromotive force for n pairs in series will be represented approximately by

$$ne(t - t_0).$$

The magnitude of the current (C) is found by dividing this by the sum of the internal and external resistances $(R_0 + R)$; and the useful work done externally per second is RC^2. It reaches a maximum when the external resistance is equal to the internal; and its amount is then

$$\frac{n^2 e^2 (t - t_0)^2}{4R_0}.$$

The value of the internal resistance R_0 depends upon the dimensions and specific resistances of the bars. Denoting the latter quantities by r_1, r_2, and taking σ_1, σ_2 to represent the areas of section, the common length being l, we have

$$R_0 = nl \left(\frac{r_1}{\sigma_1} + \frac{r_2}{\sigma_2}\right);$$

so that the external work per second is

$$\frac{ne^2 (t - t_0)^2}{4l \left(\dfrac{r_1}{\sigma_1} + \dfrac{r_2}{\sigma_2} \right)}.$$

We will now compare this with the work dissipated by ordinary conduction of heat along the bars.

If Q be the amount of heat conducted by the n pairs, r_1', r_2' the *thermal* resistances, then

$$Q = n \left(\frac{\sigma_1}{r_1' l} + \frac{\sigma_2}{r_2' l} \right) (t - t_0).$$

The fraction of this heat, supplied at temperature t, which might be converted into work by a perfect engine working between the absolute temperatures t and t_0, is $(t - t_0)/t$; so that the work dissipated per second is

$$\frac{nJ (t - t_0)^2}{tl} \left(\frac{\sigma_1}{r_1'} + \frac{\sigma_2}{r_2'} \right),$$

where J denotes the mechanical equivalent of heat.

The ratio of this to the useful work is

$$\frac{4J}{te^2} \left(\frac{r_1}{\sigma_1} + \frac{r_2}{\sigma_2} \right) \left(\frac{\sigma_1}{r_1'} + \frac{\sigma_2}{r_2'} \right),$$

independent of $(t - t_0)$, of n, and of l. It is further evident that the ratio in question does not depend upon the absolute values of the sections, or of the electrical and thermal resistances, but only upon the *ratios* of these quantities. Thus the *efficiency* of the thermopile is independent of the absolute dimensions of the bars, and even of the difference of temperatures at the junctions. The *power* is increased by diminishing the length and increasing the sections to a limit not indicated by these expressions (in which the terminal temperatures are regarded as given), and probably determined in practice by the necessity of conveying the heat to the scene of action.

The resistances being given, the ratio of sections σ_2/σ_1 is to be determined so as to make our ratio a minimum. This happens when

$$\left(\frac{\sigma_2}{\sigma_1} \right)^2 = \frac{r_2 r_2'}{r_1 r_1'};$$

and thus we get

$$\frac{4J}{te^2} \left(\sqrt{\frac{r_1}{r_1'}} + \sqrt{\frac{r_2}{r_2'}} \right)^2.$$

To turn this into numbers for the case of German silver and iron, we have the following approximate numbers in C.G.S. measure * :—

$$r_1 = 2\cdot0 \times 10^4, \qquad\qquad r_1' = 1\cdot0 \times 10^1,$$
$$r_2 = 1\cdot0 \times 10^4, \qquad\qquad r_2' = 5\cdot0 \times 10^0,$$
$$e = 3\cdot0 \times 10^3.$$

The value of J is $4\cdot2 \times 10^7$, and for t we will assume 500 (absolute measure). The use of these gives, as the ratio of the work that would be obtained by a perfect engine from the conducted heat to that actually obtained from the thermo-electric force, the number 300; from which we may conclude that the steam-engine and dynamo are not likely to be superseded by a German-silver and iron thermopile, even though considerable allowance be made for the admitted roughness of the numerical estimate.

As regards other materials, it is interesting to note that the ratio contains as divisor the *square* of the electromotive force per degree.

---

* Everett's " Units and Physical Constants "; Landolt's *Tabellen.*

130.

ON WAVES PROPAGATED ALONG THE PLANE SURFACE OF AN ELASTIC SOLID.

[*Proceedings of the London Mathematical Society*, XVII. pp. 4—11, 1885.]

IT is proposed to investigate the behaviour of waves upon the plane free surface of an infinite homogeneous isotropic elastic solid, their character being such that the disturbance is confined to a superficial region, of thickness comparable with the wave-length. The case is thus analogous to that of deep-water waves, only that the potential energy here depends upon elastic resilience instead of upon gravity*.

Denoting the displacements by α, β, γ, and the dilatation by θ, we have the usual equations

$$\rho \frac{d^2\alpha}{dt^2} = (\lambda + \mu) \frac{d\theta}{dx} + \mu \nabla^2 \alpha, \quad \&c., \quad \dots\dots\dots\dots\dots(1)$$

in which

$$\theta = \frac{d\alpha}{dx} + \frac{d\beta}{dy} + \frac{d\gamma}{dz}. \quad \dots\dots\dots\dots\dots\dots(2)$$

If α, β, γ all vary as e^{ipt}, equations (1) become

$$(\lambda + \mu) \frac{d\theta}{dx} + \mu \nabla^2 \alpha + \rho p^2 \alpha = 0, \quad \&c. \quad \dots\dots\dots\dots(3)$$

Differentiating equations (3) in order with respect to x, y, z, and adding, we get

$$(\nabla^2 + h^2) \theta = 0, \quad \dots\dots\dots\dots\dots\dots(4)$$

in which

$$h^2 = \rho p^2 / (\lambda + 2\mu). \quad \dots\dots\dots\dots\dots\dots(5)$$

---

* The statical problem of the deformation of an elastic solid by a harmonic application of pressure to its surface has been treated by Prof. G. Darwin, *Phil. Mag.* Dec. 1882. Jan. 1886. —See also Camb. Math. Trip. Ex. Jan. 20, 1875, Question IV.

Again, if we put

$$k^2 = \rho p^2/\mu, \qquad \qquad (6)$$

equations (3) take the form

$$(\nabla^2 + k^2)\, \alpha = \left(1 - \frac{k^2}{h^2}\right)\frac{d\theta}{dx}, \quad \&c. \qquad (7)$$

A particular solution of (7) is*

$$\alpha = -\frac{1}{h^2}\frac{d\theta}{dx}, \qquad \beta = -\frac{1}{h^2}\frac{d\theta}{dy}, \qquad \gamma = -\frac{1}{h^2}\frac{d\theta}{dz}; \qquad (8)$$

in order to complete which it is only necessary to add complementary terms u, v, w satisfying the system of equations

$$(\nabla^2 + k^2)\, u = 0, \qquad (\nabla^2 + k^2)\, v = 0, \qquad (\nabla^2 + k^2)\, w = 0, \ldots (9)$$

$$\frac{du}{dx} + \frac{dv}{dy} + \frac{dw}{dz} = 0. \qquad (10)$$

For the purposes of the present problem we take the free surface as the plane $z = 0$, and assume that, as functions of x and y, the displacements are proportional to e^{ifx}, e^{igy}. Thus (4) takes the form

$$(d^2/dz^2 + h^2 - f^2 - g^2)\, \theta = 0;$$

so that

$$\theta = Pe^{-rz} + Qe^{+rz}, \qquad (11)$$

where

$$r^2 = f^2 + g^2 - h^2. \qquad (12)$$

In (11), r is supposed to be real; otherwise the dilatation would penetrate to an indefinite depth. For the same reason, we must retain only that term (say the first) for which the exponent is negative within the solid†. Thus $Q = 0$, and we will write for brevity $P = 1$, or rather $P = e^{ipt}\, e^{ifx}\, e^{igy}$; but the exponential factors may often be omitted without risk of confusion, so that we may take

$$\theta = e^{-rz}. \qquad (13)$$

At the same time the particular solution becomes

$$\alpha = -\frac{if}{h^2}\, e^{-rz}, \qquad \beta = -\frac{ig}{h^2}\, e^{-rz}, \qquad \gamma = \frac{r}{h^2}\, e^{-rz}. \qquad (14)$$

For the complementary terms, which must also contain e^{ifx}, e^{igy} as factors, equations (9) become

$$(d^2/dz^2 + k^2 - f^2 - g^2)\, u = 0, \quad \&c.; \qquad (15)$$

---

* Lamb on the Vibrations of an Elastic Sphere, *Math. Soc. Proc.* May 1882.

† By discarding these restrictions we may deduce the complete solution applicable to a plate, bounded by parallel plane free surfaces; but I have not obtained any results which seem worthy of quotation.

whence, as before, on the assumption that the disturbance is limited to a superficial stratum,

$$u = Ae^{-sz}, \qquad v = Be^{-sz}, \qquad w = Ce^{-sz}, \quad \dots\dots\dots(16)$$

where

$$s^2 = f^2 + g^2 - k^2. \quad \dots\dots\dots\dots\dots\dots(17)$$

In order to satisfy (10), the coefficients in (16) must be subject to the relation

$$ifA + igB - sC = 0. \quad \dots\dots\dots\dots\dots\dots(18)$$

The complete values of α, β, γ may now be written

$$\alpha = -\frac{if}{h^2}e^{-rz} + Ae^{-sz}, \qquad \beta = -\frac{ig}{h^2}e^{-rz} + Be^{-sz}, \qquad \gamma = \frac{r}{h^2}e^{-rz} + Ce^{-sz},$$

$$\dots\dots\dots(19)$$

in which A, B, C are subject to (18); and the next step is to express the boundary conditions for the free surface. The two components of tangential stress must vanish, when $z = 0$, and these are proportional to

$$\frac{d\beta}{dz} + \frac{d\gamma}{dy}, \qquad \frac{d\gamma}{dx} + \frac{d\alpha}{dz}$$

respectively. Hence

$$sB = \frac{2igr}{h^2} + igC, \qquad sA = \frac{2ifr}{h^2} + ifC. \quad \dots\dots\dots(20)$$

Substituting from (20) in (18), we find

$$C(s^2 + f^2 + g^2)h^2 + 2r(f^2 + g^2) = 0. \quad \dots\dots\dots(21)$$

We have still to introduce the condition that the normal traction is zero at the surface. We have, in general,

$$N_3 = \lambda\theta + 2\mu\frac{d\gamma}{dz};$$

or, if we express λ in terms of μ, h, k,

$$N_3 = \mu\left\{\left(\frac{k^2}{h^2} - 2\right)\theta + 2\frac{d\gamma}{dz}\right\};$$

so that the condition is

$$k^2 - 2h^2 - 2(+r^2 + h^2sC) = 0,$$

or, on substitution for r^2 of its value from (12),

$$k^2 - 2(f^2 + g^2) - 2h^2sC = 0. \quad \dots\dots\dots\dots(22)$$

By eliminating C between (21) and (22), we obtain the equation by which the time of vibration is determined as a function of the wave-lengths and of the properties of the solid. It is

$$\{k^2 - 2(f^2 + g^2)\}\{s^2 + f^2 + g^2\} + 4rs(f^2 + g^2) = 0,$$

or, by (17),

$$\{2(f^2 + g^2) - k^2\}^2 = 4rs(f^2 + g^2). \quad \dots\dots\dots(23)$$

If we square (23), and introduce the values of r^2 and s^2 from (12), (17), we get

$$\{2\,(f^2+g^2)-k^2\}^4 = 16\,(f^2+g^2)^2\,(f^2+g^2-h^2)\,(f^2+g^2-k^2).$$

As f and g occur here only in the combination (f^2+g^2), a quantity homogeneous with h^2 and k^2, we may conveniently replace (f^2+g^2) by unity. Thus

$$k^8 - 8k^6 + 24k^4 - 16k^2 - 16h^2k^2 + 16h^2 = 0. \quad\ldots\ldots\ldots\ldots(24)$$

Since the ratio $h^2 : k^2$ is known, this equation reduces to a cubic and determines the value of either quantity.

If the solid be incompressible ($\lambda = \infty$), $h^2 = 0$, and the equation becomes

$$k^6 - 8k^4 + 24k^2 - 16 = 0. \quad\ldots\ldots\ldots\ldots\ldots\ldots(25)$$

The real root of (25) is found to be $\cdot91275$, and the equation may be written

$$(k^2 - \cdot91275)\,(k^4 - 7\cdot08725\,k^2 + 17\cdot5311) = 0.$$

The general theory of vibrations of stable systems forbids us to look for complex values of k^2, as solutions of our problem, though it would at first sight appear possible with them to satisfy the prescribed conditions by taking such roots of (12), (17), as would make the *real* parts of the exponents in e^{-rz}, e^{-sz} negative. But, referring back to (23), which we write in the form

$$(2 - k^2)^2 = 4rs,$$

or, in the present case of incompressibility, by putting $r = 1$,

$$(2 - k^2)^2 = 4s,$$

we see that we are not really free to choose the sign of s. In fact, from the complex values of k^2, viz., $3\cdot5436 \pm 2\cdot2301i$, we find

$$4s = -2\cdot7431 \pm 6\cdot8846i;$$

so that the real part of s is of the opposite sign to r, and therefore e^{-rz}, e^{-sz} do not both diminish without limit as we penetrate further and further into the solid.

Dismissing then the complex values, we have, in the case of incompressibility, the single solution

$$k^2 = \frac{\rho p^2}{\mu} = \cdot91275\,(f^2+g^2). \quad\ldots\ldots\ldots\ldots\ldots(26)$$

From (19), (20), (21), we get in general

$$h^2\alpha = if\left\{-e^{-rz} + \frac{2rs}{s^2+f^2+g^2}\,e^{-sz}\right\}, \quad\ldots\ldots\ldots\ldots(27)$$

$$h^2\beta = ig\left\{-e^{-rz} + \frac{2rs}{s^2+f^2+g^2}\,e^{-sz}\right\}, \quad\ldots\ldots\ldots\ldots(28)$$

$$h^2\gamma = r\left\{+e^{-rz} - \frac{2\,(f^2+g^2)}{s^2+f^2+g^2}\,e^{-sz}\right\}. \quad\ldots\ldots\ldots\ldots(29)$$

In the case of incompressibility, we have k^2 given by (26), and

$$r^2 = f^2 + g^2, \quad s^2 = \cdot 08725 \, (f^2 + g^2).$$

Hence

$$
\left.
\begin{aligned}
h^2\alpha &= if \{ - e^{-rz} + \cdot 5433 \, e^{-sz} \} \, e^{ipt} \, e^{ifx} \, e^{igy} \\
h^2\beta &= ig \{ - e^{-rz} + \cdot 5433 \, e^{-sz} \} \, e^{ipt} \, e^{ifx} \, e^{igy} \\
h^2\gamma &= \sqrt{(f^2 + g^2)} \, \{ e^{-rz} - 1\cdot840 \, e^{-sz} \} \, e^{ipt} \, e^{ifx} \, e^{igy}
\end{aligned}
\right\} \quad \dots\dots\dots\dots(30)
$$

If we suppose the motion to be in two dimensions only, we may put $g = 0$; so that $\beta = 0$, and

$$
\left.
\begin{aligned}
h^2\alpha/f &= i \{ - e^{-fz} + \cdot 5433 \, e^{-sz} \} \, e^{ipt} \, e^{ifx} \\
h^2\gamma/f &= \{ \ e^{-fz} - 1\cdot840 \, e^{-sz} \} \, e^{ipt} \, e^{ifx}
\end{aligned}
\right\} , \quad \dots\dots\dots\dots(31)
$$

in which

$$k = \cdot 9554 f, \quad s = \cdot 2954 f. \quad \dots\dots\dots\dots\dots(32)$$

For a progressive wave we may take simply the real parts of (31). Thus

$$
\left.
\begin{aligned}
h^2\alpha/f &= (e^{-fz} - \cdot 5433 \, e^{-sz}) \sin (pt + fx) \\
h^2\gamma/f &= (e^{-fz} - 1\cdot840 \, e^{-sz}) \cos (pt + fx)
\end{aligned}
\right\} . \quad \dots\dots\dots\dots(33)
$$

The velocity of propagation is p/f, or $\cdot 9554 \sqrt{(\mu/\rho)}$, in which $\sqrt{(\mu/\rho)}$ is the velocity of purely transverse plane waves. The surface waves now under consideration move, therefore, rather more slowly than these.

From (32), (33), we see that α vanishes for all values of x and t when $e^{(s-f)z} = \cdot 5433$, i.e., when $fz = \cdot 8659$. Thus, if λ' be the wave-length $(2\pi/f)$, the horizontal motion vanishes at a depth equal to $\cdot 1378 \lambda'$. On the other hand, there is no finite depth at which the vertical motion vanishes.

To find the motion at the surface itself, we have only to put $z = 0$ in (33). We may drop at the same time the constant multiplier (h^2/f) which has no present significance. Accordingly,

$$\alpha = \cdot 4567 \sin (pt + fx), \qquad \gamma = - \cdot 840 \cos (pt + fx), \quad \dots\dots\dots(34)$$

showing that the motion takes place in elliptic orbits, whose vertical axis is nearly the double of the horizontal axis.

The expressions for stationary vibrations may be obtained from (30) by addition to the similar equations obtained by changing the sign of p, and similar operations with respect to f and g. Dropping an arbitrary multiplier, we may write

$$
\left.
\begin{aligned}
\alpha &= -f \{ - e^{-rz} + \cdot 5433 \, e^{-sz} \} \cos pt \sin fx \cos gy \\
\beta &= -g \{ - e^{-rz} + \cdot 5433 \, e^{-sz} \} \cos pt \cos fx \sin gy \\
\gamma &= \ r \{ + e^{-rz} - 1\cdot840 \, e^{-sz} \} \cos pt \cos fx \cos gy
\end{aligned}
\right\} , \quad \dots\dots\dots(35)
$$

in which

$$r = \sqrt{(f^2 + g^2)}, \qquad s = \cdot 2954 \sqrt{(f^2 + g^2)}. \quad \dots\dots\dots\dots(36)$$

As before, the horizontal motion vanishes at a depth such that

$$\sqrt{(f^2 + g^2)}\, z = \cdot 8659.$$

We will now examine how far the numerical results are affected when we take into account the finite compressibility of all natural bodies. The ratio of the elastic constants is often stated by means of the number expressing the ratio of lateral contraction to longitudinal extension when a bar of the material is strained by forces applied to its ends. According to a theory now generally discarded, this ratio (σ) would be $\frac{1}{4}$; a number which, how-ever, is not far from the truth for a variety of materials, including the principal metals. In the extreme case of incompressibility σ is $\frac{1}{2}$, and there seems to be no theoretical reason why σ should not have any value between this and -1*.

The accompanying table will give an idea of the progress of the values of $k^2/(f^2 + g^2)$ as dependent upon λ/μ, or upon σ. It will be observed that the value diminishes continuously with λ, in accordance with a general principle†.

λ	σ	h^2/k^2	$k^2/(f^2+g^2)$	$k/\sqrt{(f^2+g^2)}$
∞	$\frac{1}{2}$	0	$\cdot 9127$	$\cdot 9554$
μ	$\frac{1}{4}$	$\frac{1}{3}$	$\cdot 8453$	$\cdot 9194$
0	0	$\frac{1}{2}$	$\cdot 7640$	$\cdot 8741$
$-\frac{2}{3}\mu$	-1	$\frac{3}{4}$	$\cdot 4746$	$\cdot 6896$

As an example of finite compressibility, we will consider further the second case of the table. From (12), (17),

$$r^2 = \cdot 7182\,(f^2 + g^2), \qquad r = \cdot 8475\,\sqrt{(f^2 + g^2)},$$
$$s^2 = \cdot 1547\,(f^2 + g^2), \qquad s = \cdot 3933\,\sqrt{(f^2 + g^2)}.$$

Hence, from (27), (28), (29), in correspondence with (30), we have

$$\left.\begin{aligned}
h^2\alpha &= if\,\{-e^{-rz} + \cdot 5773\,e^{-sz}\}\, e^{ipt}\, e^{ifx}\, e^{igy} \\
h^2\beta &= ig\,\{-e^{-rz} + \cdot 5773\,e^{-sz}\}\, e^{ipt}\, e^{ifx}\, e^{igy} \\
h^2\gamma &= \cdot 8475\,\sqrt{(f^2 + g^2)}\,\{e^{-rz} - 1\cdot 7320\,e^{-sz}\}\, e^{ipt}\, e^{ifx}\, e^{igy}
\end{aligned}\right\} \quad \ldots\ldots(37)$$

* Prof. Lamb, in his able paper, seems to regard all negative values of σ as excluded *a priori*. But the necessary and sufficient conditions of stability are merely that the resistance to com-pression ($\lambda + \frac{2}{3}\mu$) and the resistance to shearing (μ) should be positive. In the second extreme case of a medium which resists shear, but does not resist compression, $\lambda = -\frac{2}{3}\mu$, and $\sigma = -1$. The velocity of a dilatational wave is then $\frac{4}{3}$ of that of a distortional plane wave. (Green, *Camb. Trans.* 1838.) The general value of σ is $\lambda/(2\lambda + 2\mu)$.

† *Math. Soc. Proc.* June 1873, vol. IV. p. 359 [vol. I. p. 171]. *Theory of Sound*, t. I. p. 85. Lamb, *loc. cit.* p. 202.

For a progressive wave in two dimensions, we shall have

$$\left.\begin{array}{l} h^2\alpha/f = (e^{-rz} - {\cdot}5773\,e^{-sz})\sin{(pt+fx)} \\ h^2\gamma/f = ({\cdot}8475\,e^{-rz} - 1{\cdot}4679\,e^{-sz})\cos{(pt+fx)} \end{array}\right\} \quad \ldots\ldots\ldots\ldots(38)$$

At the surface,

$$\left.\begin{array}{l} h^2\alpha/f = +{\cdot}4227\sin{(pt+fx)} \\ h^2\gamma/f = -{\cdot}6204\cos{(pt+fx)} \end{array}\right\} , \quad \ldots\ldots\ldots\ldots\ldots(39)$$

so that the vertical axes of the elliptic orbits are about half as great again as the horizontal axes.

It is proper to remark that the vibrations here considered are covered by the general theory of spherical vibrations given by Lamb in the paper referred to. But it would probably be as difficult, if not more difficult, to deduce the conclusions of the present paper from the analytical expressions of the general theory, as to obtain them independently. It is not improbable that the surface waves here investigated play an important part in earthquakes, and in the collision of elastic solids. Diverging in two dimensions only, they must acquire at a great distance from the source a continually increasing preponderance.

131.

ON PROF. HIMSTEDT'S DETERMINATION OF THE OHM.

[*Philosophical Magazine*, XXI. pp. 10—13, 1886.]

As there is still some discrepancy in the values of the ohm obtained by able workers using various methods, it seems desirable to put forward any criticisms that may suggest themselves, in the hope that the causes of disturbance may thus come to be better understood. I propose accordingly to make a few remarks upon the paper of Professor Himstedt, translated in your November number, not at all implying that his results may not be as good as any other, but rather in order to raise discussion on certain points which the author may be able to treat satisfactorily when he publishes a more detailed account of his work.

The leading feature in the method of Prof. Himstedt is the use of a commutator, or separator, by which the make- and break-induced currents are dissociated, one or the other passing in a stream at equal small intervals of time through a galvanometer, by whose aid their magnitude is appreciated. The instrument works with mercury contacts. When I first considered the methods available for the solution of this problem at Cambridge in 1880, I found ready to my hand an ingenious apparatus, contrived by Prof. Chrystal for this very purpose. The contacts were effected by metallic dippers, controlled by eccentrics, and passing in and out of mercury cups. What determined me against this method*, notwithstanding its obvious advantages in respect of sensitiveness, was the recollection of unavailing attempts of my own in 1870 to make satisfactory mercury contacts with dippers carried by electrically maintained tuning-forks. Even when silver was the metal employed, the contacts were

---

* I may remark that Brillouin used a commutator of this nature in his researches on the comparison of coefficients of induction: *Thèses présentées à la Faculté des Sciences de Paris*, 1882.

uncertain, and no trustworthy galvanometer deflection could be obtained. It may be mentioned, in passing, that the object was to obtain, through the galvanometer, a stream of charges of a condenser, separated from the discharges, with a view to the determination of v (the ratio of electrical units). Dippers in electrical connexion with the body of the fork (and by means of a wire attached to the stalk with one pole of the condenser) were carried on both the upper and lower prongs. Underneath these, mercury cups were so arranged that the vibrating-fork was in contact with them alternately, but never with both at the same time. One of the cups was connected with the insulated pole of the battery and the other with the earth. The fork was driven by a current entirely insulated from it. It was found, however, that the contacts could not be made perfect, and the direct use of the fork was abandoned in favour of a commutator with platinum contacts driven by the fork*. This form of the apparatus is unsuitable as a separator of induced currents; and I was inclined to favour the observation of a single induced current with a ballistic galvanometer as carried out by Rowland, and afterwards by Glazebrook.

It is to be presumed that the contact difficulty has been overcome by Prof. Himstedt; and my principal reason for mentioning it is that I found it particularly capricious and insidious. The galvanometer indication would often remain steady for minutes together, and then suddenly change. It would be interesting to know whether Prof. Himstedt has met with any behaviour of this sort.

The next question that I wish to raise relates to the measurement by the galvanometer of a series of induced currents, each of short duration. On page 421 there is a reference to "cross magnetization" that I do not quite understand. I have myself † objected to the use of a ballistic galvanometer, on the ground of the tacit assumption that the needle at the moment of the impulse, when subject to a powerful cross-magnetizing force, retains its axial magnetization unaltered; but in the method of Prof. Himstedt the question assumes a different shape. In this case the needle stands in an oblique position, and we have to consider whether the axial magnetization does not alter under the action of a force having a *sensible axial component* ‡. In all probability Prof. Himstedt has considered this matter. It admits of a very simple test, all that is necessary being to deflect the needle into its oblique position with an external permanent magnet, and then to allow the induced currents to pass, *suppressing the*

---

* "On the Determination of the Number of Electrostatic Units in the Electromagnetic Unit of Electricity," J. J. Thomson, *Phil. Trans.* 1883, p. 719.

† *Phil. Trans.* 1882, p. 670. [Vol. II. p. 48.]

‡ "On a Permanent Deflection of the Galvanometer Needle, &c." *Brit. Assoc. Report*, 1868. *Phil. Mag.* Jan. 1877. [Vol. I. p. 310.] Chrystal, *Phil. Mag.* Dec. 1876.

interruption of the secondary contact. Both make- and break-induced cur-rents would then pass, whose mean value is zero; and any deflection of the needle under these conditions would be a sign that its magnetism fluctuated and that the evaluation of either stream alone would be vitiated.

An interesting feature in Prof. Himstedt's work is the arrangement of the primary and secondary coils, of which the former is a long solenoid embraced by the latter. The fact that as regards the secondary the induction-coefficient depends sensibly upon the number of turns only, without regard to radius, is much in its favour. Any one who has had to do with the measurement of coils will appreciate too the advantage of reducing the primary to a single layer. There are, however, disad-vantages in this arrangement which must be kept in sight. I will not dilate upon the use of a wooden core on which to wind the primary, though I should think it hardly safe. But assuming that there is no important uncertainty as to the value of R (the mean radius of the primary), though it should be remarked that it occurs in the formula as a square, nor in the data relating to the secondary, we have still to consider the factor K, expressive of the number of turns per unit length in the primary. So far as appears, the value of this quantity is obtained by simply dividing the whole number of turns, 2864, by the measured length, 135·125 cm. Now there is here a tacit assumption either that the wire is wound with perfect uniformity, or that we have to deal only with the mean value. The latter alternative is manifestly incorrect, since the central parts lying nearly in the plane of the secondary are necessarily more effective than the remoter parts. In point of fact the simplicity of this arrangement is more apparent than real, relating rather to calcula-tion than to measurement, as I have already had occasion to remark* in connexion with a somewhat similar use of a long solenoid in Mascart's determination of the electrochemical equivalent of silver. How far uni-formity was attained in the present case I have no means of judging; but where the successive turns are merely brought into contact with one another I should not expect a high degree of precision, if only because the thicknesses of the wire and silk are liable to vary. Again, it may be possible to verify the uniformity *à posteriori*, or to obtain data for the calculation of a correction. But at any rate it seems misleading to exhibit the result as determined by the average number of turns per unit length, when it really depends also upon the ratios of the rates of winding at the various parts of the length.

* *Phil. Trans.* 1884, p. 413. [Vol. II. p. 280.]

132.

ON THE CLARK CELL AS A STANDARD OF ELECTRO-MOTIVE FORCE.

[*Philosophical Transactions*, CLXXVI. pp. 781—800, 1886.]

§ 39. THE importance of a convenient standard of electro-motive force is now fully recognised. It gives the most available means of measuring currents, especially of large amount, and has been used for this purpose by several experimenters. I may refer to my investigation on the Constant of Magnetic Rotation of Light in Bisulphide of Carbon*, in which the currents were all measured by reference to a Clark cell, whose value was originally obtained by absolute measurements and verified at intervals by the silver voltameter. Clark cells are exceedingly convenient in use, and would doubtless be generally employed, could confidence be felt in their permanence, and in the equality of cells set up by different persons from the same recipe. To these questions I have given much attention; and the result of a large experience is very favourable to the trustworthiness of the cells, if reasonable precautions be observed in charging them. I believe that any one who takes the trouble to set up three or four cells and compares them occasionally, will be in possession of a standard of E.M.F. which he may trust to about $\frac{1}{1000}$th part.

The present memoir is to be regarded as supplementary to that on the Electro-chemical Equivalent of Silver, and on the Absolute Electro-motive Force of Clark Cells†, and the paragraphs are numbered accordingly. The total number of cells experimented upon is large. Of my own construction there have been about 60 of the ordinary kind (with solid zincs), and about 30 of the H-pattern (§ 28) with zinc amalgam. In addition to these some

---

* See *ante*, p. 343. [Vol. II. p. 360.]
† *Phil. Trans.* Part II. 1884. [Vol. II. p. 278.]

40 cells made by others have been compared, with very interesting results to be given later.

Before entering into details it may be convenient to summarise the principal sources of error. The E.M.F. may be too high, (1) because the paste is acid, (2) because the paste is not saturated with zinc sulphate. The first fault tends to cure itself, and is rarely found after the cells are a month old. The second is the origin of the more serious discrepancies that have been met with in commercial cells. If the E.M.F. is too low, the cause may be, (1) that the cell has become dry, in which case the drop will probably be progressive, (2) the solution is *super*-saturated with zinc sulphate, (3) the mercury is impure.

Believing that these cells are capable of affording standards of a high degree of precision, and that they ought to be in general use, I have gone into considerable detail as to the procedure which may be adopted. This may give the impression that the preparation is troublesome, but in reality the method that I propose is much simpler than those hitherto employed and thought to be necessary. To show how easy it is to set up these cells, I may refer to two large ones, contained in glass cylinders of about 4 inches diameter and provided with wooden covers by which the electrodes are carried. Enough common mercury was poured in to cover the bottom, contact being made with it by means of a platinum wire sealed in a glass tube. The jar was then filled to a height of about 4 inches with saturated solution of commercial zinc sulphate with which some mercurous sulphate had been rubbed up in a mortar. The zinc electrode was cut from ordinary sheet metal, and was suspended horizontally near the top of the liquid by a projecting tail. After the first few weeks these large cells have never deviated from the standard by much more than $\frac{1}{1000}$, and have been found very convenient for certain purposes on account of their comparatively small resistance. They have also been used for preliminary comparisons with cells whose value was unknown, in which case there was danger of more current passing than it is desirable to allow through delicate standards.

§ 40. The method followed for making the recent comparisons is the same in substance as that described in § 28. The use of a high resistance galvanometer gave a greater facility of reading, a change of $\frac{1}{10000}$ in the E.M.F. under measurement giving a motion of the spot of light which could be seen without a telescope from across the room.

The accompanying table (XIII.) gives the values of most of the older cells in continuation of that contained in the note to § 30 [vol. II. p. 331]. Cells (4), (8), (9) were, I think, left at Cambridge; (18) and (19) were observed at intervals during 1885, but the E.M.F. was found to fall. When about three parts per thousand too low, they were removed for examination,

TABLE XIII.

	Jan. 2, 1885	Jan. 26, 1885, 56°	March 2, 1885, 52°	April 7, 1885, 56°	June 4, 1885, 71°	June 29, 1885, 64°	Aug. 14, 18, 1885, 63°	Sept. 3, 4, 1885, 64·5°, 62·5°	Oct. 22, 23, 24, 1885, 57°, 57°, 56°	Nov. 23, 24, 1885, 54°, 52°	Dec. 26, 1885, 52°	April 27*, 1886, 58°
Clark 1	1·0000	1·0000	1·0000	1·0000	1·0000	1·0000	1·0000	1·0000	1·0000	1·0000	1·0000	1·0000
,, 5	1·0000	1·0002	1·0002	1·0002	1·0004	·9998	1·0000	1·0001	1·0001	·9999	·9999	·9999
,, 10	1·0002	1·0000	1·0000	1·0000	1·0001	1·0000	1·0001	...	1·0002	1·0002	1·0005	1·0003
,, 11	1·0002	1·0000	1·0002	1·0000	1·0003	1·0002	1·0001	...	1·0001	1·0001	1·0002	1·0000
,, 12	1·0004	1·0002	1·0002	1·0001	1·0002	1·0002	1·0001	...	1·0004	1·0003	1·0006	1·0001
,, 13	1·0000	·9998	·9998	1·0000	·9998	1·0000	1·0000	1·0000	1·0000	1·0001	1·0003	1·0001
,, 14	1·0000	1·0000	1·0000	·9999	·9998	1·0000	·9999	1·0005	1·0000	1·0001	1·0003	1·0000
,, 15	1·0006	1·0005	1·0005	1·0002	1·0002	1·0003	·9998	1·0000	1·0006	·9999	1·0006	1·0007
,, 16	1·0004	1·0001	1·0001	1·0002	·9998	1·0000	1·0000	1·0001	1·0001	1·0003	1·0005	1·0000
H_5	1·0000	·9998	·9998	·9998	1·0001	1·0002	·9999	1·0003	·9999	·9999	1·0001	1·0001
H_{10}	·9998	·9998	·9998	1·0002	1·0002	1·0002	1·0001	1·0001	1·0001	1·0003	1·0005	1·0003
H_{11}	·9998	·9996	·9998	1·0002	1·0002	1·0002	1·0003	1·0003	1·0001	1·0003	1·0004	1·0004
H_{12}	·9995	1·0003	1·0001	·9996	1·0002	1·0003	broken	1·0001				
a	1·0002	1·0000	·9999	1·0002	1·0004	1·0004	1·0001	1·0003	1·0003	1·0003	1·0001	1·0003
b	1·0002	1·0002	1·0000	1·0002	1·0004	1·0004	1·0001	1·0003	1·0001	1·0002	1·0004	1·0003
c	1·0000	·9998	·9998	1·0000	1·0002	1·0002	1·0000	1·0000	1·0000	1·0001	1·0004	1·0003
d	1·0000	1·0000	1·0000	1·0000	1·0002	1·0004	1·0001	...	1·0000	1·0001		
e	1·0010	1·0004	1·0008	1·0010	1·0010	1·0010	1·0004	1·0009	1·0009	withdrawn		
f	1·0002	1·0000	1·0001	1·0004	1·0006	1·0006	1·0003	1·0004	1·0004			

* Added since the reading of the paper.

and found to be dry. The water had exuded, or evaporated, through cracks in the paraffin wax. The cells of the H-pattern, H_6, H_{13} were broken in a manner to be presently explained. On the other hand some new cells of the H-pattern, $a, b, \ldots f$, are included. They are those referred to in the previous paper as having been fitted up by Mr Threlfall, and are more than a year old.

The agreement exhibited in Table XIII. is very remarkable. In many cases the cells may be depended on not to vary relatively more than 2 or 3 parts in 10,000, notwithstanding considerable changes of temperature. It is, indeed, doubtful whether even the whole of the small variations recorded are real. 1° C. influences the E.M.F. about 8 parts in 10,000, and differences of temperature of two or three-tenths of a degree may well have occurred, since the cells were variously mounted, and no particular precautions were taken beyond the avoidance of readings at times when the temperature of the room (immediately under the roof) was changing rapidly.

It may be convenient to recall that cells 1, 5, were made in Oct., 1883; 10–13 and 14–19 in May, 1884; H_5 in March, 1884; H_{10}, H_{11} also in March, 1884.

§ 41. In cells of the ordinary type the principal source of weakness is imperfect sealing at the top, due to cracks in the paraffin wax. As pointed out by Dr Alder Wright*, a better result is obtained if the whole cell be imbedded in a large mass of wax than when (as in my cells) the wax is applied merely inside the tube, above the cork sustaining the zinc. During the last year I have replaced paraffin by marine glue, which, so far as can be judged at present, may be relied upon to effect a complete seal. The procedure will be described presently more in detail.

The cause of failure in the H-cells is of a different nature. Many of the earlier cells had been found to break in the amalgam leg, and the trouble was attributed to a hardening and expansion of the contents (§ 29). Such a hardening had, in fact, been observed in one or two cases. More recent experience, however, has proved that the cause must be looked for elsewhere, several cells having failed in which no trace of solid amalgam was to be found. Nevertheless the amalgam is the cause of the trouble, for out of a large number of breakages *not one* has occurred in the leg containing pure mercury. It would appear that some alloying takes place with the platinum wire in contact with the amalgam, and that this gradually extends itself with fatal results to the part of the platinum sealed into the glass, from which place the cracks are

---

* *Phil. Mag.* July, 1883, p. 32.

always observed to radiate. It is hoped that a cure will be found in a plan, adopted for some recent cells, of melting in a little cement (marine glue has been used) so as to protect from the amalgam the part of the platinum which lies nearest to the glass; but it is too soon to speak with certainty.

§ 42. The H-form lends itself to hermetical sealing, and at one time I anticipated advantage from this course. There is, however, such a large amount of spare liquid that there is no likelihood of trouble from desiccation, even if the corks allow a little evaporation. Indeed, by withdrawing the corks a fresh supply of liquid could be introduced at any time. It happened on one occasion that an H-cell to which a large excess of salt had been added, was so far crusted up next the metallic surfaces that it began to show signs of failing E.M.F., much as if it were going dry. The mass was so compact that no impression could be made upon it with a glass rod; but it was bored through with a steel reamer, when the E.M.F. at once recovered its normal value. In such cases the accessibility is advantageous, especially for purposes of experiment. It is well, however, to avoid such a large excess of salt as was present in this case. By alternate melting and crystallisation as the temperature rises and falls, there is a tendency to aggregation, of which the cell above referred to affords an extreme example.

In the construction of cells with solid zinc electrodes, I have fallen back upon a simplified pattern—nothing more in fact than a small tube with a platinum wire sealed through its closed end. See figure.

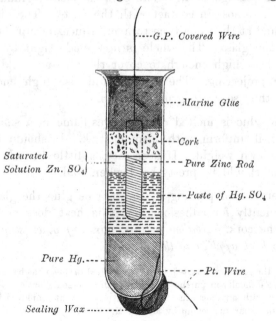

G.P. Covered Wire

Marine Glue

Cork

Saturated Solution Zn. SO₄

Pure Zinc Rod

Paste of Hg. SO₄

Pure Hg.

Pt. Wire

Sealing Wax

The only objection to this form is that the cells cannot, without precaution, be supported from underneath. Most of mine are held at the centre by a spring (cut from sheet metal) against a piece of board mounted on its edge. In this case the copper electrodes are secured in sealing-wax to the wooden stand. For single cells, when portability is desired and convenience of immersion in water or ice, it is a good plan to enclose the whole in a rather long and narrow test-tube. A little cotton wool supports the cell and prevents it from shaking about laterally. The gutta-percha covered leads pass through a piece of cork inserted near the top of the test-tube, and a little marine glue poured over the cork makes all tight. In order to give mechanical support to the platinum wire, which is liable to break where it passes through the glass, the external application of sealing-wax is recommended—a precaution applicable also to the H-cells.

§ 43. In charging the cells the first step is to pour in sufficient pure* mercury to cover the platinum effectively. The paste (of which more presently) is next introduced, with the aid of a small funnel, care being taken not to soil the sides above the proper level. The zincs, cut from rods of pure zinc, as supplied by Hopkins and Williams, and not recast, are soldered to copper wires and cleaned in the lathe. Just before use they are dipped in dilute sulphuric acid, washed in distilled water, and dried with a clean cloth or filter paper. Each zinc is mounted in a short piece of cork fitting the tube (but not too tightly), and nicked in order to allow of the passage of air. The cork is pushed gradually down until its lower face is almost in contact with the paste. The object is to leave but little air, and at the same time to avoid squeezing up the paste between the cork and the glass. The whole is now made tight by pouring marine glue over the cork high enough to cover the zinc and soldering, and leave only the wire projecting. The tube should rise high enough to receive the glue, and thus secure a good adhesion.

The marine glue is melted over the gas flame in a small pot or basin, and stirred, until uniform, with a small stick. It should be fluid enough to pour by its own weight. If necessary, a little benzole may be added, but the cement should be pretty hard when cold.

In the operation of pouring in the marine glue the glass is heated by the glue sufficiently for adhesion; but this heat does not extend appreciably below the cork. *Neither in this, nor any other stage of the process of charging, is heat applied to the paste.*

* Except when the contrary is stated, mercury distilled *in vacuo* has been used for Clark cells. There is, I believe, a difficulty in purchasing mercury thus treated; but every physical laboratory should be provided with an apparatus for this purpose. That employed by me was distilled at Cambridge in an apparatus set up by Mr W. N. Shaw.

§ 44. The earlier cells, prepared with paste, which was doubtless strongly acid, frequently gave irregular results for several weeks. Extreme cases are afforded by 15 and 16, which are shown by Table VIII. of the former paper to have been at first more than 2 per cent. too strong. Moreover, as appears from the continuation of this table in the notes [p. 331], it took nearly two months for these cells to settle down to their normal values. The cause of irregularity is to be sought rather at the mercury than at the zinc (or amalgam) electrode.

In order to examine this question, H-cells were charged with pure mercury and paste in both legs, and filled up as usual with saturated zinc sulphate solution. There should, of course, have been no E.M.F.; but the value of one of the cells was ·0041 Clark, and remained tolerably constant for several days. By stirring with a glass rod, the E.M.F. could be either increased or diminished. After some weeks the cells had come sensibly to zero, and would bear stirring (in one or both legs) without much disturbance. To another cell, which still showed irregularity, zinc carbonate was added. The E.M.F. was much reduced, and in a few days was scarcely sensible even on stirring.

When the paste is neutralised in the first instance with zinc carbonate, the irregularities are much reduced. Two cells thus prepared had an E.M.F. less than ·0001 Clark, and were scarcely affected by stirring. On Jan. 27, 1885, a piece of zinc wire was poked through the paste, so as momentarily to touch the mercury in one leg of one of these cells. A large E.M.F. was thus developed, which remained operative for half-an-hour or more; but on Jan. 28 the E.M.F. was only ·0003 Clark, and on Jan. 31 ·0002 Clark. It is clear that the mercurous sulphate has the property of freeing the mercury from the smallest contamination with zinc.

§ 45. In consequence of these observations more recent cells have been prepared with neutralised paste. This course has the advantage that the cells attain their normal values in a few days, sometimes within one day, of charging. So far as I can judge, however, there is no difference in the ultimate value whether the paste be acid or neutral. In the former case the cell probably neutralises itself by dissolution of zinc, a certain amount of gas being liberated.

It is convenient to keep a stock bottle of saturated solution of zinc sulphate. This may be prepared in a flask by mixing distilled water with about twice its weight of crystals. A little carbonate of zinc is added to neutralise free acid, and solution is effected with the aid of *gentle* heat. If time can be afforded, it is a good plan to let the solution stand, as a good deal of iron is usually deposited, even when "pure" zinc sulphate is used. The solution may then be filtered in a warm place into the stock bottle. When it is intended to charge H-cells, or prepare paste, the bottle should be exposed to a gentle warmth for a few hours, and the solution should be drawn with a pipette from near the crystals at the *bottom of the bottle*. Otherwise there is no security that the liquid used will be saturated.

To prepare paste we may rub up together in a mortar 150 gms. mercurous sulphate (as purchased), 5 gms. zinc carbonate, and as much of the saturated zinc sulphate solution as is required to make a thick paste*. Carbonic anhydride is liberated, and must be allowed a sufficient time to separate. I have found it convenient to leave the paste in the mortar for two or three days, rubbing it up at intervals with additions of zinc sulphate solution until the gas has escaped. By the addition of a small crystal, and by evaporation, we have security that the paste is saturated, and will remain so, notwithstanding such moderate elevation of temperature as the cells are expected to bear. The paste may then be transferred to a tightly-corked bottle, and, so far as my experience extends, will remain available for many months at least. Before pouring, the bottle of paste should be well shaken up.

The performance of the newer cells has been satisfactory (one irregularity will be mentioned later), and the substitution of marine glue for paraffin wax promises a longer life. A large number of observations have been recorded, but it does not appear necessary to give them here in detail.

§ 46. I have been anxious to compare with my cells some prepared by others, and have to thank many physicists for the opportunity of doing so. Dr Alder Wright was good enough to send me several of his cells, with which comparisons of especial value have been made. I shall have occasion to remark more at length upon the results obtained with these cells, and for the moment will only say that the difference between them and mine is under $\frac{1}{1000}$th part. Cells prepared by Mr Threlfall (not those previously mentioned), by Dr Fleming, and by Prof. G. Forbes also agreed well.

Dr Fleming's cells, of which six remained in my hands, were at first irregular, and even now show somewhat larger variations than the best cells are liable to. The cause is, I believe, to be found in insufficient purity of mercury, as is suggested by the appearance of the metallic surface. When the mercury is quite pure, the surface is as bright as in a thermometer bulb. I have instituted special experiments in order to ascertain the effects of impure mercury. On two occasions three cells have been charged alike in all respects, except that in the first vacuum-

---

* It is usually found that on neutralisation the mercurous sulphate turns yellow, so that the paste presents ultimately somewhat the appearance of mustard. I do not know whether the change of colour is normal, or is to be attributed to impurity.

distilled mercury was employed, in the second ordinary mercury, and in the third mercury purposely contaminated with tin-foil (probably containing also lead). The results were not very distinct; but they indicate that impurity in the mercury is apt to depress the E.M.F. (*e.g.* by $\frac{1}{2}$ per cent.), and especially to make it irregular. I am disposed to attach great importance to purity of mercury, and believe it to be more essential than purity of zinc*, although I should not recommend the use of common zinc when the purer metal can be obtained so easily. Other cells have been prepared with mercury to which a little silver and copper (filings from a silver coin) were added. After the first week or two the E.M.F. of these cells was normal (to within a $\frac{1}{1000}$th part). It is probable that metals more oxidisable than mercury are removed from it by the paste, as certainly happens in the case of zinc.

§ 47. It has been abundantly proved by v. Helmholtz, Alder Wright and other workers, that the E.M.F. of Clark's, Daniell's, and similar cells, rises as the zinc sulphate solution is diluted. In some such cells to be discussed presently (39, 40), the E.M.F. is about $1\frac{1}{2}$ per cent. higher than for a normal Clark. Dr Hopkinson, some of whose cells have been compared with mine, writes that he is pretty clear that the worst irregularities of Clark's cells are due to the zinc sulphate not being saturated. A cell 2 per cent. in excess could be made right by simply introducing crystals of $ZnSO_4$. It is evident that sufficient care has not been taken in this respect in the preparation of cells sold to the public.

§ 48. In this matter of saturation there is another danger to be encountered, to which my attention was first drawn in connexion with some cells prepared with great care by Mr Mortimer Evans, and left in my charge by Sir W. Thomson. Of these, two were in practical agreement with mine; but the other eight, though in close agreement with one another, were too low (according to my standard) by rather more than four parts per thousand. And this state of things persisted without the smallest change for two or three months, during which tests were applied at intervals.

Being anxious to examine a phenomenon to which my experience had afforded no parallel, I opened carefully one of the abnormal cells (T_2) to the extent of withdrawing the zinc. My idea was that possibly the zinc solution was *supersaturated*, in which case the E.M.F. might be expected to be too low. Attached to the zinc, however, were found what appeared to be crystals of sulphate; but in order to be on the safe side, a few particles from the stock bottle were added to the cell, and the zinc was

* Of ten cells, prepared at the same time as a set to be used in series, five were made with pure zinc and five with common sheet zinc. No difference in the performance can be detected.

replaced. After the lapse of a few hours the E.M.F. was tested, *and was found to be normal*, as it has remained ever since.

There was now little doubt but that the solution had been supersaturated, in the sense with which we are concerned. The presence of crystals is no evidence to the contrary, unless it can be proved that the crystals were those of the *normal, heptahydrated*, salt.

At this stage I wrote to Mr M. Evans to inquire whether there was anything in the history of the cells that would account for the separation into two classes, and I was informed that all the cells had been prepared originally in the same manner. The mercury was twice distilled *in vacuo*, and in other respects the greatest care had been taken. When, however, the cells came to be tested, it was found that owing to contraction all but *two* were wanting in proper contact between the mercury and platinum, and that this contact could only be restored by remelting over a water-bath the whole of the paraffin wax in which the cells were imbedded. It was by this operation, no doubt, that the solution became supersaturated. The agreement with mine of the two cells which were not heated (one of them (T_3) is referred to later) is very satisfactory as showing that the great precautions exercised by Mr Evans lead to the same E.M.F. as I have obtained with far less trouble. I may add that a second abnormal cell (T_8) moved to equality with my standards on being opened.

§ 49. With a view to the better understanding of this matter I made myself acquainted with the beautiful researches of M. Gernez* upon supersaturation, conducted principally with solutions of sulphate and of acetate of soda; and have performed parallel experiments upon sulphate of zinc. A very strong solution of this salt prepared hot, and sealed up in a glass tube, will sometimes cool without any deposit. More often it throws down an abnormal (lower) hydrate. If in this condition the tube be heated pretty rapidly in boiling water, some of the salt dehydrates further to a powder (presumably mono-hydrated); if it be allowed to cool again the inferior crystalline hydrate reforms. However long the solution stands cold over this hydrate, it is still supersaturated as regards the normal hydrate, the minutest addition of which causes the supernatant liquid to become almost solid, with needles penetrating it in all directions. The experiment has been repeated many times with less strong solutions standing in open test-tubes, charged and preserved with the simple precautions indicated by M. Gernez.

It is evident that "supersaturation" is a term without definite meaning until further explained. Gernez has shown that a solution may be super-

---

* *Annales de l'école normale*, t. III. 1866, p. 163; t. v. 1876, p. 1.

saturated with respect to one or both of two different hydrates, *i.e.*, will crystallise similarly on contact with the smallest fragment, and not super-saturated at all with respect to the anhydrous salt, the addition of which causes no effect. De Coppet* has proved that the so-called supersaturated solution may be disposed to take up a further portion of anhydrous salt; as may, indeed, be inferred from previously known facts, since in general the anhydrous salt must tend either to leave or to enter the solution, and the former alternative is excluded by the observed behaviour on first contact.

§ 50. In view of the above facts we can hardly doubt that a Clark cell, heated nearly to the temperature of boiling water and then cooled, would be likely to become supersaturated; but I thought it would be satisfactory actually to try the experiment. A normal cell of my own preparation and containing an excess of undissolved salt, was maintained for several hours at an elevated temperature, and tested after cooling. A temperature of 38° C. did not permanently alter the cell; neither did a temperature of 49°, nor one of 60°. But after an exposure to about 80° a permanent change set in. Immediately after cooling the value in terms of the standards was ·9914, but after one day's standing it settled to ·9943, close to which value it has since remained. It appears from the above that the cell probably requires to be heated sufficiently to decompose the normal hydrate, and not merely to bring all the immersed salt into solution. In the latter case there may well be solid particles within reach, which re-determine normal crystallisation on cooling.

A second experiment was tried with an old cell (*e* of Table XIII.) which contained a large excess of undissolved salt. This was of the *H*-pattern, which lends itself more conveniently to observation and experiment. On November 13 the cell was heated for several hours nearly to 100° C. After cooling a solid mass of crystals was to be seen over the metals in both legs, and it might have been supposed that the operation had been unsuccessful. On November 16, however, the E.M.F. was found to have changed (from about 1·0005) to ·9949, at which value it remained until November 25. On that day, at 5^h p.m. the corks were drawn. At 6^h no effect had been produced, and a fragment of the normal hydrate was dropped into each leg. At 6^h 45^m new crystals had formed, and the E.M.F. had risen to ·9996. A few hours later the E.M.F. was 1·0000, at which value it has since remained within two or three parts in 10,000. It was remarked that the crystalline deposit on contact with the normal hydrate was much less in amount than had been met with in experimental tubes with simple zinc sulphate. The explanation is probably that during the heating no complete diffusion of the salt was effected, so that

---

* *C. R.* LXXIII. p. 1324, 1871. See also Nicol, *Phil. Mag.* June, 1885.

after cooling supersaturation was limited to the lower layers. Both metals being at the bottom in this form of cell, the E.M.F. is independent of the condition of the upper parts.

It is worthy of note that all the "supersaturated" cells which I have tested are about 5 parts per 1000 too low. That they should give a definite E.M.F. is to be expected whenever the lower hydrate is formed. For the solution in contact with the lower hydrate is just as definite in composition as when it is in contact with the normal hydrate. It is, however, possible to have "supersaturated" solutions without formation of the lower hydrate, and then the E.M.F. would be indefinite. The deficiency may certainly be less than the 5 parts per 1000, and may probably be more in certain cases.

§ 51. In view of the possibility of error from under and over-saturation, the reader may be inclined to ask whether it would not be better to prescribe a *dilute* solution for standard cells, as is conveniently done for standard Daniell's. One advantage attending this construction is (as will presently appear) a lower temperature-coefficient. Again, we should be inclined to expect a more definite dependence upon temperature. In order to bring a saturated cell to its normal condition after warming (for example), it is necessary not merely that the whole of the contents should acquire the new temperature, but also that sufficient time should be allowed for *diffusion*. If the solution in contact with the zinc be weaker than corresponds to saturation at the altered temperature, the full loss of E.M.F. will not be experienced. In this respect the *H*-cells, in which the excess of salt rests upon the metals, would seem to have an advantage. But I cannot say that in practice I have met with the defect due to imperfect diffusion. Cells which have stood at 10° or 12° seem to acquire their new values after an hour or two's immersion in ice. The argument weighs, however, in favour of *small* cells, through which diffusion of temperature and matter can take place quickly. Such experience as I have had of cells prepared with dilute solutions, would not lead me to prefer them, even were there no difficulty in, or necessity for, a standardising. In the case of clear solutions, such as are used for Daniell's, the specific gravity is a convenient test of strength; but I do not see how a standard unsaturated paste could be accurately prepared without a good deal of trouble. Another objection to dilute solutions is the progressive alteration of E.M.F., due to evaporation, which must take place whenever the sealing is at all imperfect*.

In truth, there is no real difficulty in avoiding both under and over-saturation, if the experimenter will bear in mind the known properties of the materials with which he is dealing. The grosser errors, arising from

* Alder Wright. *Loc. cit.* p. 33.

the first cause, can only occur as the result of carelessness. As to the latter, it may be that supersaturation has sometimes entered as a consequence of excessive precautions against the admission of air. It cannot occur in the presence of the minutest fragment of the normal hydrate. Opinions may perhaps differ upon this point, but I am myself disposed to condemn the use of heat in charging the cells. If hot paste be brought into contact with hot mercury, and then closed hermetically, there must be some risk of supersaturation.

§ 52. The next question which I propose to consider is that of the temperature-coefficients of Clark cells. My observations on cell No. 1 at Cambridge, § 36, gave for the proportional fall of E.M.F. per degree Centigrade in the neighbourhood of 15° the number ·00082, so that at $t°$ C. we might take

$$E = 1\cdot435 \left\{1 - \cdot00082 \left(t - 15\right)\right\}.$$

This number is in agreement with that found by Helmholtz for saturated cells, but it differs seriously from the number (·00041) given by Alder Wright* also for saturated cells. These discrepancies have naturally led to the conclusion that the temperature-variations of Clark cells are uncertain, and Dr Fleming has insisted upon the advantage in this respect possessed by the Daniell†. If indeed it were a matter of chance whether the temperature-coefficient of a Clark were ·0008, or ·0004, the utility of these cells as standards for delicate work would be seriously impaired. A glance, however, at Table XIII.‡ will show that such uncertainty need not exist. The results of March 2 and June 4 correspond to a difference of 19° F., or about 11° C., so that if one cell had the coefficient ·0008, and another the coefficient ·0004, the change of temperature would separate them to the extent of 44 parts in 10,000, whereas the greatest change observed (perhaps not due to this cause at all) is but 5 parts. Many observations on recent cells, made in ordinary course, point in the same direction. In one or two cases there has been an apparent rise at temperatures above 65° F., indicating a drop in the temperature-coefficient relatively to that of No. 1. I have attributed this to an insufficient excess of undissolved salt in cells prepared when the weather was cold, the result of course being a failure of saturation at high temperatures, attended (as will presently appear) by a fall of temperature-coefficient.

Being desirous of clearing up, as far as possible, any questions connected with the practical use of these cells as standards, I determined to supplement the former observations with special experiments at some-

---

* *Phil. Mag.* July, 1883, p. 36.

† On the "Use of Daniell's Cell as a Standard of Electro-motive Force," *Phil. Mag.* Aug. 1885. It appears, however, to be, by a slip of the pen, that the coefficient for the Daniell is represented as only $\frac{1}{20}$ of that of the Clark. The numbers given lead to the ratio $\frac{1}{4}$.

‡ I must apologise for the Fahrenheit degrees.

what extreme temperatures, which should include as great a variety of constructions as possible. Most of the cells were so mounted that they could not well be tried at temperatures differing from that of the surrounding air, and I had to content myself with varying the temperature of the room by opening windows and burning gas. Care was taken that no great variation occurred within two or three hours of the comparisons. Under these circumstances tests were made of (1), (10) (Table XIII.); T_1 (one of the abnormal cells of Mr M. Evans'), T_3 (a normal cell of the same batch); W_{31}, W_{59}, W_{62}—three cells by Dr Alder Wright; M_{183}, a cell sold by Messrs Clark and Muirhead,—at temperatures from 47° F. to 69° F. These cells were all supposed to be saturated. There were also two prepared purposely with diluted paste—(39), (40). Besides these, two saturated cells of my own construction mounted in test-tubes, which could be immersed in water or ice, were tested from 32° F. to $67\frac{1}{2}$° F.

In order to obtain an absolute result we must have command of a standard independent of temperature-variations. At Cambridge I employed for this purpose a cell kept constantly in ice. The present observations were rather protracted, and I preferred to rely upon two cells (35), (38), mounted in test-tubes and imbedded with a thermometer in a mass of sand, itself situated in an underground recess. The variations of temperature were here very small and readily determined, so that there was no practical uncertainty on this account. The variation of the two cells relatively to one another was less than $\frac{1}{10000}$ during the whole month of observation. (Table XV.)

TABLE XIV.—Value of Cells referred to No. 1.

Temperature (Fahr.)	Date, 1885	(1)	T_1	T_3	W_{31}	W_{59}	W_{62}	M_{183}
47	Nov. 8 . .	1·0000	·9957	. .	·9991	·9991	·9993	. .
$49\frac{1}{4}$	,, 2 . .	1·0000	·9955	. .	·9993	·9999	. .	·9997
51	Oct. 31 . .	1·0000	·9957	1·0001	·9996	1·0000	. .	·9997
$56\frac{1}{2}$	,, 28 . .	1·0000	·9957	1·0003	. .	. .	. .	. .
57	,, 27 . .	1·0000	·9957	1·0001	1·0000	1·0000	. .	. .
$58\frac{1}{4}$	Nov. 4 . .	1·0000	·9954	. .	·9992	·9992	1·0000	·9996
61	Oct. 30 . .	1·0000	·9957	. .	·9997	·9993	·9999	·9980
67	Nov. 27 . .	1·0000	·9951	·9996	1·0001	·9996	·9997	1·0000
69	July 20 . .	1·0000	·9957	·9998	. .	. .	. .	. .

From the results in Table XIV., reduced to No. 1 as standard, the reader will see that there is no distinct difference of coefficient. It is interesting to note that T_1 (which there is every reason to consider supersaturated) keeps its distance from No. 1. When I first thought of supersaturation, I regarded this agreement of temperature-coefficient as an argument on the other side, not at that time recognising the probable occurrence of the lower hydrate. But what will still more arrest attention is the agreement of Dr Alder Wright's cells with No. 1. There are some irregularities, possibly dependent upon imperfect penetration of temperature through the masses of paraffin in which the cells are imbedded; but there is no distinct evidence of a lower coefficient, and certainly no such difference as that between ·0008 per degree Cent. and ·0004. The same may be said of M_{183}, which I received from Dr Alder Wright.

§ 53. The absolute variations with temperature of (1), (10), (39), (40) are shown in Table XV., in which all the electro-motive forces are expressed in terms of (38) at 51° F., a small correction being introduced to allow for the 1° or 2°, by which the actual temperature of (38), (35), may have differed from 51°. The coefficients for (1) and (10) are almost identical. The observed values for (1) agree pretty well with the formula

$$·9970 \{1 - ·000425 \, (t - 57)\},$$

as will be seen from the adjoining column calculated therefrom. This is in Fahrenheit degrees. The corresponding formula in Centigrade degrees is

$$·9970 \{1 - ·000765 \, (t - 13·9)\}.$$

It appears, therefore, that the temperature-coefficient for these cells, including those of Dr Alder Wright, is ·00077, and that the observed values are utterly irreconcilable with such a coefficient as ·0004.

For (39) and (40), prepared with unsaturated solution, the temperature-coefficient is exactly the half of that for the saturated cells. Reckoned in Fahrenheit degrees, it is ·00021, or in Centigrade degrees ·00038. The agreement with this value is shown by the column calculated from the formula

$$1·0134 \{1 - ·000212 \, (t - 57)\}.$$

The coefficient ·00038 is so close to that given by Alder Wright, viz., ·00041, that one cannot help suspecting that the cells used by him for this purpose may have been unsaturated, or possibly supersaturated (without deposit of lower hydrate).

Over the above range of temperature a linear expression represents the E.M.F. sufficiently well. When, however, the values recorded for (36), and (37) for temperatures from 32° F. to 67½° F. were plotted, a distinct curvature was apparent indicating a lower coefficient at the lower temperatures. Table XVI. exhibits the observed values, and (for comparison

TABLE XV.—In terms of (38) at 51° Fahr.

Temperature (Fahr.)	Date, 1885	(35) reduced to 51°	(1) observed	(1) calculated	(10) observed	(39) observed	(39) calculated	(40) observed
° 47	Nov. 8 . .	1·0003	1·0013	1·0013	1·0014	1·0154	1·0155	1·0152
48½	,, 1 (aft.) .	1·0003	1·0008	1·0006	. .	. .	. .	. .
49¼	,, 2 . .	1·0003	1·0004	1·0003	1·0006	1·0150	1·0151	1·0149
51	Oct. 31 . .	1·0003	·9996	·9996	·9997	1·0148	1·0147	1·0147
51	Nov. 7 . .	1·0004	·9997	·9996	1·0000	. .	. .	. .
51½	,, 18 . .	1·0004	·9993	·9993	. .	. .	. .	. .
52	,, 12 (morn.)	1·0004	·9994	·9991	. .	. .	. .	. .
52½	,, 17 . .	1·0004	·9994	·9989	. .	. .	. .	. .
53	,, 21 . .	1·0003	·9988	·9987	. .	. .	. .	. .
53½	,, 11 (even.)	1·0004	·9989	·9985	. .	. .	. .	. .
54	,, 20 . .	1·0003	·9982	·9983	. .	. .	. .	. .
56½	Oct. 28 . .	1·0003	·9971	·9972	·9973	1·0133	1·0135	1·1032
56½	Nov. 12 (aft.) .	. .	·9974	·9972	. .	. .	. .	. .
57	Oct. 27 . .	1·0003	·9970	·9970	·9972	1·0134	1·0134	1·0133
57½	Nov. 13 . .	1·0004	·9970	·9968	. .	. .	. .	. .
58¼	,, 4 . .	1·0004	·9964	·9965	·9963	1·0131	1·0131	1·0130
61	Oct. 30 . .	1·0004	·9955	·9953	·9955	1·0125	1·0125	1·0125
67	Nov. 27 . .	1·0004	·9928	·9928	·9927	1·0111	1·0113	1·0110

TABLE XVI.—In terms of (38) at 51° Fahr.

Temperature (Fahr.)	Date, 1885	(37) observed	(37) calculated	Temperature (Fahr.)	Date, 1885	(36) observed	(36) calculated
° 32	Nov. 10 . .	1·0051	1·0050	° 32	Nov. 11 (aft.) .	1·0067	1·0067
32	,, 11 (aft.) .	1·0050	1·0050	32	,, 11 (even.)	1·0066	1·0067
32	,, 11 (even.)	1·0051	1·0050	32	,, 12 (morn.)	1·0070	1·0067
40	,, 17 . .	1·0028	1·0027	32	,, 12 (even.)	1·0068	1·0067
41¾	,, 21 . .	1·0021	1·0022	40	,, 17 . .	1·0038	1·0041
47½	,, 8 . .	1·0003	1·0003	41¾	,, 21 . .	1·0028	1·0035
48¾	,, 12 (aft.) .	·9995	·9998	47½	,, 18 . .	1·0010	1·0013
49½	,, 12 (morn.)	·9996	·9996	49½	,, 13 . .	1·0003	1·0005
49½	,, 13 . .	·9993	·9996	50½	,, 2 . .	1·0003	1·0001
50½	,, 2 . .	·9997	·9992	50½	,, 10 . .	1·0001	1·0001
52½	Oct. 31 . .	·9986	·9985	53	Oct. 31 . .	·9992	·9991
52½	Nov. 18 . .	·9984	·9985	53	Nov. 18 . .	·9992	·9991
57½	Oct. 27 . .	·9966	·9966	57½	Oct. 27 . .	·9970	·9971
57½	,, 28 . .	·9967	·9966	57½	,, 28 . .	·9971	·9971
61¾	,, 30 . .	·9946	·9949	61¾	,, 30 . .	·9950	·9952
66½	Nov. 4 . .	·9926	·9928	66½	Nov. 4 . .	·9931	·9929
67¼	,, 20 . .	·9916	·9925	67¼	,, 20 . .	·9923	·9925
67½	,, 7 . .	·9922	·9924	67½	,, 7 . .	·9925	·9924

with them) numbers calculated from quadratic expressions. For (37) the expression is

$$\cdot9985\,\{1-\cdot00037\,(t-52\tfrac{1}{2})-\cdot0000025\,(t-52\tfrac{1}{2})^2\}\,;$$

the equivalent of which in Centigrade degrees is

$$\cdot9985\,\{1-\cdot000666\,(t-11\cdot4)-\cdot0000081\,(t-11\cdot4)^2\}.$$

According to this, the change for one degree Cent. is the following linear function of temperature

$$\cdot000674+\cdot0000162\,(t-11\cdot4)\,;$$

so that the temperature-coefficient ranges from ·000489 at 0° C. to ·000813 at 20° C. At 15° C. it would be ·00073.

For the other cell (36) the observed values of E.M.F. are pretty well represented by

$$\cdot9991\,\{1-\cdot00042\,(t-53)-\cdot0000028\,(t-53)^2\},$$

from which are deduced the numbers in the column headed "calculated." In Centigrade degrees this becomes

$$\cdot9991\,\{1-\cdot000756\,(t-11\cdot6)-\cdot0000091\,(t-11\cdot6)^2\}\,;$$

giving for the temperature-coefficient at $t°$ Cent.

$$\cdot000765+\cdot000018\,(t-11\cdot6).$$

At 15° the value from this formula is ·00083.

It would seem that these two cells have temperature-coefficients which differ sensibly. But this way of presenting the matter is apt to give an exaggerated impression. The difference in the coefficients indicates a separation of electro-motive forces at the rate of $\frac{1}{10000}$ only per degree Cent., so that the whole relative change for ordinary indoor variations would not exceed $\frac{1}{1000}$ of the whole. It will be seen from Tables XIV., XV. that, through a more limited range of temperature, a large number of various cells are satisfactorily represented by the coefficient (intermediate between those just found) ·00077; and I believe that the adoption of this number* for cells with *saturated solutions* can lead to no appreciable error in ordinary use. For very special purposes it will, no doubt, be desirable to protect standard cells from the larger temperature-variations, which is very easily done by keeping them in a cellar, from which well-insulated wires will convey the E.M.F. to any desired point. As to the absolute magnitude of the variations of E.M.F. with temperature, it may be worth while to recall that they are about the *double* of those experienced by German silver resistances.

* It should be mentioned that (*loc. cit.*) Dr Fleming found for a Clark cell the coefficient .00082 at 15° C.

§ 54. A good many special experiments have been tried with cells variously constructed, in order to elucidate as far as possible the behaviour of Clark's. Two of these, labelled H (1, $\frac{1}{2}$), H (1, $\frac{1}{4}$), were referred to in a short communication to the British Association at Montreal*, and have been observed at intervals since, with results which are, I think, of interest. In these cells, the legs representing the zincs were charged with a strong† but fluid amalgam of zinc and mercury; so that if the other legs had been charged with pure mercury and paste of mercurous sulphate, an ordinary Clark would have resulted. This, however, was not done. No paste at all was used, and in place of pure mercury a dilute amalgam was substituted— in H (1, $\frac{1}{2}$) one obtained by diluting the strong amalgam with its own volume of pure mercury, and in H (1, $\frac{1}{4}$) one containing one volume of strong amalgam to three volumes of pure mercury. The cells were filled up, as usual, with zinc sulphate solution.

The electro-motive forces, expressed as fractions of No. 1, are recorded in the adjoining Table (XVII.). It is clear that the E.M.F. rises rapidly with temperature—something like 30 per cent. for 15° F. The large amount of this variation, and the fact of its taking place in the opposite direction to that followed by Clark's, are very remarkable.

TABLE XVII.

Date, 1884	Temperature (Fahr.)	H (1, $\frac{1}{2}$)	H (1, $\frac{1}{4}$)	Date, 1885	Temperature (Fahr.)	H (1, $\frac{1}{2}$)	H (1, $\frac{1}{4}$)
	°				°		
July 21 .	. .	·0041	. .	Jan. 2 .	. .	·0033	·0083
,, 22 .	. .	·0041	. .	,, 14 .	. .	·0031	·0081
,, 23 .	. .	·0041	. .	Aug. 3 .	63	·0044	·0095
,, 24 .	. .	·0041	. .	Sept. 5 .	63	·0042	·0095
,, 29 .	. .	·0040	·0091	Nov. 6 .	52	·0034	·0083
,, 30 .	. .	·0040	·0091	,, 12 .	56	·0039	·0086
Aug. 6 ,,	67	·0044	·0096	,, 27 .	67	·0044	·0093
Sept. 4 .	62	·0041	·0093	Dec. 2 .	51	·0034	·0084
Oct. 8 .	63	·0040	·0093				
,, 28 .	59	·0037	·0090				
Nov. 17 .	. .	·0036	·0087				
Dec. 5 .	. .	·0037	·0089				

It would appear that, apart from changes of temperature, the cells have retained their E.M.F. tolerably constant for more than a year and a half— a result which could scarcely have been expected, in view of the very small quantities of zinc which they contain.

* *Brit. Assoc. Report*, 1884, p. 651. [Vol. II. p. 359.]
† As to what is to be regarded as strong, more facts will presently be given.

If we diminish still further the proportion of zinc in the weaker amalgam, the E.M.F. increases; until, if we omit the zinc altogether, the E.M.F. may reach ·8 or ·9 Clark. In this case, however, the E.M.F. is very unsteady.

On Feb. 18, 1885, a cell (F) of the H-pattern was charged with pure mercury in one leg, and in the other with an amalgam containing one milligram of zinc dissolved in 30 grams of mercury. No paste being used, the cell was filled up with saturated zinc sulphate solution. Under these conditions the E.M.F. was unsteady—about ·7. On Feb. 23 some paste was added to the leg containing pure mercury, on which the E.M.F. became steady at ·939 Clark, even when the amalgam was stirred with a glass rod. On Feb. 28, E.M.F. = ·935; on March 2, ·931; on March 6, ·920; on April 12, ·068; and on Dec. 2, ·0003.

Another cell (G) charged with pure mercury and paste in one leg had no zinc added to the pure mercury in the other leg, except such as it received by the passage for 3^m of the current from two large Clarks, in whose circuit 1000 ohms of wire-resistance was also included. Under these circumstances, the deposited zinc should have been about $\frac{1}{10}$ mg. The resulting E.M.F. was ·923, falling off from day to day, as in the case of (F). Nine months afterwards the value was ·0003.

From these and other experiments it appeared that an astonishingly minute proportion of zinc in the amalgam was sufficient to give the cell most of the force of a Clark. I had contemplated examining in greater detail the electro-motive character of weak amalgams of known composition, when my attention was called by Prof. Ayrton to an interesting investigation by Messrs Hockin and Taylor*, covering the same ground. In their cells zinc was opposed to the various amalgams; so that the circumstances were really the same as in the case of H (1, $\frac{1}{2}$), H (1, $\frac{1}{4}$), above discussed, the strong amalgam being equivalent to zinc. A few of their numbers may here be quoted. When the zinc was opposed to pure mercury, E.M.F. = 1·186 volt. (According to my experience this should have been uncertain.) The mercury being replaced by amalgams, the E.M.F. observed were as follows:—

					Volts
Zinc 1 part, mercury			23,600,000	parts	1·179
„	„	„	11,800,000	„	1·080
„	„	„	7,530,000	„	·655
„	„	„	5,900,000	„	·513
„	„	„	1,800,000	„	·214
„	„	„	400,000	„	·134
„	„	„	200,000	„	·124

---

* "On the Electro-motive Force of Mercury Alloys," *Journal of Society of Telegraph Engineers*, vol. VIII. 1879.

" Mercury alloyed with any number of metals takes the place in the scale of E.M.F. of the most electro-positive metal it may contain, if the amount of the electro-positive metal present is not less than about $\frac{1}{1000000}$ in weight of the mercury."

§ 55. In order to obtain any stability of E.M.F. in these cells, it seems to be necessary that the mercury should either be alloyed with a sensible, though perhaps relatively very small, portion of zinc, or else protected with mercurous sulphate. With pure mercury without paste not only is the E.M.F. variable from hour to hour, but it can scarcely be said to be definite even at a particular time—that is, it may be altered by the passage of a very small quantity of electricity, such as should have no effect whatever upon a properly constituted cell. Sometimes when the galvanometer contact was made, a throw of the needle was observed, not followed by any corresponding permanent deflection, the cell in fact behaving like a charged condenser.

In the absence of paste a very small addition of zinc to mercury gives it a definite and tolerably permanent character. If, however, there be any mercurous sulphate, even though originally in another leg of an H-cell, the mercury is gradually repurified. To this cause is to be attributed the gradual fall to zero of the E.M.F. of the cells (F) and (G) above considered.

Indeed, the whole tendency of these observations is to suggest that the action of the mercurous sulphate may be secondary, rather than primary; so that if zinc could be opposed to really pure mercury, no mercurous sulphate would be needed in a Clark cell. It may be, however, that in that case a minute quantity of mercurous sulphate would form itself spontaneously. In such cells as H $(1, \frac{1}{2})$ we are compelled to suppose that the chemical origin of the E.M.F. is the tendency of zinc and mercury to combine, or rather the tendency of two different amalgams to equalise themselves. There is here a close parallelism to the electro-motive forces due to affinity of saline solutions of different strengths, manifested, for example, when an electrode of zinc is in contact with a strong solution of zinc sulphate, and the second similar electrode is in contact with a weak solution.

Before quitting this subject, I may mention observations on what was practically a Clark cell, although prepared without zinc and without paste. On March 7, 1885, an H-cell (J) was charged with pure mercury in both legs, and filled up with zinc sulphate solution. A current from an external source of about $\frac{1}{400}$ ampère was then passed through it for $3\frac{1}{2}$ hours. The E.M.F. at subsequent dates were as follows :—

Clark

March 9	·983
March 11	·980
April 9	·978
July 7	·973
December 12	·971

It will be seen that a tolerable approximation to a Clark was obtained by this simple process. Zinc is thrown down upon the mercury in one leg, and mercury dissolved in the other by the passage of the current.

§ 56. The comparisons recorded in Table XIII. furnish abundant proof that there has been no special change in the E.M.F. of No. 1, but do not of themselves exclude the possibility of a general movement of the cells in one direction as the result of age, by which the relative values of cells of nearly the same date might remain unaffected. Comparisons with younger cells, indeed, go a long way towards negativing this supposition; but considering that all the older cells had undergone a journey (from Cambridge to Terling), I thought that a re-determination of the absolute E.M.F., by means of silver, would be a valuable confirmation of their constancy.

With the assistance of Mrs Sidgwick, two such silver determinations were effected in August, 1885, as described in § 38. Reduced to 15°, the E.M.F. of No. 1 was found on August 15 to be 1·4541 B.A. volts, and on August 19, 1·4533, giving as a mean 1·4537 B.A. volts. The value found from absolute measurements in the autumn of 1883, and the spring of 1884, was 1·4542 (§ 36). That given (note to § 37) as the result of two experiments in Nov., 1884, is 1·4534; but a slight error has been discovered in the reductions leading to the latter number. By an oversight the factor 1·00041, necessary to correct for the slight heating of the resistance by the passage of the current (§ 33), was omitted, the effect of which omission is an under-estimate of the difference of potentials at the terminals of the resistance traversed by a known current, and consequently of the E.M.F. of the cell under test. This correction being introduced, the numbers will stand thus:—

TABLE XVIII.

Date	E.M.F. of No. 1 at 15° C. in B.A. volts
Oct. 1883 to April 1884 . .	1·4542
Nov. 1884	1·4540
Aug. 1885	1·4537

The slight fall in these numbers has little significance; and we may regard the E.M.F. of No. 1 as having remained constant to within about $\frac{1}{2000}$ for nearly two years.

I believe that the E.M.F. of cells constructed with reasonable precautions, especially as to saturation, will rarely differ at ordinary temperatures by more than $\frac{1}{1000}$ part from

$$1{\cdot}454 \left\{1 - {\cdot}00077 \left(t - 15\right)\right\} \text{ B.A. volts,}$$

or (the B.A. unit of resistance being taken equal to $\cdot$9867 ohm)

$$1{\cdot}435 \left\{1 - {\cdot}00077 \left(t - 15\right)\right\} \text{ true volts.}$$

§ 57. In Table X., § 32, are given some comparisons with No. 1 Clark of a standard Daniell with equi-dense solutions of Raoult's pattern. As a mean at 16° C.,

$$\text{Daniell} = {\cdot}7703 \text{ Clark} = 1{\cdot}1046 \text{ true volt.}$$

I have never succeeded in obtaining really good results with Daniells of any construction, variations in the condition of the copper* rapidly altering the E.M.F. by two or three thousandths. The mean of a rather large number of comparisons in August, 1885, gives about $\cdot$7715 Clark at 16°, and this number was confirmed by further observations in December, 1885. The sp. gr. of the solutions was 1$\cdot$101; the zinc was amalgamated, and the copper was freshly-coated electrolytically. The effect of variations in the condition of the metals and of the solutions are discussed by Alder Wright and by Fleming.

Some observations have also been made with apparatus and solutions lent me by Dr Fleming, and arranged nearly according to the description published by him†. The variations observed with this form of Daniell were smaller than with Raoult's, perhaps in consequence of the better protection of the copper solution from the atmosphere‡. The mean value for 16° was about

$$\cdot 7674 \text{ Clark} = 1{\cdot}1004 \text{ volt,}$$

in pretty good agreement with the value found by Dr Fleming himself.

---

* It seems not unlikely that the greater part of the uncertainty of ordinary Daniells might be got rid of by substitution of silver for copper. Silver *sulphate*, however, is not sufficiently soluble. But I have not made experiments in this direction, believing that the mercury of the Clark cell is better still.

† *Phil. Mag.* August, 1885.

‡ In my experience the copper found its way to the zinc more readily than one would have expected. The amalgamated zinc became coated with a furry deposit, which struck a blue colour on solution in nitric acid and supersaturation with ammonia. In using Raoult's cells I have endeavoured to hold the copper in check by placing a loose strip of zinc in the connecting Y-tube.

In correcting for the effect of a small variation of temperature, I have allowed only for the variations of the Clark, assuming that of the Daniell to be insensible.

P.S.—May, 1886.—I have lately had the opportunity of comparing with mine five cells set up by Dr A. Muirhead, who has had great experience in these matters. All five agreed with my No. 1 to about one part in *ten thousand*. Dr Muirhead informs me that other cells, including one prepared by himself seven or eight years ago, agree closely with these.

[1900. Further remarks upon the Clark cell, with reference to a paper by Prof. Carhart, will be found in the *Electrician* for Jan. 24, 1890.]

133.

TESTING DYNAMOS.

[*Electrical Review*, XVIII. p. 242, 1886.]

In the measurements of the efficiency of dynamos by Dr Hopkinson's ingenious method*, would it not be possible to carry out the principle more fully, so as to dispense with *all* measurement of mechanical power, by introducing into the circuit a few storage cells which should supply the small percentage of energy wanted? In this way all the data could be observed electrically.

* [1900. See *El. Rev.* XVIII. p. 230, 1886. In this method one machine acting as a dynamo is connected, mechanically and electrically, to a similar machine acting as a motor, so that the power recovered from the motor is utilised to drive the dynamo in relief of the prime mover. In as much as mechanical measurements are subject to more uncertainty than electrical measurements, the approximate elimination of the former measurement is a great advantage.]

134.

THE REACTION UPON THE DRIVING-POINT OF A SYSTEM EXECUTING FORCED HARMONIC OSCILLATIONS OF VARIOUS PERIODS, WITH APPLICATIONS TO ELECTRICITY.

[*Philosophical Magazine*, XXI. pp. 369—381, 1886.]

THE object of the present communication is to prove some general mechanical theorems, which may be regarded as in some sort extensions of that of Thomson relating to the energy of initial motions. The question involved in the latter may be thus stated* :—

" Given any material system at rest. Let any parts of it be set in motion suddenly with any specified velocities possible, according to the connections of the system; and let its other parts be influenced only by its connections with these. It is required to find the motion." And the solution is " that the motion actually taken by the system is that which has less kinetic energy than any other motion fulfilling the prescribed velocity conditions." On the other hand, if the impulses are given, a theorem of Bertrand tells us that the kinetic energy is the *greatest* possible.

For our present purpose we suppose the system to be set in motion by an impulse of one particular type, which we may call the first. The impulse itself may be denoted by $\int \Psi_1 dt$, and the corresponding velocity generated by $\dot{\psi}_1$. Under any given circumstances as to constraint, the velocity and the impulse are in proportion to one another; and the resulting kinetic energy T is proportional to the square of either, being equal to $\frac{1}{2}\dot{\psi}_1\int\Psi_1 dt$. Now Thomson's theorem asserts that the introduction of a constraint can only increase the value of T when $\dot{\psi}_1$ is given. Hence, whether $\dot{\psi}_1$ be given or not, the constraint can only increase the ratio of $\frac{1}{2}T$ to $\dot{\psi}_1^2$, or of $\int\Psi_1 dt$ to $\dot{\psi}_1$. This form of the statement virtually includes both Bertrand's and Thomson's

---

* Thomson and Tait's *Natural Philosophy*, §§ 316, 317.

theorems, which are thus seen to be merely different aspects of the same truth. If the velocity be given, the impulse is a minimum in the absence of constraint. If the impulse be given, the velocity is under the same circumstances a maximum. Calling the ratio of $\int \Psi_1 dt$ to $\dot{\psi}_1$ the moment of inertia of the system when subjected to forces of the type in question, we may say that this moment can only be increased by the introduction of a constraint forcing the motion to follow a different law from that natural to it.

In close analogy to this theorem there are two others, relating to equilibrium and to steady motion resisted by viscous forces, of at least equal importance*. They may be thus stated.

Conceive a system to be displaced from stable equilibrium by a force of specified type. If the corresponding displacement be given in magnitude, the force is a minimum—or if the magnitude of the force be given, the displacement is a maximum,—when there is no constraint. Or we may say that the *stiffness* of the system, with respect to the kind of force in question, is increased by constraint. Examples, in illustration of the general proposition, are given in the papers already cited.

The third theorem depends upon the properties of the dissipation-function, and its most interesting application is to the conduction of heat and electricity. To take the latter case, if an electromotive force be applied to any system of conductors, the "resistance" to steady currents can only be increased by the imposition of a constraint, such for example as the rupture of a contact.

Hitherto we have supposed the forces to be either instantaneous or steady; and the three theorems depend upon the functions T, F, and V, expressing respectively kinetic energy, dissipation, and potential energy, only one of them being supposed to come into consideration at a time. We have now to inquire under what conditions the theorems remain intact when the impressed force is a harmonic function of the time.

As regards the first theorem, the justification for the neglect of F and V may be that they are non-existent, as in many problems of ordinary hydrodynamics. In such cases the motion is at any instant of the same character as if it had been generated impulsively from rest, and the moment of inertia is a minimum. But even when F and V are generally sensible, their influence tends to diminish as the frequency of alternation increases, and we approach at last a state of things in which they may be neglected. From this point onwards we may say that the moment of inertia is a minimum in the unconstrained condition. Thus in a system of electrical conductors subject to a rapidly periodic electromotive force, the distribution of currents

---

* *Phil. Mag.* Dec. 1874, "A Statical Theorem" [vol. I. p. 223]; March 1875, "General Theorems relating to Equilibrium and to Initial and Steady Motion" [vol. I. p. 232]. See also *Theory of Sound*, ch. IV.

is ultimately independent of the resistances, and the *self-induction* is a minimum in the absence of constraint.

In like manner, even when T and F are sensible, the motion tends to be more and more determined by V, as the frequency of the vibrations is imagined to *diminish*. An "equilibrium theory" ultimately becomes applicable, and the "stiffness" is a minimum when there are no constraints.

The theorem in which F is mainly concerned stands in a somewhat different position. If T and V are both sensible, we cannot find an extreme case, in respect of the frequency of the vibration, which shall annul their influence. If, however, V vanish, we can make F paramount by taking the period sufficiently long; and if T vanish, we can attain the same object by limiting ourselves to the case when the period is very short. If T and V both vanish, the theorem of minimum resistance in the absence of constraint holds good for all periods of vibration. In the application to a system of electrical conductors which possess resistance and induction, but no capacity for *charge* needing to be regarded, we find that while (as already stated) the induction becomes paramount when the vibrations are very rapid, on the other hand when they are very slow the distribution is determined ultimately by the resistances only. In the first case the self-induction, and in the second the effective resistance, is a minimum in the absence of constraints.

We are now prepared to enter upon the consideration of the problem which is the main subject of the present paper, viz. the behaviour of systems in which F, and one or other of the two remaining functions T and V, are sensible, but without the restriction to very rapid or to very slow motions by which the influence of the second function may be got rid of. The investigation is almost the same whether it be T or V that enters; for the sake of definiteness I will take the first alternative.

Consider then a system, devoid of potential energy, in which the coordinate ψ_1 is made to vary by the operation of the harmonic force Ψ_1, proportional to e^{ipt}. The other coordinates ψ_2, ψ_3, ... may be chosen arbitrarily, and it will be very convenient to choose them (as may always be done) so that no product of them enters into the expressions for T and V. They would be in fact the principal or normal coordinates of the system on the supposition that ψ_1 is constrained (by a suitable force of its own type) to remain zero. The expressions for T and F thus take the following forms:—

$$T = \tfrac{1}{2}a_{11}\dot{\psi}_1^2 + \tfrac{1}{2}a_{22}\dot{\psi}_2^2 + \tfrac{1}{2}a_{33}\dot{\psi}_3^2 + \cdots$$
$$+ a_{12}\dot{\psi}_1\dot{\psi}_2 + a_{13}\dot{\psi}_1\dot{\psi}_3 + a_{14}\dot{\psi}_1\dot{\psi}_4 + \cdots \cdots \cdots \cdots (1)$$
$$F = \tfrac{1}{2}b_{11}\dot{\psi}_1^2 + \tfrac{1}{2}b_{22}\dot{\psi}_2^2 + \tfrac{1}{2}b_{33}\dot{\psi}_3^2 + \cdots$$
$$+ b_{12}\dot{\psi}_1\dot{\psi}_2 + b_{13}\dot{\psi}_1\dot{\psi}_3 + b_{14}\dot{\psi}_1\dot{\psi}_4 + \cdots \cdots \cdots \cdots (2)$$

from which we get the equations of motion

$$a_{11}\ddot{\psi}_1 + a_{12}\ddot{\psi}_2 + a_{13}\ddot{\psi}_3 + \ldots + b_{11}\dot{\psi}_1 + b_{12}\dot{\psi}_2 + \ldots = \Psi_1,$$

$$a_{12}\ddot{\psi}_1 + a_{22}\ddot{\psi}_2 + b_{12}\dot{\psi}_1 + b_{22}\dot{\psi}_2 = 0,$$

$$a_{13}\ddot{\psi}_1 + a_{33}\ddot{\psi}_3 + b_{13}\dot{\psi}_1 + b_{33}\dot{\psi}_3 = 0,$$

$$\ldots\ldots\ldots\ldots\ldots\ldots\ldots\ldots\ldots\ldots\ldots\ldots\ldots$$

since there are no forces other than Ψ_1. We now introduce the supposition that the whole motion is harmonic in response to Ψ_1. Thus the above equations may be replaced by

$$(ip\,a_{11} + b_{11})\,\dot{\psi}_1 + (ip\,a_{12} + b_{12})\,\dot{\psi}_2 + (ip\,a_{13} + b_{13})\,\dot{\psi}_3 + \ldots = \Psi_1,$$

$$(ip\,a_{12} + b_{12})\,\dot{\psi}_1 + (ip\,a_{22} + b_{22})\,\dot{\psi}_2 = 0,$$

$$(ip\,a_{13} + b_{13})\,\dot{\psi}_1 + (ip\,a_{33} + b_{33})\,\dot{\psi}_3 = 0.$$

$$\ldots\ldots\ldots\ldots\ldots\ldots\ldots\ldots\ldots\ldots\ldots\ldots\ldots$$

By means of the second and following equations, $\dot{\psi}_2$, $\dot{\psi}_3$, ... are expressed in terms of $\dot{\psi}_1$. Introducing these values into the first, we get

$$\frac{\Psi_1}{\dot{\psi}_1} = ip\,a_{11} + b_{11} - \frac{(ip\,a_{12} + b_{12})^2}{ip\,a_{22} + b_{22}} - \frac{(ip\,a_{13} + b_{13})^2}{ip\,a_{33} + b_{33}} - \ldots \quad \ldots\ldots(3)$$

The ratio $\Psi_1 : \dot{\psi}_1$ is a complex quantity, of which the real part corresponds to the work done by the force in a complete period, and dissipated in the system. By an extension of electrical language we may call it the *resistance* of the system and denote it by the letter R'. The other part of the ratio is imaginary. If we denote it by $ipL'\dot{\psi}_1$, or $L'\ddot{\psi}_1$, L' will be the moment of inertia, or self-induction of electrical theory. We write therefore

$$\Psi_1 = (R' + ipL')\,\dot{\psi}_1; \quad\ldots\ldots\ldots\ldots\ldots\ldots\ldots\ldots(4)$$

and the values of R' and L' are to be deduced by separation of the real and imaginary parts of the right-hand member of (3).

Now the real part of

$$\frac{(ip\,a_{12} + b_{12})^2}{ip\,a_{22} + b_{22}}$$

$$= \frac{b_{12}^2 b_{22} + p^2 a_{12}(2\,b_{12}a_{22} - a_{12}b_{22})}{b_{22}^2 + p^2 a_{22}^2} = \frac{b_{12}^2}{b_{22}} - \frac{p^2(a_{12}b_{22} - a_{22}b_{12})^2}{b_{22}(b_{22}^2 + p^2 a_{22}^2)}; \quad \ldots\ldots(5)$$

so that

$$R' = b_{11} - \Sigma\,\frac{b_{12}^2}{b_{22}} + p^2\,\Sigma\,\frac{(a_{12}b_{22} - a_{22}b_{12})^2}{b_{22}(b_{22}^2 + p^2 a_{22}^2)}. \quad \ldots\ldots\ldots\ldots\ldots(6)$$

This is the value of the resistance as determined by the constitution of the system, and by the frequency of the imposed vibration. Each component of the latter series (which alone involves p) is of the form $\alpha p^2/(\beta + \gamma p^2)$, where α, β, γ are all positive, and (as may be seen most easily by considering its reciprocal) increases continuously as p^2 increases from zero to infinity. We conclude that as the frequency of vibration increases, the value of R' increases continuously with it. At the lower limit the motion is determined sensibly by the quantities b (the resistances) only, and the corresponding resultant resistance R' is an absolute minimum, whose value is

$$b_{11} - \Sigma \frac{b_{12}^2}{b_{22}}. \quad\quad\quad\quad\quad\quad\quad\quad\quad\quad\quad\quad(7)$$

At the upper limit the motion is determined by the inertia of the component parts without regard to resistances, and the value of R' is

$$b_{11} - \Sigma \frac{b_{12}^2}{b_{22}} + \Sigma \frac{(a_{12}b_{22} - a_{22}b_{12})^2}{b_{22}a_{22}^2}. \quad\quad\quad\quad\quad(8)$$

That the resistance in this case would exceed that expressed by (7) might have been anticipated from the analogue of Thomson's theorem; but we now learn in addition that at every stage of the transition, during which in general the motions of the various parts disagree in phase, every increment of frequency of vibration is accompanied by a corresponding increment of resistance.

Again, the imaginary part of

$$\frac{(ip\,a_{12} + b_{12})^2}{ip\,a_{22} + b_{22}}$$

$$= ip\,\frac{b_{12}(2a_{12}b_{22} - a_{22}b_{12}) + p^2a_{22}a_{12}^2}{b_{22}^2 + p^2a_{22}^2} = ip\,\frac{a_{12}^2}{a_{22}} - ip\,\frac{(a_{12}b_{22} - a_{22}b_{12})^2}{a_{22}(b_{22}^2 + p^2a_{22}^2)}, \quad\dots(9)$$

so that

$$L' = a_{11} - \Sigma\,\frac{a_{12}^2}{a_{22}} + \Sigma\,\frac{(a_{12}b_{22} - a_{22}b_{12})^2}{a_{22}(b_{22}^2 + p^2a_{22}^2)}. \quad\quad\quad\dots(10)$$

In the latter series each term is positive, and continually diminishes as p^2 increases. Hence every increase of frequency is attended by a diminution of the moment of inertia, which tends ultimately to the minimum corresponding to disappearance of the dissipative terms.

Certain very particular cases in which R' and L' remain constant do not require more than a passing allusion. If T and F are of the same form, every such quantity as $(a_{12}b_{22} - a_{22}b_{12})^2$ vanishes.

As examples of the general theorem may be mentioned the problems considered by Prof. Stokes in his well-known paper upon "The Effect of the Internal Friction of Fluids on the Motion of Pendulums*." Consider, for instance, the result for a sphere of radius a, vibrating (according to e^{ipt}) in a fluid for which the kinematic coefficient of viscosity is μ'. M' denoting the

* Camb. Trans. vol. IX. 1850.

mass of the fluid displaced by the sphere, Prof. Stokes's results may be written

$$R' = \frac{9M'}{4a} \left\{ \frac{2\mu'}{a} + \sqrt{(2\mu'p)} \right\}, \qquad L' = M' \left\{ \frac{1}{2} + \frac{9\sqrt{(2\mu')}}{4a\sqrt{p}} \right\}.$$

When p is zero, which represents uniform motion of the sphere,

$$R' = \frac{9\mu'M'}{4a^2}, \qquad L' = \infty \,*.$$

As p increases, the expressions show that, in agreement with the theorem, R' continually increases and L' continually diminishes. In fact R' tends to become infinite, and L' to assume the value $(\frac{1}{2}M')$ given by ordinary hydrodynamics, in which viscosity is not regarded.

The use of the principal coordinates would not often be advantageous when the object is a special calculation of L' and R', rather than the establishment of a general theorem. In one very important case—that of two degrees of freedom only—the question does not arise, since but one other coordinate ψ_2 enters in addition to ψ_1. Under this head we may take the problem of the reaction upon the primary circuit of the electric currents induced in a neighbouring secondary circuit. In this case the coordinates (or rather their rates of increase) are naturally taken to be the currents themselves, so that $\dot\psi_1$ is the primary, and $\dot\psi_2$ the secondary current.

In usual electrical notation we represent the coefficients of self and mutual induction by L, N, M, so that

$$T = \tfrac{1}{2}L\dot\psi_1^2 + M\dot\psi_1\dot\psi_2 + \tfrac{1}{2}N\dot\psi_2^2,$$

and the resistances by R and S. Thus

$$a_{11} = L, \qquad a_{12} = M, \qquad a_{22} = N;$$
$$b_{11} = R, \qquad b_{12} = 0, \qquad b_{22} = S;$$

and (6) and (10) become at once

$$R' = R + \frac{p^2M^2S}{S^2 + p^2N^2}, \qquad L' = L - \frac{p^2M^2N}{S^2 + p^2N^2} \quad \ldots\ldots\ldots(11, 12)$$

These formulæ were given long ago by Maxwell†, who remarks that the reaction of the currents in the secondary has the effect of increasing the effective resistance and diminishing the effective self-induction of the primary circuit.

If the rate of alternation be very slow, the secondary circuit is without influence. If, on the other hand, the rate be very rapid,

$$R' = R + \frac{M^2S}{N^2}, \qquad L' = \frac{LN - M^2}{N}.$$

---

* That the energy of the motion is infinite in this case does not appear to have been noticed.
† *Phil. Trans.* 1865; M is misprinted for M^2.

The formulæ (11) and (12) may be applied to deal with a more general problem of considerable interest, which arises when the secondary circuit acts upon a third, this upon a fourth, and so on, the only condition being that there must be no mutual induction except between immediate neighbours in the series. Thus a_{13}, a_{14}, a_{24}, ... (or, as we should here call them, M_{13}, M_{14}, M_{24}, ...) are supposed to vanish, as would usually happen in experiment. For the sake of distinctness we will limit ourselves to four circuits.

In the fourth circuit the current is due *ex hypothesi* only to induction from the third. Its reaction upon the third, for the rate of alternation under contemplation, is given at once by (11) and (12); and if we use the complete values applicable to the third circuit under these conditions, we may thenceforth ignore the fourth circuit. In like manner we can now deduce the reaction upon the secondary, giving the effective resistance and self-induction of that circuit under the influence of the third and fourth circuits; and then, by another step of the same kind, we may arrive at the values applicable to the primary circuit, under the influence of all the others. The process is evidently general; and we know by the theorem that, however numerous the train of circuits, the influence of the others upon the first must be to increase its effective resistance and diminish its effective inertia, in greater and greater degree as the rapidity of alternations increases.

In the limit, when the rapidity of alternation increases indefinitely, the distribution of currents is determined by the induction-coefficients irrespective of resistance, and it is of such a character that the currents are alternately opposite in sign as we pass along the series*.

As another example under the head of two degrees of freedom, we will take the case of two electrical conductors in parallel. It is not necessary to include the influence of the leads outside the points of bifurcation. Provided there be no mutual induction between these parts and the remainder, their induction and resistance enter into the result by simple addition.

Under the operation of resistance only, the total current $\dot{\psi}_1$ would divide itself between the conductors R and S in the parts

$$\frac{S\dot{\psi}_1}{R+S}, \quad \text{and} \quad \frac{R\dot{\psi}_1}{R+S}.$$

We may conveniently take the second coordinate $\dot{\psi}_2$ so that the currents in the two conductors are

$$\frac{S}{R+S}\dot{\psi}_1 + \dot{\psi}_2, \quad \text{and} \quad \frac{R}{R+S}\dot{\psi}_1 - \dot{\psi}_2,$$

$\dot{\psi}_1$ still representing the total current.

* See a paper, "On some Electromagnetic Phenomena considered in connection with the Dynamical Theory," *Phil. Mag.* July 1869. [Vol. I. p. 1.]

Thus,

$$F = \tfrac{1}{2} R \left(\frac{S}{R+S} \, \dot{\psi}_1 + \dot{\psi}_2 \right)^2 + \tfrac{1}{2} S \left(\frac{R}{R+S} \, \dot{\psi}_1 - \dot{\psi}_2 \right)^2$$

$$= \tfrac{1}{2} \dot{\psi}_1{}^2 \frac{SR}{R+S} + \tfrac{1}{2} \dot{\psi}_2{}^2 (R+S) ;$$

and if L, M, N be the induction-coefficients of the two conductors,

$$T = \tfrac{1}{2} \dot{\psi}_1{}^2 \frac{LS^2 + 2MSR + NR^2}{(R+S)^2}$$

$$+ \dot{\psi}_1 \dot{\psi}_2 \frac{(L-M) S + (M-N) R}{R+S} + \tfrac{1}{2} \dot{\psi}_2{}^2 (L - 2M + N).$$

Accordingly,

$$a_{11} = \frac{LS^2 + 2MSR + NR^2}{(R+S)^2}, \qquad a_{12} = \frac{(L-M) S + (M-N) R}{R+S},$$

$$a_{22} = L - 2M + N ;$$

$$b_{11} = \frac{SR}{R+S}, \qquad b_{12} = 0, \qquad b_{22} = R + S ;$$

and thus by (6), (10),

$$R' = \frac{SR}{R+S} + \frac{p^2}{R+S} \frac{\{(L-M) S + (M-N) R\}^2}{(R+S)^2 + p^2 (L - 2M + N)^2}, \quad \dots\dots(13)$$

$$L' = \frac{LS^2 + 2MSR + NR^2}{(R+S)^2} - \frac{\{(L-M) S + (M-N) R\}^2}{(R+S)^2 (L - 2M + N)}$$

$$+ \frac{\{(L-M) S + (M-N) R\}^2}{(L - 2M + N) \{(R+S)^2 + p^2 (L - 2M + N)^2\}}. \quad \dots\dots(14)$$

It should be remarked that $(L - 2M + N)$ is necessarily positive, representing twice the kinetic energy of the system when the current in the first conductor is $+1$ and in the second -1.

Of the three terms in (14) the second and third cancel one another when p vanishes, and when p is very great the third term tends to disappear. The first and second terms together may be put into the form

$$\frac{LN - M^2}{L - 2M + N}, \quad \dots\dots\dots\dots\dots\dots\dots(15)$$

independent (as it should be) of the resistances. In this $(LN - M^2)$ is necessarily positive, but may be relatively small when the wires are wound together. The energy of the system is then very small, when the currents are so rapid that their distribution is determined by induction.

There is an interesting distinction to be noted here dependent upon the manner in which the connections are made. Consider, for example, the case of a bundle of five contiguous wires wound into a coil, of which three wires connected in series (so as to give maximum self-induction) constitute one of the branches in parallel, and the other two, connected similarly in series, constitute the other branch. There is still an alternative in respect to the manner of connection of the two branches. If steady currents would circulate opposite ways (M negative), the total current is divided into two parts in the ratio of $3 : 2$, in such a manner that the more powerful current in the double wire nearly neutralizes at external points the magnetic effects of the less powerful current in the triple wire, and the total energy of the system is very small. But now suppose that the connections are such that steady currents would pass the same way round in both branches (M positive). It is evident that the condition of minimum energy cannot be satisfied if the currents are in the same direction, but requires that the smaller current in the triple wire should be in the opposite direction to the larger current in the double wire. In fact the ratio of currents must be $3 : - 2$; so that (as on the same scale the total current is 1) the component currents in the branches are both numerically greater than the total current which is divided between them. And this peculiar feature becomes more and more strongly marked the nearer L and N approach to equality*.

When there are several conductors in parallel, the results would in general be very complicated. When, however, there is no mutual induction between the various members, simplification occurs. If the currents be denoted by $\dot{\psi}_1$, $\dot{\psi}_2$, $\dot{\psi}_3 \dots$, the difference of potentials at the common terminals is

$$E = (ipL_1 + R_1)\,\dot{\psi}_1 = (ipL_2 + R_2)\,\dot{\psi}_2 = \dots,$$

so that

$$\frac{E}{\dot{\psi}_1 + \dot{\psi}_2 + \dots} = \frac{1}{\Sigma\,(ipL + R)^{-1}}.$$

But if R' and L' be the effective resistance and self-induction respectively of the combination,

$$\frac{E}{\dot{\psi}_1 + \dot{\psi}_2 + \dots} = R' + ipL',$$

so that

$$\frac{1}{R' + ipL'} = \Sigma\,\frac{1}{R + ipL}. \qquad \dots\dots\dots\dots\dots\dots(16)$$

Now

$$\Sigma\,\frac{1}{R + ipL} = \Sigma\,\frac{R - ipL}{R^2 + p^2L^2};$$

or, if we write

$$\Sigma\,\frac{R}{R^2 + p^2L^2} = A, \qquad \Sigma\,\frac{L}{R^2 + p^2L^2} = B, \dots\dots\dots\dots(17)$$

$$\Sigma\,\frac{1}{R + ipL} = A - ipB = \frac{1}{(A + ipB)/(A^2 + p^2B^2)}.$$

Hence

$$R' = \frac{A}{A^2 + p^2B^2}, \qquad L' = \frac{B}{A^2 + p^2B^2} \dots\dots\dots\dots(18)$$

---

* The reader who is interested in this subject is referred to my papers in the *Phil. Mag.* July 1869, June 1870, " On Some Electromagnetic Phenomena," &c.   [Vol. I. p. 1, p. 14.]

Equations (17) and (18) contain the solution of the problem.   When $p = 0$,

$$R' = \frac{1}{\Sigma(R^{-1})}, \qquad L' = \frac{\Sigma(LR^{-2})}{(\Sigma R^{-1})^2} . \quad \dots\dots\dots\dots\dots(19)$$

When $p = \infty$,

$$R' = \frac{\Sigma(RL^{-2})}{(\Sigma L^{-1})^2}, \qquad L' = \frac{1}{\Sigma(L^{-1})} . \quad \dots\dots\dots\dots\dots(20)$$

These examples will suffice.

The relation between Ψ_1 and $\dot{\psi}_1$ expressed in (4) may be exhibited in another way in terms of the phase difference (ϵ) and the ratio of maxima. Thus if

$$\Psi_1 = P e^{i\epsilon} \dot{\psi}_1,$$

we have

$$P = \sqrt{(R'^2 + p^2 L'^2)}, \qquad \tan \epsilon = pL'/R' . \dots\dots\dots\dots\dots(21)$$

As p increases from 0 to ∞, ϵ usually ranges from 0 to $\frac{1}{2}\pi$.   At first sight it might appear probable that every increment of p would involve an increment of ϵ, but this seems not to be generally true.   For consider a case in which

$$a_{11} = 0, \qquad a_{12} = 0, \qquad a_{13} = 0, \dots$$

so that by (10)

$$pL' = \Sigma \frac{a_{22} b_{12}^2 p}{b_{22}^2 + p^2 a_{22}^2} .$$

Here pL' begins (as usual) at zero and ends at zero.   During part of the range, therefore, it falls; and thus since R' rises throughout, it follows that ϵ does not rise throughout.

It may be worth while to remark that in some cases, where we cannot deal with phases, we are concerned principally with the value of $\sqrt{(R'^2 + p^2 L'^2)}$, a quantity which practical electricians are then tempted to call the resistance of the system.   This temptation should be overcome, and the name reserved for R', on which depends the amount of energy dissipated.   It must be admitted, however, that a name for $\sqrt{(R'^2 + p^2 L'^2)}$ is badly required. Perhaps it might be called the "throttling."   [Heaviside's term is *impedance*.]

The corresponding theorem in cases when T vanishes is deduced in a similar manner with use of the potential energy,

$$V = \tfrac{1}{2} c_{11} \psi_1^2 + \tfrac{1}{2} c_{22} \psi_2^2 + \tfrac{1}{2} c_{33} \psi_3^2 + \dots + c_{12} \psi_1 \psi_2 + c_{13} \psi_1 \psi_3 + c_{14} \psi_1 \psi_4 + \dots .$$

Thus, if we write

$$\Psi_1 = \mu' \psi_1 + R' \frac{d\psi_1}{dt} \dots \quad \dots\dots\dots\dots\dots(22)$$

we find

$$\mu' = c_{11} - \Sigma \frac{c_{12}^2}{c_{22}} + p^2 \Sigma \frac{(b_{12} c_{22} - b_{22} c_{12})^2}{c_{22}(c_{22}^2 + p^2 b_{22}^2)}, \quad \dots\dots\dots\dots(23)$$

$$R' = b_{11} - \Sigma \frac{b_{12}^2}{b_{22}} + \Sigma \frac{(b_{12} c_{22} - b_{22} c_{12})^2}{b_{22}(c_{22}^2 + p^2 b_{22}^2)} . \quad \dots\dots\dots\dots(24)$$

As p^2 increases, the "stiffness" (represented by μ') increases, and the "resistance" diminishes.

After what has been said it will not be necessary to occupy space with illustrations of the present theorem. Indeed its applications seem to afford less interest. It is curious that here, again, the easiest examples would be taken from electricity, although the principle itself is one of general mechanics. These (relating to the periodic charge and discharge of condensers through high resistances) may be left to the reader who wishes to pursue the subject further. The application to the theory of the conduction of heat may also be noticed.

When the three functions T, F, and V are all sensible, it is not generally possible to make the transformation to sums of squares upon which our process was founded. There are, however, special cases in which the same transformation which is required to simplify T and V is successful also as regards F. Among these are of course to be reckoned cases in which F does not appear, and those where there is but one other coordinate besides ψ_1. Assuming that b_{23}, b_{34}, ... vanish, we have

$$\frac{\Psi_1}{\psi_1} = c_{11} - p^2 a_{11} + ip\, b_{11} - \frac{(c_{12} - p^2 a_{12} + ip\, b_{12})^2}{c_{22} - p^2 a_{22} + ip\, b_{22}} - \cdot \quad \dots\dots\dots(25)$$

If we put

$$\Psi_1 = \mu' \psi_1 + R' \frac{d\psi_1}{dt}, \quad\dots\dots\dots\dots\dots\dots(26)$$

we obtain the values of μ' and R' by writing in (23) and (24) throughout $c_{11} - p^2 a_{11}$, $c_{22} - p^2 a_{22}$, $c_{12} - p^2 a_{12}$, ... for c_{11}, c_{22}, c_{12}, ... respectively.

A simpler case, which may be worth special mention, arises when all the coefficients b_{12}, b_{13}, ... vanish. We have then

$$\mu' = c_{11} - p^2 a_{11} - \Sigma \frac{(c_{12} - p^2 a_{12})^2 (c_{22} - p^2 a_{22})}{(c_{22} - p^2 a_{22})^2 + p^2 b_{22}{}^2}, \quad\dots\dots\dots\dots(27)$$

$$R' = b_{11} + \Sigma \frac{(c_{12} - p^2 a_{12})^2 b_{22}}{(c_{22} - p^2 a_{22})^2 + p^2 b_{22}{}^2}. \quad\dots\dots\dots\dots(28)$$

This case, when there are two degrees of freedom, is considered in my book on the *Theory of Sound*, § 117.

If all the frictional coefficients b_{22}, b_{33}, ... disappear, we have

$$R' = b_{11}, \quad\dots\dots\dots\dots\dots\dots\dots\dots\dots\dots\dots\dots(29)$$

and

$$\mu' = c_{11} - p^2 a_{11} - \Sigma \frac{(c_{12} - p^2 a_{12})^2}{c_{22} - p^2 a_{22}}. \quad\dots\dots\dots\dots\dots\dots(30)$$

Whenever, during its increase, p approaches and passes through one of the values proper to the free vibrations of the system supposed to be vibrating under the condition that ψ_1 is constrained by a suitable force Ψ_1 to remain zero, μ' rises to $-\infty$ and passes through to $+\infty$.

135.

ON THE SELF-INDUCTION AND RESISTANCE OF STRAIGHT CONDUCTORS.

[*Philosophical Magazine*, XXI. pp. 381—394, 1886.]

IN connection with the experimental results of Professor Hughes*, I have recently been led to examine more minutely the chapter in Maxwell's *Electricity and Magnetism* (vol. II. ch. xiii.), in which the author calculates the self-induction of cylindrical conductors of finite section. The problems being virtually in two dimensions, the results give the ratio $L : l$, where L is the coefficient of self-induction, and l the length considered. And since both these quantities are linear, the ratio is purely numerical. In some details the formulæ, as given by Maxwell, require correction, and in some directions the method used by him may usefully be pushed further. The present paper may thus be regarded partly as a review, and partly as a development of Maxwell's chapter.

The problems divide themselves into two classes. In the first class the distribution of the currents is supposed to be the same as it would be if determined solely by resistance, undisturbed by induction; in particular the density of current in a cylindrical conductor is assumed to be uniform over the section. The self-induction calculated on this basis can be applied to alternating currents, only under the restriction that the period of the alternation be not too small in relation to the other circumstances of the case. If this condition be not satisfied, the investigation must be modified so as to include a determination of the distribution of current. A problem of this class considered by Maxwell (§ 689) relates to the " Electromotive Force required to produce a Current of Varying Intensity along a Cylindrical Conductor†."

---

* Inaugural Address to the Society of Telegraph Engineers, January 1886.

† That some of the results arrived at experimentally by Hughes might be attributed to unequal distribution of current in the conductors was pointed out by Prof. Forbes in the course of a discussion which followed the delivery of Prof. Hughes's address.

In connection therewith another problem of the same class will here be treated, in which the mathematical conditions are simpler, and the results more readily apprehended.

In § 685 Maxwell takes the problem of two cylindrical conductors, the first of which conveys the outgoing and the second the (numerically equal) return current. The external radii are a_1, a_1'; the internal radii a_2, a_2'; b the distance between the centres. A possible difference in the magnetic quality is contemplated, the permeabilities for the material composing the cylinders being denoted by μ, μ', and that of the intervening space by μ_0.

The first correction I have to note relates merely to a slip of the pen. The result (22) should run

$$\frac{L}{l} = 2\mu_0 \log \frac{b^2}{a_1 a_1'} + \tfrac{1}{2}\mu \left[\frac{a_1^2 - 3a_2^2}{a_1^2 - a_2^2} + \frac{4a_2^4}{(a_1^2 - a_2^2)^2} \log \frac{a_1}{a_2} \right]$$
$$+ \tfrac{1}{2}\mu' \left[\frac{a_1'^2 - 3a_2'^2}{a_1'^2 - a_2'^2} + \frac{4a_2'^4}{(a_1'^2 - a_2'^2)^2} \log \frac{a_1'}{a_2'} \right].$$

As printed in Maxwell's book the square brackets are omitted. This error does not affect the following formula (23), in which the cylinders are supposed to be solid. By putting a_2, a_2' equal to zero, we get

$$\frac{L}{l} = 2\mu_0 \log \frac{b^2}{a_1 a_1'} + \tfrac{1}{2}(\mu + \mu').$$

It must, however, be remarked that in the derivation of (22) Maxwell appears to have overlooked the effect of the matter composing one conductor in disturbing the lines of induction due to the current in the other. On this account the formula is correct only when the permeabilities μ are all equal, and the results cannot be applied to iron wires without reservation. It would seem, however, that the error is of small importance when the wires are distant. The application to wires in contact, contemplated in § 688, will hold good only for the non-magnetic metals.

If we write c^2 for $(a_1^2 - a_2^2)$, so that c is the radius of the solid cylinder of equal sectional area, we have in (22),

$$\frac{a_1^2 - 3a_2^2}{a_1^2 - a_2^2} + \frac{4a_2^4}{(a_1^2 - a_2^2)^2} \log \frac{a_1}{a_2}$$
$$= \frac{3c^2 - 2a_1^2}{c^2} - \frac{2(a_1^2 - c^2)^2}{c^4} \log \left(1 - \frac{c^2}{a_1^2}\right)$$
$$= \frac{2}{3}\frac{c^2}{a_1^2} + \text{terms in } \frac{c^4}{a_1^4}.$$

Hence, when the thickness of the cylinders is relatively small,

$$\frac{L}{l} = 2\mu_0 \log \frac{b^2}{a_1 a_1'} + \frac{\mu}{3}\frac{c^2}{a_1^2} + \frac{\mu'}{3}\frac{c'^2}{a_1'^2}.$$

If b, c, c' be given, the self-induction diminishes with increase of a_1, a_1', especially when μ, μ' are much greater than μ_0.

When μ is constant throughout, the "geometric mean distance" (§§ 691, 692) may conveniently be introduced. If A_1, A_2 be the areas occupied by the outgoing and return currents, we have

$$\frac{L}{l} = 4\mu_0 \left[\log R_{A_1 A_2} - \tfrac{1}{2} \log R_{A_1} - \tfrac{1}{2} \log R_{A_2}\right],$$

where R_{A_1}, R_{A_2} are to be understood as in (5), (6), (9) § 692.

For two circular areas,

$$\log R_{A_1 A_2} = \log b, \quad \log R_{A_1} = \log a_1 - \tfrac{1}{4}, \quad \log R_{A_2} = \log a_1' - \tfrac{1}{4},$$

if a_1, a_1' be the radii and b the distance between the centres; so that

$$\frac{L}{l} = 2\mu_0 \log \frac{b^2}{a_1 a_1'} + \mu_0,$$

as before.

In § 692 the value of R_A is given for rectangles and circular rings. For an ellipse of semi-axes a and b Prof. J. J. Thomson gives*

$$\log R = \log \frac{a+b}{2} - \tfrac{1}{4}, \quad \dots\dots\dots\dots\dots\dots\dots(1)$$

in which of course the case of the circle is included.

It is evident that for a given area R is least when the figure is circular. In that case we have

$$R = \cdot 4393 \sqrt{(A)}. \quad \dots\dots\dots\dots\dots\dots\dots\dots(2)$$

In the case of the square,

$$R = \cdot 44705 \sqrt{(A)}. \quad \dots\dots\dots\dots\dots\dots\dots\dots(3)$$

For the ellipse [of eccentricity e], Prof. Thomson's result leads to

$$R = \cdot 4393 \sqrt{A} \cdot (1 + \tfrac{1}{32} e^4 + \dots), \quad \dots\dots\dots\dots\dots\dots(4)$$

showing the small effect of moderate eccentricity when the area is given.

As examples of very elongated forms we may take the ellipse or the rectangle. In the latter case the value approximates to that applicable to a line given by Maxwell (5) § 692. If the length be a,

$$R = a\, e^{-\tfrac{3}{2}} = A b^{-1} e^{-\tfrac{3}{2}}, \quad \dots\dots\dots\dots\dots\dots\dots(5)$$

increasing without limit for a given area as b decreases.

It has been pointed out that Maxwell's result (22) § 685 is not rigorous, unless μ be constant. In order to put a case in which the lines of induction

---

* In a private letter.

follow a simple law in spite of the presence of iron, we may suppose that the conductors are co-axal cylindrical shells. The outgoing current of total strength C travels in the interior cylinder of radii a_2, a_1; the return current of strength $-C$ in the outer cylinder of radii a_2', a_1'.

In Maxwell's notation we have the equations

$$\frac{dH}{dr} = -\mu\beta, \qquad \beta r = 4\pi \int_0^r wr\,dr,$$

so that

$$\iint H\,w\,dx\,dy = 2\pi \int_0^\infty H\,wr\,dr = \tfrac{1}{2}\int H\,d\,(\beta r)$$

$$= \tfrac{1}{2}[H\beta r] + \tfrac{1}{2}\int_0^\infty \mu\beta^2 r\,dr.$$

Now βr vanishes both at zero and infinity, so that we may take

$$\iint H\,w\,dx\,dy = \tfrac{1}{2}\int_0^\infty \mu\beta^2 r\,dr, \quad\quad\quad\quad\quad\ldots\ldots\ldots\ldots\ldots\ldots(6)$$

in which β represents the magnetic force, everywhere perpendicular to r.

In the integration from 0 to ∞ there are five regions to be considered. In the first, from 0 to a_2, there is no magnetic force. In the second, from a_2 to a_1, the magnetic force depends upon the total current travelling through the strata which are internal with respect to the point in question. In terms of the total current C we have

$$\beta = \frac{2C}{a_1^2 - a_2^2}\left(r - \frac{a_2^2}{r}\right). \quad\quad\quad\quad\ldots\ldots\ldots\ldots\ldots\ldots(7)$$

The permeability is here supposed to be μ.

In the third region, between the cylinders, the permeability is μ_0, and the magnetic force is given by

$$\beta = \frac{2C}{r}. \quad\quad\quad\quad\quad\quad\quad\quad\ldots\ldots\ldots\ldots\ldots\ldots(8)$$

Within the second cylinder the permeability is μ', and

$$\beta = \frac{2C}{r} - \frac{2C}{a_1'^2 - a_2'^2}\left(r - \frac{a_2'^2}{r}\right). \quad\quad\ldots\ldots\ldots\ldots\ldots\ldots(9)$$

In the fifth region, from a_1' to ∞, $\beta = 0$.

Effecting the integrations, as indicated in (6), we obtain the value of

$$\iint H\,w\,dx\,dy,$$

which gives $2T/l$; and (if L be the coefficient of self-induction) $T = \tfrac{1}{2}LC^2$. The result is

$$\frac{L}{l} = 2\mu_0 \log\frac{a_2'}{a_1} + \frac{2\mu}{a_1^2 - a_2^2}\left\{\frac{a_1^2 - 3a_2^2}{4} + \frac{a_2^4}{a_1^2 - a_2^2}\log\frac{a_1}{a_2}\right\}$$

$$+ \frac{2\mu'}{a_1'^2 - a_2'^2}\left\{\frac{a_2'^2 - 3a_1'^2}{4} + \frac{a_1'^4}{a_1'^2 - a_2'^2}\log\frac{a_1'}{a_2'}\right\}. \quad\ldots\ldots\ldots(10)$$

Perhaps the most interesting application of the general result is to trace the diminution of self-induction as the two currents are brought into closer and closer proximity. Let us suppose that the intervening space is reduced without limit, so that $a_2' = a_1$. Suppose further that $\mu' = \mu$, and that both conductors have the same sectional area πc^2, so that

$$a_1{}^2 - a_2{}^2 = a_1{}'^2 - a_2{}'^2 = c^2.$$

Under these circumstances we have

$$\frac{L}{l} = \frac{2\mu}{c^4}\left\{ -r^2 c^2 + (r^2 - c^2)^2 \log \frac{r}{\sqrt{(r^2 - c^2)}} + (r^2 + c^2)^2 \log \frac{\sqrt{(r^2 + c^2)}}{r} \right\}, \quad \ldots(11)$$

in which r is written for the radius of the common surface.

If c is small in comparison with r, (11) becomes

$$\frac{L}{l} = \frac{2\mu c^2}{3 r^2}\left\{ 1 + \frac{1}{10}\frac{c^2}{r^2} + \ldots \right\}, \quad \ldots\ldots\ldots\ldots\ldots\ldots(12)$$

showing that when the sectional areas are given, the self-induction diminishes without limit as the radius (r) increases.

If b denote the thickness of the walls, we have ultimately $c^2 = 2br$, and

$$\frac{L}{l} = \frac{4\mu b}{3r}. \quad \ldots\ldots\ldots\ldots\ldots\ldots\ldots\ldots\ldots(13)$$

If the material composing the conductors be soft iron, the self-induction will be several hundred times greater than in the case of copper or other non-magnetic metal.

I now pass on to § 689, in which Maxwell solves a problem of the second class, relative to the self-induction of a cylindrical conductor, regard being had to the disturbance from uniformity in the distribution of the current over the section, due to induction. I will introduce the permeability μ, which in this question Maxwell treats as unity. His equations (14), (15), thus become

$$\mu C = -\left(\alpha\mu \frac{dT}{dt} + \frac{2\alpha^2 \mu^2}{1^2 . 2^2}\frac{d^2 T}{dt^2} + \ldots + \frac{n\alpha^n \mu^n}{1^2 . 2^2 \ldots n^2}\frac{d^n T}{dt^n} + \ldots \right),$$

$$AC - S = T + \alpha\mu \frac{dT}{dt} + \frac{\alpha^2 \mu^2}{1^2 . 2^2}\frac{d^2 T}{dt^2} + \ldots + \frac{\alpha^n \mu^n}{1^2 . 2^2 \ldots n^2}\frac{d^n T}{dt^n} + \ldots ;$$

where α, equal to l/R, represents the conductivity (for steady currents) of unit of length of the wire.

If $\phi(x)$ denote the function

$$1 + x + \frac{x^2}{1^2 . 2^2} + \ldots + \frac{x^n}{1^2 . 2^2 \ldots n^2} + \ldots, \quad \ldots\ldots\ldots\ldots\ldots(14)$$

these equations may be written

$$\frac{dS}{dt} = A\frac{dC}{dt} - \phi\left(\alpha\mu\frac{d}{dt}\right)\frac{dT}{dt}, \qquad C = -\alpha\phi'\left(\alpha\mu\frac{d}{dt}\right)\frac{dT}{dt}.$$

Moreover, if E denote "the electromotive force due to other causes than the induction of the current upon itself,"

$$\frac{E}{l} = \frac{dS}{dt}.$$

To apply this to periodic currents following the harmonic law, where all the functions are proportional to e^{ipt}, we may replace d/dt by ip. Hence, eliminating dT/dt, we get

$$\frac{\alpha E}{lC} = \frac{E}{RC} = ip\alpha \cdot A + \frac{\phi(ip\alpha\mu)}{\phi'(ip\alpha\mu)}. \qquad \ldots\ldots\ldots\ldots\ldots(15)$$

To interpret (15), we have to separate the real and imaginary parts of ϕ/ϕ'. Now if x be small, we get in ascending powers of x by ordinary division,

$$\frac{\phi(x)}{\phi'(x)} = 1 + \frac{1}{2}x - \frac{1}{12}x^2 + \frac{1}{48}x^3 - \frac{1}{180}x^4 + \frac{13}{8640}x^5 - \ldots; \quad \ldots\ldots(16)$$

so that

$$\frac{\phi(ip\alpha\mu)}{\phi'(ip\alpha\mu)} = 1 + \frac{1}{12}p^2\alpha^2\mu^2 - \frac{1}{180}p^4\alpha^4\mu^4 + \ldots$$
$$+ i\left\{\frac{1}{2}p\alpha\mu - \frac{1}{48}p^3\alpha^3\mu^3 + \frac{13}{8640}p^5\alpha^5\mu^5 - \ldots\right\}. \quad \ldots\ldots\ldots(17)$$

Thus, if we write

$$E = R'C + ipL'C, \qquad\ldots\ldots\ldots\ldots\ldots\ldots(18)$$

we find

$$R' = R\left\{1 + \frac{1}{12}p^2\alpha^2\mu^2 - \frac{1}{180}p^4\alpha^4\mu^4 + \ldots\right\}$$
$$= R\left\{1 + \frac{1}{12}\frac{p^2l^2\mu^2}{R^2} - \frac{1}{180}\frac{p^4l^4\mu^4}{R^2} + \ldots\right\}, \qquad\ldots\ldots\ldots(19)$$

$$L' = l\left\{A + \mu\left(\frac{1}{2} - \frac{1}{48}p^2\alpha^2\mu^2 + \frac{13}{8640}p^4\alpha^4\mu^4 - \ldots\right)\right\}$$
$$= l\left\{A + \mu\left(\frac{1}{2} - \frac{1}{48}\frac{p^2l^2\mu^2}{R^2} + \frac{13}{8640}\frac{p^4l^4\mu^4}{R^4} - \ldots\right)\right\}, \qquad\ldots\ldots(20)*$$

in which R' represents the effective resistance and L' the effective self-induction.

If the rate of alternation be very slow, so that p is small, these equations give (as was to be expected), $R' = R$, and

$$L' = l\{A + \tfrac{1}{2}\mu\}, \qquad\ldots\ldots\ldots\ldots\ldots\ldots\ldots(21)$$

* For the case $\mu = 1$, (19) and (20) follow readily from Maxwell's equation (18) § 690.

representing the self-induction for steady or slowly alternating currents. If we include the next terms, we see that as the frequency increases, the self-induction begins to diminish. At the same time the resistance begins to increase.

These results are merely very special cases of a general law*, from which we may learn that as the frequency of alternation gradually increases from zero to infinity, there is a steady rise of resistance and accompanying fall of self-induction. The application of the general idea to the present case is very simple. At slow rates of alternation the distribution of current, being such as to make the resistance a minimum, is uniform over the section; and this distribution, since it involves magnetization of the outer parts of the cylinder, leads to considerable self-induction, especially in iron. On the other hand, when the rate of alternation is very rapid, the endeavour is to make the self-induction a minimum irrespective of resistance. This object is attained by concentration of the current into the outer layers. The magnetization of the conductor is thus more and more avoided, but of course at the expense of increased resistance. We may gather from the general argument, what (19) and (20) in their actual forms do not tell us, that as p increases without limit, R' also becomes infinite, while the part of L' depending upon the magnetization of the conductor tends to zero.

The increase of resistance proper (not merely of the " throttling" due to the combined effect of resistance and self-induction) in iron wires of moderate diameter subjected to varying currents, is one of the most striking of Prof. Hughes's results. So far as I am aware, neither Maxwell nor any other theorist had anticipated that the alteration of resistance would be important under such circumstances †.

In order to see under what conditions the alteration of resistance (and of self-induction) would become sensible, we have to examine the value of

$$\frac{1}{12} \frac{p^2 l^2 \mu^2}{R^2}.$$

We will take first the case of an iron wire of ·4 centim. diameter. The specific resistance of iron in C.G.S. measure is about 10^4; so that

$$\frac{R}{l} = \frac{10^4}{\pi \times ·04}.$$

Thus, if $p = 2\pi/\tau$, τ being the complete period,

$$\frac{1}{12} \frac{p^2 l^2 \mu^2}{R^2} = \frac{5·2 \mu^2}{10^{10} \tau^2}. \quad\ldots\ldots\ldots\ldots\ldots\ldots\ldots(22)$$

---

* See preceding article. [Art. 134.]

† In the paper referred to I have quoted Maxwell's calculation of increased resistance and diminished self-induction due to the operation of currents in a *secondary* circuit.

The value of μ is more difficult to assign. It must be remembered that it is only in a very rough sense that μ can be treated as a constant at all. For small degrees of magnetization and for solid iron we may perhaps take $\mu = 300$ [100 would be a better value]. In the case of the hissing sounds of a microphone-clock, working by a scraping contact, τ must be less than $\frac{1}{1000}$ second. Taking it, however, at this value, we get

$$\frac{1}{12} \frac{p^2 l^2 \mu^2}{R^2} = 47 \text{ nearly,}$$

which shows that under these circumstances the resistance and self-induction are entirely different from what they would be for slow rates of alternation.

We will now consider the case of copper, where $\mu = 1$. The specific resistance may be taken to be 1640. If a be the radius in centimetres,

$$\frac{1}{12} \frac{p^2 l^2 \mu^2}{R^2} = \frac{\pi^4}{3} \frac{a^4}{1640^2 \tau^2} = \frac{1 \cdot 2\, a^4}{10^5 \tau^2}. \quad\text{...................}(23)$$

From an alternate-current machine we may have currents of period ·01 second (100 positive and 100 negative pulses per second). In such a case our fraction becomes $\cdot 12\, a^4$. This shows that for diameters of 1 centim. and over, the augmentation of resistance in the mains of an alternate-current system will rapidly become of commercial importance. A remedy may be found in the use of a more elongated section, or in subdivision of the main into a number of detached parts.

In physical experiments, such as those in which absolute resistance is determined by the method of the revolving coil*, (23) may be neglected, as we may see by supposing $a = \frac{1}{10}$, $\tau = \frac{1}{4}$.

The ultimate form of (15), when p is very great, may be arrived at analytically. The series (14) may be then replaced by

$$\frac{1}{2\sqrt{\pi}} \frac{e^{2\sqrt{x}}}{x^{\frac{1}{4}}}, \quad\text{.................................}(24)$$

from which we find $\phi(x) = x^{\frac{1}{2}} \phi'(x)$, or

$$\frac{\phi(ip\alpha\mu)}{\phi'(ip\alpha\mu)} = \sqrt{(\tfrac{1}{2}p\alpha\mu)} \cdot (1 + i). \quad\text{.......................}(25)$$

Accordingly, the limiting values of resistance and self-induction are given by

$$R' = R\sqrt{(\tfrac{1}{2}p\alpha\mu)} = \sqrt{(\tfrac{1}{2}pl\mu R)}, \quad\text{................................}(26)$$

$$L' = l\left\{A + \frac{\sqrt{\mu}}{\sqrt{(2p\alpha)}}\right\} = l\left\{A + \sqrt{\left(\frac{\mu R}{2pl}\right)}\right\}, \quad\text{................}(27)$$

* See for example *Phil. Trans.* 1882. "Experiments to Determine the Value of the B.A. Unit of Resistance in Absolute Measure." [Vol. II. p. 38.]

the first of which increases without limit with p, while the second tends to the finite value lA.

In the preceding problem the return current is supposed to be at a distance. As an example in which the self-induction of the whole circuit may become small, it would be natural to imagine the currents to travel in co-axal cylindrical shells, the interval between which might be considered to diminish indefinitely. The interest of the solution would, however, centre in the extreme case arrived at by supposing the radii of the cylinder to be great in comparison with the thickness of the walls; and if we limit ourselves to this from the first, the analysis will be a good deal simplified.

Neglecting then the curvature, we treat the walls as plane, and the width of the strips (corresponding to the circumference of the cylinders) as infinite; so that our functions, while remaining, as hitherto, independent of z (measured parallel to the axes of the cylinders), now become also independent of the second rectangular coordinate y, and may be treated as functions only of the time and of x, the coordinate measured perpendicularly to the walls. The problem is thus the distribution of currents in a circuit composed of two parallel infinitely long, infinitely wide, and equally thick strips, one of which conveys the outgoing and the other the return current. The thickness of each conducting strip will be denoted by b, and that of the intervening insulating layer by $2a$. The origin of x may conveniently be taken at the central point, in the middle of the insulating layer.

We might commence with the investigation of steady currents; but it will be sufficient to regard them as alternating (e^{ipt}). The results applicable to steady currents can always be deduced by simply putting p equal to zero.

Assuming, then, that the component currents u, v vanish, as well as the components of magnetic force γ, α, we have, in Maxwell's notation, the equations

$$\rho w = -\frac{d\Psi}{dz} - \frac{dH}{dt} = -\frac{d\Psi}{dz} - ipH, \quad\quad\quad\text{(28)}$$

$$d\beta/dx = 4\pi w, \quad\quad\quad \mu\beta = -dH/dx; \quad\quad\quad\text{(29)}$$

so that

$$w = \frac{1}{4\pi}\frac{d\beta}{dx} = -\frac{1}{4\pi\mu}\frac{d^2H}{dx^2}, \quad\quad\quad\text{(30)}$$

and

$$\frac{d^2H}{dx^2} = \frac{4\pi\mu}{\rho}\frac{d\Psi}{dz} + i \cdot \frac{4\pi\mu p}{\rho} \cdot H. \quad\quad\quad\text{(31)}$$

We will now apply (31) to that conducting strip which lies on the positive side of the origin. Since, by hypothesis, $d\Psi/dz$, the rate at which the potential varies, is independent of x, the solution for regular periodic motion may be written

$$H = \frac{i}{p}\frac{d\Psi}{dz} + A\,e^{m(x-a)} + Be^{-m(x-a)}, \quad \ldots\ldots\ldots\ldots(32)$$

$$\mu\beta = -mA\,e^{m(x-a)} + mBe^{-m(x-a)}, \quad \ldots\ldots\ldots\ldots(33)$$

in which A and B are constants, so far arbitrary, and

$$m^2 = \frac{4\pi\mu pi}{\rho}. \quad \ldots\ldots\ldots\ldots\ldots\ldots(34)$$

One relation between A and B is supplied by the condition that the magnetic force β must vanish at the external surface, where $x = a + b$. Hence

$$A\,e^{mb} = Be^{-mb}. \quad \ldots\ldots\ldots\ldots\ldots(35)$$

If C be the total current corresponding to width y, we have

$$\frac{C}{y} = \int_a^{a+b} w\,dx = \frac{1}{4\pi}(\beta_{a+b} - \beta_a) = -\frac{1}{4\pi}\beta_a = \frac{m}{4\pi\mu}(A - B),$$

by (33); so that

$$A - B = \frac{4\pi\mu C}{my}. \quad \ldots\ldots\ldots\ldots\ldots(36)$$

These equations determine A and B.

Another condition is afforded by the consideration that, on account of the symmetry, H must vanish when $x = 0$, or $H_0 = 0$. Within the insulator, whose permeability we take to be unity, $w = 0$, so that by (29) $\beta = \beta_a$, and

$$H_a - H_0 = -\int\beta\,dx = -\beta_a a.$$

Hence from (32), by equating the values of H_a,

$$\frac{4\pi aC}{y} = \frac{i}{p}\frac{d\Psi}{dz} + A + B. \quad \ldots\ldots\ldots\ldots(37)$$

Now by (35), (36),

$$A + B = -\frac{4\pi\mu C}{my}\frac{e^{mb} + e^{-mb}}{e^{mb} - e^{-mb}},$$

so that

$$-\frac{d\Psi}{dz} = ipC\frac{4\pi a}{y} + \frac{4\pi\mu ipC}{my}\frac{e^{mb} + e^{-mb}}{e^{mb} - e^{-mb}}. \quad \ldots\ldots\ldots\ldots(38)$$

In (38) the first term represents a part of the effective self-induction, and contributes nothing to the effective resistance. This self-induction, per unit length, is simply $4\pi a/y$, and is independent of p. The second term, being neither wholly real nor wholly imaginary, contributes both to self-induction and to resistance. If, separating the real and imaginary parts of the right-hand member of (38), we write

$$-l\frac{d\Psi}{dz} = R'.C + L'.ipC; \quad \ldots\ldots\ldots\ldots(39)$$

then R' represents the resistance and L' the self-induction of length l of the conductor measured parallel to z.

From (34)

$$m = \sqrt{\left(\frac{2\pi\mu p}{\rho}\right)} \cdot (1+i) = q(1+i), \quad \dots\dots\dots \dots\dots\dots (40)$$

if we write for brevity

$$q = \sqrt{\left(\frac{2\pi\mu p}{\rho}\right)}. \quad \dots\dots\dots\dots\dots\dots\dots (41)$$

The general expressions for R' and L' are somewhat complicated. If the rate of alternation be slow, p, and with it m and q, are small. In this case (38) may be written approximately

$$-\frac{d\Psi}{dz} = ipC\,\frac{4\pi a}{y} + \frac{\rho C}{by}\left(1 + i\,\frac{4\pi\mu pb^2}{3\rho}\right);$$

so that

$$R' = \frac{\rho l C}{by}, \quad \dots\dots\dots\dots\dots\dots\dots\dots\dots (42)$$

in accordance with Ohm's law, and

$$L' = \frac{4\pi a l}{y} + \frac{4\pi}{3}\frac{\mu b l}{y}. \quad \dots\dots\dots\dots\dots\dots (43)$$

If $a = 0$, the first term in (43) disappears, and we get a simplified result which should agree with one found previously (13). We may compare them by replacing y with $2\pi r$. The apparent discrepancy that the self-induction by (13) is twice as great as by (43) depends merely upon a slight difference in the way of reckoning. The result in (13) refers to a *double* length l, one part from the outgoing and the other from the return conductor.

At the other extreme, when p is very great, a simple result again applies. In that case $(e^{mb} + e^{-mb})/(e^{mb} - e^{-mb})$ may be replaced by unity, and (38) becomes

$$-\frac{d\Psi}{dz} = ipC\,\frac{4\pi a}{y} + (1+i)\,\frac{C\sqrt{(2\pi\mu p\rho)}}{y}.$$

Thus

$$R' = \frac{l\sqrt{(2\pi\mu p\rho)}}{y}, \quad \dots\dots\dots\dots\dots\dots\dots (44)$$

$$L' = \frac{4\pi a l}{y} + \frac{l\sqrt{(2\pi\mu\rho)}}{y\sqrt{p}}. \quad \dots\dots\dots\dots\dots (45)$$

These formulæ show that the resistance increases without limit with p, being proportional to $\sqrt{p}$, and that the self-induction diminishes towards the limit $4\pi a l/y$. If a be zero, that is if the insulating layer be infinitely thin, the self-induction diminishes without limit as p increases, being proportional to $p^{-\frac{1}{2}}$. Another important point is that when p is great enough, the values of

R' and L' are independent of b, the thickness of the strips. The meaning of this is, of course, that, under such circumstances, the currents concentrate themselves more and more towards the inner parts, in the endeavour to diminish the effective self-induction.

The distribution of current in the extreme case, where it is not limited by the thinness of the strips, is readily expressed. We have in general

$$w = \frac{mC}{y} \frac{e^{m\,(x-a-b)} + e^{-m\,(x-a-b)}}{e^{mb} - e^{-mb}}, \quad \dots\dots\dots\dots(46)$$

becoming, when $b = \infty$,

$$w = \frac{mC}{y} e^{-m\,(x-a)} ; \quad \dots\dots\dots\dots\dots(47)$$

or, on introducing the value of m from (40),

$$w = \frac{\sqrt{2}\,.\,qC}{y} e^{-q\,(x-a)} \left\{ \cos\left(\tfrac{1}{4}\pi - qx + qa\right) + i \sin\left(\tfrac{1}{4}\pi - qx + qa\right) \right\}. \quad \dots\dots(48)$$

The thickness through which the current is important is found from

$$q\,(x-a) = 1,$$

or

$$x - a = \sqrt{\left(\frac{\rho}{2\pi\mu p}\right)}, \quad \dots\dots\dots\dots\dots(49)$$

diminishing as p increases.

It should be remarked in conclusion that when a is very small and l very great, there may be a sensible accumulation of electricity upon the inner surfaces of the strips acting as plates of a condenser. In such a case u will no longer vanish, w will become a function of z, and our results will require modification.

136.

ON THE COLOURS OF THIN PLATES.

[Edinburgh Transactions, XXXIII. pp. 157—170, 1886.]

Introduction.

THE first impression upon the mind of the reader of the above title will probably be, that the subject has long since been exhausted. The explanation of these colours, as due to interference, was one of the first triumphs of the Wave Theory of Light; and what Young left undone was completed by Poisson, Fresnel, Arago, and Stokes. And yet it would be hardly an exaggeration to say that the colours of thin plates have never been explained at all. The theory set forth so completely in our treatises tells us indeed how the composition of the light reflected depends upon the thickness of the plate, but what will be its colour cannot, in most cases, be foretold without information of an entirely different kind, dealing with the chromatic relations of the spectral colours themselves. This part of the subject belongs to Physiological Optics, as depending upon the special properties of the eye. The first attempt to deal with it is due to Newton, who invented the chromatic diagram, but his representation of the spectrum is arbitrary, and but a rough approximation to the truth. It is to Maxwell that we owe the first systematic examination of the chromatic relations of the spectrum, and his results give the means of predicting the colour of any mixed light of known composition. Almost from the time of first reading Maxwell's splendid memoir, I have had the wish to undertake the task of calculating from his data the entire series of colours of thin plates and of exhibiting them on Newton's diagram. The results are here presented, and it is hoped may interest many who feel the fascination of the subject and will be pleased to see a more complete theory of this celebrated series of colours.

The diagram [below] explains many things already known from observation, such as the poverty of the blue of the first order and of the green of the second order. For good blues we must look to the second and third orders, and for good greens to the third and fourth. The point in which the diagram disagrees most with descriptions by former observers, *e.g.* Herschel, relates to the precedence of the reds of the first and second orders. The first red has usually been considered inferior, but the reason appears to lie in its feeble luminosity and consequent liability to suffer from contamination of white light. This and other questions are further discussed in the sequel.

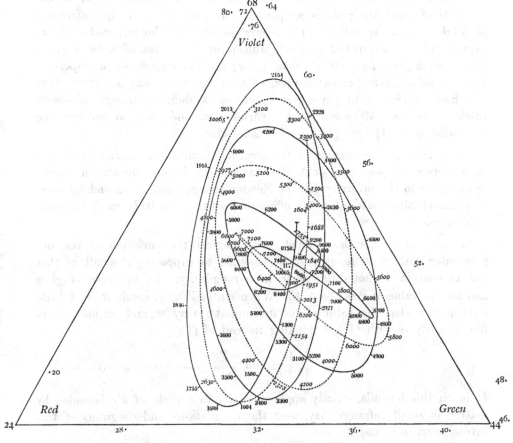

The complementary colours, best obtained with the use of polarised light, are also calculated and exhibited on a diagram.

§ 1. The calculation, according to Young and Poisson, of the amount of light of given wave-length (λ) reflected from a thin plate is given in all treatises on physical optics. If D be the thickness, β the obliquity of the ray within the plate, $1 : e$ the ratio in which the amplitude is altered

in one reflection, then for the intensity of light in the reflected system we
have

$$\frac{4e^2 \sin^2(\pi V/\lambda)}{(1-e^2)^2 + 4e^2 \sin^2(\pi \overline{V}/\lambda)} \quad \dots\dots\dots\dots\dots\dots(1)$$

in which the intensity of the original light is taken to be unity, and V is
written for $2D \cos \beta$. The colours exhibited in white light are to be found
by combining the chromatic effects of all the rays of the spectrum.

When, as in Newton's rings, the thickness of the plate varies from point
to point, there is a series of colours determined by supposing D to vary in
the above expression. This series is not absolutely independent of the
material of which the plate is composed, even if we disregard the differences
of brightness corresponding to the occurrence of e^2 in the numerator of our
expression. On account of retarded propagation, the value of λ for a given
ray is less in glass, for instance, than in air; and in consequence of dispersion
there is no accurate proportionality, so that we cannot say absolutely that
a definite thickness in glass corresponds to a definite, though different,
thickness in air. Moreover, since e varies from one body to another, the
denominator of (1) changes its value somewhat.

It is evidently impracticable to carry out calculations strictly applicable
to all cases. If we take for λ the wave-length in air, we obtain results
appropriate to the ordinary case of Newton's rings; and in extending them
to plates of other material, we in effect neglect the relatively small influence
of dispersion.

Again, we may without much error neglect the variation of the denominator with wave-length, which amounts to supposing e^2 small, or that
the two media do not differ much in refrangibility. In the case of glass
and air the value of e^2 is about $\frac{1}{25}$. When $\sin^2(\pi V/\lambda)$ is small, it is of little
consequence what the value of the denominator may be, and we may there-
fore identify it with $(1 + e^2)^2$, taking instead of (1),

$$\frac{4e^2}{(1 + e^2)^2} \sin^2 \frac{\pi V}{\lambda} . \quad \dots\dots\dots\dots\dots\dots(2)$$

It is on this formula, strictly applicable only to a plate of air bounded by
matter of small refrangibility, that the calculations and diagrams of this
investigation are based.

§ 2. The colours of Newton's scale are met with also in the light
transmitted by a somewhat thin plate of doubly-refracting material, such
as mica, the plane of analysis being perpendicular to that of primitive
polarisation. To this case also our calculations are applicable, if we neglect
the dispersion, and (as is usual) the light transmitted after two or more
reflections at the surfaces of the plate.

If the analyser be turned through 90°, a new series of colours is exhibited complementary to the first series. The purity of the colours, as regards freedom from admixture with white, is greatest when the principal section of the crystal is inclined at 45° to the plane of polarisation, and it is in this case also that the colours of the first series attain their maximum brightness. If we represent the first series by $\sin^2(\pi V/\lambda)$, the second series in the case referred to will be represented by $\cos^2(\pi V/\lambda)$. It should be noticed that the colours of Newton's rings seen by transmitted light are complementary to those seen by reflection; but the scale of colours is far more dilute than that obtainable as above with the aid of double refraction.

The colours of the first series are met with also in other optical experiments, e.g., at the centre of the illuminated patch, when light issuing from a point passes through a small round aperture in an otherwise opaque screen *.

§ 3. In order to be able to calculate the colour of any given mixture of light, it is necessary to know the exact chromatic relations of the spectral rays themselves. This is precisely the question investigated by Maxwell†. Selecting three rays as standards of reference, he expresses the colours of other rays in terms of them. The actual observations in all cases consisted in matches of two whites, one the original white which had not undergone prismatic analysis, the other a white compounded of three rays,—first of the three standard rays themselves, then of two standard rays in combination with a fourth ray which it was desired to express in terms of the standards. The auxiliary white was then eliminated.

The three points selected were at 24, 44, and 68 of the scale to which the spectrum was referred. "I chose these points, because they were well separated from each other on the scale, and because the colour of the spectrum at these points does not appear to the eye to vary very rapidly, either in hue or in brightness, in passing from one point to another. Hence, a small error of position will not make so serious an alteration of colour at these points, as if we had taken them at places of rapid variation; and we may regard the amount of the illumination produced by the light entering through the slits in these positions as sensibly proportional to the breadths of the slits.

"(24) corresponds to a bright scarlet about one-third of the distance from C to D; (44) is a green very near the line E; and (68) is a blue, about one-third of the distance from F to G."

A specimen observation is given:—

"Oct. 18, J. $18 \cdot 5\,(24) + 27\,(44) + 37\,(68) = W.$

* Airy's *Tract on Optics*, § 79.
† "On the theory of Compound Colours," *Phil. Trans.* 1860.

This equation means that on the 18th of October the observer J (myself) made an observation in which the breadth of the slit X was 18·5, as measured by the wedge, while its centre was at the division (24) of the scale; that the breadths of Y and Z were 27 and 37, and their positions (44) and (68); and that the illumination produced by these slits was exactly equal, in my estimation as an observer, to the constant white W.

"The position of the slit X was then shifted from (24) to (28), and when the proper adjustments were made, I found a second colour-equation of this form—

$$\text{Oct. 18, J.} \quad 16\,(28) + 21\,(44) + 37\,(68) = W.$$

Subtracting one equation from the other, and remembering that the figures in brackets are merely symbols of position, not of magnitude, we find

$$16\,(28) = 18\cdot5\,(24) + 6\,(44),$$

showing that (28) can be made up of (24) and (44), in the proportion of 18·5 to 6.

"In this way, by combining each colour with two standard colours, we may produce a white equal to the constant white. The red and yellow colours from (20) to (32) must be combined with green and blue, the greens from (36) to (52) with red and blue, and the blue from (56) to (80) with red and green."

The values employed in the present paper are those of Maxwell's second observer K (whose vision in the region of the line F was more normal than his own)*, and are given in his table No. VI. For our purpose they require some extension, especially at the violet end. Thus the equivalents of (16), (84), (88), (92), (96), (100), are obtained by a graphical extrapolation from the curves given by Maxwell. The adjoining Table I. is deduced from his with some reduction, in order to exhibit the value, in terms of the three standards, of the illumination due to the unit width of slit in each case. It will be seen that the extrapolation at the upper end of the spectrum is necessary in order to make up anything like the full total of (68).

The colour produced by combining all the light which passed the prisms from (16) to (100) is the white of the apparatus. Its equivalent in terms of the standards is given by

$$W' = 3\cdot973\,(24) + 6\cdot520\,(44) + 6\cdot460\,(68).$$

It differs a little from the standard white of the original matches, i.e.,

$$W = 18\cdot6\,(24) + 31\cdot4\,(44) + 30\cdot5\,(68),$$

not only in consequence of omission of some extreme red and violet, but probably also on account of absorption by the prisms.

* It is understood that K represents Mrs Maxwell. In these matters a woman's observations are generally to be preferred to a man's, as less liable to irregularities of the kind described in *Nature*, Nov. 17, 1881. [Vol. I. p. 542.]

The colours of the spectrum were exhibited by Maxwell in Newton's manner, and are reproduced on our diagram [p. 499], in which each colour is represented by the centre of gravity of three weights at the corners of an equilateral triangle, the magnitudes of the weights being taken proportional to the quantities of (24), (44), and (68) required to compound the colour, so that the corners themselves represent the standard colours.

TABLE I.

Scale	Wave-length	Colour	(24)	(44)	(68)
16	2580	red	+ ·140	...	...
20	2450	red	·420	+ ·009	+ ·063
24	2328	scarlet	1·000	...	...
28	2240	orange	1·155	·360	− ·006
32	2154	yellow	·846	·877	·005
36	2078	yellow-green	·484	1·246	·032
40	2013	green	+ ·127	1·206	− ·008
44	1951	green	...	1·000	...
48	1879	bluish-green	− ·063	·759	+ ·085
52	1846	blue-green	·055	·506	·282
56	1797	greenish-blue	·050	·340	·495
60	1755	blue	·047	·190	·753
64	1721	blue	− ·033	·033	·905
68	1688	blue	...	...	1·000
72	1660	indigo	+ ·019	·006	·944
76	1630	indigo	·025	+ ·016	·693
80	1604	indigo	·005	− ·028	·479
84	1580	...	...	...	·333
88	1560	...	...	...	·208
92	1540	...	...	...	·146
96	1520	...	...	...	·083
100	1500	...	...	...	·042
			+ 3·973	+ 6·520	+ 6·460

The wave-lengths are given in Fraunhofer's measure (in terms of the Paris inch*). The scale is such that for D, $\lambda = 2175$, and for F', $\lambda = 1794$.

The fact that the spectrum colours lie, roughly speaking, upon two sides of the triangle [p. 499], indicates that all pure oranges and yellows can be made up by a mixture of pure red and pure green, and in like manner that all varieties of pure blue and blue-green can be compounded of pure violet and pure green. If, as there is reason to believe, the curve representing the spectrum is slightly rounded off at the green corner, this means

* 1 Paris inch = 2·7070 cm.

that the *same* spectrum green is not available for both pure yellows and pure blues. The green lying most near the corner gives with red yellows, and with violet blues, which are somewhat less saturated than the corresponding colours of the spectrum.

§ 4. The colours of thin plates are to be calculated in accordance with (2) from Table I., as white was calculated, but with introduction throughout of the factor $\sin^2(\pi V/\lambda)$. For each thickness of plate V is constant, but an integration over the spectrum is required. Table II. gives a specimen

TABLE II.

	$\sin^2(\pi V/\lambda)$		
	$V=1846$	$V=3600$	$V=6800$
16	·607	·896	·828
20	·490	·991	·420
24	·367	·980	·060
28	·275	·892	·013
32	·188	·737	·225
36	·118	·553	·572
40	·066	·379	·863
44	·028	·219	1·000
48	·003	·068	·865
52	·000	·024	·702
56	·007	·000	·391
60	·026	·026	·148
64	·052	·081	·023
68	·084	·164	·008
72	·119	·256	·089
76	·164	·373	·264
80	·209	·483	·467
84	·255	·589	·667
88	·297	·678	·817
92	·342	·762	·932
96	·389	·840	·994
100	·440	·904	·989

of the values of the factors, and may be considered to represent the brightness, at various points, of the spectrum that would be formed by analysing the light reflected. The three retardations given correspond to the reds of the first and second orders, and to the green of the fourth order. In actual calculation these numbers would not occur (nor indeed those of Table I.), but would be represented by their logarithms.

From the necessity of determining a large number of points, the calculations ran to great length. They have not been performed throughout in duplicate, but have been so far re-examined as to exclude any error which could appreciably affect the diagram. In many cases neighbouring points verify one another to a sufficient degree of accuracy.

TABLE III.—*First Series.*

V	(24)	(44)	(68)	V	(24)	(44)	(68)
0	·77	1·65	2·28	5200	2·69	4·72	1·42
1006·5	3·82	6·46	5·87	5300	3·06	4·30	1·95
1300	3·75	5·07	2·79	5400	3·33	3·78	2·68
1500	3·01	3·20	·82	5600	3·51	2·75	4·15
1604	2·51	2·23	·27	5800	3·20	2·01	5·03
1688	2·04	1·49	·18	6000	2·53	1·77	4·93
1755	1·67	1·01	·27	6200	1·75	2·11	3·97
1846	1·20	·53	·69	6400	1·09	2·82	2·74
1951	·75	·26	1·63	6600	0·76	3·56	1·85
2013	·49	·26	2·25	6700	0·74	3·87	1·71
2154	·13	·67	3·81	6800	0·83	4·12	1·75
2328	·09	1·82	5·44	6900	1·00	4·26	1·99
2630	·99	4·44	5·87	7000	1·26	4·32	2·41
2927	2·59	5·95	3·37	7100	1·55	4·27	2·91
3100	3·29	5·77	1·71	7200	1·86	4·13	3·41
3300	3·78	4·68	·59	7400	2·45	3·69	4·17
3400	3·81	4·04	·58	7600	2·83	3·16	4·48
3500	3·74	3·17	·93	7800	2·93	2·76	4·02
3600	3·50	2·40	1·59	8000	2·72	2·56	3·24
3800	2·68	1·26	3·42	8200	2·35	2·62	2·45
4000	1·67	·93	5·08	8400	1·89	2·85	2·28
4200	·79	1·48	5·71	8600	1·52	3·17	2·52
4400	·29	2·68	5·02	8800	1·32	3·47	2·98
4600	·35	3·96	3·43	9000	1·34	3·65	3·69
4800	·91	4·91	1·86	9200	1·53	3·73	4·04
4900	1·33	5·13	1·31	9400	1·83	3·67	3·84
5000	1·79	5·17	1·03				

The final results, expressed as before in terms of the standards (24), (44), (68), are exhibited in Table III. In the first column are to be found the values of V (expressed in the same measure as λ). Thus, when $V = 1688$, the illumination vanishes at the point (68) on Maxwell's scale, for which $\lambda = 1688$. If the compound light reflected from a plate of this thickness were analysed by the prism, the centre of a dark band would be found at (68). Although the extinction is absolute at only one point, still the neighbouring region, which naturally contributes most of the colour-component (68), is very obscure, and thus the total of this component reaches only ·178, while the two other components are present in fair quantity. The resulting colour is a good orange.

As V increases, the dark band moves down the spectrum. When $V = 1951$, the centre of the band is at (44); thus nearly all the green is eliminated, and the colour is a rich purple. Again, when $V = 2328$, the centre of the band is at (24), the resulting colour is a rich blue. This band then moves out of the visible spectrum; but a new one presently makes its appearance, and begins to invade the spectrum from the violet end. When $V = 2 \times 1688$ or 3376, the ray (68) is again extinguished, and the colour is the yellow of the second order. For higher values of V, there may be two or more dark bands simultaneously, as appears in Table II., when $V = 6800$.

§ 5. Any sequence of colours may conveniently be represented on Newton's diagram, in the manner adopted by Maxwell for the particular sequence found in the spectrum. Such a curve would represent, for example, the colours of an absorbing medium, as the thickness traversed varies from nothing to infinity. In all such cases the curve starts from the point white, and ends at the point representative of that ray of the spectrum to which the medium is most transparent. For many coloured media the curve would not depart widely from a straight line ruled outwards from white to a point on one of the sides of the triangle. But when the medium is dichromatic, as for example a solution of chloride of chromium, the curve might start in one direction and ultimately come round to another. Thus in the case referred to the course of the curve from white would be towards the middle of the blue side of the triangle, then after a good progress in that direction it would bend round through yellow, and ultimately strike the triangle at a point near the red corner representative of the extreme visible rays at the lower end of the spectrum. The principal object of the present investigation was to exhibit in a similar manner upon Newton's diagram the curve of the colours of thin plates. To find the point corresponding to the retardation 1688, we imagine weights proportional to the numbers 2·04, 1·49, ·18 to be situated at the three angular points of the triangle, and construct the centre of gravity of such weights. This point represents the colour due to retardation 1688.

§ 6. The diagram [p. 499] embodies the results of Table III., so far as the *quality* of the effects is concerned. When the thickness, or retardation (V), is infinitely small, the amount of light reflected of course vanishes, but the *colour* approaches a limit, found by combining the constituents in quantities proportional to λ^{-2}, the limit of $\sin^2(\pi V / \lambda)$. This limiting blue of the first order would be the blue of the sky, according to the theory which attributes the light to reflection from thin plates of water in the form of bubbles. The blue of the sky is, however, really a much richer colour than this, and corresponds more nearly to that calculated on the supposition that the disturbance is due to spheres, or masses of other shape, small in all their dimensions relatively to the wave-lengths of light. According to this view, the colour is that found by taking the components of white light proportionally to λ^{-4}, instead of λ^{-2}*.

The curve, starting thus from a definite point, takes a nearly straight course in the direction of white (W'), which it passes a little upon the green side. The white of the first order on Newton's scale is thus somewhat greenish, as must obviously be the case when we consider that it arises when the maximum reflection is in the green or yellow portion of the spectrum, so that the red and blue must be relatively deficient; but the deviation from white is very small, and is not usually recognised. After leaving white the curve passes through the yellow, and approaches pretty close to the side of the triangle at a point representing the D-line in the orange†. The retardation is here 1688. The colour then reddens, but makes no approach to the spectrum reds lying near the corner of the triangle. Passing rapidly through the purple "transition tint," it becomes bluer, until it attains the magnificent blue or violet of the second order, in the neighbourhood of $V = 2328$. At this point there is a good approach to the corresponding spectrum colour, although the latter lies here a little outside the triangle. Leaving blue the colour rapidly deteriorates, becoming greener, but nowhere attaining a good green. The best yellow of the second order at 3400 is nearly as pure as the best of the first order, but inclines less to orange. The reds of the second order are even less pure than those of the first, but the inferiority diminishes as we approach the second transition-tint in the purple. The blue of the third order at 4200 is much inferior to the corresponding colour of the second order, but gradually acquires a superiority as it becomes greener near 4400. The blue-greens which follow, and the full greens from 4800 to 5000, are splendid colours, beyond comparison superior to the corresponding colours of the second order, but yet falling far short of the spectrum colours near (44). On the other hand, in the third order the yellows are not so pure as in the first and second orders, and there is even less approach to red, although a better show is made in the purple at 6000. In the transition from this purple to green, the blue falls short even of the blue of the first order, but the green at 6800 is very fine, sensibly equal to one of the greens of the third order. It will be remarked that in the fourth order greens there is little variety, the direction both on the outward and on the backward course being nearly in a line through white. On the return to white, which is very closely approached, a contrary curvature sets in, so that the earlier reds are more

* See several papers by the Author, published in the *Philosophical Magazine*, "On the Light from the Sky, its Polarisation and Colour," Feb. 1871, April 1871; "On the scattering of Light by Small Particles," June 1871; "On the Electro-Magnetic Theory of Light," August 1881, &c. [Vol. I. Arts. 8, 9, 74.]

† The points 20, 24, 28, ... on the diagram, represent the spectrum colours as determined by Maxwell.

blue than the later. The curve then bends round on the yellow side of white, until it attains a rather feeble blue-green at 9000.

§ 7.   It will be interesting to compare the diagram with descriptions by previous writers of Newton's scale of colours.   In his article on Light in the *Encyclopædia Metropolitana* (1830), Sir John Herschel says:—"The colours, whatever glasses be used, provided the incident light be white, always succeed each other in the same order; that is, beginning with the central black spot as follows:—

"First ring, or first order of colours,—*Black, very faint blue, brilliant white, yellow, orange, red.*

"Second ring, or second order,—*Dark purple or rather violet, violet, blue, green* (very imperfect, a yellow-green), *vivid yellow, crimson-red.*

"Third ring, or third order,—*Purple, blue, rich grass-green, fine yellow, pink, crimson.*

"Fourth ring, or fourth order,—*Green* (dull and bluish), *pale yellowish-pink, red.*

"Fifth ring, or fifth order,—*Pale bluish-green, white, pink.*

"Sixth ring, or sixth order,—*Pale blue-green, pale pink.*

"Seventh ring, or seventh order,—*Very pale bluish-green, very pale pink.*

"After these the colours become so pale that they can scarcely be distinguished from white.

"On these we may remark, that the green of the third order is the only one which is a pure and full colour, that of the second being hardly perceptible, and of the fourth comparatively dull and verging to an apple-green; the yellow of the second and third orders are both rich colours, but that of the second is especially rich and splendid; that of the first being a fiery tint passing into orange. The blue of the first order is so faint as to be scarce sensible, that of the second is rich and full, but that of the third much inferior; the red of the first order hardly deserves the name—it is a dull brick-colour; that of the second is rich and full, as is also that of the third; but they all verge to crimson, nor does any pure scarlet or prismatic red occur in the whole series."

Herschel's observations were made in the usual way with glass lenses,—a course convenient in respect of measurement of thicknesses, but incapable of doing justice to the colours, in consequence of the contamination with white light reflected at the upper surface of the upper plate and at the lower surface of the lower plate. The latter reflection should at any rate be got rid of by using a glass, either opaque, or blackened at the hind surface.

§ 8. For his description Newton used the soap-bubble, " because the Colours of these Bubbles were more extended and lively than those of the Air thin'd between two Glasses, and so more easy to be distinguished." He takes the colours in the reverse order, beginning with large retardations. I give his description as nearly as may be in his own words, but adapted to the more convenient notation followed by Herschel:—

" The red of the fourth order was also dilute and dirty, but not so much as the former three; after that succeeded little or no yellow, but a copious green (fourth order), which at first inclined a little to yellow, and then became a pretty brisque and good willow-green, and afterwards changed to a bluish colour; but there succeeded neither blue nor violet.

" The red of the third order inclined very much to purple, and afterwards became more bright and brisque, but yet not very pure. This was succeeded with a very bright and intense yellow, which was but little in quantity and soon changed to green; but that green was copious and something more pure, deep and lively than the former green. After that followed an ex- cellent blue of a bright sky colour (third order), and then a purple, which was less in quantity than the blue, and much inclined to red.

" The red of the second order was at first a very fair and lively scarlet, and soon after of a brighter colour, being very pure and brisque, and the best of all the reds. Then after a lively orange followed an intense bright and copious yellow, which was also the best of all the yellows; and this changed first to a greenish-yellow and then to a greenish-blue; but the green between the yellow and the blue was very little and dilute, seeming rather a greenish-white than a green. The blue which succeeded became very good, and of a very bright sky-colour, but yet something inferior to the former blue; and the violet was intense and deep, with little or no redness in it, and less in quantity than the blue.

" In the last red appeared a tincture of scarlet next to violet, which soon changed to a brighter colour, inclining to an orange; and the yellow which followed was at first pretty good and lively, but afterwards it grew more dilute, until by degrees it ended in perfect whiteness*."

§ 9. Some small discrepancies in the descriptions of Newton and Herschel probably depend upon ambiguities in the use of colour names. In the rings of high order what Newton calls blue, Herschel describes as bluish- green. Both observers remark upon the poverty of the green of the second order, but the diagram shows that it is superior to that of the fifth order. Neither Newton nor Herschel seem to have done full justice to the green of the fourth order, which at its best rivals closely the corresponding colour of the third order. My own observations are in accordance with the teaching of the diagram, which shows, moreover, that as we depart from

* Newton's *Opticks*, 1704, book II. p. 21.

retardation 6800 the colour of the fourth order rapidly deteriorates by admixture with white, while the colours of the third order in the neighbourhood of 4800 retain their purity as they change in hue.

One discrepancy between the diagram and the above descriptions will at once strike the reader. According to the diagram, the red and purple of the first order are superior to those which follow, whereas Herschel says that the red of the first order hardly deserves the name. Judged by the standard of the spectrum red at (24), this criticism would apply to them all; but the question is as to the relative merits of the various reds. The explanation depends upon considerations of brightness, of which the curve takes no account. If we refer to Table III., we see that at 1846 the red component is 1·20, but that at the corresponding point for the red of the second order (between 3600 and 3800) it rises to about 3·0. The deficiency of brightness in the first order goes a long way by itself to explain the apparent inferiority, for dark red gives rather the impression of brown; but if there is the slightest admixture of white light, the comparison is still more unfair. It would be useless, for example, to take the colours from an air-plate between lenses. The feebly luminous red of the first order is then drowned in a relatively large proportion of white light, which tells much less upon the brighter, though less pure, red of the second order. This complication does not arise when soap-films are employed, and the red of the first order is evidently much improved; but the rapidity of transition at this part of the scale renders observation difficult. The best comparison that I have been able to make is with the aid of a beautiful mica combination kindly lent me by Rev. P. Sleeman. When this is examined in a dark room between crossed nicols, and lighted brilliantly from a part of the sky near the sun, the red of the first order is seen in great perfection, and I had no difficulty in believing it to be superior to that of the second order. It is not very easy to bring the rivals into juxtaposition under equal brightnesses; but there is, I think, no reason to doubt that the first order would come off victorious. The composition of the lights will be understood by reference to Table II.

§ 10. The only colours which can be said to make any approach to spectrum purity are the yellows of the first two orders, and the blue and green-blue of the second and third orders respectively. There is a corresponding difficulty in obtaining good greens by absorption. To do so it is necessary that the transmitted spectrum should terminate at two pretty well marked points; in the case of red the difficulty is much less, all that is requisite being that the transmission should increase rapidly as the refrangibility of the light diminishes.

Besides the absolute brightness, there are two other circumstances which may influence the estimation of the colours of thin plates as normally presented. It is probable that in some cases the colours are much affected by contrast with their neighbours. To this cause we may attribute the difficulty in observing the transition between the reds and blue-greens of the fourth and higher orders. As the nearly neutral transition-tint is approached from either side, the effect upon the eye is improved by contrast, so as largely to compensate for the increasing poverty of the real colour. Much, again, depends upon the rapidity with which differences occur with varying retardation. When Newton speaks of the yellow of the second order as copious, he refers (I imagine) rather to the width of the band than to the brightness of the light. The diagram gives important information on this subject also. Compare, for example, in the first order, the change from 1500 to 1755, with that from 1755 to 1846 or 1951. The rapidity of the change in the latter interval is the foundation of the usefulness of the "transition-tint" in polarimetric work. If we wish to compare the rates of progress in different orders, we must distinguish according as we contemplate sensitiveness to small absolute, or to small relative, variations of retardation.

§ 11. The points of intersection of the curve are of interest, as corresponding to colours obtainable with two different thicknesses. The first that presents itself is the yellow, common to the first and second orders. The table shows that the latter is the brighter. In the second and third orders the similar colours differ but little in brightness. One occurs in the blue and another in the greenish-yellow. Nor is there much difference of brightness between the otherwise nearly identical greens of the third and fourth orders. It follows that if observers are able to distinguish in all cases which order of colours they are dealing with, it must be by reference to a sequence, rather than by estimation of a single colour.

§ 12. With respect to the absolute retardations or thicknesses at which the various colours are formed, careful observations have been made by Reinold and Rücker*. For comparison with their results I will take the green of the fourth order at 6800. In air at perpendicular incidence, this answers to a thickness of 3.40×10^{-5} Paris inches, or 9.19×10^{-5} cm. The numbers in their Table (p. 456), Column V., are

Green, ...	8·41
„	8·93
Yellow-green, ...	9·64

so that the agreement is pretty good. I would remark in passing that the diagram does not recognise a yellow-green of this order; but the appearance of such may perhaps be explained by contrast.

§ 13. The series of colours complementary to those of Table III. are found by subtraction of the numbers there given from those representative

* " On the Electrical Resistance of Thin Liquid Films, with a Revision of Newton's Table of Colours," *Phil. Trans.* 1881.

of white, viz., 3·97, 6·52, 6·46, respectively. The resulting numbers are exhibited in Table IV., in which the first entry for zero retardation corresponds to the full white*.

TABLE IV.—*Second (Complementary) Series.*

V	(24)	(44)	(68)	V	(24)	(44)	(68)
0	3·97	6·52	6·46	5200	1·28	1·80	5·04
1006·5	·15	·06	·59	5300	·91	2·22	4·51
1300	·22	1·45	3·67	5400	·64	2·74	3·78
1500	·96	3·32	5·64	5600	·46	3·77	2·31
1604	1·46	4·29	6·19	5800	·77	4·51	1·43
1688	1·93	5·03	6·28	6000	1·44	4·75	1·53
1755	2·30	5·51	6·29	6200	2·23	4·41	2·49
1846	2·77	5·99	5·77	6400	2·89	3·70	3·72
1951	3·22	6·26	4·83	6600	3·22	2·96	4·61
2013	3·48	6·26	4·21	6700	3·23	2·65	4·75
2154	3·84	5·85	2·65	6800	3·14	2·40	4·71
2328	3·88	4·70	1·02	6900	2·97	2·26	4·47
2630	2·98	2·08	·59	7000	2·72	2·20	4·05
2927	1·38	·57	3·09	7100	2·42	2·25	3·55
3100	·68	·75	4·75	7200	2·11	2·39	3·05
3300	·19	1·84	5·87	7400	1·52	2·83	2·29
3400	·16	2·48	5·88	7600	1·14	3·36	1·98
3500	·23	3·35	5·53	7800	1·04	3·76	2·44
3600	·47	4·12	4·87	8000	1·25	3·96	3·22
3800	1·29	5·26	3·04	8200	1·63	3·90	4·01
4000	2·30	5·59	1·38	8400	2·08	3·67	4·18
4200	3·19	5·04	·75	8600	2·45	3·35	3·94
4400	3·68	3·84	1·44	8800	2·65	3·05	3·48
4600	3·62	2·56	3·03	9000	2·63	2·87	2·77
4800	3·06	1·61	4·60	9200	2·44	2·79	2·42
4900	2·64	1·39	5·15	9400	2·15	2·85	2·62
5000	2·18	1·35	5·43				

The curve representative of this series of colours on Newton's diagram is given by the dotted line in the figure, so far as the tabulated numbers permit. It starts from the point White, and passes rapidly through a whitish-yellow to a very dark red and purple at $V = 1006·5$. This part of the curve can not be drawn from the tabulated data,—a defect of no great consequence, for the quantity of light being so insignificant, its quality is of little interest. From $V = 1300$ onwards the curve is pretty well determined.

It will be seen that the two series of colours are of pretty much the same general character. The green at 5800 in the second series compares favourably with the greens of the third and fourth orders in the first series.

* In comparing with Table III., it should be remembered that the numbers there given under the head of $V=0$ are relative only, the true values being infinitely small.

137.

NOTES, CHIEFLY HISTORICAL, ON SOME FUNDAMENTAL PROPOSITIONS IN OPTICS.

[*Philosophical Magazine*, XXI. pp. 466—476, 1886.]

It is little to the credit of English science that the fundamental optical theorems of Cotes and Smith should have passed almost into oblivion, until rediscovered in a somewhat different form by Lagrange, Kirchhoff, and von Helmholtz. Even now the general law governing apparent brightness seems to be very little understood, although it has acquired additional importance in connection with the theory of exchanges and the second law of Thermodynamics. In seeking the most natural basis for the law of magnifying, usually attributed to Lagrange, I was struck with the utility of Smith's phrase "apparent distance," which has never been quite forgotten, and was thus induced to read his ch. v. book ii.*, founded upon Cotes's "noble and beautiful theorem." I think that it may be of service to present a re-statement, as nearly as may be in his own words, of the more important of the laws deduced by Smith, accompanied by some remarks upon the subject regarded from a more modern point of view.

The general problem is thus stated:—

"To determine the apparent distance, magnitude, situation, degree of distinctness and brightness, the greatest angle of vision and visible area, of an object seen by rays successively reflected from any number of plane or spherical surfaces, or successively refracted through any number of lenses of any sort, or through any number of different mediums whose surfaces are plane or spherical. With an application to Telescopes and Microscopes."

* Smith's *Compleat System of Opticks*, Cambridge, 1738. French translations were published by P. Perzenas, Avignon, 1767, and by Duval Leroy, Brest, 1767.

It is divided into three propositions, of which the first is :—

" Having the focal distances and apertures of any number of lenses of any sort, placed at any given distances from one another and from the eye and object, it is required to find the apparent distance, magnitude, situation, degree of distinctness and brightness of the object seen through all the lenses; together with the greatest angle of vision and visible area of the object, and the particular aperture which limits them both."

Apparent distance means the distance at which the object would have to be placed so as to appear by direct vision of the same apparent magnitude as through the lenses.

Fig. 1.

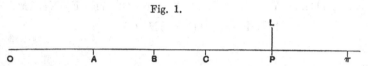

Let PL (fig. 1) be an object viewed by the eye at O through any number of lenses placed at A, B, C, ... whose focal distances are the lines a, b, c, ... and whose common axis is the line $OABCP$. In the standard case the lenses are supposed to be concaves. Then if $O\pi$ be the apparent distance,

$$O\pi = OP + \frac{OA \cdot AP}{a} + \frac{OB \cdot BP}{b} + \frac{OC \cdot CP}{c} + \frac{OA \cdot AB \cdot BP}{ab}$$
$$+ \frac{OA \cdot AC \cdot CP}{ac} + \frac{OB \cdot BC \cdot CP}{bc} + \frac{OA \cdot AB \cdot BC \cdot CP}{abc}.$$

The statement is for three lenses, but the law of formation of the terms is general. If a, b, c are infinite, we fall back on direct vision, and $O\pi = OP$. If any of the lenses are convex, the focal distances of such lenses must be looked upon as negative.

" § 262. Corol. 1. While the glasses are fixt, if the eye and object be supposed to change places, the apparent distance, magnitude, and situation of the object will be the same as before. For the interval OP being the same, and being divided by the same glasses into the same parts, will give the same theorem for the apparent distance as before." This is a proposition of the utmost importance, from which follows without much difficulty :—

" § 263. Corol. 2. When an object PL is seen through any number of glasses, the breadth of the principal pencil where it falls on the eye at O, is to its breadth at the object-glass C, as the apparent distance of the object to its real distance from the object-glass; and consequently in Telescopes, as the true magnitude of the object to the apparent."

True magnitude here means apparent magnitude as seen directly, and the theorem is identical with that enunciated more than half a century later by Lagrange*.

In the investigation of apparent brightness the extreme ray $PtsrK$ from the central point P of the image is considered (fig. 2):—

Fig. 2.

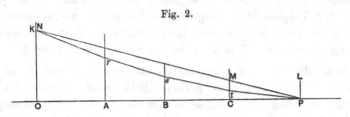

"If OK be not less than ON, the area of the pupil will be totally inlightened by the pencil which flows from P. Let $PtsrN$ be a ray of that pencil, cutting the object-glass Ct in t; and supposing the glasses were removed, let an unrefracted ray PMN cut the line Ct in M. Then the quantity of refracted rays which fall upon the line NO is to the quantity of unrefracted rays which would fall upon it as the angle CPt to the angle CPM, that is as the apparent magnitude of the line NO seen from P, to the true. And therefore, by turning the figure round the axis OP, the quantity of refracted rays which fill the pupil is to the quantity of unrefracted rays which would fill it (as the apparent magnitude of any surface at O seen from P, to the true; or as the apparent magnitude of any surface at P seen from O to the true; and consequently) as the apparent magnitude of the least surface, or physical point P, to the true; that is as the picture of the point P formed upon the retina by those refracted rays, to its picture formed by the unrefracted rays. These pictures of the point P are therefore equally bright and cause the appearance of P to be equally bright in both cases. Now let the pupil be larger than the greatest area inlightened at O by the pencil that flows from P; and supposing a smaller pupil equal to this area, we have shown that the pictures of P made upon the retina by refracted and unrefracted rays would be equally bright; and consequently each of them would be less bright than when the larger pupil is filled with unrefracted rays, in the same proportion as the smaller pupil, or area inlightened by the refracted rays, is less than the larger pupil, inlightened by unrefracted rays."

"§ 267. Corol. 6. It is evident that an object seen through glasses may appear as bright as to the naked eye; but never brighter, even though all the incident light be transmitted through the glasses."

* " Sur une loi générale d'Optique," *Mémoires de l'Académie de Berlin*, 1803.

Smith's splendid work was evidently unknown to Helmholtz, when, after establishing the law of apparent brightness, he remarks*:—

"Diese Folgerung ist schon von Lagrange gezogen worden. Leider hat er den zweiten Fall, der gerade bei starken Vergrösserungen der gewöhnliche ist, nicht besprochen, nämlich den, wo das in die Pupille eindringende Strahlenbündel diese nicht ganz ausfüllt. Das mag nicht wenig zu der Vergessenheit beigetragen haben, in welche seine wichtige Abhandlung gefallen ist." It is indeed astonishing that a theorem of such obvious importance should need to be discovered three times†.

The first advance upon the position attained by Smith is to be found in Kirchhoff's celebrated (but perhaps little read) memoir, "Ueber das Verhältniss zwischen dem Emissionsvermögen und dem Absorptionsvermögen der Körper für Wärme und Licht‡." The Theory of Exchanges renders it evident that the law of apparent brightness must have an even higher generality than Smith had claimed for it. No limitation can be admitted to systems of optical surfaces centred upon an axis, to which the rays are supposed to be but slightly inclined. Kirchhoff's investigation is founded upon Hamilton's characteristic function T, which we may here take to represent the reduced optical distance $(\int \mu ds)$ along a ray between any two points. At the extremities O, O', of the central ray of a pencil undergoing any number of reflections and refractions, planes are drawn perpendicular to the final directions of the ray, and in these planes rectangular coordinate axes x_1, y_1, x_2, y_2 are taken. T expresses the reduced distance between a point x_1, y_1, in the first plane and a point x_2, y_2 in the second plane, as a function of these four variables. In Smith's

* *Pogg. Ann.* Jubelband, 1874. p. 566.

† In his biography of Lagrange (*Suppl. Enc. Brit.*; Young's Works, vol. II. p. 575) Young, with some want of appreciation, thus refers to the memoir "On a General Law of Optics, 1803":—"A demonstration of the foundation of the method long since used by English opticians for determining the magnifying powers of telescopes of all kinds, which form an image of the object-glass beyond the eye-glass, by measuring the diameter of that image. The author hazards, in this paper, the very singular assertion, that the illumination of the object must be the same in all telescopes whatever, notwithstanding the common opinion that it depends upon the magnitude of the object-glass; and his reasoning would be correct, if the pupil of the eye were always less than the image of the object-glass in question; since, as he observes, the density of the light in this image is always inversely as the magnifying power; but he forgets to consider that the illumination on the retina, when the whole pencil is taken in, is in the joint ratio of the density and the extent; a consideration which justifies the common opinion on this subject, and shows that a most profound mathematician may be egregiously mistaken in his conclusions, if he proceeds to calculate upon erroneous grounds. It deserves, however, to be remembered that the brightness of any given angular portion of a magnified image must always be somewhat less than that of an equal portion of the object seen by the naked eye, because it can be no greater if the pencil fills the pupil, and will be less in proportion as the pencil is smaller than the pupil, besides the unavoidable loss of light as the refracting surfaces."

In his *Théorie des Lunettes*, Berlin, 1778, Lagrange himself refers to Smith's *Optics*, Ch. v. Book II.

‡ *Pogg. Ann.* t. CIX. p. 275 (1860).

terminology Kirchhoff's result may be thus stated:—The inverse square of the apparent distance between O and O' is

$$\frac{d^2T}{dx_1\,dx_2}\,\frac{d^2T}{dy_1\,dy_2} - \frac{d^2T}{dx_1\,dy_2}\,\frac{d^2T}{dx_2\,dy_1};$$

the apparent distance being the same whether O' be seen from O, or O be seen from O'.

A nearly similar investigation in terms of the corpuscular theory is to be found in Thomson and Tait's *Natural Philosophy* (1867), §§ 326, 327, where the authors emphasize strongly the optical importance of their conclusions. "The most obvious optical application of this remarkable result is, that in the use of any optical apparatus whatever, if the eye and the object be interchanged without altering the position of the instrument, the magnifying power is unaltered." And again, "Let the points O, O' be the optic centres of the eyes of two persons looking at each other through any set of lenses, prisms, or transparent media arranged in any way between them. If their pupils are of equal sizes in reality, they will be seen as similar ellipses of equal apparent dimensions by the two observers."

It will be remarked that in general the "magnifying power" varies in different directions, a circle being seen as an ellipse. This phrase and that of "apparent distance" may be understood in an extended sense. Thus the apparent distance may be taken to mean the distance at which the object, if seen direct, would present the same angular *area* as when seen through the lenses &c.

The demonstration of the law of apparent distance upon the basis of Hamilton's characteristic function, is of course satisfactory, and perhaps indispensable to a complete investigation; but it is not unimportant to remark that the essential part of the law is really included in the vastly more general reciprocal theorem, established in the first instance by v. Helmholtz for vibrations in a uniform gaseous medium*, and capable of extension to all vibrating systems, even though subject to dissipative forces†.

Let Ψ denote the intensity of a radiant source at O, ψ the corresponding amplitude of luminous vibration at O'. By the doctrine of rays, the energy transmitted across any section of a pencil is the same, and thus if σ_1 be the area of the cross sections at O' of a thin pencil of rays from O whose angular magnitude is there ω_1, we have

$$\sigma_1\psi^2 = H\omega_1\Psi^2,$$

where H is an absolute constant.

---

* *Theorie der Luftschwingungen in Röhren mit offenen Enden.* Crelle, Bd. LVII. 1860.
† *Proc. Math. Soc.* June 1873 [vol. I. p. 179]; *Theory of Sound*, vol. I. §§ 107, 108, 109.

In like manner if σ_2 be the area at O of a pencil starting from O' with angular opening ω_2,

$$\sigma_2 \psi'^2 = H \omega_2 \Psi'^2,$$

where Ψ' now measures the source at O', and ψ' the luminous amplitude at O. But by the reciprocal theorem

$$\psi' : \Psi' = \psi : \Psi,$$

and thus

$$\frac{\omega_1}{\sigma_1} = \frac{\omega_2}{\sigma_2};$$

or the "apparent distance" of O' from O is the same as of O from O'.

If we now assume the reciprocal character of apparent distance, there is no difficulty in deducing the law of apparent brightness in a perfectly general manner. For consider the whole light received over a small area σ_1 at O' (perpendicular to the ray) from a small luminous area σ_2 at O (also perpendicular to the ray). If, as before, ω_1 denote the angular opening at O of the pencil which corresponds to σ_1, this light is proportional to $\omega_1 \sigma_2$. But $\omega_1 \sigma_2 = \omega_2 \sigma_1$, if ω_2 be the apparent magnitude of σ_2 seen from O'. Hence the whole quantity of light received by σ_1 is proportional to $\omega_2 \sigma_1$; so that if σ_1 be given, representing, for example, the area of the pupil, the whole light received at O' from a small area σ_2 at O is proportional to the apparent magnitude of that area. In other words the apparent brightness is constant.

In this way of regarding the matter the law of apparent brightness becomes a deduction from the general reciprocal theorem. The argument may, of course, be reversed, so as to exhibit the reciprocal character of apparent distance as a consequence of the law respecting brightness. And this view of the subject may perhaps commend itself to those who appreciate the independent evidence for the law of brightness derived from the theory of enclosures as based upon the second law of thermodynamics.

In any case the law connecting magnifying power with the section of the pencil follows as an immediate consequence. If σ_2 be given, its apparent magnitude, as seen from O', is given by

$$\omega_2 = \sigma_2 \frac{\omega_1}{\sigma_1};$$

and is, therefore, inversely proportional to the section at O' of a pencil of given angular magnitude issuing from O. This principle is of great use in the design of optical instruments The application to the telescope is fully stated by Smith. By means of it the one-dimensional magnifying power of prisms, so placed that the emergence is more nearly grazing than the incidence, is readily traced*.

* See "Investigations in Optics," *Phil. Mag.* January 1880. [Vol. I. p. 455.]

There is, of course, no limit, either in telescopes or in microscopes, to the magnifying which may be obtained by sufficiently diminishing the section of the emergent pencil; but, as is now well known, the resolving power cannot thus be indefinitely augmented. It is interesting to note that Smith was aware of the fact, though he could have no knowledge of the reasons for it. He points out that in examining objects of great intrinsic brightness, there would appear to be advantage in diminishing the aperture of the object-glass. For by this means the disturbing influence of aberration, both spherical and chromatic, would be mitigated. "But* in reality it is quite otherwise; and that for two reasons. First because the minute parts...may be better discerned when all the light remains in the telescope than when it is reduced to $\frac{1}{100}$ part, though not in the same proportion. The other reason is that when the aperture is too much contracted the outlines that circumscribe the pictures in the eye become confused, which is carefully to be minded, and also what are the limits of this confusion. This is certain, that as the aperture is contracted the slender pencils or cylinders of rays that emerge from the eye-glass into the eye are also contracted in the same proportion. Now if the breadth of one of these pencils be less...than $\frac{1}{60}$ or $\frac{1}{72}$ part of an inch, the outlines of the pictures are spoiled, for some unknown reason in the make of the eye....For by looking through a hole in a thin plate, narrower than $\frac{1}{5}$ or $\frac{1}{6}$ of a line, the edges of objects begin to appear confused, and so much the more as the hole is made narrower." If we assume that a given apparent definition requires a given diameter of emergent pencil, it follows that the resolving power of telescopes is proportional to aperture.

The theory of the microscopic limit has been much discussed in recent years, and has been placed upon a satisfactory basis by Abbe, and by v. Helmholtz, who treats the subject in his usual masterly style. But I think that due credit has not been given to Fraunhofer, whose argument appears substantially correct. In his discussion on gratings Fraunhofer† remarks:—"Es ist nicht wohl denkbar, dass die Politur, welche wir durch Kunst auf Glas etc. hervorbringen können, mathematisch vollkommen sey. Besteht diese Politur aus *Unebenheiten*, welche, in Hinsicht ihrer Entfernung von einander, kleiner als ω sind [ω denotes the wave-length], so sind sie sowohl für durchfahrendes als zurückgeworfenes Licht ohne Nachtheil, und es können dadurch keine Farben irgend einer Art entstehen; auch wäre es durch kein Mittel möglich diese Unebenheiten sichtbar zu machen." And here he appends the very pregnant note:—"Man kann daraus schliessen, was möglicher Weise durch Mikroskope noch zu sehen ist. Ein mikroskopischer Gegenstand z. B.,

---

* Book II. Ch. 7, p. 144. Smith is here expounding the views of Huyghens (*Dioptrica*).
† Gilbert, *Ann.* 1823, p. 337.

dessen Durchmesser = ω ist, und der aus zwei Theilen besteht, kann nicht mehr als aus zwei Theilen bestehend erkannt werden. Dieses zeigt uns eine Gränze des Sehevermögens durch Mikroskope."

Fraunhofer's views did not commend themselves to Herschel[*], who regarded the "alleged limit to the powers of microscopes" as not "following from the premises." It so happens that I can give an independent opinion upon the persuasiveness of Fraunhofer's reasoning, for I had occasion in 1870, in connection with my own work upon the reproduction of gratings, and before the publication of the investigations of Abbe and v. Helmholtz, to consult his writings, when the note above quoted attracted my attention and fully convinced me of the general truth of the doctrine of the microscopic limit. It seemed evident, at any rate, that two radiant points, separated by only a small fraction of the wave-length, could not be optically distinguished.

It is worthy of notice that while Fraunhofer speaks of the whole wave-length, modern investigation fixes the half wave-length rather, as the limit of microscopic vision. It seems, however, that, on his own principles, Fraunhofer should have arrived at the latter result; for a grating whose period is equal to the wave-length can show colours when sufficiently inclined[†]. It is easy to see that when the angular aperture of a microscope is nearly 180°, a displacement of the radiant point amounting to half a wave-length, perpendicularly to the line of vision, will entail on one side of the pencil an acceleration and on the other a retardation of that amount, so that at the original focal point upon the retina the phases will now range just over a complete period. The displacement of half a wave-length corresponds therefore to something rather less than the half width of the best possible image of a mathematical point.

A definite limit to an operation such as visual resolution (involving in some degree a mental judgment) is, of course, not to be expected. For the microscope, the purely physical question of the distribution of light in the ultimate image of a mathematical point cannot be definitely solved without some assumption as to the manner in which the light is radiated in different directions. Even if the radiation were uniformly distributed, it does not follow that the light emerging from the eye-piece consists of ordinary plane waves, equally intense over the area of section. It would seem, indeed, that such a uniform distribution of rays is inconsistent with good definition, except at the very centre of the field of the microscope. For, in accordance with principles already discussed, it implies at all points an equal magnifying power, which however is to be reckoned always in relation to an object supposed to be perpendicular to the initial

[*] *Enc. Met.* "Light," § 758, 1830.
[†] Fraunhofer seems to have gone wrong in his formula for the case of oblique incidence.

direction of the ray. But inasmuch as the extreme rays of the pencil start in a direction oblique to the axis, it is evident that in relation to a given external object different parts of the system have different magnifying powers. In order that the efficient magnifying power (in the radial direction) may be alike for the whole of the emergent pencil, the rays must be concentrated towards the outer parts according to a law which will be obvious. In this way the resolving power might perhaps come to be a little greater than that estimated by v. Helmholtz, the concentration of rays towards the circumference playing somewhat the same part as a central stop.

Much more might be said upon this subject, but probably without results of practical importance. My main purpose has been to emphasize fundamental optical principles which have met with strange neglect, and to show how much excellent work had been done in this direction by some early writers.

P.S. Reference should have been made above to an interesting paper of Clausius*, in which the author develops very fully the analytical theory of radiation, as based upon Hamilton's function. One general theorem, previously established with rather less generality by v. Helmholtz, may here be noticed. The angles of the cones formed by the pencil of rays at an object and at its image (supposed to be astigmatic) stand simply in the inverse ratio of the areas of the corresponding elements of object and image.

In terms of apparent distance the argument may be put thus. Consider a cross section of the pencil at any intermediate point P. Then the squares of the apparent distances of P from the object and image are as the solid angles of the cones. Again, consider a pencil of rays passing through P, which must mark out corresponding portions of object and image, and it is evident that these areas are also in the ratio of the squares of the same apparent distances. Hence the proposition, which is thus seen to be intimately connected with the notion of apparent distance.

*Pogg. Ann. t. cxxi. p. 1, (1864).

138.

ON THE INTENSITY OF LIGHT REFLECTED FROM CERTAIN SURFACES AT NEARLY PERPENDICULAR INCIDENCE.

[*Proceedings of the Royal Society*, XLI. pp. 275—294, 1886.]

IN the present communication I propose to give an account of a photometric arrangement presenting some novel features, and of some results found by means of it for the reflecting power of glass and silver surfaces. My attention was drawn to the subject by an able paper of Professor Rood[*], who, in giving some results of a photometric method, comments upon the lack of attention bestowed by experimentalists upon the verification, or otherwise, of Fresnel's formulæ for the reflection of light at the bounding surfaces of transparent media. It is true that polarimetric observations have been made of the *ratio* of the intensities with which the two polarised components are reflected; but even if we suppose (as is hardly the case) that these measurements are altogether confirmatory of Fresnel's formulæ, the question remains open as to whether the actual intensity of each component is adequately represented. This doubt would be set at rest, were it shown that Young's formulæ for perpendicular incidence (to which Fresnel's reduce), viz., $(\mu - 1)^2/(\mu + 1)^2$, agrees with experiment.

Professor Rood's observations relate to the effect of a plate of glass when interposed in the course of the light. He measures, in fact, the transmission of light by the plate, and not directly the reflection. No one is in a better position than myself for appreciating the advantages of this course from the point of view of experiment. In the first place, the incidence can easily be made strictly perpendicular, in which case no question arises of a separate treatment of the two polarised components of ordinary light. And, what is much more important, the interposition of the plate leaves the course of the light unchanged, and thus allows

---

[*] *Amer. Journ. Sci.* vol. XLIX. 1870 (March); vol. L. 1870 (July).

the alteration of intensity to be determined in an accurate manner with the simplest arrangements.

On the other hand, the measurement of the transmitted, instead of the reflected light, is open to grave objection on more than one ground. It may be doubted whether the influence of *absorption* is altogether negligible, even when the thickness of the plate is as small as that mentioned by Professor Rood, viz., 1·67 mm. But the feature which strikes me most unfavourably is the necessary magnification of error, when we deduce the proportion of light reflected from the observed loss of light transmitted. The transmitted light is about 91 per cent.; and thus an error arising from the neglect of absorption, or from imperfect matches, amounting, say, to 1 per cent., leads to a relative error of more than 10 per cent. in the estimated reflection. The importance of this consideration may be illustrated by Professor Rood's actual results. In the first case recorded by him the observed transmission was 91·440 as against the theoretical 91·736. The difference 0·296 is indeed very small reckoned upon the transmitted light; but if we translate the results into terms of the reflected light, they present a different appearance. On the supposition that the whole loss in the transmitted light is due to reflection, we get for the intensity of the reflected light 8·560, which is to be compared with the theoretical 8·264. The difference 0·296 is now some 3½ per cent., and is thus by no means insignificant. In the other case given by Professor Rood the discrepancy is greater still, amounting to 7 per cent. It may be remarked that in both cases the amount of the reflection appears to be in excess of that given by Young's formula. But the cause may lie in the assumption that the whole failure of transmission is due to reflection. And whatever the explanation may be, we can hardly agree with Professor Rood when he concludes that these experiments show "that the reflecting power of glass with the above index of refraction, *conforms in the closest manner to the predictions of theory.*"

In the hope of being able to deal directly with the reflected light, I made a great many trials of various devices during the spring of 1885, but without finding anything satisfactory. Indeed, at one time, I had almost come round to the opinion that the difficulties of measuring the reflected light were so great that Professor Rood had shown a wise discretion in declining to face them, and that after all the best results would perhaps be reached through measurements of the transmitted light, checked by the use of plates of different thicknesses so as to eliminate absorption*. If, indeed, we give up the perpendicular incidence, the objection founded upon the relatively small quantity of light reflected may be met; for at an incidence of 70°, about half the light (polarised in the

* I was not aware until lately that Sir John Conroy was at work in this direction.

plane of incidence) is reflected. In such an experiment it would of course be necessary to determine accurately the angle of incidence.

The difficulties referred to have their origin in the necessary alteration in the course of the light by the act of reflection. The direct and reflected light cannot be interchanged in any simple manner, and the shift necessary to bring about the substitution may easily lead to systematic error. In the apparatus (presently to be described) to which I was finally led, the difficulty seems to be fairly overcome so far as regards the accuracy of the results, but at the cost of several tiresome adjustments, impeding the ready trial and interchange of various reflectors.

My apparatus differs in several respects from that generally used for photometric purposes. Before describing it in detail, it may be worth while to indicate some of the considerations which led me to design it.

The photometers in ordinary use may be said to depend upon the principle of diffusion. If the illuminating candle, or lamp, be drawn back from the screen to double the original distance, the brightness of the screen as perceived by the eye is supposed to be quartered. This implies that (within certain limits) the brightness of the screen is independent of the apparent magnitude of the source of light (the total radiation being given), or that the light diffused by the screen in a particular direction (towards the eye) is independent of the direction of incidence. Reciprocally, the light incident in a different direction is supposed to be diffused through a considerable angle with some approach to uniformity. There is no doubt that with proper arrangements this condition may be satisfied with sufficient accuracy for practical purposes. My object in formulating it is to show that the use of a diffusing screen in photometry is *necessarily* attended by an enormous reduction of light.

For our present purpose this loss of light is a serious matter. Weakened to $\frac{1}{25}$ by reflection from glass, the light of an ordinary candle or lamp is hardly sufficient to illuminate a diffusing screen properly, unless placed so close that measurement of the distances becomes uncertain. The difficulty might perhaps be got over by the use of incandescent electric lamps, but such were not at my command. When, as in Sir John Conroy's experiments*, the reflecting surfaces under test are metallic, or when (as above suggested) the observation relates to the transmission of light by an oblique plate of transparent material, the illumination given by a lamp may be adequate.

In my apparatus all the reflections are *regular*, and there is no further loss of light than the characters of the surfaces entail. An incidental advantage is that the accurate *flatness* of surfaces demanded by methods in which illumination is inferred from distance, is here unnecessary. The

* *Roy. Soc. Proc.* vol. xxxv. 1883, p. 26.

apparatus was first set up during the summer of 1885; but the glasses then at my disposal were not good enough, and when the parallel glass mirrors, &c., necessary for satisfactory working came into my possession, the season was so far advanced that I decided to postpone operations until the following summer.

Description of Apparatus.

The light is admitted into the room through a pane of finely-ground glass fitted into the shutter. All other light is carefully excluded, and the walls and ceiling are blackened—an almost indispensable provision. The ground glass carefully cleaned* is illuminated not only by the direct light of the sky, but also by light from above reflected at a large mirror. The reception of light through a large angle not only favours the aggregate brightness and tends to moderate the changes due to passing clouds, but it makes the uniformity of the field more independent of the evenness with which the glass is ground. Under these circumstances, and when there is no sunshine, direct or reflected, falling upon the ground glass, the latter may be looked upon as a tolerably uniform source of diffused light. This uniformity, however, is not relied upon; but the arrangements are so made that the parts of the field compared are contiguous or identical, and are seen by rays which leave at the same angle. It will be convenient first to describe generally the course of the light, and afterwards the manner in which the adjustments were effected.

Proceeding from the ground glass (A), fig. 1, the light falls upon the transparent plates, B, B', at which nearly equal parts are reflected. These plates are of worked glass about 6 inches by 4 inches, and are placed at the polarising angle. By this means are obtained two beams of polarised light of nearly equal, and of constant relative, brightness. The light transmitted by both plates is stopped at a screen, and takes no further part.

On the right-hand side the light reflected at B is again reflected by a mirror of worked glass, silvered behind at C, and assumes the direction CDF. D and F are alternative positions of the same mirror (also of worked glass, silvered behind). When the glass under test E is in use, the shifting mirror is in the position D, and the light follows the course $CDEFH$. In the contrary case, there is one reflection instead of two, and the ray takes the finally identical course $CDFH$. At H this central ray is reflected at the extreme edge of a speculum of silver-on-glass in the direction HI, to a small observing telescope, which is focussed upon this edge.

---

* Strong sulphuric acid is an excellent detergent for this purpose.

By adjustments that will presently be explained, it is secured that the reflections at D and F shall take place under the same angle, and

Fig. 1.

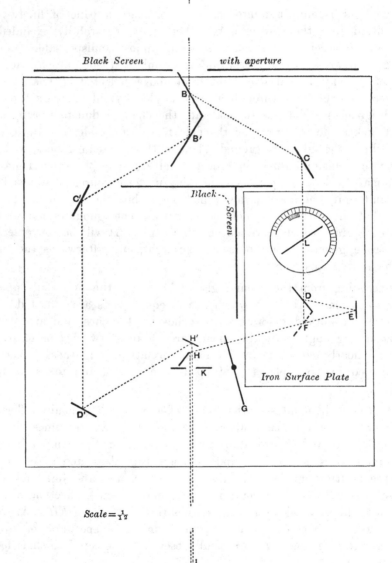

therefore with the same (moderate) loss of light; but when E is in use the brightness is diminished some twenty times. To compensate this in the other position the disk G is introduced. It consists of a blackened disk of tin, from which (along a sufficient length measured radially) a sector is cut out, so that when the disk is caused to revolve the view is cut off and blackness substituted for about nineteen-twentieths of the whole time. When the speed exceeds about twenty-five revolutions per second, there is no perceptible flicker, and the light is seen with a simple diminution of brightness. The idea of the method is so to adjust the angular opening that the effects of the glass under test and of the disk shall be equal.

The two brightnesses last considered can only be seen successively. They are separately tested with a comparison light, reflected at B' from the same primary beam. After reflection at a silvered mirror C', and then at a blackened glass D', this light falls upon a silver-on-glass speculum at H', and passes thence to the observing telescope at I. In setting up the apparatus a control over the brightness of the comparison light is obtained by varying the angle of incidence upon D'.

In order that the line of division between the two fields, as seen from I, may be quite sharp, it is necessary that the final reflector H be a speculum. To obtain a reflecting surface, perfect up to the very edge, a piece of silvered glass is carefully cut (on the glass side) with a diamond. If the operation is properly performed, the silver is left undisturbed, and when the plate is inclined, as at H, no part of the glass substratum is visible.

In adjusting the apparatus the object aimed at is to cause the central ray, ABB', issuing from A on the ground glass, to assume ultimately the position HI, whether it proceed by the course on the left, or by either of the alternative courses on the right. As may be supposed, this is more easily said than done. All the reflectors require to be adjusted so as to be perpendicular to the plane in which the central ray is to travel. This plane is conveniently taken horizontal, so that every reflector has to be vertical*.

The central ray is defined by diaphragms at A, B, B', points in the same horizontal straight line. At A the light is admitted through a small aperture only. At B, B' the holes are cut in thin cardboard screens held in definite positions up to the glasses. It was found convenient to have them rather large—about half an inch in diameter. In setting up the

---

* The levelling of the reflectors was effected with the aid of the straight edge of a long board, adjusted until it coincided with the prolongation of its image. My assistant, Mr Gordon, is expert at this adjustment of the edge to perpendicularity with the reflecting surface. The verticality of the latter is then tested by the application of a spirit-level to the edge of the board.

apparatus the glasses B, B', C are readily put into position, accuracy being required only in the levelling. The line CDF is now defined, and the next step is the more difficult one of fixing the two positions for the stand carrying the mirror D. This stand is (like all the others) provided with levelling screws, but these must not be used in passing from the one position to the other. A heavy metal surface plate was laid down upon the table as the support for this stand, and carefully levelled. Under these circumstances the mirror if vertical in one position will remain vertical even though displaced; and this remains true, even though the feet of the stand do not rest immediately upon the plate, but upon small flat buttons of metal of uniform thickness, and perforated with equal holes, by which the feet of the stand are guided to definite positions. When the adjustments are complete, these buttons are fastened to the surface plate by dropping cement round their edges.

The position F may now be chosen at convenience, and without any particular care except in the levelling. The central ray, as fixed by the diaphragms, should fall near the middle of the surface. The other positions would also be somewhat arbitrary were it not for the necessity of securing the same angle of incidence and reflection in the two positions. To assist in this a small frame of brass wire is provided, carrying two pointers, and so arranged that it can always be placed in an absolutely definite position with respect to the mirror. By means of hooks it makes two contacts with the back, and two with the upper edge of the mirror. Of the other two contacts required to make up the necessary six, one is with the lower part of the face of the mirror, and the other with one of its vertical edges. Of the pointers, one (in the path of the incident ray) leads upwards, and the other (in the path of the reflected ray) leads downwards. By bending them suitably their extremities may be brought into the path of the central ray, so that when the eye is placed in such a position (H) as to see the central point A in the middle of the (apparently elliptical) aperture $B*$, this central point is just enclosed between the barely meeting pointers. By so choosing the second position, D, that this condition is again satisfied to an eye looking along ED, we secure not only the same angle of reflection but the use (for the central ray) of the same part of the glass. In making the adjustment we may first bring the pointer on which the incident ray strikes into the already determined line CD, and then rotating the apparatus about the vertical through this point, bring the second pointer to coincide with the reflected ray.

We have now to consider how to fix the position of E, the reflector under examination. Replacing the shifting mirror into its first position F,

---

* Auxiliary lighting, with a candle or otherwise, is sometimes necessary in order to see these apertures properly.

we mark the line of the central reflected ray, FH, by needles, standing up from the table and as far apart as convenient. Transferring the shifting mirror to the position D, we have so to place E that the reflected ray shall coincide with the same line as before. For this purpose not only must the azimuth of E be correct, but its plane must be brought into the intersection of the already determined lines DE, FE. A levelled slab of glass is provided, on which to rest the feet of the stand carrying E. The mirror is now brought into a vertical plane, and may then be shifted on the slab without loss of this adjustment. The remaining double adjustment is best made systematically. By rotation about *any* vertical axis, the central ray may be caused to pass over one of the needles. If it fails to pass over the other, the axis of rotation must be shifted backwards and forwards until a suitable rotation allows satisfaction of both conditions. The ray now follows in both cases the course FGH, and the mirror H, with the sharp edge, may next be pushed in so as just to catch the ray in question and send it to the observing telescope (half of a small opera-glass) at I.

The adjustments for the auxiliary light on the left-hand side are a simpler matter. All the mirrors being levelled, the central ray is brought to the point H', in the prolongation of IH. Nothing then remains but to turn the final (vertical) mirror round H' until the reflected ray coincides with HI. When the eye looks in along this line, the bright spot should be seen in the same position from both mirrors.

To guard against accidental displacements, the movable pieces were usually secured with a little sealing-wax. A diaphragm at K limits the field of view, and is so placed that the aperture is bisected by the division line H. It is not necessary to do more than allude to various screens employed to cut off stray light and render the room as dark as possible.

The principal trouble experienced, that of making and retaining the adjustments, is connected with the rather large scale of the apparatus, which made it difficult to use a single levelled bed for all the movable pieces. The question is thus suggested, what is it that fixes the absolute scale? And the rather unexpected answer must be—the diameter of the pupil of the eye, which is the only linear quantity concerned*.

In order to understand this it is necessary to bear in mind that although in describing the adjustments we speak of a single ray only, we are of necessity really dealing with a complete beam. The observation of a match requires that the two parts of the field of view have finite angular magnitudes, and from every point of the field there must proceed a pencil of rays limited by the pupil, or by the telescope. If all these

---

* The wave-length of light may be regarded here as infinitely small.

rays are to be treated as sensibly parallel during their passage through the apparatus, certain limitations must be observed. For easy observation the field of view should subtend at the eye an angle of not less than a degree, so that if no telescope be employed the defect of parallelism must exceed this amount. The linear scale of the apparatus is not thus fixed, however, for we might suppose the eye (armed when necessary with a focusing lens) to approach without limit the final mirrors. But if we do this we increase the defect of parallelism due to the aperture of the eye. It is true that we may elude the objection by contracting proportionally the effective aperture, but only at an expense of brightness, which cannot usually be afforded. In accordance with a universal rule, full brightness requires that the aperture of the eye be filled with light. In this way we see how it is that the aperture of the eye controls the size of the apparatus.

The employment of a telescope introduces a certain modification, which it may be worth while to state somewhat fully, as the principle is of general application. The extreme angle between the rays of the beam may be regarded as made up of two parts: (1) the angle subtended at the object-glass by the aperture in the diaphragm (K) near the final mirrors (upon which the telescope is focused); (2) the angle subtended by the object-glass at the diaphragm. If

a = diameter of pupil,

b = diameter of aperture in diaphragm,

r = distance between telescope and diaphragm,

m = magnifying power of telescope,

α = angular diameter of field of view presented to the eye,

then $$\frac{mb}{r} = \alpha,$$

and the extreme angle between the rays of the beam

$$= \frac{b}{r} + \frac{ma}{r} = \frac{\alpha}{m} + \frac{a}{b} \cdot \alpha.$$

We may here regard α and a as given beforehand; and we see that with a given b the first term may be reduced without limit by increasing m, and that then the defect of parallelism is proportional to a, the diameter of the pupil. If m and b can both be increased without limit, we may approach as nearly as we please to a state of things in which all the rays concerned are parallel. The preservation of full brightness throughout is already secured by the supposition that the effective aperture of the object-glass is ma.

The reasoning set forth above shows at any rate that the size of the apparatus cannot be reduced below a certain point, but I do not affirm that mine was not unnecessarily large. In addition to its other advantages, the use of a telescope gives facilities for obtaining a good focus upon the division line, an adjustment of great importance for the easy recognition of small differences of brightness.

The necessarily finite magnitude of the field of view involves a certain imperfection in this, and probably in other methods of photometry. We can indeed secure that the lights seen in immediate juxtaposition come from the same part of the ground glass, but a corresponding perfection of adjustment does not apply to other parts of the field. If we suppose ourselves to be looking through the telescope at the ground glass, the part seen to the right of the division line really lies to the right on the ground glass. On the other side there is a distinction, according to the two positions of the shifting mirror. When the revolving disk is in use, the circumstances on the right-hand side of the apparatus correspond to those on the left, and thus the part of the field seen to the left of the division line really comes from the left on the ground glass. The ground glass is thus seen much as if it were looked at directly, in spite of the separation of the light into two parts following distinct courses. On the other hand, when the additional reflector (under examination) is brought into play, there is another inversion, and the part of the ground glass seen to the left comes really from the right of the central line. In this case, therefore, it is the same part of the ground glass which is seen in both final mirrors. The distinction here pointed out would be of no consequence if the field were absolutely uniform, or if it were possible to compare the parts seen in immediate juxtaposition, without regard to the parts a little further removed. But if the original field vary slightly in brightness from right to left, it will be a question how far the eye would select for the match *continuity* of brightness across the division line, or how far it would demand equality in the *average* brightnesses of the two parts presented.

It now remains to describe certain accessories. During the observations it is necessary to have some means of varying the relative brightnesses of the two parts of the field without removing the mirrors or altering the width of the slit in the revolving disk. For this purpose a plate of glass (*L*), capable of rotation about a vertical axis, was introduced into the path of the light on the right-hand side of the apparatus (between the second and third reflections). As the angle of incidence upon this plate increases, a greater proportion of the light is reflected and thrown away, and a less proportion is transmitted to the eye.

The observation consists in varying the azimuth of this plate until the match is satisfactory, after which the obliquity of the plate is measured. The transmission by the plate at the measured obliquity can then be found approximately from Fresnel's formula*. It may, perhaps, be objected that the use of this formula assumes the very thing that the experiments were principally intended to test; but the objection is evaded, almost if not altogether, when the aperture in the disk is so nearly adjusted to the ideal width that the oblique plate comes to take nearly the same azimuth for both sets of readings, *i.e.*, with and without the use of the mirror under examination. The use of the formula to allow for a small out-standing difference of obliquities can lead to no appreciable error. If on a first trial a large difference be found, a corrected aperture is calculated with the aid of Pickering's table, and the disk readjusted or replaced.

A fixed oblique plate has sometimes been used on one or other side of the apparatus in order to effect a rough adjustment of the brightness, and to bring the necessary obliquity of the rotating plate to a convenient amount (30°—60°). This was less trouble than a readjustment of the mirrors on the left, with an alteration in the angle of incidence upon the black glass D'.

In taking an observation the adjustment of the relative brightnesses was facilitated by a device which may now be described. If the attempt be made to secure an absolute match between the two parts of the field in view, a doubt is apt to arise as to whether the disappearance of the division line is due to the success of the adjustment or to fatigue of the eyes, leading, as in my case it very rapidly does, to imperfect focusing. This difficulty is less felt when the adjustment is under the immediate control of the observer, who can then satisfy himself of the sensitiveness of his eye by making the necessary displacement; but in the present experiments (on account of the distance of the telescope) it was con-venient to employ an assistant. A glass plate, perpendicular to the path of the light, and attached to a sort of pendulum, was therefore provided on the left-hand side, in such a manner that by pulling and letting go a string it could be introduced or withdrawn at pleasure. The effect of the plate would be to stop some 8 per cent. of the light, and the adjust-ment was so made that with glass *in* the (apparent) right of the field was as much too dark as it was too bright when the glass was *out*. The difference of brightness, amounting according to the above estimate to 4 per cent., was always fully apparent, and probably no setting more than 2 per cent. in error would be allowed to pass, giving, as such would do, a difference of 6 per cent. on the one side and of 2 per cent. upon

---

* A convenient table is given by Pickering, *Phil. Mag.* vol. XLVII. 1874, p. 129. If A be the proportion of light reflected at a single surface, the transmission through a *transparent* plate is given by $(1-A)^2 + (1-A)^2 A^2 + (1-A)^2 A^4 + \ldots\ldots = (1-A)/(1+A)$. The whole reflection is thus $2A/(1+A)$, from which Pickering's table is calculated. An erratum may be noted. For 65° the reflection should be 39·6, not 38·4, the value of A being supposed to be $\sin^2(\theta - \theta_1)/\sin^2(\theta + \theta_1)$, while $\sin \theta = 1\cdot55 \sin \theta_1$. There are some other minor inaccuracies.

the other*. Since the auxiliary light is eventually eliminated, it makes
no difference, of course, whether we take for the comparison the full
light, or the mean of the lights with and without the interposition of
the plate.

A 2 per cent. error in single settings may lead to a 4 per cent. error
in the comparison of the effects of the reflector and of the disk; and
accordingly (since this may occur in either direction) an 8 per cent. discrepancy in the results is *possible*. This, however, would be very unlikely,
and with a two- or three-fold repetition of the individual settings would
be practically out of the question.

The revolving disks, used to diminish the light on the right-hand side
of the apparatus in about the same degree as by the mirror under test,
were cut from tin plate, about 9 inches in diameter, and carefully centred.
The angular apertures were finally calculated from measurements of the
chord of the arc, and of the radius. It is important that the disks be
thoroughly blackened, in view of the assumption that no light reaches the
eye except during the passage of the aperture. Here is one reason why
it is desirable to keep the room as dark as possible. The disk should
also be properly balanced. On one occasion a curious and at first puzzling
effect was observed. The division line, which should present no visible
width, sensibly widened, appearing sometimes darker than the nearly
balanced adjoining fields, and sometimes, though more rarely, appearing
relatively bright. The explanation is to be found in a vibration of the
mirror, whose edge forms the division line, in a horizontal direction, perpendicular to the line of sight, the vibration being communicated from
the revolving wheel through the floor to the table upon which the mirrors
stood. It is evident that if the two lights under comparison were equal,
not merely on the average, but at every moment of time, such a movement of the mirror would have no disturbing influence, and could not
make the division line visible. But it is otherwise when one of the lights
is intermittent, and the vibrations of the mirror are executed (as here
they must be) in the same period. For suppose that at the moment
when the division line is advanced, so as to invade still further the field
from the back mirror, the light is reaching the eye through the aperture
in the disk. In this case the parts near the edge of the vibrating mirror
will be sending to the eye the full light due to this part of the field.
During the remainder of the vibration, no light should reach the eye,
but if this mirror retreats, the back mirror sends its continuous light
from the same apparent place, so that when the angular opening in the
disk is small, it is possible for the part of the field over which the
division line vibrates to present an almost doubled brightness, combining

* The accuracy of the settings falls much short of that attained by Professor Rood.

in fact the illumination of the two parts of the field. A different phase relation may evidently lead to an abnormal diminution of brightness in the same region. These effects disappeared when the disk was better balanced.

Prism of Crown Glass (I).

In ordering a glass for the purpose of determining the reflecting power of a surface, a prism was preferred to a plate, both on account of the easier separation of the reflections from the front and back surface, and also because the refractive index could be determined more readily. During the observations the hind surface was coated with black varnish, the effect of which, however, in annulling the second reflection, was far from complete.

With this glass, carefully cleaned (but not repolished), six sets of observations were made, four by myself and two by Mrs Sidgwick. Each set consisted of three or four settings with the glass in operation, and about the same number with substitution of the revolving disk. The following is a set of readings by myself on August 7th, 1886 :—

Face of Prism (I).

Reflection.	Revolving disk.
°	°
39·0	40·3
38·2	42·7
37·8	43·0
39·1	42·0
—	43·5
Mean ... 38·5	42·3

The angles here given are the obliquities of the adjustable glass plate used to graduate the intensity. According to Pickering's table, calculated from Fresnel's formula ($\mu = 1·55$), the effect of this plate at $38·5°$ would be to reflect 15·7 per cent. of the light incident upon it. The light transmitted is therefore 84·3 per cent. In like manner the light transmitted by the plate at an obliquity of $42·3°$ is 82·5 per cent.

In order to complete the calculation of the reflecting power of the glass surface, we must know the proportion of light transmitted by the revolving disk. Measurements gave for the chord of aperture of this disk in fiftieths of an inch 45·0, corresponding to a radius of 174·25. The angle of aperture is thus $14° 50' = 14·83°$. Accordingly the factor expressing the reduction of light by use of the disk is—

$$14·83 / 360 = 0·04119.$$

The reflection from the face of the glass prism is thus—

$$\frac{82{\cdot}5}{84{\cdot}3} \times 0{\cdot}04119 = 0{\cdot}0403.$$

The following is a summary of the results obtained at this time :—

Face of Prism (I).

Lord Rayleigh.		Mrs Sidgwick.	
Aug. 4 ...... 0·0411	....	Aug. 4 ...... 0·0405	
„ 5 ...... 0·0413	....	„ 5 ...... 0·0413	
„ 7 ...... 0·0403	....	—	
„ 9 ...... 0·0413	....	—	
Mean ...... 0·0410	....	Mean ... 0·0409	

Final mean .... 0·04095.

Since this number is very nearly the same as that (0·04119) due to the disk alone, we see that the result is scarcely at all dependent upon the correctness of the assumed effect of the oblique plate.

It now remains to make comparison with the reflection as given by Fresnel's formula, viz.:—

$$\frac{\sin^2 (\theta - \theta')}{\sin^2 (\theta + \theta')}, \qquad\qquad\qquad (A)$$

where $\sin \theta / \sin \theta' = \mu$, and μ is the refractive index.

The index was determined in the usual way from the angle of the prism (i) and from the minimum deviation (D). The value of i was found to be 9° 50′. The minimum deviation of soda light on one side was 5° 5′, and on the other 5° 4½′. Thus $D = 5°\ 4\frac{3}{4}′$, and

$$\mu = \frac{\sin \frac{1}{2} (D + i)}{\sin \frac{1}{2} i} = 1{\cdot}5141.$$

The angle of incidence (θ), which was the same in all the observations with the various reflectors, was measured by determining the angle (2θ) between the incident and reflected ray. This measurement does not require great precision, for a change of a whole degree in the value of θ would alter the reflection about 1 per cent. only. I found—

$$\theta = 13°\ 52′.$$

With these values of θ and μ, we find—

$$\frac{\sin^2 (\theta - \theta')}{\sin^2 (\theta + \theta')} = 0{\cdot}04514,$$

about 10 per cent. in excess of the reflection actually observed.

In order to satisfy myself that the deficiency of reflection was real and permanent, this prism was remounted after a thorough cleaning, and further observations were taken, as summarised in the following table. The revolving disk was the same as before:—

Prism of Crown Glass (I), Remounted.

	Lord Rayleigh.	Mr Gordon.	Mean.
Aug. 24	0·04085	0·04100	0·0409
„ 25, morn.	0·04183	0·04199	0·0419
„ 25, even.*	0·03950	0·04030	0·0399
„ 26	0·04190	0·04170	0·0418
Mean ...	0·04102	0·04125	0·04113

The difference between 0·04113 and the mean previously found, viz., 0·04095, has no significance.

In consequence of the detection of a greatly augmented reflection from another glass surface (II, below), as the result of a repolish with putty powder, this surface also was submitted to similar treatment. Immediately afterwards, on August 30th, a much increased reflection was observed, the numbers by two observers being—

0·0481, 0·0472; mean, 0·0476.

The disk gave, as before, a transmission 0·04119; so that the numbers for the repolished face depend too much upon the assumed effect of various obliquities of the inclined plate to be fully trustworthy, even were they sufficiently numerous to guard against accidental errors. But they proved, unequivocally, a considerable increase in reflecting power as the result of the repolish.

In view of these results, a new disk was prepared of angular aperture about $17\frac{1}{4}°$, and, consequently, with a transmission equal to 0·04763. The numbers obtained with this are shown in the following table:—

Prisms of Ground Glass (I) Repolished.

	Lord Rayleigh.	Mr Gordon.
Aug. 30, aft.	0·0461	0·0451
„ 31, morn.	0·0451	0·0454
„ 31, aft.	0·0448	0·0447
Mean	0·0453	0·0451

Final mean = 0·0452.

* The considerable discrepancy shown in this set of readings was probably caused by insufficiency of light.

The observed result now agrees remarkably well with that calculated from Fresnel's formula; but unfortunately it depends more (for about 5 per cent. of its value) than could be wished upon the use of the oblique plate.

Prism of Crown Glass (II).

So soon as it appeared that the reflection from the face of prism (I) fell so much short of what was to be expected in accordance with Fresnel's theory, I tried another prism whose surface was still older than that of (I). The event proved a still more marked deficiency. With the aid of the disk giving transmission 0·04119, the following numbers were obtained:—

Prism (II), before Repolishing.

	Lord Rayleigh.	Mrs Sidgwick.	Mr Gordon.
Aug. 26	0·0349	—	0·0344
„ 27	0·0342	0·0350	—

Mean ...... 0·0346.

Although somewhat dependent upon the assumed effect of the oblique plate, this number is far too low to be consistent with anything obtainable from Fresnel's formula with an admissible index*. This circumstance suggested a repolishing of the surface, which, however, was superior to that of (I), so far as could be judged from close inspection in a favourable light. The repolishing was executed by Mr Gordon by means of a disk of wood charged with putty powder and mounted in the lathe. Observation now demonstrated a remarkable improvement in the reflecting power, as the following numbers will show:—

Prism (II), after Repolishing.

	Mrs Sidgwick.	Lord Rayleigh.	Mr Gordon.
Aug. 28, morn. ...	0·0491	0·0488	—
„ 28, aft.	—	0·0479	0·0473
„ 30, morn. ...	—	0·0484	0·0481

Mean ...... 0·0483.

Here again too much depends upon the oblique plate, the transmission of the disk being only 0·04119; but there can be no doubt of an increase

---

* If e^2 be the proportions of light reflected at incidence θ, then Fresnel's formula is equivalent to

$$\tan \theta_1 = \frac{1-e}{1+e} \tan \theta,$$

by which θ_1 is found. The index is then given by $\sin \theta / \sin \theta_1$. In the present case

$$e^2 = 0·0346, \quad e = 0·1860, \quad (1-e)/(1+e) = 0·08140 / 1·1860,$$

so that, since $\theta = 13° 52'$, $\theta_1 = 9° 37'$, $\mu = 1·434$.

in the reflection of something like 30 per cent. If we may argue from the number obtained from prism (I) after repolishing, under nearly similar circumstances, viz., 0·0476, we may conclude that the true reflecting power of this prism is about 0·0460.

Altogether the evidence favours the conclusion that recently polished glass surfaces have a reflecting power differing not more than 1 or 2 per cent. from that given by Fresnel's formula; but that after some months or years the reflection may fall off from 10 to 30 per cent., and that without any apparent tarnish.

The question as to the cause of the falling off, I am not in a position to answer satisfactorily. Anything like a disintegration of the surface might be expected to reveal itself on close inspection, but nothing of this kind could be detected. A superficial layer of lower index, formed under atmospheric influence, even though no thicker than $\frac{1}{100000}$ inch, would explain a diminished reflection. Possibly a combined examination of the lights reflected and transmitted by glass surfaces in various conditions would lead to a better understanding of the matter. If the superficial film act by diffusion or absorption, the transmitted light might be expected to fall off. On the other hand, the mere interposition of a transparent layer of intermediate index would entail as great an *increase* in the transmitted as falling off in the reflected light. There is evidently room here for much further investigation, but I must content myself with making these suggestions.

Plate Glass Silvered Behind.

This glass was silvered chemically by the milk-sugar process, and by transmitted light showed the sky of a normal deep blue colour. The film was not polished. In determining the efficiency of this and other good reflectors, the black glass mirror D' was replaced by one silvered behind. The first trial without a revolving disk gave for the reflecting power 0·82. This result, of course, depended entirely upon the assumed influence of various obliquities of the adjusting plate. A disk was therefore prepared with two opposite projecting teeth, in which the ratio of aperture to circumference turned out on careful measurement to be 0·8230. This number, therefore, represents the transmission of light by the disk. Using this disk I found the following values for the reflecting power of the mirror for light incident upon it at an angle of 13° 52′:—

Aug. 11 .... 0·823
„ 12 .... 0·833

Mean .... 0·828

This result relates, like all the others, to light polarised in the plane of incidence. Mirrors of this kind are durable, and not being exposed to tarnish are more convenient than specula, whenever the double reflection is not objectionable. The high reflecting power is a satisfactory feature.

Silver-on-Glass Speculum.

This was the silver side of the same glass as the last, polished with wash-leather and a little rouge. The milky film was not perfectly removed. Four observations, not over concordant, probably in consequence of variation of reflecting power at different parts of the surface, gave—

Lord Rayleigh.	Mrs Sidgwick.	Mr Gordon.
Aug. 14 ... 0·902	Aug. 16 ... 0·933	Aug. 14 ... 0·920
„ 16 ... 0·895		

Mean ... 0·912.

The surface was then repolished and remounted with the following results:—

Lord Rayleigh.	Mr Gordon.
Aug. 18 ...... 0·950	Aug. 18 ...... 0·952
„ 19 ...... 0·938	„ 19 ...... 0·911
—	„ 21 ...... 0·938
Mean ...... 0·944	Mean ...... 0·934

Mean ...... 0·939.

The increase in efficiency may have been due to a more careful selection of the best polished central part as much as to actual improvement in the polish of the speculum as a whole. The transmission of the disk used with this surface is 0·9105.

Sir John Conroy* found an even higher number (0·975) as the reflecting power of silver films for light polarised in the plane of reflection, and incident at 30°.

Mirror of Black Glass.

A plate of opaque glass has the advantage that the influence of the hinder surface is eliminated without more ado; but, on the other hand, it lends itself less readily to determinations of index. The following results were obtained with such a plate:—

	Mrs Sidgwick.	Lord Rayleigh.
July 29	0·0580	—
„ 30	0·0581	0·0570
„ 31	0·0583	0·0572
Aug. 2	0·0574	0·0578
„ 3	0·0581	0·0577
Mean	0·0580	0·0574

Final mean ...... 0·0577.

* Roy. Soc. Proc. vol. XXXVII. 1884, p. 38.

During these observations a disk was employed giving transmission 0·0577, so that in this case the final result is absolutely independent of the effect of the adjustable oblique plate. It will be observed that the separate results obtained by Mrs Sidgwick and by myself differ, even in the means, by 1 per cent. This is not the only instance in which the errors have presented a suspiciously systematic appearance; but the differences being always small could not be submitted to any satisfactory examination. It rarely happened, for instance, that Mrs Sidgwick and I could find definite fault with each other's settings.

When these results were first obtained, I thought that they would turn out to be too high for agreement with Fresnel's formula, supposing that the index of the glass was low. A subsequent measurement of the specific gravity, however, gave reason for suspecting that the glass might be flint, a conclusion confirmed by determinations of the refractive index.

These were made by two methods: (1) by observation of the polarising angle in air, (2) by observation of the angle at which total reflection sets in when the mirror is immersed in bisulphide of carbon. The first is, perhaps, the simpler in respect of experimental arrangements, but it is open to the objection that the inference of the refractive index from the polarising angle is somewhat theoretical.

The black glass was mounted upon the turntable of an ordinary goniometer. In the focus of the collimator was placed a wire, seen dark in a bright field of view. Various positions of the turntable were then tried, such that on rotating a Nicol held at the eye the dark patch appeared to pass somewhat to the right or to the left of the collimator wire. After each observation the web of the telescope was set to coincidence with the collimator wire, and a reading taken. Success depends in some degree upon the use of a suitable light. Sunshine diffused through ground glass answered the purpose very well.

Right.		Left.		Central.
64° 25′	....	63° 56′	....	64° 7′
16	....	40	....	64 0
17	....	55	....	—
14	....	55	....	—

The table gives a set of circle readings. In the first column the patch was to the right of the collimator wire, in the second to the left, and in the third there was no appreciable deviation. We may, therefore, take as the reading for the polarising angle 64° 5′, with a probable error not exceeding 3′ or 4′. The reading for a direct setting of the telescope upon the collimator wire was $-5'$, so that the polarising angle is $\frac{1}{2}(180-64° 10')=57° 55'$. Whence according to Brewster's law—

$$\mu = \tan 57° 55' = 1\cdot5952.$$

This relates to white light.

To find the index of refraction by the method of total reflection, the mirror was mounted vertically in a small tank of plate glass, cemented with glue and treacle, and containing bisulphide of carbon. The mirror E, as shown in fig. 2, was parallel to one of the sides of the tank, and

Fig. 2.

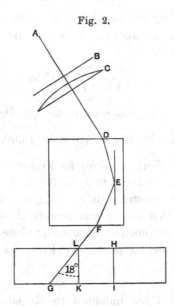

a cover was provided to check evaporation. A uniform field of homogeneous light could be obtained from a salted spirit lamp, A, with the aid of a plate of ground glass, B, and a collimating lens, C. The eye looking in along such directions as GF, is able to mark with considerable accuracy the direction in which total reflection begins. By the aid of plumb-lines, &c., this direction and that of the face of the mirror (seen from above) were marked upon a board, and it appeared that the angle GLK, between the face of the mirror and the direction GLF of the first totally reflected emergent light was 18°.

A beautiful variation in the experiment may be made by replacing the spirit lamp with a candle, and subsequently analysing the reflected light by a direct vision prism. For this purpose a screen carrying a slit should be interposed as near F as conveniently may be. As the incidence of the light upon the black glass becomes more grazing total reflection sets in, but first at the violet end of the spectrum. When the eye is looking nearly in the right direction, the spectrum appears to be covered by a veil proceeding from the red end up to a point dependent upon the precise direction of the light. By slightly shifting the eye, the veil may be made to reach any desired part of the spectrum, and then we know for what ray total reflection is just commencing. By bringing the veil to touch the soda line (rendered visible with the aid of the spirit lamp), precisely the same direction was found as had previously been marked out with use of homogeneous light. It would be possible in this way to determine with considerable accuracy the dispersive powers of opaque bodies.

The angle of $18°$, being measured in air, is not the complement of the true angle of reflection. If we take $1\cdot630$ as the index of CS_2 for soda light, we find for this angle

$$\sin^{-1}\left(\frac{\sin 18°}{1\cdot630}\right) = 10°\ 56';$$

whence for the index of the glass relative to soda light,

$$\mu = 1\cdot630 \times \cos 10°\ 56' = 1\cdot600.$$

The amount of reflection according to Fresnel's formula, with an incidence of $13°\ 52'$ and an index $1\cdot600$, is $0\cdot05726$, a little *less* than that actually observed. The agreement is as good as could be expected, but it should be noticed that this mirror was merely cleaned and not repolished with putty powder. If repolishing were to produce as much effect in this case as upon the acute-angled prism (I), Fresnel's formula would be left considerably in arrear*.

P.S. Nov. 9, 1886.—I am indebted to Mr Glazebrook for a determination of the refractive index of the prism of crown glass II. He finds $\mu = 1\cdot5328$. The introduction of this into Fresnel's formula ($\theta = 13°\ 52'$) gives for the reflecting power $0\cdot0477$.

---

* Some of the results here given were communicated to the British Association at Birmingham, where also was read a paper by Sir John Conroy on the same subject.

139.

NOTES ON ELECTRICITY AND MAGNETISM.
I. ON THE ENERGY OF MAGNETIZED IRON.

[*Philosophical Magazine*, XXII. pp. 175—183, 1886.]

THE splendid achievements of the last ten years in the practical application of Magnetism have given a renewed impetus to the study of this subject
which is sure to bear valuable fruit. Especially to be noted are two memoirs
recently published in the *Philosophical Transactions of the Royal Society*, by
Prof. Ewing*, and by Dr Hopkinson†, in which are detailed very important
data derived from laborious experiment, accompanied by much interesting
and suggestive comment.

The results of observation are usually expressed, after the example of
Rowland and Stoletow, in the form of curves showing the relation between $\mathfrak{B}$
and $\mathfrak{H}$, the magnetic induction and the magnetizing force. It may be well
here to recall the convention in accordance with which $\mathfrak{H}$ is measured. At
any point in air, the magnetic force is defined in an elementary manner, and
without ambiguity, but when we wish to speak of magnetic force in iron,
further explanation is needed. The continuity of the iron is supposed to be
interrupted by an infinitely thin crevasse in the interior of which we imagine
the measurement to be effected. If the crevasse is parallel to the direction
of magnetization, the force thus found is denoted by $\mathfrak{H}$, and is independent of
free magnetism on the walls of the crevasse. If, however, the crevasse be
perpendicular to the lines of force, there is a full development of free
magnetism ($\mathfrak{I}$) upon the walls, and the interior force is now $\mathfrak{B}$, equal to
$\mathfrak{H} + 4\pi\mathfrak{I}$. In the estimation of $\mathfrak{H}$ (as well as of $\mathfrak{B}$) the influence of all free
magnetism, not dependent upon the imaginary interruption of continuity, is
of course to be included. On this account the value of $\mathfrak{H}$ in the interior,

---

* "Experimental Researches in Magnetism," vol. CLXXVI. Part II. p. 523.
† "Magnetization of Iron," *ibid.* p. 455.

and even at the centre, of a bar of iron placed in an otherwise uniform magnetic field, is greatly reduced, unless the length of the bar be a very large multiple of the diameter.

Experiment shows that the relation of $\mathfrak{B}$ to $\mathfrak{H}$ is not of a determinate character. In a cycle of operations, during which $\mathfrak{H}$ is first increased, and is afterwards brought back to its original value, the induction $\mathfrak{B}$ is always greater on the descending than on the ascending course. This phenomenon, which is exemplified familiarly by the retention of magnetism in a bar after withdrawal of the magnetizing force, is called by Ewing *hysteresis*. The

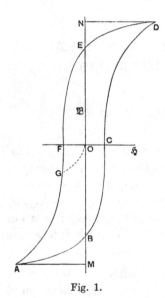

Fig. 1.

accompanying curve $ABCDEFGA$ (fig. 1) is copied from one given by him as applicable to very soft iron, conducted round a cycle from strong negative to strong positive magnetization and back again. The "residual magnetism" or "retentiveness" (OE) amounts to a large fraction (sometimes to 93 per cent.) of the maximum.

The work spent in carrying the iron round a magnetic cycle is represented by $-\int \mathfrak{J}\,d\mathfrak{H}$, as was first shown by Warburg*, who supposes the magnetic force operative upon the soft iron to be due to permanent magnets, and variable with their position. The work required to carry the permanent magnets through the proposed cycle of motions is then proved to have the above written value, applicable to the unit of volume of the soft iron. If $\mathfrak{J}$ were proportional to $\mathfrak{H}$, or even related to it in any determinate manner, the integral would vanish; but on account of hysteresis it has a finite value.

So long as we limit our attention to complete cycles, we may write indifferently $-\int \mathfrak{J}\,d\mathfrak{H}$, or $-(4\pi)^{-1}\int \mathfrak{B}\,d\mathfrak{H}$, since $\int \mathfrak{H}\,d\mathfrak{H}$ vanishes. Again, under the same restriction,

$$-\int \mathfrak{J}\,d\mathfrak{H} = +\int \mathfrak{H}\,d\mathfrak{J} = \frac{1}{4\pi}\int \mathfrak{H}\,d\mathfrak{B}.$$

When, however, we wish to consider incomplete cycles, especially with reference to the behaviour of soft iron, it is more suitable to take $\mathfrak{J}$ as independent variable. We are led naturally to this form if we suppose that, as in the more important practical applications, the varying magnetizing force is due to an electric current, upon which the magnetized iron reacts inductively†.

* *Wied. Ann.* XIII. p. 141 (1881).

† Hopkinson, *l.c.* p. 466.

In order to avoid the question of free polarity, we may consider, first, a ring electro-magnet with an iron core, of length l and section σ. If n be the number of windings, the whole inductive electromotive force is $n\sigma \cdot d\mathfrak{B}$, and the element of work is $n\sigma C \cdot d\mathfrak{B}$, C being the current at the moment in question. But

$$l\mathfrak{H} = 4\pi nC;$$

so that the element of work is, per unit of volume of iron,

$$\frac{1}{4\pi}\mathfrak{H}\,d\mathfrak{B}.$$

If we express $\mathfrak{B}$ in terms of $\mathfrak{H}$ and $\mathfrak{J}$, we have

$$\frac{1}{4\pi}\mathfrak{H}\,d\mathfrak{B} = \frac{1}{4\pi}\mathfrak{H}\,d\mathfrak{H} + \mathfrak{H}\,d\mathfrak{J},$$

of which the latter part is specially due to the iron. In practice the former part is small, and the distinction between

$$\frac{1}{4\pi}\mathfrak{H}\,d\mathfrak{B} \text{ and } \mathfrak{H}\,d\mathfrak{J}$$

may often be disregarded.

But it is by no means a matter of indifference whether we take $\mathfrak{B}\,d\mathfrak{H}$ or $\mathfrak{H}\,d\mathfrak{B}$. The difference between the two modes of reckoning may be exemplified in the case of iron already nearly "saturated," and exposed to an increasing force. Here $\int\mathfrak{B}\,d\mathfrak{H}$ is large, while $\int\mathfrak{H}\,d\mathfrak{B}$ is small; so that the latter corresponds better with the changes which we suppose to be taking place in the iron, as well as to the circumstances of ordinary practice.

Let us now consider a little more closely the cycle of Fig. 1. From A to B, $\mathfrak{H}$ is negative, while $d\mathfrak{B}$ is positive; so that along AB the inductive electromotive force is in aid of the current, and work is received *from* the iron of amount represented by the area ABM. From B to D, $\mathfrak{H}$ is positive as well as $d\mathfrak{B}$, and work represented by $BDNB$ may be supposed to be put *into* the iron. From D to E, work, represented by NED, is received from the iron, and from E to A work, represented by AME, is expended. From this we see that not only is work, represented by the area $ABCDEA$, dissipated in the complete cycle, but that at no part of the cycle is there more than an insignificant fraction of work recovered. The case is not one of a storing of energy recoverable with a small relative loss, but rather one of almost continuous dissipation.

And here the question is forced upon us, whether it is true, as is usually supposed, that the strong residual magnetism at E is really a store of energy. From the fact that the magnetism may be got rid of by very moderate tapping, we may infer, I admit, that some energy is necessarily dissipated in passing from the magnetized to the unmagnetized condition, but the dissipation may be exceedingly small; and the argument is not conclusive, since the mechanical energy of the vibrations may be involved in the process. If we attempt to demagnetize the iron in a straightforward manner by the application of a reversed force, following the course indicated by *EFGO*, then, so far from recovering, we actually expend energy—that, namely, represented by the area *EFGOE*. For practical purposes, at any rate, it would seem that magnetized iron cannot be regarded as the seat of available energy.

The opposite opinion, which is widely entertained, appears to depend upon insufficient observance of the distinction, vital to this subject, between closed and unclosed magnetic circuits. It is not disputed that available energy accompanies the magnetization of a short bar of iron, but this is in virtue of the free polarity at the ends. The work stored is in fact that which might be obtained, were the bar flexible, by allowing the ends to approach one another, under their mutual attraction. When this operation is finished, so that the bar has become a ring, there is no longer any work to be got out of it, though it remains magnetized.

In further illustration of this matter, reference may be made to some interesting observations by Elphinstone and Vincent* on closed magnetic circuits. As is well known, the armature of a horseshoe electromagnet remains strongly attracted after cessation of the battery-current. If, even after a considerable interval of time, the coils of this electromagnet were connected with those of a second electromagnet also provided with an armature, and the first armature were then violently pulled away, attraction set in and persisted between the second armature and its electromagnet, the magnetism of the original circuit being as it were transferred to the second. Or, if a galvanometer were substituted for the second electromagnet, a deflection followed the forcible withdrawal of the armature. In these experiments the necessary energy is obtained, not from the magnetism of the closed circuit, but from the work done in opening it, that is in pulling away the armature. [1900. A criticism, subsequently (*Phil. Mag.* XXII. p. 469) withdrawn, upon a remark of Prof. Ewing, is omitted.]

When we know, as from Prof. Ewing's results, the behaviour of a given sample of iron under the influence of various forces $\mathfrak{H}$ *actually operative*, we can deduce by means of Poisson's theory the magnetism assumed by ellipsoids of any shape in response to any uniform *external force* $\mathfrak{H}'$. If $\mathfrak{I}$ be the magnetization parallel to the axis of symmetry (2c), the demagnetizing effect of $\mathfrak{I}$ is $N\mathfrak{I}$, where N is a numerical constant, a function of the eccentricity (e)†. When the ellipsoid is of the ovary or elongated form,

$$a = b = c \sqrt{(1 - e^2)},$$

$$N = 4\pi \left(\frac{1}{e^2} - 1 \right) \left(\frac{1}{2e} \log \frac{1+e}{1-e} - 1 \right),$$

---

* *Proc. Roy. Soc.* vol. XXX. p. 287 (1880).

† Maxwell's *Electricity and Magnetism*, § 438.

becoming in the limiting case of the sphere $(e = 0)$

$$N = \tfrac{4}{3}\pi;$$

and at the other extreme of elongation assuming the form

$$N = 4\pi \frac{a^2}{c^2}\left(\log\frac{2c}{a} - 1\right).$$

If the ellipsoid is of the planetary form,

$$a = b = \frac{c}{\sqrt{(1 - e^2)}}\;*,$$

and

$$N = 4\pi\left(\frac{1}{e^2} - \frac{\sqrt{(1 - e^2)}}{e^3}\sin^{-1}e\right).$$

In the case of a very flattened planetoid $(e = 1)$, N becomes in the limit equal to 4π.

The force actually operative upon the iron is formed by subtracting $N\mathfrak{J}$ from that externally imposed, so that

$$\mathfrak{H} = \mathfrak{H}' - N\mathfrak{J};$$

and if from experiments on very elongated ellipsoids $(N = 0)$ we know the relation between $\mathfrak{H}$ and $\mathfrak{J}$, then the above equation gives us the relation between $\mathfrak{H}'$ and $\mathfrak{J}$ for any proposed ellipsoid of finite elongation. If we suppose that $\mathfrak{H}$ is plotted as a function of $\mathfrak{J}$, we have only to add in the ordinates $N\mathfrak{J}$, proper to a straight line, in order to obtain the appropriate curve for $\mathfrak{H}'$.

As an example, let us apply this method to deduce the behaviour of the soft iron of Ewing's Fig. 2, when made into an ellipsoid whose polar axis is fifty times the equatorial axis, and carried round a cycle through strong positive and strong negative magnetism. We have

$$N = \frac{4\pi}{50^2}\{\log_e 100 - 1\} = 4\pi \times \cdot001442.$$

The curve ABC (Fig. 2), traced from Prof. Ewing's, gives the relation

Fig. 2.

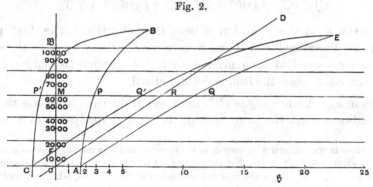

* There is here a slight variation from Maxwell's notation.

35—2

between $\mathfrak{H}$ and $\mathfrak{B}$, the latter of which we may identify with $4\pi\mathfrak{I}$*. The equation of the straight line is

$$\mathfrak{H} = N\mathfrak{I} = \cdot001442 \times 4\pi\mathfrak{I};$$

and with allowance for the different scales adopted for ordinates and abscissæ, is represented on the diagram by OD. In order to find the points Q, Q' appropriate to the ellipsoid (50 : 1) from P, P', we have merely to measure PQ, $P'Q'$ equal to RM. We thus obtain the curve $AQEQ'FC$, on which the points of zero magnetization are the same as on the original curve†. We see that a much stronger field is now required to produce the higher degrees of magnetization, and that there is less hysteresis—the magnetic state is more nearly a definite function of the external field. A similar construction might be used reversely to pass from observed results relative to ellipsoids of moderate elongation to the curve appropriate to ellipsoids of infinite elongation, on which alone we can base our views of the real character of magnetic media.

Prof. Ewing has traced by experiment the influence of various degrees of elongation on the magnetism of cylindrical rods. Results of this kind are exhibited in his Fig. 3, but they are not strictly comparable with those obtained above, not only because the latter relate to ellipsoids, but also on account of the different character of the magnetic operations represented. His curves begin at a condition of zero field and zero magnetization.

The work expended in producing a small change of magnetization of the ellipsoid, acted upon by a uniform field, is $\mathfrak{H}'d\mathfrak{I}$ simply per unit of volume. This we may see, perhaps most easily, by supposing the iron to be replaced by an electric current of equal magnetic moment. The element of work done then depends upon the coefficient of mutual induction M of the two circuits, and M may be regarded as the number of lines of force due to the original current which pass through the fictitious circuit. The whole work is thus

$$\int \mathfrak{H}'d\mathfrak{I} = \int \mathfrak{H}\, d\mathfrak{I} + N\int \mathfrak{I}\, d\mathfrak{I} = \int \mathfrak{H}\, d\mathfrak{I} + \tfrac{1}{2}N\mathfrak{I}^2,$$

if we reckon from the condition of zero magnetization. The first part is that already considered, and shown to be almost entirely wasted; the second, which in most cases of open magnetic circuits is much the larger, is completely recovered when the iron is demagnetized.

Thus in Fig. 2, since $QQ' = PP'$, the areas of the two curves are the same, which indicates that the same amount of work is dissipated in a complete

---

* The curve is symmetrical with respect to O as centre, and $\mathfrak{H}$ is measured in c.g.s. units.

† Dr Hopkinson (*loc. cit.* p. 465) has already applied this method to the determination of the particular point F, indicative of the residual magnetism in the ellipsoid, when the external force is withdrawn.

cycle. But the work absorbed during one part and restored during the remainder of the cycle is much greater in the case of AEC, corresponding to the ellipsoid of moderate elongation.

The coefficient N reaches its maximum when the ellipsoid is very oblate. In this case

$$\int \mathfrak{H}'d\mathfrak{J} = \int \mathfrak{H}\, d\mathfrak{J} + 2\pi \mathfrak{J}^2,$$

which is applicable to large plates magnetized perpendicularly to their surfaces. This is the form to which the iron must be reduced in order that a given magnetization of a given volume may store* the largest amount of energy. In this case the energy is nearly all recoverable; but we must remember that no practicable field would carry the magnetization very far.

In the theory of alternating currents the neighbourhood of iron is often treated as if its only effect were to increase the self and mutual induction of the circuits. A writer conversant with experiment usually guards himself by a reference to the currents induced in the iron considered as a conductor. The latter effect may be in great measure eliminated by a proper subdivision of the iron, with intervening non-conducting strata; but a glance at Fig. 2 shows at once that, apart altogether from internal currents, the influence of the iron is of a more complicated character. If the curve connecting $\mathfrak{J}$ (or $\mathfrak{B}$) and $\mathfrak{H}$ were a straight line, the same on the upward as on the down-ward course, then the presence of iron would simply increase the self-induction. When the iron constitutes a closed magnetic circuit, this is very far from being true. Indeed it would be nearer the mark to say that the iron increases the apparent *resistance* of the electric circuit, leaving the self-induction unchanged. In so far as the curve of Fig. 2 can be identified with an ellipse, the reaction of the iron can be represented as equivalent to a change in the apparent resistance *and* self-induction of the circuit. Which of the two is the more important depends somewhat upon the other circum-stances of the case; but with closed electromagnets the magnetic work dissipated during the period (corresponding to increased resistance) is always greater than the work spent during one part and recovered during the remainder of the period (corresponding to increased self-induction). On this account, the resistance of an iron wire to variable currents is greater than to steady currents, even though the current be constrained to be uniformly distributed over the section. In the absence of such constraint, the resist-ance undergoes a further increase in consequence of the tendency of the current to concentrate itself towards the exterior†. In general both causes must cooperate to produce an apparent increase of resistance to variable currents.

* It is not meant here to imply that the energy is resident *in* the iron.

† " On the Self-induction and Resistance of Straight Conductors," *Phil. Mag.* vol. XXI. (1886), p. 388. [Vol. II. p. 492.]

When the magnetic circuits are open, as with bars of iron of moderate length, the reaction of the iron manifests itself mainly as increased self-induction. This happens also in the case of closed magnetic circuits, when the magnetic changes are very small.

In general, since the curve of Fig. 2 differs widely from an ellipse, the reaction of the iron cannot be fully represented as equivalent to a change in the resistance and self-induction of the magnetizing circuit. In any case of strict periodicity the reaction may, however, be analyzed, in accordance with Fourier's theorem, into harmonic components with periods which are sub-multiples of the original period. The neighbourhood of iron may thus introduce overtones into what would otherwise be a simple sound.

140.

NOTES ON ELECTRICITY AND MAGNETISM.

II. THE SELF-INDUCTION AND RESISTANCE
OF COMPOUND CONDUCTORS.

[*Philosophical Magazine*, XXII. pp. 469—500, 1886.]

In his inaugural address to the Society of Telegraph Engineers[*], and in a subsequent communication to the Royal Society[†], Prof. Hughes has described a series of interesting experiments, which have attracted a good deal of attention in consequence both of the official position and known experimental skill of the author. Some of the conclusions which he advances can hardly be sustained, and have met with severe criticism at the hands of Weber, Heaviside, and others. There are certain other points raised by him, or suggested by his work, which seem worthy of consideration ; and I propose in the present paper to give an account of some investigations, mainly experimental, carried on during the summer months, which may, I hope, tend to settle some controverted questions.

Prof. Hughes's first apparatus consists of a Wheatstone's quadrilateral, with a telephone in the bridge, one of the sides of the quadrilateral being the wire or coil under examination, and the other three being the parts into which a single German-silver wire is divided by two sliding contacts. If the battery-branch be closed, and a suitable interrupter be introduced into the telephone-branch, balance may be obtained by shifting the contacts. *Provided that the interrupter introduces no electromotive force of its own*[‡], the balance indicates the proportionality of the four resistances. If P be the unknown resistance of the conductor under test, Q, R the resistances

---

[*] *Journ. Tel. Eng.* vol. XV. (1866), p. 1.

[†] *Proc. Roy. Soc.* vol. XL. (1886), p. 451.

[‡] This condition is not always satisfied. With the reed-interrupter (see below) a loud sound may sometimes be heard, although the battery-branch be open.

of the adjacent parts of the divided wire, S that of the opposite part (between the sliding contacts), then, by the ordinary rule, $PS = QR$; while Q, R, S are subject to the relation

$$Q + R + S = W,$$

W being a constant. If now the interrupter be transferred from the telephone to the battery-branch, the balance is usually disturbed on account of induction, and cannot be restored by any mere shifting of the contacts. In order to compensate the induction, another influence of the same kind must be introduced. It is here that the peculiarity of the apparatus lies. A coil is inserted in the battery and another in the telephone-branch, which act inductively upon one another, and are so mounted that the effect may be readily varied. The two coils may be concentric and relatively movable about the common diameter. In this case the action vanishes when the planes are perpendicular. If one coil be very much smaller than the other, the coefficient of mutual induction M is proportional to the cosine of the angle between the planes. By means of the *two* adjustments, the sliding contact and the rotating coil, it is usually possible to obtain a fair silence.

In his address Prof. Hughes interpreted his observations on the basis of an assumption that the self-induction of P was represented by M, irrespective of resistance, and that the resistance to variable currents could (as in the case of steady currents) be equated to QR/S. In the discussion which followed I pointed out that this was by no means generally true, and I gave * the following formulæ as applicable to the case in which the only sensible induction among the sides of the quadrilateral is the self-induction L of the conductor P :—

$$QR - SP = p^2ML, \quad\dotfill (1)$$
$$M(P + Q + R + S) = SL. \quad\dotfill (2)$$

The electrical vibrations are here supposed to follow the harmonic law, with frequency $p/2\pi$. " It will be seen that the ordinary resistance balance $(SP = QR)$ is departed from. The change here considered is peculiar to the apparatus, and, so far as its influence is concerned, it does not indicate a real alteration of resistance in the wire. Moreover, since p is involved, the disturbance depends upon the rapidity of vibration, so that in the case of ordinary mixed sounds silence can be attained only approximately. Again, from the second equation we see that M is not in general a correct measure of the value of L. If, however, P be very small, the desired condition of things is approached; since, by the construction of the apparatus, $Q + R + S$ is constant (say W), and if P be small enough S does not differ much from W, *i.e.* most of the wire forming the three sides of the combination is devoted to the member opposite to P."

* *Journ. Tel. Eng.* vol. xv. p. 54.

The formulæ are easily proved. Since there is no current through the bridge, there must be the same current (x) in P and one of the adjacent sides (say) R, and for a like reason the same current y in the sides Q and S. The difference of potentials at time t between the junction of P and R and the junction of Q and S may be expressed by each of the three following equated quantities:—

$$Qy - Px - L\frac{dx}{dt} = -M\frac{d(x+y)}{dt} = Rx - Sy \; ;$$

from which the required results are obtained by elimination of the ratio $x : y$, and introduction of the supposition that all the quantities vary harmonically with frequency $p/2\pi$.

The inadequacy of Prof. Hughes's original interpretation has been remarked upon also by Prof. Weber* and by Mr O. Heaviside†, who have obtained the corrected formulæ. I give them here because I agree with Prof. Weber that this form of apparatus possesses distinct advantages. As he points out, if P be known, the application of (2) really presents no difficulty, and allows of L being readily found in terms of M.

There are many cases in which we may be sure beforehand that P (the effective resistance of the conductor, or combination of conductors, to the variable currents) is the same as if the currents were steady, and then P may be regarded as known. There are other cases, however,—some of them will be treated below,—in which this assumption cannot be made; and it is impossible to determine the unknown quantities L and P from (2) alone. We may now fall back upon (1). By means of the two equations, P and L can always be found in terms of the other quantities. But among these is included the frequency of vibration; so that the method is only practically applicable when the interrupter is such as to give an absolute periodicity. A scraping contact, otherwise very convenient, is thus excluded‡; and this is undoubtedly an objection to the method.

My own experiments have been made with three different forms of apparatus. The first was constructed upon the model of that originally described by Hughes, and still to be preferred for some purposes. The others will be described in due course; but it will be convenient to consider first those parts which are common—the interrupters and the induction-compensators.

The Interrupters.

When regular vibrations are not required, a scraping contact interrupter is the least troublesome. Mine is of the roughest possible construction. It

* *El. Rev.* April 9, 1886; July 9, 1886.
† *Phil. Mag.* August, 1886.
‡ A toothed-wheel interrupter, as usually employed, does not give a regular vibration of the period corresponding to the passage of a tooth.

is driven by a small jet of water issuing from a glass nozzle in communication with a tap, and impinging upon blades bent in a piece of tin plate
and revolving about a vertical axis. The upper part of the axis carries
a small cylinder of roughened iron, against which a brass spring lightly
presses. As in Hughes's apparatus, the scraping contact is periodically
broken altogether by a projecting finger, which during part of the revolution
pushes back the brass spring. This is a point of some importance, for a
faint scraping sound is far better heard and identified when thus rendered
intermittent. The apparatus stands in the sink, so that the water scattered
from the revolving blades runs away without giving trouble. The pressure
exercised by the contact-spring requires readjustment from day to day if
the loudest sound is wanted.

But for many of the most interesting experiments a scraping contact
is unsuitable. Prof. Hughes has found, indeed, that in some cases the
natural pitch of the telephone-plate is predominant; so that the vibration,
as it reaches the ear, is not quite so mixed as might have been expected
from its origin. When, however, the induction and resistance under
observation are rapidly varying functions of the frequency of vibration,
it is evident that no sharp results can be obtained without an interrupter
giving a perfectly regular electrical vibration. With proper appliances an
absolute silence, or at least one disturbed only by a slight sensation of
the octave of the principal tone, can be obtained under circumstances
where a scraping contact would admit of no approach to a balance at all.

A thoroughly satisfactory interrupter of this kind has not, to my knowledge, been constructed. Tuning-forks, driven electromagnetically with
liquid or solid contacts, answer well so long as the frequency required
does not exceed 128 or 256 per second; but here we desire frequencies of
from 500 to 2000. My experiments have been made with harmonium-
reeds as interrupters, the vibrating tongue making contact once during
each period with the slightly rounded end of a brass or iron wire, which
can be advanced exactly to the required position by means of a screw cut
upon it. Blown with a well regulated wind, such reeds have given good
results even up to 2000 (complete) vibrations per second; but they are
often capricious and demand frequent readjustment. The reed which I
have usually employed makes about 1050 vibrations per second, and
answered its purpose fairly well. Hitherto I have not been able to satisfy
myself as to the cause of the falling off in efficiency, which often sets in
suddenly, and persists until cured by a readjustment. Another objection
to this interrupter is the simultaneous production of loud aerial sounds,
which must be prevented from reaching the ear of the observer at the
telephone by several interposed doors.

[1900. An interrupter which has given very good results is described in
Theory of Sound, vol. II. p. 456.]

The Induction-Compensators.

Two instruments, similar in all respects, were made by my assistant Mr Gordon, much after the pattern employed by Prof. Hughes. In each there is a small coil mounted so that one diameter coincides with a diameter of a larger coil, and movable about that diameter. The mutual induction M between the two circuits depends upon the position given to the smaller coil, which is read off by a pointer attached to it and moving over a graduated circle. The circles are so divided that the reading (θ) would* be zero when the *axes* of the coils were coincident, or the planes parallel. In this position M is arithmetically a maximum (M_0); and we consider its algebraic sign to be positive. At 90°, when the axes are at right angles, $M = 0$. At 180° M would be negative, and of the same arithmetic value (M_0) as at 0°.

The coils are wound upon boxwood rings, and in each there are 45 convolutions. The mean diameters are about 3 inches and $1\frac{1}{2}$ inch.

Some of the earlier experiments were interpreted by a theory of the compensator, which I knew at the time to be very rough. If the small coil be treated as infinitely small, then

$$M = M_0 \cos \theta.$$

On the same supposition we have, from the roughly measured dimensions,

$$M_0 = 60,000 \text{ centim.}$$

The law of the simple cosine was found to lead to considerable anomalies; and when at a later date (August 19) I carried out my intended calibrations, some very curious results revealed themselves.

The best arrangement for calibration and for determination of the constant of the instruments is to institute distinct primary and secondary circuits. The former included a battery, a scraping contact [p. 553], and the two outer (larger) coils of the compensators. The latter included the two inner coils and a telephone. The precise procedure will depend upon whether we can assume the exact equality of the two compensators. In that case we may introduce, and retain during the observations, another pair of induction-coils, one of course in the primary and the other in the secondary circuit, and of such power as to produce a displacement of about 30° of the compensator. Thus, while in the absence of the additional coils, balance would be obtained when *both* compensators stand at 90°, their introduction would lead to such readings as 90°, 60°; 100°, 70°; 110°, 80°; &c. By this means various parts of the scale of one compensator can be compared with non-corresponding parts of the other; and this is sufficient if the two are similar.

* The position is mechanically unattainable.

This method was used; but it is perhaps better to arrange so that each compensator is calibrated independently of the other. In this case alternate readings are taken with and without the cooperation of the additional coils; and the equivalent induction is found for each compensator at various parts of its scale. The following set of readings will give an idea of the *modus operandi*.

Additional coils in		Additional coils out	
Reading of I.	Reading of II.	Reading of I.	Reading of II.
92°	30°	92°	42¼°
102½	42¼	102½	53
113	53	113	64
124½	64	124½	75½
135½	75½	135½	86½
148½	86½	148½	98

It will be seen that the adjustment is made alternately on the two compensators. Thus in the second compensator the steps $30°-42\frac{1}{4}°$, $42\frac{1}{4}°-53°$, $53°-64°$, &c. have all the same value, whatever may be the construction of the first compensator, which indeed need not be graduated at all. In like manner, the steps from $92°-102\frac{1}{2}°$, $102\frac{1}{2}°-113°$, &c. on the first compensator have an equal value.

An examination of these and other results, not here recorded, leads to the unexpected conclusion that from 40° to 140°, *i.e.* through a range of 100° about the perpendicular position, the scale of induction does not differ appreciably from the scale of degrees. From 30° to 40°, or from 140° to 150°, the induction is something like a tenth part less than that corresponding to 10° in the neighbourhood of 90°. Within the whole mechanical range of the instruments, from 30° to 150°, there could scarcely be an error of 2 per cent. in assuming M proportional to the angle measured from perpendicularity, *i.e.* $(\frac{1}{2}\pi - \theta)$, or, say, θ'.

The general explanation of this very convenient property is not difficult to understand; since for high values of θ' the approximation over the whole circumference of coils of not very unequal diameters must lead to a more rapid increase of M than if the smaller coil were very small; and it is conceivable that for some particular ratio of diameters the increase may just so much exceed that represented by $\sin \theta'$, as to correspond nearly to θ'. I was desirous, however, of explaining this very peculiar relation more completely, and have therefore developed the theory for the case of a ratio of 2 : 1 (nearly that of my apparatus) on the basis of formulæ

given by Maxwell. The details of the calculation are given in the form
of an appendix [p. 577].

It may suffice here to say that the experimental result is abundantly
confirmed; and that reason is found for the conclusion that the proportionality of induction to angle would be even better maintained if the
diameter of the smaller coil were increased from ·50 to ·55 of that of the
larger. The non-mathematical reader may be content to accept this proportionality over most of the range of the actual instruments upon the
experimental evidence.

The absolute value of the induction-coefficient corresponding to each
degree of the compensators was determined at the time of the calibration
by comparison with the calculable induction-coefficient between two coils
wound in measured grooves cut on the surface of a wooden cylinder. These
coils contained respectively 21 and 22 convolutions; and the induction-
coefficient for the mean windings is found to be 277·3 centim. by a calculation of which it is not necessary to record the details. Hence, for the
actual coils,

$$M = 21 \times 22 \times 277\cdot3 = 1\cdot281 \times 10^5 \text{ centim.}$$

The obvious procedure for the comparison would be to combine the compensators without additional coils, so as to obtain a balance at the telephone
when both stand at 90°, and then to observe the displacement or displacements necessary when the standard cylinder coils are introduced. In my
case, however, the range of induction provided by the compensators was
insufficient to balance the standard, if used in this way, even when displacements were made in such (opposite) directions as to cooperate in
changing the total induction. An additional pair of coils was therefore
introduced, for the purpose, as it were, of shifting the zero, of which nothing
required to be known, since they remained connected whether or not the
standard coils were in operation. With this modification, balance could be
attained in both cases.

With standard coils in, one pair of readings was 130°, 43½°; and with
standard coils out, 50°, 127½°. The connections were such that when there
was no external change, the corresponding readings would move in the
same direction [p. 556]; and thus the number of degrees of one compensator
equivalent to the standard is

$$130 - 50 + 127\tfrac{1}{2} - 42\tfrac{1}{2}, \text{ viz. } 165.$$

Accordingly, every degree of either compensator, not too far removed from
the middle of the range, represents 776·3 centim. of induction-coefficient.

The maximum coefficient, when $\theta = 0$, according to Table II. (Appendix),
would be about 56100 centim.

In view of the statement* "that the coefficient of mutual induction is

* [Hughes] *Proc. Roy. Soc.* vol. xl. p. 468.

less in iron than in copper wires," I may mention an experiment made with the aid of one of the compensators, in which the effect of the substitution of iron for copper is directly examined. The mutual induction measured is that between two circuits, one of which was composed of the two copper coils of 21 and 22 convolutions spoken of above connected in series; and the other of a single turn of wire situated midway between, and lying in a shallow scratch or groove on the wooden cylinder, by which its position was accurately defined. The arrangements being the same as in the determination of the constant of the compensator, the value of the double induction (obtained by reversal) between the circuit of a single turn of copper wire and the circuit of 43 turns was 40°·7. The single turn of copper wire was now replaced by a turn of iron wire of equal diameter and bedded in the same scratch, with the result that the double induction was 40°·6, the same value being obtained whether the iron wire were included in the primary circuit with the battery and interrupter, or in the secondary with the telephone. Care had, of course, to be exercised in the disposition of the leads, in consequence of the use of a single turn only for one of the circuits. So far as the experiment could show, the induction is absolutely the same, whether the single turn be of iron or of copper.

To return now to the bridge arrangement, the following are a few examples of the use of the original form of apparatus. The scale of the wire readings was in $\frac{1}{50}$ inch, the whole length $(Q + R + S)$ being 1960. In ohms the resistance of the whole wire is 4·00. The interrupter was the "reed," making about 1050 (complete) vibrations per second. Thus $p = 2\pi \times 1050$. The first case is that of a helix of insulated copper wire, without core of any kind. To get a balance the compensator had to be placed at 54°, so that $M = 36°$, each degree representing 776 centim. The resistances also necessary were

$$Q = 610, \ R = 190; \ \text{therefore} \ S = 1160.$$

They are expressed in scale-divisions, the value of each of which is

$$\frac{4·00 \times 10^9}{1960} = 2·04 \times 10^6 \ \frac{\text{centim.}}{\text{sec.}}.$$

If, as we are almost entitled to do, we assume that the resistance P to variable currents is the same as that readily found with use of steady currents, viz. 87·3 scale-divisions, we may at once deduce

$$L = M\frac{P + Q + R + S}{S} = \frac{36 \times 776 \times 2833}{1160} = 68200 \ \text{centim.}$$

We will, however, dispense with this assumption. Eliminating L between the two equations, we get for the determination of P,

$$P\left\{1 + \frac{p^2 M^2}{S^2}\right\} = \frac{Q \cdot R}{S}\left\{1 - \frac{p^2 M^2 (Q + R + S)}{S \cdot Q \cdot R}\right\}.$$

In the fractions containing M^2 the resistances must be expressed in absolute measure. We find

$$\frac{p^2M^2}{S^2} = \frac{4\pi^2 \times 1050^2 \times 36^2 \times 776^2}{1160^2 \times 10^{12} \times 2\cdot04^2} = \cdot0061 ;$$

$$\frac{p^2M^2(Q+R+S)}{S.Q.R} = \frac{4\pi^2 \times 1050^2 \times 36^2 \times 776^2 \times 1960}{1160 \times 610 \times 190 \times 10^{12} \times 2\cdot04^2} = \cdot1189 ;$$

so that

$$P = \cdot876 \frac{Q.R}{S} ,$$

differing some 12 per cent. from the value (QR/S) given by the usual formula. Inserting the values of Q, R, S, we have

$$P = 87\cdot5 \text{ scale-divisions.}$$

This is the effective resistance to variable currents of the frequency in question.

With steady currents the readings were

$$Q = 557, \quad R = 190, \quad S = 1213 ;$$

so that

$$P_0 = \frac{557 \times 190}{1213} = 87\cdot3.$$

The resistance to variable currents, calculated by the correct formulæ with knowledge of the frequency of vibration, is thus almost identical with the value found with steady currents; whereas if we were to ignore the disturbance of the ordinary resistance rule by induction, we should erroneously conclude that the resistance to variable currents was some 12 per cent. higher than to steady currents.

Of other experiments made with this coil I will only mention one. When a stout copper rod was inserted, the circumferential secondary currents induced in it altered the readings with variable currents to

$$Q = 660, \quad R = 190 ; \quad M = 29\tfrac{1}{2}°.$$

The effective self-induction is evidently diminished, and the effective resistance increased in accordance with the universal rule*. The precise values may be obtained from our two fundamental equations, in the manner exemplified above.

If the foregoing experiment (with the copper core) be attempted with the scraping-contact interrupter, giving a mixed sound, no definite balance is obtainable.

---

* See equations (8), (10), (11), (12), *Phil. Mag.* May 1886, pp. 373—375. [Vol. II. p. 479.]

The second example that I shall give is of a wire of soft iron about $1\frac{1}{2}$ metre long and 3·3 millim. diameter. Here with variable currents from the reed-interrupter, of the same period as before,

$$Q = 178, \quad R = 190, \quad S = 1592; \quad M = 8 \times 776 \text{ centim.};$$

from which we find

$$P = ·985 \frac{Q.R}{S} = 20·93 \text{ scale-divisions.}$$

In the present case the ordinary simple rule (QR/S) would lead to an error of $1\frac{1}{2}$ per cent. only.

The resistance to steady currents is given by

$$P_0 = \frac{100 \times 190}{1670} = 11·38.$$

We may conclude that the effective resistance to variable currents of this frequency (1050) is 1·84 times the resistance to steady currents.

A long length of wire from the same hank was examined later by another method [p. 568], and gave for the ratio in question 1·89.

In some of his experiments* Prof. Hughes found that it made but little difference to the self-induction of an iron wire, whether it was arranged as a compact coil of several turns, or as a single wide loop. The question is readily examined with the present form of apparatus; for, since the resistance is not altered, the compensator readings give an accurate relative measure of the self-induction. A hank of nineteen convolutions of insulated soft iron wire required for balance 25°·8; but when opened out into a single (approximately circular) loop, the reading was only 11°·2, a much greater difference than that mentioned by Hughes.

A better experiment may be made with a coil of doubled wire. The length previously used was divided into halves, which were tied together closely with cotton thread and bent into a compact coil of 9 convolutions and about $4\frac{1}{2}$ inches diameter. When the two wires were connected in series, in such a manner that the direction of electric circulation was the same in both, the self-induction was represented by 24°·2; but when one wire was reversed, the self-induction fell to 9°·2, the large difference depending entirely upon the mutual induction between the two iron wires.

I had intended to apply this apparatus to investigate the self-induction of wires of various materials and diameters, formed into single circular loops, but the subject has been so ably treated by Prof. Weber as to render further work unnecessary, at least as regards the non-magnetic metals.

* *Proc. Roy. Soc.* vol. XL. p. 457.

That the circle is the proper standard form for accurate measurement cannot be doubted; but the effect of magnetic quality is shown most markedly when the wires compared are of given length and diameter, and doubled so as to form single close loops. For a total length $2l$ of copper wire, the self-induction is smallest when the wires are just in contact, and then *

$$L = 3.772\,l.$$

In practice some interval is required for insulation, so that the coefficient of l may perhaps be taken to be 4. To iron wires the theory is not strictly applicable†, but we may probably assume without serious error

$$L = l(4 \log 2 + \mu) = l(2.772 + \mu),$$

μ being the *permeability*. Prof. Hughes finds for the ratio of iron to copper under these circumstances 440 : 18‡; according to which we should have

$$\frac{3 + \mu}{4} = \frac{440}{18},$$

or $\mu = 95$, in approximate agreement with values found by other methods.

Although the original apparatus of Hughes is capable of very good results, and is especially suitable when the wires under test are in but short lengths, the fact that induction and resistance are mixed up in the measurements is a decided drawback, if it be only because the readings require for their interpretation calculations not readily made upon the spot. The more obvious arrangement is one in which both the induction and resistance of the branch containing the subject under examination are in every case brought up to the given totals necessary for a balance. To carry this out conveniently we require to be able to add self-induction without altering resistance, and resistance without altering self-induction, and both in a measurable degree. The first demand is easily met. If we include in the circuit the *two* coils of an induction-compensator, connected in series, the self-induction of the whole can be varied in a known manner by rotating the smaller coil. For the self-induction of the instrument, used in this manner, may be regarded as made up of the constant self-inductions of the component coils taken separately, and of twice the positive or negative mutual induction between them. The first part, in consequence of its constancy, need not be regarded; and thus every degree (within the admissible range) may be taken as representing 2×776.3, or 1552.6 cm., of self-induction.

The introduction, or removal, of resistance without alteration of self-induction cannot well be carried out with rigour. But in most cases the

---

* Maxwell's *Electricity and Magnetism*, §§ 686, 688.
† *Phil. Mag.* May 1886, p. 383. [Vol. II. p. 487.]
‡ *Loc. cit.* p. 457.

object can be sufficiently attained with the aid of a resistance-slide of thin German-silver wire. It may be in the form of a nearly close loop, the parallel out-going and return parts being separated by a thin lath of wood. A spring of stout brass wire making contact with both parts short-circuits a greater or less length of the bight.

In the Wheatstone's quadrilateral, as arranged for these experiments, the adjacent sides R, S are made of similar wires of German silver of equal resistance ($\frac{1}{2}$ ohm). Being doubled they give rise to little induction, but the accuracy of the method is independent of this circumstance. The side P includes the conductor, or combination of conductors, under examination, an induction-compensator, and the resistance-slide. The other side, Q, must possess resistance and self-induction greater than any of the conductors to be compared, but need not be susceptible of ready and measurable variations. But, as a matter of fact, the second induction-compensator was used in this branch, and gave certain advantages in respect of convenience. Sometimes also a rheostat was included; but during a set of comparisons the condition of this branch was usually maintained constant, the necessary variations being made in P. In order to avoid mutual induction between the branches, P and Q were placed at some distance away, being connected with the rest of the apparatus by leads of doubled wire.

It will be evident that when the interrupter acts in the battery branch, balance can be obtained at the telephone in the bridge only under the conditions that both the aggregate self-induction and resistance in P are equal to the corresponding quantities in Q. Hence when one conductor is substituted for another in P, the alterations demanded at the compensator and in the slide give respectively the changes of self-induction and of resistance.

In this arrangement the induction and resistance are well separated, so that the results can be interpreted without calculation. During the month of July a large number of observations on various combinations of conductors were effected, but the results were not wholly satisfactory. There seemed to be some uncertainty in the determination of resistance, due to the inclusion of the two movable contacts of the resistance-slide in one of the sides (P) of the quadrilateral*. I therefore pass on to describe a slight modification by means of which much sharper measurements were attainable.

In order to get rid of the objectionable movable contacts, some sacrifice of theoretical simplicity seems unavoidable. We can no longer keep Q (and therefore P when a balance is attained) constant; but by reverting to the arrangement adopted in a well-known form of Wheatstone's bridge, we cause the resistances taken from P to be added to Q, and *vice versâ*. The transferable resistance is that of a straight wire of German silver, with which one

---

* Prof. Hughes appears also to have met with this difficulty in his second apparatus.

telephone terminal makes contact at a point whose position is read off on a divided scale. Any uncertainty in the resistance of *this* contact does not influence the measurements.

Fig. 1.

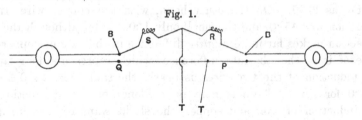

The diagram shows the connection of the parts. One of the telephone terminals goes to the junction of the ($\frac{1}{2}$ ohm) resistances R and S, the other to a point upon the divided wire. The branch P includes one compensator (with coils connected in series), the subject of examination, and part of the divided wire. The branch Q includes the second compensator (replaceable by a simple coil possessing suitable self-induction), a rheostat, or any resistance roughly adjustable from time to time, and the remainder of the divided wire. The battery branch, in which may also be included the interrupter, has its terminals connected, one to the junction of P and R, the other to the junction of Q and S. When it is desired to use steady currents, the telephone can of course be replaced by a galvanometer.

In this arrangement, as in the other, balance requires that the branches P and Q be similar in respect both of self-induction and of resistance. The changes in induction due to a shift in the movable contact may usually be disregarded, and thus any alteration in the subject (included in P) is measured by the rotation necessitated at the compensator. As for the resistance, it is evident that (R and S being equal) the value of any additional conductor interposed in P is measured by twice the displacement of the sliding contact necessary to regain the balance.

The position of the contact was read to tenths of an inch; and, since the actual resistance per inch was ·0246 ohm, a displacement of that amount represents ·0492 ohm. To save unnecessary reductions, the resistance of any conductor will usually be expressed in terms of the contact displacement caused by its introduction, just as the self-induction is expressed in degrees of the compensator.

In order to compare the behaviour of iron and copper, two double coils were prepared as nearly similar as conveniently could be. The iron coil was that already spoken of [p. 560]. The resistance of each wire was ·9 ohm, and the diameter ·032 inch. In the double copper coil the resistance of each wire was ·1 ohm, and the diameter ·037 inch. Each coil consisted of 9 (double) convolutions, of diameter about $4\frac{1}{2}$ inches.

The two iron wires being connected in series, the large self-induction (when the current circulated the same way in both wires) was found to be 65°·1; the small self-induction (when the directions of circulation were different) was 23°·1. On the other hand, with the copper wires the large self-induction was 45°·0, and the small only 1°·0. Thus, although the manner of connection makes far more *relative* difference in the case of copper than in the case of iron, the absolute difference, which represents four times the mutual induction of the two wires, is nearly the same, viz. 44°·0 for copper and 42°·0 for iron. There is here no evidence of any distinction in the mutual induction of iron and copper, the slight want of agreement being easily attributable to different degrees of closeness of approach in the two cases. The readings for resistance were sensibly the same, whether the currents were steady (balance being tested by a galvanometer), or were variable with a frequency of 1050 per second. They were also unaffected by the reversal of one of the wires.

With the same pair of double coils an interesting experiment may be made by observing the effect of closing the second wire upon the apparent resistance and self-induction of the first. To steady currents the resistance of one of the copper wires was 1·75, unaltered by closing the circuit of the other wire. With secondary open, the same resistance was found to apply to periodic currents of frequency 1050; but when the secondary was closed the resistance rose to 2·67. On the other hand, the closing of the secondary reduced the self-induction from 11°·2 to 4°·7. It will be instructive to compare these results with Maxwell's formulæ:—

$$R' = R + \frac{p^2 M^2 S}{S^2 + p^2 N^2}, \qquad L' = L - \frac{p^2 M^2 N}{S^2 + p^2 N^2}; \quad \ldots\ldots\ldots(3, 4)$$

which we may do by means of a value of M (the mutual induction) deduced from the previous experiment, in which the wires were connected in series. Thus

$$M = 11°·0 = 11·0 \times 1553 \text{ centim.}$$

From the present experiment,

$$R = S = 1·75 \text{ inches of slide} = 1·75 \times ·0492 \times 10^9 \frac{\text{centim.}}{\text{sec.}},$$

$$L = N = 11°·2 \text{ of compensator} = 11·2 \times 1553 \text{ centim.},$$

$$p = 2\pi \times 1050;$$

so that

$$\frac{p^2 M^2}{R^2 + p^2 L^2} = ·60.$$

Thus, according to the formulæ, the resistance R' to the periodic currents should be

$$R' = R + ·60R = 2·80 \text{ inches of slide.}$$

This compares with the 2·7 actually found. In like manner
$$L' = L - ·60L = 4°·5 \text{ of compensator,}$$
agreeing as well as could be expected with the observed value 4°·7.

Similar experiments were made on the double coil of iron wire. With secondary open, the resistance of one wire to steady currents was 16·79. To periodic currents of frequency 1050 the resistance was just perceptibly greater, viz. 16·85, which increased a little further (17·15) when the secondary was closed. The closing of the secondary left the self-induction sensibly unaffected at 21°·2. The much slighter influence of the secondary here observed is due mainly to the higher resistance of the iron as compared with copper. A calculation carried out as before gives

$$\frac{p^2 M^2}{R^2 + p^2 L^2} = ·016,$$

agreeing pretty well with the proportional change observed in the resistance. The corresponding change in self-induction would be barely sensible.

In the case where the primary and secondary circuits are similar ($S = R$, $N = L$), Maxwell's general formulæ may be written in the form

$$\frac{R' - R}{R} = \frac{L - L'}{L} = \frac{p^2 M^2}{R^2 + p^2 L^2}; \quad \dots\dots\dots\dots\dots\dots(5)$$

and we may note two extreme cases. When p is small, or, more fully expressed, when the period of vibration is long in comparison with the time-constant of either circuit, viz. L/R, the reaction of the secondary currents is of small importance. On the other hand, when p is large, the right-hand member of (5) approaches to the form M^2/L^2; and this again does not differ much from unity when the two circuits consist of a double coil of non-magnetic wire. Under such circumstances the reaction of the secondary tends to destroy the self-induction and to double the resistance of the primary.

Being desirous of investigating an example approximating to the second extreme, I prepared a double coil of stouter wire than the preceding. The diameter was about ·08 inch, and the length of each wire was 318 inches. There were 20 (double) turns, so that the mean diameter of the coil, wound as compactly as possible, was about 5 inches. The resistance of each wire was about ·05 ohm.

The coefficient of mutual induction was determined by comparison of the self-induction (L) of one wire with that of the two wires connected oppositely in series, viz. ($2L - 2M$). In this way it appeared that

$$M = 43°·1 = 43·1 \times 1553 \text{ centim.}$$

The interrupter was the reed, of frequency 1050.

Observation showed that the closing of the secondary diminished the self-induction from 44°·4 to 3°·4. The resistance to steady currents was ·92 inch. The resistance to the periodic currents was ·97 with secondary open, and 1·74 with secondary closed.

Taking then

$$L = 44\cdot4 \times 1553 \text{ centim.}, \quad R = \cdot97 \times \cdot0492 \times 10^9 \frac{\text{centim.}}{\text{sec.}},$$

we get

$$\frac{p^2M^2}{R^2+p^2L^2} = \frac{10^{17} \times 1\cdot951}{10^{17} \times \cdot023 + 10^{17} \times 2\cdot071} = \cdot932.$$

According to the formula, therefore,

$$L' = \cdot068\,L = \cdot068 \times 44\cdot4 = 3^\circ\cdot0,$$

which is to be compared with the observed 3°·4.

The application of the formula to the calculation of R' is somewhat embarrassed by the observed difference of resistances to steady and to periodic currents when the secondary was open, of which the theory takes no account. It is true that the difference is small, but it appeared to lie outside the limits of error of observation. It is not accounted for merely by the tendency of periodic currents to adhere to the outer parts of a conducting cylinder. If this observation stood alone, one would be inclined to attribute the discrepancy to some action, whether electro-magnetic or electro-static, of the neighbouring secondary, even though open; but, as we shall have occasion to notice, a similar tendency of the resistance to increase when periodic currents are substituted for steady ones is to be observed in cases where no such explanation is available. The effect was, however, too small to be investigated further without some modification in the apparatus, or in the nature of the conductors submitted to examination.

If we take, as found for the periodic currents, $R = \cdot97$, we get

$$R' = 1\cdot93 \times \cdot97 = 1\cdot87,$$

instead of 1·74 as observed. On the other hand, if we take $R = \cdot92$, we get

$$R' = 1\cdot93 \times \cdot92 = 1\cdot77.$$

The next experiment was contrived to illustrate the behaviour under periodic currents of a system composed of two conductors connected in parallel. The general theoretical formulæ are given in a former paper[*]; but the more special case selected for experiment was one in which the mutual induction of the two conductors (M) and the self-induction of one of

[*] *Phil. Mag.* May 1886, formulæ (13), (14), p. 377. [Vol. II. p. 482.]

them (N) can be neglected. The formulæ for the resistance and self-induction of the combination then reduce to

$$R' = \frac{SR}{R+S}\left[1 + \frac{S}{R}\frac{p^2L^2}{(R+S)^2 + p^2L^2}\right], \quad \dots\dots\dots\dots\dots(6)$$

$$L' = \frac{S^2L}{(R+S)^2 + p^2L^2}, \quad \dots\dots\dots\dots\dots\dots\dots\dots\dots(7)$$

in which $SR/(R+S)$ represents the resistance to steady currents $(p = 0)$. The peculiar features of the arrangement are brought out most strongly by taking a case in which S (the resistance of the induction-less component) is great compared with R. It is then obvious that steady, or slowly alternating, currents flow mainly through R, and accordingly that the resistance and self-induction of the combination approximate to R and L respectively. Rapidly alternating currents, on the other hand, flow mainly through S, so that the resistance of the combination approximates to S, and the self-induction to zero. These common-sense conclusions are of course embodied in the formulæ.

The conductors combined in parallel were (1) the coil of stout copper [p. 563] with its two wires permanently connected in parallel so as to give maximum self-induction (L), and (2) a moderate length of somewhat fine brass wire. With steady currents the resistances were

$$R = {\cdot}45, \qquad S = 2{\cdot}29, \qquad R_0' = {\cdot}35.$$

It had been expected that the resistances R, S, of the separate conductors would have been sensibly the same whether tested by steady or by periodic currents; but the resistances in the latter case tended always to appear higher. Thus with the same reed as interrupter,

$$R = {\cdot}52, \qquad S = 2{\cdot}33, \qquad L = 43^{\circ}{\cdot}7, \qquad N = {\cdot}3^{\circ};$$

and for the combination,

$$R' = 2{\cdot}04, \qquad L' = 3^{\circ}{\cdot}0.$$

These results of observation illustrate satisfactorily the general behaviour of the combination to periodic currents of high frequency, and they agree fairly well with the formulæ. According to these, if we take the values of R and S as observed with periodic currents, we have

$$R' = 2{\cdot}16, \qquad L' = 2^{\circ}{\cdot}61.$$

The altered distribution of current under the influence of induction, and consequent increase of resistance, exemplified in the above examples, is an extreme case of what may happen to a sensible extent within a simple conducting cylinder, especially of iron, when the diameter is not very small in relation to the frequency of electrical vibration. In order to avoid magnetizing the material of the conductor, the current tends to confine itself to the outer strata, in violation of the condition of minimum resistance. Prof. Hughes has already given examples of this effect; but they are difficult to compare with theory in consequence of his employment of a vibration of indefinite pitch. The following observations were made with the usual reed interrupter, giving about 1050 vibrations per second.

A somewhat hard Swedish-iron wire, 10·03 metres long, and 1·6 millim. diameter, was first examined. The resistance to steady currents was 10·3, and to the variable currents given by the reed, 12·0. The wire was then softened in the flame of a spirit-lamp, after which the resistance to steady currents was 10·4, and to variable currents 12·1. Expressed in ohms, the resistance to steady currents is

$$10\cdot4 \times \cdot0492 = \cdot51 \text{ ohm.}$$

From these data we may deduce an approximate value of the magnetic permeability (μ) of the material for circumferential magnetization. For if l be the length, R the resistance to steady currents, $p/2\pi$ the frequency of vibration, we have for the resistance (R') to variable currents the approximate expression *

$$R' = R\left\{1 + \frac{1}{12}\frac{p^2 l^2 \mu^2}{R^2} - \frac{1}{180}\frac{p^4 l^4 \mu^4}{R^4} + \dots\right\};$$

so that for the rough determination of μ we may take in the present example,

$$\frac{1}{12}\frac{p^2 l^2 \mu^2}{R^2} = \frac{1\cdot7}{10\cdot4}.$$

The result is $\mu = 108.$

A more accurate use of the formula would bring out a sensibly higher value; but it is hardly worth while to pursue the matter, inasmuch as any deduction of μ from the small observed difference of resistance (1·7) is necessarily subject to considerable error.

In order to get better materials for a determination of μ by this method, a stouter wire of Swedish iron was next tested, 18·34 metres in length and 3·3 millim. in diameter. The metal was rather hard. The resistance to steady currents was found to be 4·7, and to the variable currents from the reed 8·9. These are, as usual, in terms of the scale of the apparatus. The absolute resistance to steady currents

$$R = \cdot230 \times 10^9 \frac{\text{centim.}}{\text{sec.}}.$$

In this example, the change of resistance (in the ratio 1·89 : 1) is so great that no use can be made of the approximate formula quoted above, but we

---

* *Phil. Mag.* May 1886, equation (19), p. 387. [Vol. II. p. 491.]

must revert to the original series. In the notation employed in the paper referred to, if $\phi(x)$ denote the function*,

$$1 + x + \frac{x^2}{1^2 \cdot 2^2} + \cdots + \frac{x^n}{1^2 \cdot 2^2 \cdots n^2} + \cdots,$$

the resistance to variable currents (R'), and the self-induction (L'), are given by

$$\frac{R'}{R} + ip\,\frac{L'}{R} = \frac{ipl}{R}\,A + \frac{\phi\,(ipl\mu/R)}{\phi'\,(ipl\mu/R)}; \qquad \cdots\cdots\cdots\cdots\cdots\cdots(8)$$

so that the *real* part of the fraction ϕ/ϕ' gives the ratio R'/R. By calculation from the series I find

$$\frac{\phi\,(i \times 5{\cdot}2365)}{\phi'\,(i \times 5{\cdot}2365)} = \frac{-4{\cdot}5893 + i \times 1{\cdot}5171}{-1{\cdot}0297 + i \times 1{\cdot}6662} = 1{\cdot}8906 + i \times 1{\cdot}5859,$$

in which the first term on the right agrees sufficiently nearly with the observed value of R'/R. We may conclude that

$$pl\mu/R = 5{\cdot}2365,$$

whence

$$\mu = 99{\cdot}5.$$

In order to give an idea of the degree of accuracy with which μ is determined by the observed value of R'/R, it may be worth while to record another numerical result, viz. :—

$$\frac{\phi\,(i \times 5{\cdot}6815)}{\phi'\,(i \times 5{\cdot}6815)} = 1{\cdot}9596 + i \times 1{\cdot}6544.$$

In these calculations it is assumed that the increase in R, observed when variable currents are substituted for steady ones, is due simply to a less favourable distribution of current over the section. If there were sensible hysteresis in the magnetic changes, R would be still further increased. I believe, however, that under such magnetizing forces as were at play in these experiments, there is no important hysteresis, and that μ may be treated as sensibly constant.

The increased self-induction and resistance of an iron wire, due to its magnetic quality, are doubtless disadvantages from a telephonic point of view. If found serious they may be mitigated, as Prof. Hughes has shown, by the use of a stranded wire, in which the circumferential magnetic circuits are interrupted. There has been some confusion, I think, in connection with the notion of "retardation." If we had the means of observing the passage of signals at various points of a long cable, we should find them not merely

---

* The relation of ϕ to Bessel's function of order zero is expressed by

$$\phi\,(x) = J_0\,(2i\sqrt{x}).$$

retarded (which would be of no consequence) as we recede from the sending end, but also attenuated. The amplitude of a periodic, *e.g.* telephonic, current sent into a cable becomes less and less as the distance increases. Nothing of the kind can happen in a well-insulated iron wire of negligible electrostatic capacity. Its resistance and self-induction may oppose the entrance of a current, but whatever current there is at any moment at the sending end of the wire must exist unimpaired throughout its whole extent.

I will now record a few experiments as to the effect of an iron core upon the apparent self-induction and resistance of an encompassing helix. The wire was wound in one layer upon a glass tube; the total number of turns is 205, occupying a length of 28·6 centim. The length of the wires forming the cores was 24·1 centim. The results given are the differences of the readings obtained with and without the cores, so that the resistance and self-induction of the helix itself are not included. The interrupter was the same reed as in previous experiments.

A comparison was made of the effect of a solid iron wire 1·2 millim. in diameter and of two bundles of wires of similar iron (drawn from the same specimen) of equal aggregate section and weight. One bundle contained 7 wires, and another 17. The results were:—

	1 wire	7 wires	17 wires
Resistance............	1·3	0·3	0·2
Self-induction	13°	18°	18°

showing that when the wire was undivided the secondary currents developed in it increased the apparent resistance of the helix by 1·3, and diminished the apparent self-induction.

A similar experiment was tried with a stouter wire, 3·3 millim. in diameter (from the same hank as the length of 18·34 metres treated as a conductor). In the hard condition the self-induction due to this was $24\frac{1}{2}°$, and the resistance 3·8; numbers altered to $28\frac{1}{2}°$ and 4·4 respectively by softening with a spirit-flame. The effect of a bundle of thirty-five soft wires of the same iron and of equal aggregate section was 84° of self-induction and 1·6 of resistance *.

There is nothing surprising in the conclusion, forced upon us by the observations, that the magnetic effects of iron rods 3·3 millim. in diameter

---

* It may be worth while to remark that in these experiments no approach to a balance could be obtained when a scraping contact interrupter was used. With the reed there was complete silence, or at most a slight perception of the octave. The failure of the scraping contact is due, of course, to the mixed character of the vibration, and to the fact that the adjustments necessary for balance vary rapidly with pitch.

are seriously complicated by the formation of induced internal electric currents. As I have shown on a former occasion*, the principal time-constant of a cylinder of radius a, specific resistance ρ, and permeability μ, is given by

$$\tau = \frac{4\pi\mu a^2}{(2 \cdot 404)^2 \rho}.$$

This means that circumferential currents started and then left to themselves would occupy a time τ in sinking to $1/e$ of their initial magnitude. Whether the effects of such currents will be important or not depends upon the relative magnitudes of τ and of the period of the magnetic changes actually in progress. In the present case, with

$$\mu = 100, \qquad \rho = 9827, \qquad 2a = \cdot 33,$$

the value of τ is about $\frac{1}{2000}$ of a second, that is, about *half* the period of the actual electrical vibration.

The theory of an infinite conducting cylinder exposed to periodic longitudinal magnetic force (Ie^{ipt}) has been given by Lamb†, who finds for the longitudinal magnetic induction at any distance r from the axis

$$c = \frac{J_0(kr)}{J_0(ka)} \mu I e^{ipt}, \quad\dots\dots\dots\dots\dots\dots\dots..(9)$$

where

$$k^2 = - 4\pi\mu ip/\rho. \quad\dots\dots\dots\dots\dots\dots\dots(10)$$

When the changes are infinitely slow, c reduces to $\mu I e^{ipt}$, as should evidently be the case.

A more complete solution was worked out a little later by Oberbeck‡ including what is required for our present purpose, viz. the value of

$$2\pi \int_0^a c\, r\, dr.$$

In terms of the function ϕ previously used [p. 569], (8) becomes

$$c = \mu I e^{ipt} \frac{\phi(ip\mu . \pi r^2/\rho)}{\phi(ip\mu . \pi a^2/\rho)}, \quad\dots\dots\dots\dots\dots\dots(11)$$

whence is readily deduced

$$2\pi \int_0^a c\, r\, dr = \pi a^2 . \mu I e^{ipt} \frac{\phi'(ip\mu . \pi a^2/\rho)}{\phi(ip\mu . \pi a^2/\rho)}, \quad\dots\dots\dots\dots(12)$$

where

$$\phi'(x) = 1 + \tfrac{1}{2} x + \tfrac{1}{3} \frac{x^2}{1^2 . 2^2} + \tfrac{1}{4} \frac{x^3}{1^2 . 2^2 . 3^2} + . \quad\dots\dots\dots\dots(13)$$

* *Brit. Assoc. Report*, 1882, p. 446. [Vol. II. p. 129.]

† *Math. Soc. Proc.* Jan. 1884, vol. XV. p. 141.

‡ *Wied. Ann.* vol. XXI. (1884), p. 672. There seems to be some error in the way in which the magnetic constant appears in Oberbeck's solution (47). According to it (as I understand) a copper core would be without effect.

The mathematical analogy between this problem and that of the variation of a longitudinal electrical current in a cylindrical conductor has been pointed out by Mr Heaviside*, who has also given the full solution of the latter. Maxwell's investigation, somewhat further developed in my paper†, relates principally to that aspect of the question with which experiment is best able to deal, viz. the relation between the total current at any moment and the corresponding electromotive force.

That the argument in ϕ, ϕ' is the same in (12) as in (8) will be evident, when it is remembered that R in (8) denotes the resistance of unit length of the cylinder; so that

$$\frac{l}{R} = \frac{\pi a^2}{\rho}.$$

Hence, if we may assume that the material is isotropic, the same numerical results are applicable to a given wire in both problems. But from this point the analogy fails us. What we require here to express is the ratio of the total magnetic induction to the external magnetizing force, and not the inverse relation, corresponding in the other problem to the expression of the electromotive force in terms of the total current. The experimental results are the reaction of the core upon the magnetizing circuit, expressed as alterations of apparent self-induction and resistance. Now if m be the number of turns per unit length in the magnetizing helix and C the current (proportional to e^{ipt}), we have

$$I\,e^{ipt} = 4\pi mC; \quad\quad\quad\quad\quad\quad\quad\quad\quad\quad\quad(14)$$

and for the electromotive force (E) due to the change of magnetic induction in the core, reckoned per unit length,

$$E = m\frac{d}{dt}\left\{2\pi\int_0^a c\,r\,dr\right\} = 4m^2\pi^2 a^2\mu\,.\,ipC\,.\,\phi'/\phi. \quad\quad(15)$$

In order to interpret this, we must separate the real and imaginary parts of ϕ'/ϕ. If we write

$$\phi'/\phi = P - iQ,$$

then the part of E which is in the same phase as dC/dt is $4m^2\pi^2 a^2\mu\,.\,ipC\,.\,P$; and the part which is in the same phase as C is $4m^2\pi^2 a^2\mu\,.\,pC\,.\,Q$. The first manifests itself as an increase of self-induction, and the second as an increase of resistance. If $\rho = \infty$, $P = 1$, $Q = 0$.

What we require to know for our present purpose is the effect of *introducing* the core; and to obtain this we must subtract any part of E which remains when we put $\rho = \infty$, $\mu = 1$. Calling this E_0, we have

$$E_0 = 4m^2\pi^2 a^2\,.\,ipC,$$

and

$$E - E_0 = 4m^2\pi^2 a^2\left\{ipC\,(\mu P - 1) + \mu pC\,.\,Q\right\}.$$

---

* *Phil. Mag.* August 1886, p. 118.

† *Phil. Mag.* May 1886, p. 386. [Vol. II. p. 491.]

Thus if δL, δR be the apparent augmentations of self-induction and resistance in the helix due to the introduction of the core, reckoned per unit length,

$$\delta L = 4m^2\pi^2a^2(\mu P - 1), \qquad \delta R = 4m^2\pi^2a^2\mu \, . \, pQ. \quad \dots\dots\dots(16)$$

From the calculation already made for the purposes of the other problem, we have

$$\frac{\phi'(i \times 5\cdot2365)}{\phi(i \times 5\cdot2365)} = (1\cdot8906 + i \times 1\cdot5859)^{-1} = \cdot31047 - i \times \cdot26044 \, ;$$

so that for the stout iron wire of $3\cdot3$ millim. diameter and $\mu = 99\cdot5$,

$$P = \cdot31047, \qquad Q = \cdot26044.$$

With these values the effects δL, δR of the core of $3\cdot3$ millim. diameter may be calculated; but no very good agreement with observation is to be expected, since the conditions of infinite length, isotropy, &c., were but inadequately satisfied. Inserting in (16) $m = 205/28\cdot6$, $a = \cdot165$, $\mu = 99\cdot5$, we get

$$\delta L = 1650, \qquad \delta R = 10^6 \times 9\cdot436.$$

These are expressed in absolute measure, and reckoned per unit length of core. To obtain numbers comparable with the experimental readings, we must multiply by $24\cdot1$ (the length of the core), and reduce δL by division by 1553, and δR by division by $10^9 \times \cdot0492$. The result is

$$(\delta L) = 25^\circ\cdot6, \qquad (\delta R) = 4\cdot6 \, ;$$

which agree moderately well with the observed values, viz.

$$(\delta L) = 24\tfrac{1}{2}^\circ, \qquad (\delta R) = 3\cdot8.$$

If the material composing the core were non-conducting, $P = 1$, and

$$\delta L_0 = 4m^2\pi^2a^2(\mu - 1).$$

The ratio of the actual effect to that which would be got from the same aggregate section of a bundle of wires, infinitely thin and insulated from one another, is thus

$$\frac{\delta L}{\delta L_0} = \frac{\mu P - 1}{\mu - 1},$$

of which the numerical value in the present example is $\cdot303$. The corresponding ratio of observed effects for the solid wire (softened), and for the bundle of 35 wires of the same aggregate section was

$$28\tfrac{1}{2}/84 = \cdot339.$$

The general result of these experiments is to support the conclusion arrived at by Oberbeck that the action of iron cores, submitted to periodic magnetizing forces of feeble intensity, can be calculated from the usual simple theory, provided we do not leave out of account the induced internal currents which often play a very important part. Oberbeck's observations were made with the electrodynamometer, and with rather low frequencies of vibration—about one tenth of that used in most of the observations here recorded.

We have seen in several examples that the self-induction of a combination of conductors, being a function of the pitch, admits of an indefinite series of values; and the question suggests itself to which (if any) of these corresponds the value obtained by galvanometric observation of the transition from a state of things in which all the currents are zero to one in which they have steady values under the action of a constant electromotive force. In the ordinary theory of Maxwell's method for determining self-induction from the throw of the galvanometer-needle in a Wheatstone's bridge (a resistance-balance having been already secured), the conductor under test is supposed to be simple. The general case of an arbitrary combination of conductors can only be treated by a general method. An investigation founded upon the equations of my former paper* shows that the result which would be obtained by Maxwell's method corresponds to the self-induction of the combination for *infinitely slow* vibrations.

We have supposed that the behaviour of the compound conductor is not influenced by electrostatic phenomena; otherwise the representation of the part of the electromotive force in the same phase as dC/dt as due merely to self-induction would be unnatural. So far as experiment is concerned, we have no means of distinguishing between an effect dependent upon dC/dt and one dependent upon $\int C dt$, for the phase of both is the same. We may contrast two extreme cases—(1) a simple conductor with resistance and self-induction, (2) a simple condenser with resisting leads. In the first case the electromotive force at the terminals is written

$$L \cdot ip\, C + R \cdot C;$$

in the second

$$- \mu' \cdot ip^{-1} C + R \cdot C,$$

where μ' represents the "stiffness" of the condenser. If we persisted in regarding the imaginary part in the second case as due to (negative) self-induction, we should have to face the fact that the coefficient becomes infinite as p diminishes without limit.

A number of combinations in which the induction of coils is balanced by condensers are considered by Chrystal in his valuable memoir on the differential telephone†.

---

* *Phil. Mag.* May 1886, p. 372. [Vol. II. p. 477.] The analysis may be simplified by choosing the first type so as to correspond to steady flow. The coefficients b_{12}, b_{13} ..., as well as the final values of $\dot{\psi}_2$, $\dot{\psi}_3$... are then zero, and the result may be expressed,

$$\int \Psi_1 \, dt = a_{11} \dot{\psi}_1 + b_{11} \int \dot{\psi}_1 \, dt.$$

† *Edinburgh Transactions*, 1879.

In a paper* already referred to I have shown that when two conductors in parallel exercise a powerful reciprocal induction, very curious results may follow the application of a periodic electromotive force. I have lately submitted the matter to experimental test, by which theoretical anticipations have been fully confirmed.

The two conductors in parallel were constructed out of the three wires of a heavy and compact triple coil of copper wire† mounted in a mahogany ring, which has been in my possession for many years. Of these wires two are combined in series (with maximum self-induction) to constitute one of the branches in parallel. The other branch is the third wire of the triple coil, so connected that steady currents would circulate the same way round them all. The variable currents were obtained from a battery and scraping-contact apparatus [p. 553], connected directly. Under these conditions, if the intermittence be rapid enough, the currents distribute themselves in the two branches so as nearly to neutralize one another's magnetizing-power: and this requires that the current in the single wire should be of about twice the magnitude of the current in the double wire, *and in the opposite direction.* If we call these currents 2 and −1, the current in the mains must be +1.

As may be seen from formula (13)‡, such a state of things leads to a high equivalent resistance for the system; and the question might be investigated on this basis with the apparatus already described. I preferred, however, to examine directly whether it were true that the current in one of the branches exceeded that in the main; and this could be readily done by "tapping" with the telephone. For this purpose the two branches and the main were led through short lengths of similar German-silver wire to the junction, composed of a copper plate to which the wires were soldered (Fig. 2). One

Fig. 2.

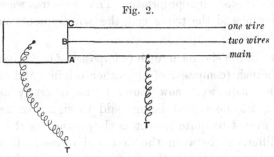

telephone terminal was soldered to the plate; the other was brought into contact with some point of the German-silver wire carrying the current to be observed. It is evident that if the three alternating currents were of

* *Phil. Mag.* May 1886, p. 378. [Vol. II. p. 482.]
† The three wires were wound on *together.*
‡ *Loc. cit.* p. 377. [Vol. II. p. 482.]

equal magnitude, sounds of like loudness would be heard at equal distances from the copper plate, whichever of the wires was touched; and, further, that the distances required to produce equal sounds are inversely as the magnitudes of the corresponding currents.

A moment's observation proved that the currents in A and B were about equal, and that in C much greater. Numerical estimates are best made with the aid of a second observer, who does not see what contacts are being tried. My assistant considered that about $6\frac{1}{2}$ inches of B and about $3\frac{1}{2}$ inches of C were required to give the same loudness as 6 inches of A. This agrees with the approximate theory as well as could be expected.

If the single wire be reversed, then, according to theory (resistances of German-silver wires neglected), the distribution should be much the same as of steady currents under the sole influence of resistance; that is, the currents in the branches should be as $+2$ to $+1$, so that on the same scale the main current would be $+3$. According to this the equivalent lengths of the German-silver wires would be 6, 9, 18. The numbers actually found by experiment were 6, 8, $17\frac{1}{2}$.

In the first part of this experiment the current in *one* of the branches is greater than in the main; but I wished to examine a case where *both* parts of the divided current exceeded the whole. This could be done with a fivefold coil, as described in the previous paper; but such was not ready to hand. In default thereof a common double coil, belonging to a large electro-magnet, was enveloped with a single layer of extra wire, which was combined in series with one of the original wires. This arrangement is less favourable than one in which the two branches are in close juxtaposition throughout; but I thought that with the aid of an iron core it could be made to answer the purpose. Such a core was provided in the form of a bundle of fine wires, solid iron being obviously inappropriate. The two wires were now connected in parallel and replaced the triple coil, the arrangements in other respects remaining unchanged.

The currents in the shorter branch (composed of one original coil simply), in the longer branch (composed of the other original and of the additional coil), and in the main were now found to be inversely as the measured distances ·9, 1·3, 2·3, no regard being paid to sign, viz. as 1·11, ·77, ·43. These numbers cannot be quite correct as they stand, for the third should be equal to the difference between the first and second. If we suppose the second and third to be correct, the first would have to be 1·20 instead of 1·11. Such an error as this may easily occur in estimating the equality of sounds heard successively; and there can be no doubt that the smaller branch current largely exceeded the main current*.

* These experiments were described before the British Association at Birmingham, September 3, 1886.

APPENDIX.—*The Induction-Compensators* [p. 557].

For the mutual induction-coefficient between two circular circuits, subtending angles α_1, α_2 at the point of intersection of their axes (lines through their centres and perpendicular to their planes), and distant c_1, c_2 from that point, Maxwell gives*

$$M = 4\pi^2 \sin^2\alpha_1 \sin^2\alpha_2 \, c_2 \left\{ \tfrac{1}{2} \frac{c_2}{c_1} Q_1'(\alpha_1) Q_1'(\alpha_2) Q_1(\theta) \right.$$

$$\left. + \ldots + \frac{1}{i(i+1)} \frac{c_2{}^i}{c_1{}^i} Q_i'(\alpha_1) Q_i'(\alpha_2) Q_i(\theta) + \ldots \right\}, \quad \ldots\ldots\ldots(17)$$

the angle between the axes being denoted by θ. $Q_i \ldots$ denote Legendre's coefficients (more usually represented by P_i), and the *dash* indicates differentiation with respect to μ. In our present application the circuits are *concentric*, so that $\alpha_1 = \alpha_2 = \tfrac{1}{2}\pi$, and c_1, c_2 are equal to their radii. Moreover $\{Q_i'(\tfrac{1}{2}\pi)\}^2$ vanishes if i be even; while if i be odd $(2n+1)$ we have

$$\{Q'_{2n+1}(\tfrac{1}{2}\pi)\}^2 = \frac{3^2 . 5^2 . 7^2 \ldots (2n+1)^2}{2^2 . 4^2 . 6^2 \ldots (2n)^2} ; \quad \ldots\ldots\ldots\ldots(18)$$

so that

$$M \frac{c_1}{4\pi^2 c_2{}^2} = \tfrac{1}{2} Q_1(\theta) + \frac{1}{3.4} \frac{3^2}{2^2} \left(\frac{c_2}{c_1}\right)^2 Q_3(\theta) + \frac{1}{5.6} \frac{3^2 . 5^2}{2^2 . 4^2} \left(\frac{c_2}{c_1}\right)^4 Q_5(\mu) + \ldots$$

$$+ \frac{1}{(2n+1)(2n+2)} \cdot \frac{3^2 . 5^2 . 7^2 \ldots (2n+1)^2}{2^2 . 4^2 . 6^2 \ldots (2n)^2} \left(\frac{c_2}{c_1}\right)^{2n} Q_{2n+1}(\theta), \quad \ldots\ldots(19)$$

which is what we have to calculate for various values of θ on the supposition that $c_2 = \tfrac{1}{2} c_1$.

The following are the values of $Q_{2n+1}(\theta)$ at intervals of 10°. It is unnecessary for our purpose to go further than Q_7.

TABLE I.

θ	$Q_1(\theta)$	$Q_3(\theta)$	$Q_5(\theta)$	$Q_7(\theta)$
90°	·00000	·00000	·0000	·0000
80	+ ·17365	− ·24738	+ ·2810	− ·2834
70	+ ·34202	− ·41301	+ ·3281	− ·1486
60	+ ·50000	− ·43750	+ ·0898	+ ·2231
50	+ ·64279	− ·30022	− ·2545	+ ·2854
40	+ ·76604	− ·02523	− ·4197	− ·1006
30	+ ·86603	+ ·32476	− ·2233	− ·4102
20	+ ·93969	+ ·66488	+ ·2715	− ·1072
10	+ ·98481	+ ·91057	+ ·7840	+ ·6164
0	+1·00000	+1·00000	+1·0000	+1·0000

From these the values of (19) were computed. They are shown in Table II., together with the sines of θ' and the differences for each step of 10°.

* *Electricity and Magnetism*, § 697.

TABLE II. $c_2^2 = \cdot 25\, c_1^2$.

θ	θ'	Induction	Diffs.	Sin θ'	Diffs.
90°	0°	·0000		·00000	
80	10	·0769	·0769	+ ·17365	·1736
70	20	·1538	·0769	+ ·34202	·1684
60	30	·2304	·0766	+ ·50000	·1580
50	40	·3058	·0754	+ ·64279	·1428
40	50	·3786	·0728	+ ·76604	·1232
30	60	·4460	·0674	+ ·86603	·1000
20	70	·5029	·0569	+ ·93969	·0737
10	80	·5416	·0387	+ ·98481	·0451
0	90	·5559	·0143	+1·00000	·0152

The column headed Induction gives the value of

$$\tfrac{1}{2} Q_1(\theta) + \frac{1}{3 \cdot 4}\, \frac{3^2}{2^2} \cdot \frac{1}{4}\, Q_3(\theta) + \frac{1}{5 \cdot 6}\, \frac{3^2 \cdot 5^2}{2^2 \cdot 4^2} \cdot \frac{1}{4^2}\, Q_5(\theta) + \dots$$

It will be seen that for moderate values of θ' the differences are very nearly constant, far more so than the differences of sin θ', which latter would apply to the induction on the supposition of a very small interior coil. The results of the experimental calibration are thus confirmed and explained.

An inspection of the table suggests that the proportionality to θ' might be improved yet further if the value of c_2/c_1 were a little increased. The following numbers calculated for a twenty per cent. increase of c_2^2/c_1^2, viz. for $c_2 = \cdot 54772\, c_1$, confirms the idea. Such a proportion, applicable to the *mean* radii of the coils, might well be designedly chosen.

TABLE III. $c_2^2 = \cdot 3\, c_1^2$.

θ	Induction	Diff.
90°	·0000	
80	·0752	·0752
70	·1508	·0756
60	·2268	·0760
50	·3025	·0757
40	·3769	·0744
30	·4479	·0710
20	·5099	·0620
10	·5532	·0433
0	·5695	·0163

The numbers in the column headed Induction are the values of

$$\tfrac{1}{2} Q_1(\theta) + \frac{1}{3 \cdot 4}\, \frac{3^2}{2^2}\, (\cdot 3)\, Q_3(\theta) + \frac{1}{5 \cdot 6}\, \frac{3^2 \cdot 5^2}{2^2 \cdot 4^2}\, (\cdot 3)^2\, Q_5(\theta) + \dots$$

The last two entries are liable to a small error from omission of Q_9, &c.

141.

NOTES ON ELECTRICITY AND MAGNETISM.—III. ON THE BEHAVIOUR OF IRON AND STEEL UNDER THE OPERATION OF FEEBLE MAGNETIC FORCES.

[*Philosophical Magazine*, XXIII. pp. 225—245, 1887.]

THE question whether or not iron responds proportionally to feeble magnetic forces is of interest not only from a theoretical point of view, but from its bearing upon the actual working of telephonic instruments. Considerable difference of opinion has been expressed concerning it, several of the best authorities inclining to the view that a finite force is required to start the magnetization. Prof. Ewing remarks*:—"As regards the hysteresis which occurs when the magnetism of soft iron is changed, my experiments confirm the idea already suggested by other observers, that when the molecular magnets of Weber are rotated they suffer, not first an elastic and then a partially non-elastic deflection as Maxwell has assumed, but a kind of frictional retardation (resembling the friction of solids), which must be overcome by the magnetizing force before deflection begins at all." In a subsequent passage† Prof. Ewing treats the question as still open, remarking that though his curves suggest that the initial value of k (the susceptibility) may be finite, they afford no positive proof that it is not initially zero, or even negative.

My attention was first called to the matter about a year and a half ago in connection with the operation of iron cores in the coils of an induction-balance. Experiment showed that iron responded powerfully to somewhat feeble forces; and I endeavoured to improve the apparatus in the hope of being able thus to examine the subject more thoroughly. Two similar long helices were prepared by winding fine insulated wire

* *Phil. Trans.* 1886, p. 526, § 5.
† *L.c.* § 61.

upon slender glass tubes. These were connected in series with a battery, a resistance-box, and a microphone-clock, so as to constitute a primary circuit. The secondary consisted of a large quantity of copper wire, mounted upon a bobbin, through the opening in which both primary coils were inserted. The circuit of the secondary was completed by a telephone. When neither primary coil contained a core, silence at the telephone could readily be obtained. The iron cores used were those described in Part II.*; and it was found that all of them (including the bundle of seventeen very fine wires) disturbed the silence until the resistance was so far increased that the magnetizing force was less than about $\frac{1}{50}$ of the earth's horizontal force (H). Moreover, there was no indication that the absence of audible effect under still smaller magnetizing forces was due to any other cause than the want of sensitiveness of the apparatus.

I did not pursue the experiments further upon these lines, because calculation showed that the feeble magnetization of a piece of iron could more easily be rendered evident directly upon a suspended needle (the magnetometric method), than indirectly by the induction of currents in an encompassing coil connected with a galvanometer. Nearly all the results to be given in this paper were obtained by a form of the magnetometric method specially adapted to the inquiry whether or not the magnetization of iron continues proportional to the magnetizing force when the latter is reduced to the uttermost.

The magnetizing-spiral first used was one of those already referred to. It consists of a single layer of fine silk-covered copper wire wound on a glass tube and secured with shellac varnish (A, Fig. 1). The total length of the spiral is 17 centim., its diameter is about ·6 centim., and the windings are at the rate of 32 per centim. The resistance is about $5\frac{1}{2}$ ohms.

The magnetometer was simply a small mirror backed by steel magnets (B), and suspended from a silk fibre, as supplied by White for galvano-meters. It was mounted between glass plates at about 2 centim. distance from the magnetizing-spiral. The earth's force was compensated by steel magnets, which also served to bring the mirror perpendicular to the helix in spite of the influence of residual magnetism in the iron core. The deflec-tions were read in the manner usual with Thomson's galvanometers, by the motion of a spot of light thrown upon a scale after reflection by the mirror. The division is in millimetres, and with the aid of a lens a displacement of $\frac{1}{10}$ of a division can usually be detected with certainty.

The direct effect of the magnetizing-spiral upon the suspended needle was compensated by a few turns of wire C, 7 centim. in diameter,

* *Phil. Mag.* December 1886, p. 490. [Vol. II. p. 570.]

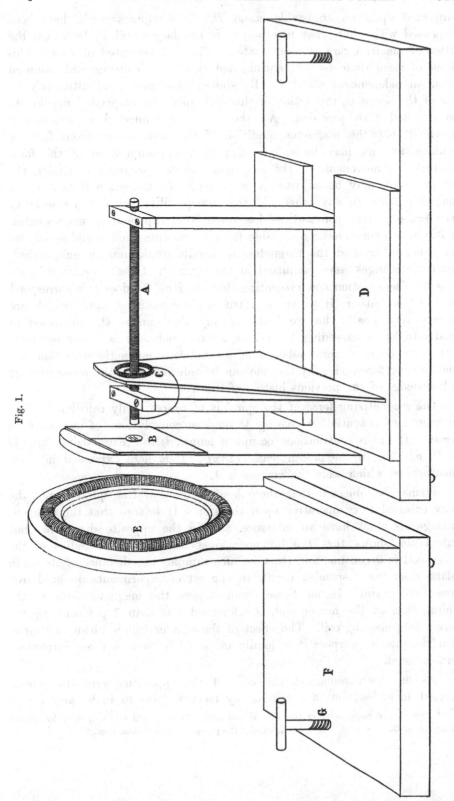

Fig. 1.

supported upon an adjustable stand D. This adjunct might have been dispensed with; but what is essential is the larger coil E, by which the effect of the *iron core* is compensated. This coil consisted of 74 convolutions, of mean diameter 18 centim., tied closely with string, and mounted upon an independent stand F. By sliding this stand, and ultimately by use of the screw G, the action of this coil upon the suspended needle can be adjusted with precision. All the coils are connected in series; and provided that the magnetic condition of the iron under given force is definite, matters may be so arranged that the imposition of the force produces no movement of the suspended needle, or, more generally, the compensation may be adjusted so as to suit the transition from any one magnetic force to any other. If the susceptibility (k) and permeability $\mu (= 4\pi k + 1)$ were constant, as has often been supposed in mathematical writings, the compensation suitable for any one transition would serve also for every other, and the magnetometer-needle would remain undisturbed, whatever changes were permitted in the strength of the magnetizing current[*]. The question now presenting itself is, How far does this correspond to fact? or, rather, How far is it true for magnetizing forces which are always very small? for we know already that, under the operation of moderate forces exceeding (say) 1 or 2 c.g.s., not only is μ not constant, but there is no definite relation at all between magnetic induction and magnetizing force, whereby the one can be inferred from the other without a knowledge of the previous history of the iron.

The magnetizing force of the spiral is of course easily calculated. The difference of potential in passing through n convolutions of current C is $4\pi nC$. If the n convolutions occupy a length l, the magnetizing force is $4\pi C \cdot n/l$; or, in the present case, $128\pi C$. C is here expressed in c.g.s. measure, on which scale the ampère is ·1.

It may be objected that the magnetic force of the spiral is not the only external force operative upon the iron. It is true that the compensating-coils must have an influence, and in the opposite direction. But calculation shows that the influence must be small. The radius of the large coil is 9 centim., and (to take an example) the distance of its mean plane from the suspended needle in one set of experiments on hard iron was 13·6 centim. Under these circumstances the magnetic force in the spiral, even at the nearer end, is influenced less than 2 per cent. by the large compensating-coil. The effect of the smaller coil is about the same. For the present purpose it is hardly worth while to take these corrections into account.

As has been remarked, the coils of the apparatus were always connected in series; but a reversing-key (serving also to make and break)

---

[*] The idea of compensating the iron is not new. The method was employed by Koosen (*Pogg. Ann.* Bd. LXXXV. S. 159, 1852) to exhibit the phenomena of " saturation."

was introduced so as to allow of the reversal of the compensating-coil in relation to the others. In one position of the key (−) the action of the coil and of the magnetized iron are opposed; in the other (+) the actions conspire. When the currents to be used were not exceedingly small, the whole apparatus was in simple circuit with a Daniell cell and such resistance-coils as were necessary. Exclusive of the cell and of the added resistances, the whole resistance was $7\frac{1}{2}$ ohms.

As an example, I will now give the details of some observations on December 6 made to test the behaviour of unannealed Swedish iron wire. The diameter of the wire is 1·6 millim.; it is from the same hank as a piece used in the experiments of Part II.* The compensating-coil was adjusted until it made no difference whether the key was open or closed (−), the additional resistance being 1000 ohms. In stating the result it will for the present be sufficient to give the German-silver resistances, that of the apparatus and of the battery being relatively of no importance. The corresponding current is about 10^{-4} c.g.s., and the strength of the magnetic field in the spiral is given by

$$128\,\pi C = ·04 \text{ c.g.s.}$$

We shall have a better idea of this if we recall that, on the same system of measurement, $H = ·18$; so that the force in action is about $\frac{1}{5}$ of that which the earth exercises horizontally.

When the resistance was altered to 11,000 ohms, the compensating-coil of course remaining undisturbed, contact (−) produced no visible motion, showing that the same compensation is suitable for the much smaller force. But at this point we require to be assured that the absence of disturbance is not due merely to want of sensitiveness. The necessary information is afforded at once by making reversed contact (+), which (with 11,000 ohms) gave a swing of 57 divisions.

To diminish the magnetizing force still further, a shunting arrangement was adopted. The current from the Daniell was led through 10,000 ohms and then through a box capable of providing resistances from 1 to 1000· The circuit of the apparatus included another coil of 10,000 ohms, and its terminals were connected to those of the box. The battery-current was thus about ·0001 ampère, or 10^{-5} c.g.s. If a be the (unplugged) resistance in the box, the E.M.F. at the terminals of the apparatus-circuit is $a \times 10^{-4}$ volts; and the current C through the magnetizing helix and compensating-coil is $a \times 10^{-9}$ c.g.s.

When $a = 1000$ ohms, (−) gave no visible deflection, while (+) caused a swing of 5 divisions.

* *l.c.* p. 488.   [Vol. II. p. 568.]

At this stage recourse was had to the "method of multiplication" in order to increase the sensitiveness*. A pendulum was adjusted until its swings were synchronous with those of the suspended needle. It was then easy to make and break contact in such a way as to augment the swing due to any outstanding force. Thus, when $a = 1000$, the swing was increased by the use of the timed contacts and ruptures (+) until it measured 26 divisions instead of 5 only. But a similar series of operations with reversed currents (−) caused no swing amounting to $\frac{1}{10}$ division; so that we may consider the compensation proved to be still perfect to about 1 per cent.

In applying the method to still smaller forces we cannot avoid a loss of sensitiveness. With $a = 100$, (+) gave 3 divisions, while the effect of (−) remained insensible. The correctness of the compensation is thus verified to about 6 per cent. of the separate effects. Had the iron, even at this stage, refused to accept magnetization, the fact would have manifested itself by the equality of the swings obtainable in the two ways, (+) and (−), of making the connections.

In the last case mentioned the current was 10^{-7} C.G.S., and the magnetic force was 4×10^{-5} C.G.S. We may therefore regard the proportionality of magnetic induction to magnetic force over the range from $\frac{1}{5} H$ to $\frac{1}{5000} H$ as an experimental fact. In view of this, neither theory nor observation gives us any reason for thinking that the proportionality would fail for still smaller forces.

Quite similar results have been obtained with steel. On December 13 a piece of drill steel (unannealed) was examined, the delicacy of the apparatus, as evidenced by the (+) effect, being about the same as in the above experiments on hard Swedish iron. No failure of proportionality could be detected with forces ranging from about $\frac{1}{5} H$ to $\frac{1}{10000} H$.

Annealed iron is a much less satisfactory subject. With unannealed

---

* The advantage of the method of multiplication seems to be hardly sufficiently appreciated. It is not merely that the effect is presented to the eye in a magnified form. That object can be attained by optical appliances, and by diminishing the directive force upon the suspended parts, whether by using a nearly astatic system of needles, or by compensating the field. For the most part these devices augment the unavoidable disturbances (which exhibit themselves by a shifting zero) in the same proportion as the effect to be measured, or at any rate rendered apparent. The real ultimate impediment to accuracy of measurement is almost always the difficulty of distinguishing the effect under examination from accidental disturbances, and it is to overcome this that our efforts should be directed. The method of multiplication is here of great service. The desired effects are largely magnified, while the disturbances, which are not isoperiodic with the vibrations of the needle, remain unmagnified, and therefore fall into the background.

It is obvious that, in order to secure this advantage, the vibrations must not be strongly damped. No doubt a highly damped galvanometer-needle is often convenient, and sometimes indispensable. But it seems to be a mistake to use it where a null method is applicable, and when the utmost delicacy is required. In such a case the inertia of the needle, and the forces both of restitution and of damping, should all be made small.

iron and steel the compensation for small forces may be made absolute, so that neither at the moment of closing the circuit nor afterwards is there any perceptible disturbance.   This means that (so far as the magnetometer-needle can decide) the metal assumes instantaneously a definite magnetic condition which does not afterwards change.   But soft iron shows much more complicated effects.   The following observations were made upon a piece of Swedish iron (from the same hank as the former) annealed in the flame of a spirit-lamp.   When an attempt was made to compensate for the imposition of a force equal to $\frac{1}{5}H$, no complete balance could be obtained.   When the coil was so placed as to reduce as much as possible the instantaneous effect, there ensued a drift of the magnometer-needle represented by about 170 divisions of the scale, and in such a direction as to indicate a continued increase of magnetization.   Precisely opposite effects followed the withdrawal of the magnetizing force.   The settling down of the iron into a new magnetic state is thus shown to be far from instantaneous.   On account of the complication entailed by the free swings of the needle, good observations on the drift could not be obtained with this apparatus; but it was evident that, whilst most of the anomalous action was over in 3 or 4 seconds, the final magnetic state was not attained until after about 15 or 20 seconds*.

The operation of feebler forces was next examined, rather with the expectation of finding the drift reduced in relative importance.   But the imposition of $\frac{1}{50}H$ was followed by a drift of 13 or 14 divisions, no very small fraction of the whole action; as was seen from the observation that the (+) effect was now 300 divisions, of which 150 are due to the iron. With 20,000 ohms in circuit, giving a force equal to $\frac{1}{100}H$, the drift was 6 or 7 divisions.   By still further diminishing the force the drift could be reduced to insignificance; but it appeared to maintain its proportion to the instantaneous effect.   Apart from the complication due to the drift, the magnetization was proportional to magnetizing force from $\frac{1}{10}H$ to $\frac{1}{5000}H$ or less†.

The question now presents itself, What is the actual value of the permeability which has been proved to be a definite constant for small forces?   In consequence, however, of the nearness of the operative pole to the suspended needle in the preceding experiments, no moderately accurate value of μ can be deduced.   But the observations described in Part II. are

---

* Prof. Ewing (*loc. cit.* § 52) describes " a time lag in magnetization," especially noticeable in the softest iron and at points near the beginning of the steep part of the magnetization-curve. It should have been stated that my apparatus was very firmly supported, and, being situated underground, was well protected from vibration.   The drift or creeping did not appear to be due to this cause.

† The results here set forth were announced in a discussion following Prof. Hughes's address to the Society of Telegraph Engineers on February 11, 1886, *Journ. Tel. Eng.* xv. p. 39, on the strength of preliminary experiments tried towards the close of 1885.

sufficient to show that the constant permeability for hard iron has some such value as 90 or 100, the forces then operative being within the prescribed limits. The fact that the initial value of μ is so large is obviously of great theoretical and practical importance. Further evidence will be brought forward presently in connection with observations made with an arrangement better suited to an absolute determination.

Too definite a character must not be ascribed to the above-mentioned limit of $\frac{1}{5} H$. Below this point the deviations from the law of proportionality, though mathematically existent, are barely sensible. In order to understand this, it is well to consider what happens when the limit is plainly exceeded. If a force of the order H be imposed, the compensating-coil (adjusted for small forces) appears to be overpowered, and a large deflection occurs. If the force be now removed, the recovery is incomplete, indicating that the iron retains residual magnetism. Subsequent applications and removals of the force produce a nearly regular effect, and always of such a character as to prove that the magnetic changes in the iron *exceed* those demanded by the law of proportionality. As might be expected, the excess varies as the *square* of the force; and thus, when the force is small enough, it becomes insignificant, and the law of proportionality expresses the facts of the case with sufficient accuracy. But the precise limit to be fixed to the operation of the law depends necessarily upon the degree of accuracy demanded.

The readings with and without the force being tolerably definite, it would of course be possible, by pushing in the compensating-coil, to bring about an adjustment in which the application or removal of the force causes no deflection. But this state of things must be carefully distinguished from the compensation obtainable with very small forces, in that it is limited to one particular step in the magnitude of force. If we try a force of half the magnitude, we find the compensation fail. Not only so, but the reading will be different under the same force according as we come to it from the one side or from the other. The curve representing the relation between force and magnetization is a loop of finite area.

Except for the purpose of examining whether the whole magnetization is assumed instantaneously (absence of drift), there is little advantage in the compensation being adjusted for the extreme range under trial. It is usually better to retain the adjustment proper to very small forces. Even though it fails to give a complete compensation, the coil offers an important advantage, which will presently appear; and its use diminishes the displacement to be read upon the scale.

We have seen that when the forces are very small there is a definite relation between force and magnetization, of such a character that one is

proportional to the other: the ratio k (the susceptibility) is a definite
constant. When, however, certain limits are exceeded there is no fixed
relation between the quantities; and if k is still to be retained, it requires
a fresh definition. It is not merely that k, as at first defined, ceases to be
constant, but rather that it ceases to exist. Upon this point the verdict
of experiment is perfectly clear. There is no curved line by which the
relation between force and magnetization can be unambiguously expressed,
and which can be traversed in both directions. As soon as the line ceases
to be straight, it ceases also to be single. I have thought it desirable to
emphasize this point, because the term " magnetization-function," introduced
by Dr Stoletow, rather suggests a different conclusion.

The curves given by Stoletow and by Rowland in their celebrated
researches are not exactly magnetization-curves in the more natural sense ;
that is to say, they do not exhibit fully the behaviour of a piece of iron
when subjected to a given sequence of magnetic forces. But a number
of such curves have been drawn by Ewing which afford all necessary
general information. Among these we may especially distinguish the
course followed by the iron in passing from strong positive to strong
negative magnetization and *vice versâ*, and that by which iron starting
from a neutral condition first acquires magnetization under the action of
a force constantly increasing.

Attention is called by Ewing to the *loops* which are formed when the
forces are carried round a (not very small) cycle of any kind. " Every
loop in the diagram shows that when we reverse the *change* of magnetizing
force from increment to decrement, or *vice versâ*, the magnetism begins
to change very gradually relatively to the change of $\mathfrak{H}$ (the force), no
matter how fast it may have been changing in the opposite direction
before. So much is this the case that the curves, when drawn to a scale
such as that of the figure, appear in all cases to start off tangent to the
line parallel to the axis on which $\mathfrak{H}$ is measured whenever the change
of $\mathfrak{H}$ is reversed in sign."

The question here raised as to the direction of the curve, after the
force has passed a maximum or minimum, is one of great importance. If
it were strictly true that this direction were parallel to the axis, it would
follow generally that iron in any condition of magnetization would be
uninfluenced by small periodic variations of magnetic force; for example,
that in many telephone experiments iron would show no magnetic properties. The experiments already detailed prove that when the whole force
and magnetization are small (they were not actually evanescent) very
sensible proportional changes of magnetization accompany small changes
of force, the ratio being such as to give a permeability not much inferior
to 100. Nothing is easier than to show that this conclusion is not limited

to very small mean forces and magnetizations. As regards the latter, we may apply and remove a force (say) of $5H$. By this operation the iron is left in a different magnetic condition, and the zero-reading of the magnetometer is altered, probably to the extent of driving the spot of light off the scale. But if we bring the needle back with the aid of external magnets, we can examine, as before, the effect of imposing a small force (under $\frac{1}{5}H$). If this be in the opposite direction to the previous large force, it will produce, in spite of the compensating-coil, a very sensible effect; for in this case the movement from 0 to $-\frac{1}{5}H$ is in continuation of the previous movement from $5H$ to 0. But subsequent applications and removals of $\frac{1}{5}H$ produce no visible effect upon the needle, as would have happened from the first had the small force operated in the positive direction. We may conclude, then, that the compensation for small forces suitable when the iron is nearly free from magnetization is not disturbed by the presence of considerable residual magnetism.

To examine the action of a small increment or decrement, when the total force is relatively large, we must either introduce a second magnetizing helix or effect the variation of current otherwise than by breaking the circuit. I found it most convenient simply to vary the resistance taken from the box, so arranging matters that the small alteration of current required could be effected by the insertion or removal of a single plug. The corresponding change of current is obtained by inspection of a table of reciprocals; and it was readily proved that within the admissible range of the apparatus the compensation was just as effective whether a step (not exceeding $\frac{1}{5}H$) was made from zero or from a force (say) of $5H$, 20 or 30 times as great as the increment or decrement itself. It need scarcely be repeated that there is an exception as regards the first step, in the case where it is in the same direction as the large movement preceding it.

At this stage the original magnetizing-coil, having been arranged for the investigation of the smallest forces, was replaced by another at a greater distance from the suspended needle. When the magnetization of the iron in its various parts fails to vary in strict proportion to the force, the effective pole is liable to shift its position; and this is an objection to the horizontal arrangement adopted in the earlier experiments. The helix was therefore placed vertically, the lower end of the iron core being a trifle below the level of the magnetometer-needle. The upper pole was at such a distance as to give but little relative effect. The length of the new helix, wound like the other upon a glass tube, is about 30 centim. The windings are in four layers, at the rate altogether of 65 per centim.; so that (under the same current) the magnetizing force is about twice as great as before. The resistance is 4·75 ohms.

A large number of observations have been made upon a core of rather hard Swedish iron, 3·30 millim. in diameter. The same compensating-coil as before was found suitable, and the arrangements were unaltered, except that an additional reversing-key was introduced, by which the poles of the Daniell cell could be interchanged. The total resistance of the circuit, independently of the box, was 7 ohms. The length of the core—or, rather, of the part exposed to the magnetizing force*—being about 100 diameters, is scarcely sufficient for an accurate determination; but from the observed position necessary for the compensating-coil we can get at least a rough estimate of the susceptibility for small forces. Thus, on December 28th, there was compensation for small forces when the distances of the needle from the mean plane of the compensating-coil and from the operative pole of the iron core were respectively 17·2 centim. and 9·3 centim. The magnetic force at the needle, due to unit current in the compensating-coil, is

$$\frac{2\pi \times 74 \times 9^2}{\{9^2 + 17\cdot2^2\}^{\frac{3}{2}}} = 5\cdot15.$$

The magnetizing force in the interior of the helix for unit current is

$$4\pi \times 65 = 817.$$

If k be the susceptibility, the strength of the pole is

$$\tfrac{1}{4}\pi \times \cdot330^2 \times 817 \times k;$$

and since the distance of this from the needle is 9·3 centim., we have, to determine k,

$$k = \frac{5\cdot15 \times 9\cdot3^2}{\tfrac{1}{4}\pi \times \cdot330^2 \times 817} = 6\cdot36;$$

so that

$$\mu = 1 + 4\pi k = 81.$$

This is probably an underestimate.

In order to obtain results comparable with those of Stoletow and Rowland, the iron was submitted to a series of cycles of positive and negative force. According to Ewing, the behaviour is simplest when the iron is first treated to a process of "demagnetization by reversals." This was effected *in situ* as a preliminary to the experiments of January 4th, the resistance in the box being increased by small steps from a few ohms to a thousand ohms; while at each stage the battery was reversed several times. It must be remarked, however, that the iron was all the while under the influence of the earth's vertical force; so that the resulting condition was certainly not one of demagnetization. But even as thus carried out, the operation was probably advantageous as obliterating the influence of the previous history of the iron core.

---

* At the upper end the iron projected beyond the coil.

The compensation was in the first place adjusted so that no displacement could be detected, whether the resistance was infinity or 2007 ohms*. This, of course, was in the position of the reversing-key denoted (−). When the iron and the compensating-coil acted in the same direction (+), the displacement was 8 divisions.

In Table I. the first column gives the total resistance of the circuit in ohms, and the second gives the reciprocals of the first, numbers proportional to the current or magnetizing force. Repetitions of a cycle are shown on the same horizontal line, for greater convenience of comparison. Thus the first application of current +197 gave the reading 242; a second application, after the cycle +197, 0, −197, 0, gave 241½. After two of these cycles had been completed, the current +326 gave the reading 245. To the readings as entered a small correction to infinitely small arcs has been applied. The letters R, L in the first column indicate the alternative positions of the battery reversing-key. It will be seen that very nearly the same numbers are obtained on repetition of a cycle, and that even the first application of an increased force gives a normal result.

The first question which suggests itself is the law connecting the magnitude of a current with the alteration of magnetization caused by its reversal. The quantities under consideration are exhibited in Table II., where the first column gives the current (x) and the second column the displacement (y) due to reversal. The relation between x and y is well expressed by the formula

$$y = -\cdot0053x + 1\cdot072x^2, \dots\dots\dots\dots\dots\dots\dots (1)$$

of which the whole of the second member is shown in column 5, and the two parts separately in columns 3 and 4. Column 6 gives the differences between the observed displacements and those calculated from the formula; they do not much exceed the errors of observation.

It will of course be borne in mind that the magnetization exhibited here is additional to the part rendered latent by the compensating-coil, and that the existence of the small linear term may be attributed to a defective adjustment of that coil. The calculated value of y for the step from infinite resistance to 2007 ohms, which is one quarter of the first step in the table (from $1007R$ to $1007L$), is

$$y = -\cdot13 + \cdot06 = -\cdot07 \text{ division.}$$

This is the step for which the coil was adjusted; and the difference between the calculated and observed (zero) value of y is perhaps as small as could have been expected. It is fair to conclude that, if the compensating-coil

---

* For greater delicacy, recourse was had to the "method of multiplication," assisted by a pendulum, as already described.

TABLE I.—Jan. 4, 1887.

Resistance	Current	Corrected Readings			
∞	0	240			
1007 *R*..........	+ 099	241			
∞	0	241			
1007 *L*..........	− 099	240			
∞	0	240			
507 *R*	+ 197	242	241½		
∞	0	241¼	241		
507 *L*	− 197	238½	238¼		
∞	0	239	239		
307 *R*	+ 326	245	245		
∞	0	......	243		
307 *L*	− 326	235	235		
∞	0	237	237		
207 *R*	+ 483	250½	250¼		
∞	0	246	246		
207 *L*	− 483	228	228		
∞	0	232½	232½		
107 *R*	+ 934	283¼	283¾	284	
∞	0	264½	264½	265	
107 *L*	− 934	195¼	195¼	195¼	
∞	0	214	214	213¼	
87 *R*............	+1149	306½	307½	307	
∞	0	276¾	276¾	277	
87 *L*............	−1149	172¼	171¼	171¼	
∞	0	201½	201	201	
177 *R*	+ 565	......	238½		
77 *R*	+1298	325¾	325¾		
177 *R*	+ 565	......	315½		
∞	0	286¼	286		
177 *L*	− 565	......	237½		
77 *L*	−1298	151½	150¾		
177 *L*	− 565	......	160		
∞	0	190½	188¾		
167 *R*	+ 599	......	......	232½	232¼
67 *R*	+1493	352¾	353	353½	353¾
167 *R*	+ 599	......	......	337½	338¾
∞	0	299	299½	300	301½
167 *L*	− 599	......	......	241½	
67 *L*	−1493	121¾	121¼	121¼	
167 *L*	− 599	......	......	136	
∞	0	174¾	174½	173½	

could have been perfectly adjusted for a very small step (the actual step was scarcely small enough), the uncompensated effects visible with larger currents would have been expressible by a quadratic term simply.

The currents (x) given in the tables are reduced to C.G.S. measure when divided by 10^6. On the same system the magnetizing force is $8\cdot2 \times 10^{-4} \times x$; so that the force due to the strongest current referred to in the table is $1\cdot2$ C.G.S., or about $7H$. When the current is reversed, the change of magnetic force is of course the double of this quantity.

In extending the definition of susceptibility to cases in which the force is not very small, we might proceed in more than one way. If we take the ratio of the change of magnetization to change of force when the force is reversed, we are following good authorities; and we get a definition which is at any rate consistent with the definition necessary when small forces are concerned. The values of k for different forces are not given by a direct comparison of the numbers in Table II., since the magnetometer-scale is arbitrary; but we may find for what force the susceptibility is (for example) the double of that applicable to infinitely small forces.

TABLE II.

Current, x	Displacement, y	$\cdot0053\,x$	$1\cdot072\,x^2$	$-0\cdot0053\,x$ $+1\cdot072\,x^2$	Diff.
99	1	0·52	1·05	0·5	+0·5
197	3¼	1·0	4·2	3·2	0
326	10	1·7	11·4	9·7	+0·3
483	22¼	2·6	25·0	22·4	−0·2
934	88½	4·9	93·7	88·8	−0·3
1149	136	6·1	141·5	135·4	+0·6
1298	174	6·9	180·6	173·7	+0·3
1493	231	7·9	238·9	231·0	0

For this purpose we must note that the conjoint effect of the magnetization due to current 50, simply applied or removed, and of the compensating-coil, was 8 divisions, of which half is due to each cause. The effect of the coil for a *reversal* of current 50 is thus 8 divisions, and being proportional to the current can be deduced for any other case. At the bottom of the table, where the current is 1493, the displacement rendered latent by the coil is thus about 240 divisions; and since at this point the uncompensated displacement is nearly of the same amount, we see that the value of k (as above defined) is here doubled. Thus, if $\mathfrak{H}$ denote the magnetizing force in C.G.S. measure, we have

$$k = 6\cdot4\,(1 + \cdot8\mathfrak{H}).$$

The *form* of the relations of k to $\mathfrak{H}$ for small forces is pretty accurately demonstrated by the observations. On the other hand, the reduction to absolute measure is rather rough*—a point of less consequence, inasmuch as the constants may be expected to vary according to the sample and condition of the iron.

The observations in Table I. give a good deal more than the extreme range of magnetization due to the reversal of a force. In all cases the two residual magnetizations (when the force is zero) are recorded; while in the two latter, where the range is greatest, further intermediate points are included. The results are plotted in Fig. 2, where it will be seen that

Fig. 2.   Unannealed Iron.   (Jan. 4)

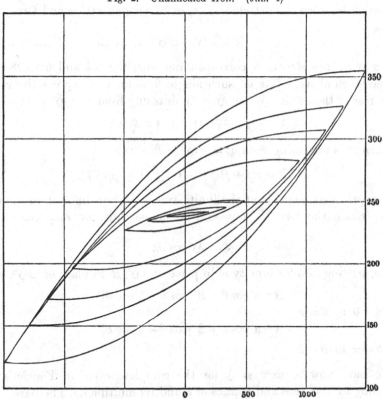

the curves start backwards in a horizontal direction after a maximum or minimum of force. Special observations (not recorded in the table) were directed to this point. Neither at the maxima nor at the zeros of force was there any evidence of failure of compensation when a small backward movement was made.

The curves do not differ much from parabolas; and in other cases, where the applied magnetic forces were all of one sign, I have found that after a

---

* In all probability the number 6·4, applicable when $\mathfrak{H} = 0$, is too small.

large movement in one direction, the curve representing a backward movement coincides somewhat closely with a parabola whose magnitude is nearly the same under different circumstances, and which is placed so that its axis is vertical and vertex coincident with the point where the backward movement commences. The reader will not forget that to obtain the real curves fully expressing the relation between magnetization and force, we must add the effect, proportional to the force, rendered latent by the compensating-coil.

On the basis of this parabolic law we may calculate the influence of hysteresis in the magnetization of iron upon the apparent self-induction and resistance of the magnetizing-coil, when periodic currents of moderate power are allowed to pass. If we reckon from the mean condition, we may express the relation between the extreme changes of magnetization and force by the formula

$$\mathfrak{J}' = \alpha \mathfrak{H}' + \beta \mathfrak{H}'^2, \quad \dots\dots\dots\dots\dots\dots(2)$$

where α and β are constants, corresponding with the 6·4 and 6·4 × ·8 of the example given above. But no such single formula can express the relation for the rest of the cycle. When $\mathfrak{H}$ is diminishing from $\mathfrak{H} = \mathfrak{H}'$ to $\mathfrak{H} = -\mathfrak{H}'$,

$$\mathfrak{J} = \alpha \mathfrak{H} + \beta \mathfrak{H}'^2 \{1 - \tfrac{1}{2}(1 - \mathfrak{H}/\mathfrak{H}')^2\};$$

but when $\mathfrak{H}$ is increasing from $\mathfrak{H} = -\mathfrak{H}'$ to $\mathfrak{H} = \mathfrak{H}'$,

$$\mathfrak{J} = \alpha \mathfrak{H} + \beta \mathfrak{H}'^2 \{-1 + \tfrac{1}{2}(1 + \mathfrak{H}/\mathfrak{H}')^2\}.$$

These expressions coincide at the limits $\mathfrak{H} = \pm \mathfrak{H}'$, but differ at intermediate points. Since the force is supposed to be periodic, we may conveniently write

$$\mathfrak{H} = \mathfrak{H}' \cos \theta;$$

whence, putting also for brevity α' in place of $\alpha \mathfrak{H}'$, β' in place of $\beta \mathfrak{H}'^2$, we get

$$\mathfrak{J} = \alpha' \cos \theta + \beta' \{\cos \theta + \tfrac{1}{2} \sin^2 \theta\}$$

from $\theta = 0$ to $\theta = \pi$,

$$\mathfrak{J} = \alpha' \cos \theta + \beta' \{\cos \theta - \tfrac{1}{2} \sin^2 \theta\}$$

from $\theta = \pi$ to $\theta = 2\pi$.

We have now to express $\mathfrak{J}$ for the complete cycle in Fourier's series proceeding by the sines and cosines of θ and its multiples. The part

$$\alpha' \cos \theta + \beta' \cos \theta,$$

being the same in the two expressions, is already of the required form. For the other part we get

$$\pm \tfrac{1}{2} \sin^2 \theta = B_1 \sin \theta + B_3 \sin 3\theta + B_5 \sin 5\theta + \dots, \quad \dots\dots\dots(3)$$

where only odd terms appear, and B_n is given by

$$B_n = -\frac{4}{\pi n (n^2 - 4)}. \quad \dots\dots\dots\dots\dots\dots(4)$$

Thus

$$\Im = (\alpha' + \beta') \cos \theta + \beta' \left\{ \frac{4}{3\pi} \sin \theta - \frac{4}{15\pi} \sin 3\theta - \frac{4}{105\pi} \sin 5\theta - ... \right\} ....(5)$$

If the range of magnetization be very small, β' vanishes, and the influence of the iron upon the enveloping coil is merely to increase its self-induction; but if β' be finite, the matter is less simple. The terms in $\sin 3\theta$, $\sin 5\theta$, &c., indicate that the response of the iron to a harmonic force is not even purely harmonic, but requires higher components for its expression. If we put these terms out of account as relatively small, we must still regard the phase of $\Im$ as different from that of $\mathfrak{H}$. The term in $\sin \theta$ will show itself as an apparent increase in the *resistance* of the coil, due to hysteresis, and independent of that which may be observed even with very small forces as a consequence of induced currents in the interior of the iron. The augmentation of resistance now under consideration may be expected to be insensible when the extreme range of magnetizing force does not exceed one-tenth of the earth's horizontal force.

In the absolute determination [p. 589] of the susceptibility to very small forces of the hard Swedish iron wire (3·30 millim. diameter), the length (about 100 diameters) was scarcely sufficient for an accurate estimate. Similar experiments on a thinner wire (1·57 millim. diameter) of the same quality of iron gave $k = 6\cdot85$, corresponding to $\mu = 87$. This is in the hard-drawn condition. After annealing the same piece of wire gave a higher result, but in this case the observation is complicated by the assumption of the magnetic state occupying a sensible time. The susceptibility applicable to the final condition is as high as 22·0, more than three times as great as before annealing. But a lower number would better represent the facts, when the small magnetic force is rapidly periodic; and it may even be that under forces of frequencies such as occur in telephonic experiments, most of the difference due to annealing would disappear. Such a conclusion is suggested by the slight influence of annealing in the experiment described in Part II.*, where is determined the increment of resistance of an iron wire due to the concentration of a variable current in the outer layers. But the matter is one requiring further examination under better experimental conditions.

The sensitiveness of the magnetometer-needle in the experiments directed to prove the constancy of susceptibility to small forces, suggests the inquiry whether iron should be used when the object is purely galvanometric. An attempt to produce a sensitive galvanometer by hanging a mirror and needle between the pointed pole-pieces of a large electromagnet, arranged as in diamagnetic experiments, was not very successful. A better result was obtained with an astatic needle system, and an electromagnet on a much smaller scale. This was of horseshoe form, the core being of hard Swedish iron wire 3·3 millim. diameter. The insulated copper wire was in three layers, of resistance

---

* *Phil. Mag.* Dec. 1886, p. 488. [vol. II. p. 568.]

·34 ohm, and the total weight of the electromagnet was 283 grams. It was held so as to embrace the upper needle system. When the time of swing from rest to rest was 4 seconds, the movement due to a current of about $\frac{1}{20,000}$ ampère was 100 divisions. The zero was steady enough to allow a displacement of half a division to be detected with tolerable certainty in each trial; so that, as actually used, the arrangement was sensitive to a current of $\frac{1}{4} \times 10^{-6}$ ampère. The addition of a similar electromagnet embracing the lower needle system, and connected in series, would double the sensitiveness, and raise the resistance to ·68 ohm. A galvanometer thus constructed, and of resistance equal to 1 ohm, would show a current of 10^{-7} ampère. Using finer wire, we might expect an instrument of 100 ohms to show a current of 10^{-8} ampère, and so on.

For comparison with the above I tried, in as nearly as possible the same way, the sensitiveness of a good Thomson astatic galvanometer of resistance 1·3 ohm. With an equal time of vibration, a current of $\frac{1}{20,000}$ ampère produced a movement of 300 divisions. The zero was perhaps a little steadier than before; but it will be seen that the sensitiveness was of the same order of magnitude. In both cases, by taking precautions and by using repetition, the delicacy might have been increased, probably tenfold.

The experiments show that there is no difficulty in constructing a galvanometer of high sensitiveness upon these lines. According to theory, with ideal iron of permeability 100, it should be possible to attain a much higher degree of sensitiveness than without iron. But the tendency to retain residual magnetism would certainly be troublesome, and probably neutralize in practice most of the advantage arising from the higher permeability, which allows of windings more distant from the needles being turned to good account. Another inconvenience may be mentioned. If the iron poles are brought at all close to the needles, there is a strong tendency to instability at moderate angles of displacement.

Experiments already described proved conclusively that the response of iron and steel to small periodic magnetic forces is not affected by the presence of a constant force, or of a residual magnetization, of moderate intensity. At the same time it appeared in the highest degree probable that the independence was not absolute, and that the response to a given small change of force would fall off as the condition of "saturation" is approached, even though we admit, in accordance with recent evidence, that saturation is attainable only in a very rough sense. The question was too important to be left undecided, but it was difficult to deal with by the magnetometric method. If the arrangement is sensitive enough to allow the effect of the small force to be measured with reasonable accuracy, it is violently disturbed by the occurrence of high degrees of magnetization. Moreover it is undesirable to depend so much, as in this method, upon what may happen near the free extremities of the iron rod, where the magnetic forces must vary rapidly. The " ballistic method," in which the changes of magnetization are indicated by the throw of a galvanometer-needle in connection with a secondary coil embracing the central parts of the rod, has the great advantage for this purpose, that the reading is independent of the stationary condition of the iron. In the first experiments by this method the magnetizing helix was similar to one already described [p. 588]; and the small, as well as the large, alterations of force were effected by varying the resistance of the circuit. By suitably choosing the resistances from a box, the small alterations of current could be obtained with sufficient suddenness by the simple introduction or removal of a plug, and were taken of the same order of magnitude at different parts of the scale. A comparison of effects (with the aid of a table of reciprocals) proved that a pretty strong total force* or magnetization did not interfere much with the response of the iron to a given force of small magnitude.

This arrangement did not well allow of the investigation being pushed further so as to deal with stronger magnetizing forces. If, with the view of increasing the current, we cut down the german-silver resistance too closely, the estimate of total resistance depends too much upon the battery, and the current becomes uncertain. This difficulty is evaded by the use of a double wire—one conveying the strong current, of which the measurement does not require to be very exact; the other conveying the weak current, of which the effect at different parts of the scale is to be examined.

In order to obtain a satisfactory ratio of length to diameter, without the loss of sensitiveness that would accompany a diminution in the section of the iron, a helix was prepared of length 59·6 centim. It was wound upon a glass tube with a double wire in three layers, the whole number of turns of each wire being 1376. The magnetizing force due to unit current in one wire is therefore

$$4\pi \times 1376/59\!\cdot\!6 = 290\!\cdot\!1.$$

The resistance of each wire is 3·2 ohms; and thus when two Grove cells are used in connection with one of the wires, a current of about an ampère (·1 C.G.S.) can be commanded. Smaller currents were obtained by the insertion of resistances from a box.

Although the secondary coil, connected with a delicate galvanometer, contained a large number of convolutions, the sensitiveness was insufficient to allow of the small magnetizing force being taken as low as would otherwise have been desirable. It was obtained by means of the second wire of the helix, which was included in the circuit of a Daniell cell and 200 ohms from a resistance-box. When the circuit was completed (or broken) at a key, the force brought into operation, or removed, was

$$\frac{290\!\cdot\!1}{2040} = \cdot 14 \text{ C.G.S.}$$

---

* Up to about 6 c.g.s.   The iron was unannealed Swedish, 3·3 millim. in diameter.

In making a series of observations it was usual, after each alteration of the strong magnetizing force, to apply and remove the small magnetizing force several times before attempting to take readings.

The results obtained by this method were of a pretty definite character. The small force produced a constant effect upon a wire of unannealed Swedish iron, 3·3 millim. in diameter, until the large force was increased from 0 to about 5 C.G.S. At about 10 C.G.S. the effect of the small force fell off 5 per cent. The highest force used, about 29 C.G.S., reduced the effect to about 60 per cent. of its original amount. On complete removal of the force due to the Grove cells, there was but a partial recovery of effect, doubtless in consequence of residual magnetization. After the wire had been removed from the helix and well shaken, the small force was found to have recovered its full efficiency.

The wire was then annealed and submitted anew to a similar series of operations. The magnetization due to the alternate application and removal of the small force was found to be at first, i.e. in the absence of a constant force, *twice* as great as before*.

The increase, however, is not long maintained, a steady force of 2 C.G.S. being already sufficient to cause a marked falling off (of about 20 per cent.). Under the operation of 29 C.G.S., the effect of the small force fell to about $\frac{1}{6}$ of its original amount. Removal from the helix and shaking in a zero field sufficed to restore the wire to its initial condition.

Similar experiments upon an annealed wire of " best spring steel " showed no sensible change of effect when the steady force was varied from 0 to about 16 C.G.S. In this case the ratio of length to diameter was about 300.

We may now regard it as established :—

That in any condition of force and magnetization, the susceptibility to small periodic changes of force is a definite, and not very small, quantity, independent of the magnitude of the small change.

That the value of the susceptibility to small changes of force is approximately independent of the initial condition as regards force and magnetization, until the region of saturation is approached.

* It should here be remembered that any part of the change of magnetization which lags behind for more than a second or two, fails to manifest itself fully in the indications of the galvanometer.

END OF VOL. II.

CAMBRIDGE : PRINTED BY J. AND C. F. CLAY, AT THE UNIVERSITY PRESS.

Printed in the United States
By Bookmasters